Advances in Studies of Heterogeneities in the Earth's Lithosphere: The Keiiti Aki Volume II

Edited by
Yehuda Ben-Zion
William H.K. Lee

2006

Birkhäuser Verlag
Basel · Boston · Berlin

Reprint from Pure and Applied Geophysics
(PAGEOPH), Volume 163 (2006) No. 2/3

Editor's

Yehuda Ben-Zion
University of Southern California
Department of Earth Sciences
Los Angeles CA, 90089-0740
USA

e-mail: benzion@terra.usc.edu

William H.K. Lee
US Geological Survey
345 Middlefield Road MS 977
Menlo Park CA 94025-3591
USA

e-mail: lee@usgs.gov

A CIP catalogue record for this book is available from the Library of Congress, Washington D.C., USA

Bibliographic information published by Die Deutsche Bibliothek:
Die Deutsche Bibliothek lists this publication in the Deutsche Nationalbibliographie; detailed bibliographic data is available in the internet at <http://dnb.ddb.de>

ISBN 3-7643-7579-5 Birkhäuser Verlag, Basel – Boston – Berlin

Part of Springer Science+Business Media
Printed on acid-free paper produced from chlorine-free pulp
Printed in Germany

ISBN 10: 3-7643-7579-5
ISBN 13: 978-7643-7579-9

e-ISBN-3-7643-7585-X

9 8 7 6 5 4 3 2 1

Keiiti Aki at the gala celebtrating his receiving the Bowie medal of the American Geophysical Union, December 16, 2004.

PURE AND APPLIED GEOPHYSICS
Vol. 163, No. 2–3, 2006

Contents

Pure appl. geophys. 163 (2006) 259–266
0033–4553/06/030259–8
DOI 10.1007/s00024-005-0029-3

Pure and Applied Geophysics

Introduction

During the December 2004 annual meeting of the American Geophysical Union (AGU), Professor Keitii (Kei) Aki was awarded the Bowie Medal, which is the highest honor bestowed annually by the AGU. To celebrate this occasion, a scientific symposium in honor of Kei was organized by Yehuda Ben-Zion, Eystein Husebye, John McRaney, Haruo Sato, and Willie Lee under the title "Quantitative Seismology: Advances in Studies of Fine-scale Heterogeneities in a Deforming Earth". Since Kei studied (and was instrumental in developing) most aspects of quantitative seismology, including source, propagation and site effects over the entire range of observed frequencies, it was difficult to pick a subset of topics that would form a reasonably focused scientific session. The title was finally chosen to reflect a set of problems that was central to Kei's interests, while having a somewhat different focus than those of two earlier 2000 symposia that commemorated Kei's retirements from academia.

The description of the 2004 AGU session in honor of Aki is as follows: *In the past 25 years, there have been tremendous advances in many aspects of quantitative seismology using physics-based approaches. New fronts were made possible by many factors: Large data sets with unprecedented resolution and bandwidth; innovative computational techniques; new theoretical models for seismic phenomena; and integrative research accounting for broad ranges of space and time scales. These advances were stimulated by many pioneering papers, and were aided strongly by the publication of the two-volume treatise "Quantitative Seismology: Theory and Methods" by Keiiti Aki and Paul Richards in 1980. This session will focus on integrative physics-based seismological studies, with an emphasis on the nature of fine-scale heterogeneities and earthquakes. Contributions from theory, observations, and applications are welcome.* The citation for the Bowie medal (Appendix 1) was given by Tom Jordan.

In contrast to the two earlier 2000 symposia honoring Kei Aki, which had only invited participants, the 2004 AGU meeting provided a venue that allowed anyone who wished to contribute or attend to do so. The special AGU session and a related follow-up banquet were highly successful, both in terms of covering current state-of-the-art research results and celebrating lasting fundamental contributions of Kei Aki. The scientific session received an overwhelming response of 85 contributions. These were given over four oral sessions which included an opening speech by Paul Richards (Appendix 2), an introduction and response by Kei at the beginning and the end of the oral sessions (Appendix 3), and 27 papers. The remaining 58 papers were presented in a poster session. Many participants wrote "tributes" to Kei, some of which were read at the banquet. The tributes provide interesting windows into aspects of Kei's personal and scientific interactions with students and colleagues and are posted at http://www.iris.edu/seismo/quakes/1964niigata/.

The present volume has 13 papers on topics related to the theme of the special 2004 AGU session in honor of Kei, which is a small fraction of the 85 presentations that were given in the meeting. The papers belong generally to three broad categories: Studies related to dynamic earthquake rupture, studies associated with imaging of earthquake locations and structures using coherent (P and S) seismic phases, and studies associated with imaging source and structure properties using scattered (coda) seismic waves. In the first group of papers, CHEN and ZHANG provide analytical and numerical results on dynamic rupture along a dipping fault in a 3-D elastic half space, using a recently-developed boundary integral formulation that accounts for the free surface and numerical regularization of the associated singular integral kernels. The calculations show that waves reflected from the surface can affect significantly the rupture growth, especially at shallow depth. A resolution test suggests that the method provides accurate results with larger grid size than is required by some other computational methods of dynamic rupture. DOR *et al.* present detailed geological mapping of rock damage in the structure of several faults of the San Andreas system in southern California. The results show strong damage asymmetry across the faults that is correlated with available information on seismic velocity structures at seismogenic depth. The observations are compatible with theoretical results on dynamic rupture along material interfaces in the fault zone structures, with preferred propagation direction associated with the velocity structure.

In the second group of papers, RICHARDS *et al.* outline the double-difference technique for relative earthquake locations, using arrival-time data obtained by waveform cross correlations and catalog phase picks, and illustrate the method with applications to regions with different size and different density of seismicity and stations. They conclude that the double-difference method and accurate arrival-time data obtained by waveform cross correlations can improve considerably the precision of relative locations, and that routine application of the double-difference and waveform cross correlation techniques is now possible at local and regional scales. ZHANG and THURBER review a recently-developed double-difference tomography method for simultaneous derivation of earthquake locations and velocity structure, based on a mixture of absolute and more accurate differential arrival times. The material includes theory, implementations, and applications of the method to synthetic and observed data at several scales. The results indicate that double-difference tomography provides generally more accurate results than standard tomography. GOT *et al.* review an alternative treatment of double-difference tomography that incorporates probabilistic optimization of results, and discuss effects of various errors in data and model parameters. Application of the method to the structure of the Kilauea volcano in Hawaii resolves internal velocity variations, and comparison to results of several other algorithms suggests that the approach is advantageous for imaging heterogeneous structures, especially when the data quality is not very high. GAUTIER *et al.* perform tomographic inversion of local earthquake arrival times for crustal structure and earthquake locations in the actively rifting Gulf of Corinth. The velocities image the rift basin and the hypocenters image a regional-scale detachment fault.

In the third group of papers, SNIEDER reviews theoretical concepts governing changes of coda waves due to small localized perturbations in the medium properties or source location, and describes how to estimate the resulting mean and variance of the travel-time perturbations. CAMPILLO reviews theoretical results on properties of coda waves in asymptotic multiple-scattering regime, and how to construct Green functions from coda waves and ambient seismic noise. The results are illustrated with applications of imaging velocity structures in Alaska and California. WEGLER *et al.* use the theory of radiative transfer to model the transport of seismic energy in 2-D and 3-D acoustic random media. The theory accounts correctly for the direct wave front, envelope broadening caused by multiple forward scattering, and late coda caused by multiple wide-angle scattering. However, for very heterogeneous media the radiative transfer results differ from those of the full wave equation. ZENG develops a set of scattered wave energy equations that include scattered surface waves and conversions of body waves to surface wave scattering. Numerical results show that scattered wave energy can be approximated well by body-wave scattering at earlier times and short distances, but at large distances scattered surface waves dominate the scattered body waves at surface stations.

MAEDA *et al.* find that the direct and scattered fundamental-mode Rayleigh waves are dominant in the earlier part of each coda envelope, however higher modes become dominant at lapse times later than 20,000–35,000 sec. The authors conclude that higher mode waves are uniformly distributed at large lapse time because of large velocity dispersion and/or scattering, and they dominate over the fundamental mode waves because of smaller attenuation in the lower mantle. PENG and BEN-ZION analyze evolving properties of coda waves generated by clusters of repeating earthquakes in the overlapping aftershock zones of the 1999 Izmit and Duzce earthquakes on the North Anatolian fault. The results show abrupt co-mainshock reduction of seismic velocities in the shallow crust, followed by gradual logarithmic-like recoveries. The decay rates of the repeating events in individual clusters are compatible with the Omori's power law relation for regional aftershocks. NISHIGAMI estimates a 3-D distribution of relative scattering coefficients in the source region of the 2004 Mid-Niigata Prefecture earthquake using inversion analysis of coda envelopes. An area with weaker scattering is found along the mainshock fault plane that appears to be related with the large slip area during the mainshock rupture. Hypocenters of the mainshock and major large aftershocks with M 5–6 tend to be located close to stronger scattering areas.

While editing this volume in honor of Kei Aki, we were greatly saddened (along with many others) to receive the news that Kei passed away on May 17, 2005. Several obituaries to Kei were already published and a few others are in preparation. A brief material in memory of Kei prepared by us is given at the end of this volume, followed by Kei's publication list.

Acknowledgments

We thank Shamita Das, Eystein Husebye, Tom Jordan, John McRaney, Paul Richards, Haruo Sato and the many other participants in the symposium and banquet for making the activities successful. We are grateful to the referees of the papers for critical reviews that improved the scientific quality of the volume. These include Brad Aagaard, Rasool Anooshehpoor, Niren Biswas, Michel Campillo, Judi Chester, Bob Crosson, Mamado Diallo, Eric Dunham, Mike Fehler, Phillipe Geubelle, Jean-Luc Got, Jeanne Hardebeck, Egill Hauksson, Michael Korn, Harold Magistrale, Martin Mai, Kevin Mayeda, Hisashi Nakahara, Zhigang Peng, Malcolm Sambridge, Haruo Sato, Peter Shearer, Roel Snieder, Paul Richards, Stephanie Ross, Justin Rubinstein, Toshiro Tanimoto, Cliff Thurber, Uli Wegler, Cecily Wolfe, and Yuehua Zeng.

Yehuda Ben-Zion
Department of Earth Sciences
University of Southern California
Los Angeles, CA 90089-0740, USA

William H. K. Lee
U. S. Geological Survey (Retired) MS 977,
345 Middlefield Road
Menlo Park, CA 94025, USA

Appendix 1: Citation for Kei Aki, 2004 Recipient of the Bowie Medal
by Thomas H. Jordan

Kei Aki's scientific research has expanded the frontiers of seismology for fifty years. He pioneered the electronic processing of seismic data to infer earth structure and properties of the earthquake source. Much of what we know about large earthquakes follows from his work. He was the first to measure seismic moment (for the 1964 Niigata earthquake); Aki moment has since replaced Richter magnitude as the fundamental measure of earthquake size. He discovered the fundamental scaling laws for seismic spectra and interpreted these laws in terms of physical models, such as the barrier theory of heterogeneous rupture. This work contributed significantly to both the basic understanding of the rupture process and the practical understanding of strong ground motions.

He elucidated the scattering and attenuation processes that govern the propagation of high-frequency seismic waves, from which he derived new methods for measuring earthquake size, scattering distributions, and intrinsic attenuation parameters. He demonstrated, for instance, that the intrinsic attenuation must decrease at high frequencies. He developed many novel approaches for describing aspects of seismic waves, including free-oscillation splitting, Gaussian beams, and boundary integral methods. In each case, he and his students successfully used these techniques to extract new types of information from seismograms.

He pioneered travel-time tomography as a means to study lithospheric structure beneath dense seismic arrays, publishing seminal papers almost a decade ahead of its widespread application to global seismology. He established new seismological perspectives on volcanic processes, including the relationship of seismicity and harmonic tremor to magma injection and eruption. He detected fault-zone guided waves and used their propagation characteristics to constrain the width and elasticity of the damage zones around faults.

As impressive as this abbreviated list might sound (more topics could easily be added), it fails to communicate Kei's true impact within the geoscience community. He literally wrote the book: *Quantitative Seismology*, which he co-authored with Paul Richards in 1980, is the most influential textbook and reference manual in the history of the field. As a teacher and mentor, he trained many bright students in his quest to understand the active Earth, producing over 50 Ph.D.'s who now occupy key positions in seismology worldwide. His success in guiding young scientists stems in part from the depth of his understanding, but also from his remarkable personal qualities—charm, wit, and a deep respect for the harmony and poetry of the natural world.

Kei's quiet leadership in seismology has demonstrated the subtlety and power of unselfish cooperation in research. He has held many positions—president of the Seismology Section of AGU, president of the Seismological Society of America, chair of the NAS Committee on Seismology—but his greatest leadership achievement was the creation of the Southern California Earthquake Center in 1991. As its founding director, Kei articulated a vision for SCEC in which the investigations by disciplinary working groups would be woven together into a system-level "master model" for earthquake hazard and risk in Southern California. The master-model concept led to many advances in seismic hazard analysis, such as the incorporation of GPS data into long-term earthquake forecasting, and it continues to guide the growing SCEC collaboration.

Some scientists loom so large in their fields that we must mark their impact with special honors. Kei Aki, the 2004 recipient of the Bowie Medal, is one of our giants!

Appendix 2: Introductory Remarks by Paul Richards, at the Aki Symposium, 8 am Dec. 13, 2004

We shall shortly begin our scientific sessions on the subject of "Quantitative Seismology: Fine-scale Heterogeneities in a Deforming Earth," and it is remarkable that 85 papers are included in these sessions, which therefore extend over two days.

The principal organization was done by Willie Lee, with some help from Yehuda Ben-Zion, Eystein Husebye, John McRaney, Haruo Sato, and Ru-Shan Wu — certainly not be by me — and it is very clear from the list of speakers and poster presenters that many of us have a close association, as a colleague or co-author or student, with Professor Keiiti Aki.

It is a great pleasure to be with him here, at the AGU meeting which later this week will recognize his accomplishments formally with the award of the Bowie Medal.

In geophysics, we understand that planet Earth is our laboratory. But few of us have seen our laboratory, and worked with it from so many perspectives, as Kei Aki.

For thirty years he lived in Japan, where he made the decision to pursue geophysics in 1949 and where he received his Bachelor's and Ph.D. degrees from the University of Tokyo. In 1960 he began his work in the United States with a post-doc appointment at Caltech. After a brief return to the University of Tokyo, he took up a professorship in M.I.T at the instigation of Frank Press.

For 18 years at M.I.T., where he had 36 Ph.D. students, he developed major new ideas on surface waves (interpreted via moment tensors), on source scaling, on tomography, on coda, on strong ground motions, on volcano seismology, and much besides.

In about 1984, I remember the excitement of Ta-Liang Teng, telling me that "Kei Aki is coming to California," and Kei himself telling me that he wanted to live in a place where earthquakes are recognized as a common experience.

In California, at the University of Southern California for 14 years where he had many more Ph.D. students, Kei promoted integration of scientific information about earthquakes and its public transfer, as the founding science director of the Southern California Earthquake Center. At the Center, for example, input from earthquake geologists was used together with the fault model of quantitative seismology, to generate output useful for earthquake engineers. In this work, the concept of seismic moment, which Kei had introduced with his classic paper in 1966 on the Niigata earthquake of 1964, was central to unifying information from plate tectonics, geology, geodesy, and historical and instrumental seismology. The public transfer of the integrated information was made in the form of probabilistic estimates of earthquake hazards.

Since 2000 Kei has been based in Réunion Island, living a life that few of us can imagine, accumulating seismological observations of an active volcano, from very close up. He travels the world — his laboratory — as a scientist who still is fascinated by the phenomena our planet exhibits – such as volcanoes that send out infrasound and seismic signals. They are influenced by the solid Earth, by the liquid magma, and by the atmosphere. Solid, fluid, gas. All three media contribute to signals recorded from volcanoes.

And today, in December 2004, he is with us here, in San Francisco. It has been a long and fantastic journey, full of remarkable accomplishments. Kei, we welcome you and invite your comments on the session about to begin, and the perspectives you have developed over many years.

Appendix 3: Introduction and Response by Keiiti Aki

Thank you, Paul, for your kind words.
Ladies and gentlemen:

I am Aki of Aki and Richards. Last time I attended an AGU meeting was 10 years ago, and I am very happy to have an opportunity to speak here again. The principal organizer of this symposium, Willie Lee, who is unfortunately absent today because of illness, asked me to summarize my work in the past 50 years as an introduction to the symposium.

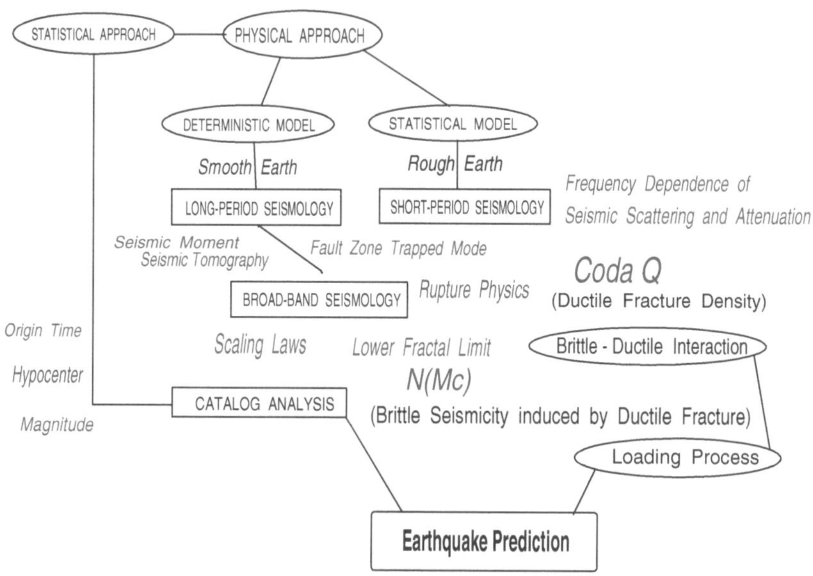

The first slide shown above summarizes my 50 years' work. It starts with earthquake prediction and ends with earthquake prediction. My first paper written in English, published in the Japanese Journal JPE in 1954, and my last paper (so far) published in EPS, the successor of JPE, in 2004, are both on earthquake prediction.

The first paper is based on statistical approach and was motivated by Norbert Wiener's "Cybernetics" published in 1948, which was the first scientific book I read seriously. After applying his prediction method to the earthquake catalog data (t,x,y,z, and M), I realized that the data available then were poor and Wiener's method based on the view of the world as a stationary linear system was too simplistic to incorporate the physics of earthquake.

So, I decided to take a physical approach to go beyond the catalog data. I took two paths; deterministic modeling and stochastic modeling. The deterministic approach required simple models of earth structures and earthquake processes and was only possible for long-period waves because short-period waves suffer from complex details of structures and processes. It started as "Long Period Seismology," producing useful concepts such as the seismic moment. It has grown to "Broad-Band Seismology" in which the upper limit of the applicable frequency range has been pushed upward steadily.

Nevertheless, the approach is basically eliminating high-frequency waves that suffer from small-scale heterogeneities. The initially smooth Earth model remains smooth because of the elimination of high-frequency waves from the data.

On the other hand, the stochastic modeling approach accepts the existence of small-scale heterogeneities and tries to find their statistical properties. It cannot define the heterogeneities specifically in time and space, but can demonstrate their existence.

In this symposium we have Raul Madariaga as an invited speaker for the deterministic approach, and Michel Campillo for the stochastic approach.

I mentioned at the beginning about my latest paper published in EPS. This paper describes my current view of earthquake and volcano precursors reached after 50 years' of travel along deterministic and stochastic modeling paths. I now see some light at the end of tunnel, and I like to encourage you to take multiple paths to achieve your goal.

Thank you for your attention.

Comments at the End of the Symposium

You all agree that it was a great symposium with uniformly excellent papers. Needless to say, I am very much pleased and impressed by new developments as well as over-all maturity attained in Seismology. It makes a great contrast to one of the first western AGU meetings I attended in 1962. It was held in the beautiful Stanford campus and the seismology session was chaired by Leon Knopoff. It was attended by half a dozen people; only the speakers were there.

I like to thank Willie Lee, Yehuda Ben-Zion, Eystein Husebye, Haruo Sato and John McRaney for organizing the symposium.

Thank you all for your participation.

Pure appl. geophys. 163 (2006) 267–299
0033–4553/06/030267–33
DOI 10.1007/s00024-005-0020-z

Pure and Applied Geophysics

Modelling Rupture Dynamics of a Planar Fault in 3-D Half Space by Boundary Integral Equation Method: An Overview

XIAOFEI CHEN,[1] and HAIMING ZHANG[1]

Abstract—In this article, we first reviewed the method of boundary integral equation (BIEM) for modelling rupture dynamics of a planar fault embedded in a 3-D elastic half space developed recently (ZHANG and CHEN, 2005a,b). By incorporating the half-space Green's function, we successfully extended the BIEM, which is a powerful tool to study earthquake rupture dynamics on complicated fault systems but limited to full-space model to date, to half-space model. In order to effectively compute the singular integrals in the kernels of the fundamental boundary integral equation, we proposed a regularization procedure consisting of the generalized Apsel-Luco correction and the Karami-Derakhshan algorithm to remove all the singularities, and developed an adaptive integration scheme to efficiently deal with those nonsingular while slowly convergent integrals. The new BIEM provides a powerful tool for investigating the physics of earthquake dynamics. We then applied the new BIEM to investigate the influences of geometrical and physical parameters, such as the dip angle (δ) and depth (h) of the fault, radius of the nucleation region (R_{asp}), slip-weakening distance (D_c), and stress inside (T_i) and outside (T_e) the nucleation region, on the dynamic rupture processes on the fault embedded in a 3-D half space, and found that (1) overall pattern of the rupture depends on whether the fault runs up to the free surface or not, especially for strike-slip, (2) although final slip distribution is influenced by the dip angle of the fault, the dip angle plays a less important role in the major feature of the rupture progress, (3) different value of h, δ, R_{asp}, T_e, T_i and D_c may influence the balance of energy and thus the acceleration time of the rupture, but the final rupture speed is not controlled by these parameters.

Key words: Dynamic rupture, boundary integral equation, half space, Green's function.

1. Introduction

Dynamic processes of earthquake sources are complex at all scales. For example, the complexities may come from the heterogeneous distribution of tectonic load (YAMASHITA, 1976; MIKUMO and MIYATAKE, 1978; MIYATAKE, 1980a,b; DAY, 1982b; CHEN *et al.*, 1987), complex geometry of the fault system in large scale (tens to hundreds of kilometers) (KOLLER *et al.*, 1992; HARRIS and DAY, 1993; AOCHI *et al.*, 2000; TADA *et al.*, 2000; AOCHI and FUKUYAMA, 2002), and the existence of barrier in small scale (hundreds of meters to several kilometers) (CHEN *et al.*, 1987; DUNHAM

[1]Laboratory of Computational Geodynamics, School of Earth and Space Sciences, Peking University, Beijing, 100871, China. E-mail: xfchen@pku.edu.cn

et al., 2003). With the increase of near-field observation data in both amount and quality, one can map rupture processes by conducting kinematic inversion (e.g., HARTZELL and HEATON, 1983; IWATA *et al.*, 2000). Although such inversions indeed help us to see the complexity of earthquake ruptures, they cannot provide quantitative understanding of physical processes of the ruptures. An effective approach to understand the physics of earthquake sources is the dynamic modelling.

Since the pioneering work of KOSTROV (1964, 1966), numerous studies have been focused on the dynamic modelling of earthquake ruptures by using a variety of methods, including the finite-difference method (FDM), the finite-element method (FEM) and the boundary integral equation method (BIEM). During this course, Prof. K. Aki and his students have made a remarkable contribution (e.g., MADARIAGA, 1976; DAS and AKI, 1977; DAS, 1980, 1981; DAS and KOSTROV, 1987; COCHARD and MADARIAGA, 1994; CHEN and AKI, 1996; MADARIAGA and COCHARD, 1996; FUKUYAMA and MADARIAGA, 1995, 1998; MADARIAGA *et al.*, 1998; MADARIAGA and OLSEN, 2000).

The FDM is firstly introduced to earthquake source dynamics by MADARIAGA (1976) and ANDREWS (1976). Since it is flexible for heterogeneous medium and stress distribution, it is widely used by numerous authors since the 1980s (e.g., DAY, 1982a,b; MIYATAKE, 1992; HARRIS and DAY, 1993; MIKUMO and MIYATAKE, 1995; OLSEN *et al.*, 1997; MADARIAGA *et al.*, 1998; CRUZ-ATIENZA and VIRIEUX, 2004). However, it seems to be limited to a vertical or horizontal fault geometry due to a regular grid distribution. Recently, ZHANG *et al.* (2004) extended the traditional FDM to include a dipping fault. Another numerical method, the FEM, is very flexible in treating problems with irregular geometry. Since huge computation is required to achieve an adequate resolution, this method is not widely used in dynamic rupture modelling until parallel computing technique was well developed recently (OGLESBY *et al.*, 1998, 2000a,b, 2001a,b; AAGAARD *et al.*, 2001).

Dynamic rupture on the fault within the earth is a typical boundary problem. Our concern is the slip and stress history on the fault. The semi-analytic BIEM is very suitable for solving such problems with stress concentration on the tip of the crack. It is now becoming one of the most favorable tools for studying the dynamic rupture processes of complex earthquake fault systems (e.g., KOLLER *et al.*, 1992; COCHARD and MADARIAGA, 1994; FUKUYAMA and MADARIAGA, 1995, 1998; TADA *et al.*, 2000; AOCHI *et al.*, 2000; AOCHI and FUKUYAMA, 2002). The boundary integral equations (abbreviated to BIEs hereafter) can be established as long as the Green's function is known. Thus the complexity of the problem-solving depends heavily on the specific form of the Green's function. Since the Green's function for elastic full space has a simple closed analytic form (see, e.g., AKI and RICHARDS, 2002), the corresponding BIEM for the full-space problem is less complex, thus the dynamic rupture problems of a complex earthquake fault system in an elastic full space have been extensively investigated to date (e.g., DAS and AKI, 1977; DAS, 1980, 1981; KOLLER *et al.*, 1992; COCHARD and MADARIAGA, 1994; FUKUYAMA and MADARIAGA, 1995, 1998; AOCHI *et al.*, 2000). In reality, the full-space model is not adequate to interpret the dynamic rupture process of an earthquake, because most of the disastrous earthquakes either rupture the ground surface or occur in shallower depth, for which the influence of the ground surface (a traction-free surface) may play an important role and cannot be neglected. Owing to the complexity of Green's function for elastic half space, the BIEM had not been rigorously extended to the problem of half space until the recent breakthrough we have made (ZHANG, 2004; CHEN and ZHANG, 2004; ZHANG and CHEN, 2005a,b).

In the first part of this paper, we briefly summarize the methodology of our recent work (ZHANG and CHEN, 2005a,b) in which the BIEM was extended to elastic half space and thus the influence of ground surface can be accurately included. The essential advancements of our new method are to use the half-space Green's function and to resolve the corresponding mathematical difficulties. Then, in the second part, we shall apply the new method to investigate the influences of different parameters on the dynamic rupture processes of a planar fault in 3-D elastic half space. These parameters include geometrical parameters, i.e., dip angle (δ) and depth (h) of the fault and radius of the nucleation region (R_{asp}), and physical parameters, i.e., slip-weakening distance (D_c), and stress inside (T_i) and outside (T_e) the nucleation region. Although inhomogeneous distribution of the physical parameters can be easily incorporated, we assume that all the parameters are homogeneous throughout this study to capture the basic physics of the dynamic ruptures.

2. *Boundary Integral Equations*

For simplicity, we consider the dynamic rupture on a rectangular planar fault with a depth h ($h \geq 0$) and a dip angle δ ($0° \leq \delta \leq 90°$) (Fig. 1). The long side of the

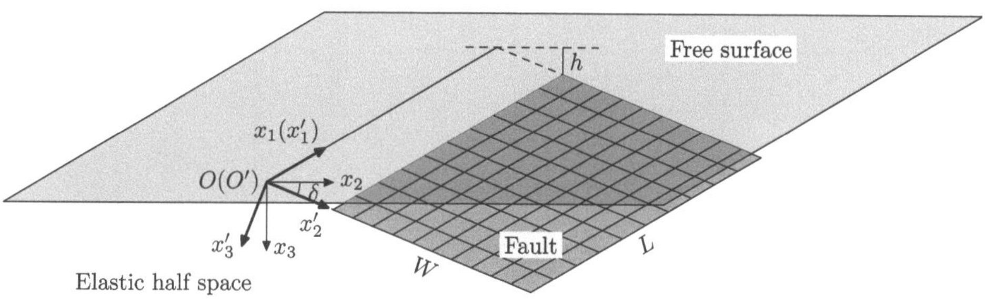

Figure 1
Geometry model of this study. The fault is simplified as a dipping rectangle with length L, width W, and a dip angle δ ($0° \leq \delta \leq 90°$). $h \geq 0$ is the distance between the upper bound of the fault and the free surface. $Ox_1x_2x_3$ and $O'x'_1x'_2x'_3$ are coordinate systems built on the free surface and the fault, where axes x_1 and x_2 are in the free surface, and axes x'_1 and x'_2 are in the fault plane. Origins O and O' are concurrent, and axes x_1 and x'_1 are concurrent, which are consistent with the intersection line of the fault and the free surface.

fault is parallel to the free surface. We assume that only shear slip occurs, the slip is only allowed in the direction of the stress drop, and no back slip is allowed. As usually takes place in dynamic modelling (e.g., FUKUYAMA and MADARIAGA, 1998; MADARIAGA *et al.*, 1998), we initiate the rupture by giving an artificial "kick."

2.1 Fundamental Boundary Integral Equations

We start from the stress representation theorem, which can be directly derived from the displacement representation theorem (see, e.g., AKI and RICHARDS, 2002). The stress representation theorem for an internal surface in frequency domain is

$$\tilde{\tau}_{kl}(\boldsymbol{x}, \omega) = c_{klnm} \iint_{\Sigma} \Delta\tilde{u}_i(\boldsymbol{\xi}, \omega) c_{ijpq} \nu_j \partial_m \tilde{G}_{np,q}(\boldsymbol{x}, \boldsymbol{\xi}, \omega) \, \mathrm{d}S(\boldsymbol{\xi}), \tag{1}$$

where $\boldsymbol{x}$ and $\boldsymbol{\xi}$ are the field and source points, respectively, Σ is a fault plane which consists of two opposite surfaces Σ^+ and Σ^-, ω is the circular frequency, $\Delta\tilde{u}_i = \tilde{u}_i|_{\Sigma^+} - \tilde{u}_i|_{\Sigma^-}$ is the slip on the fault, $\tilde{G}_{np}$ is the Green's function for half space, c_{ijpq} and ν_j are the elastic constants and the unit vector normal to the fault, respectively. $(\tilde{\cdot})$ denotes quantity in frequency domain. Spatial derivative with respect to x_i and ξ_i are marked as $\partial_i(\cdot)$ and $(\cdot)_{,i}$, respectively. Einstein's summation convention over repeated indices is assumed. Quantities in the free-surface and the fault coordinate systems are marked as $(\cdot)$ and $(\cdot)'$, respectively.

It is observed that the field point $\boldsymbol{x}$ in equation (1) locates inside volume V (Fig. 2(a)), but not on the fault plane. However, to solve the dynamic rupture problem by BIEM, we have to establish the basic equation which links the stress with the slip on fault, accordingly the field point $\boldsymbol{x}$ must be on fault plane Σ. It seems to be

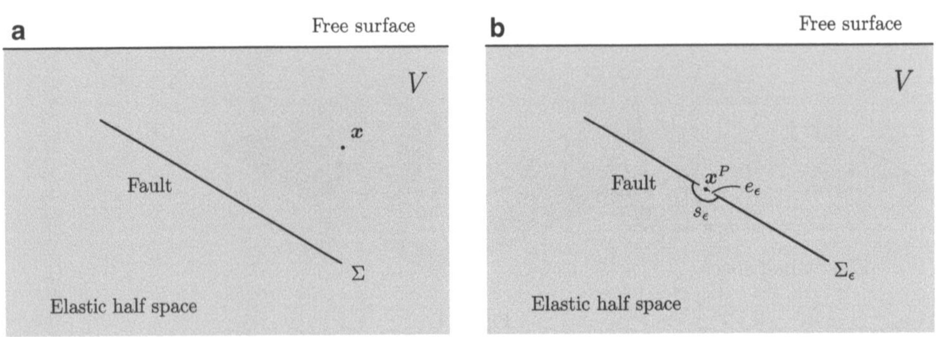

Figure 2
Sketch for changing a boundary point to an inner point. (a) Elastic half space with an inner planar fault Σ. $\boldsymbol{x}$ is an inner point of volume V. (b) Elastic half space with an inner fault Σ_ϵ, which is composed of the original planar plane without an approximation of the original boundary point $\boldsymbol{x}^P$, $\Sigma - e_\epsilon$ and a half rotary ellipsoid s_ϵ. The original boundary point $\boldsymbol{x}^P$ now becomes an inner point.

contradictory. In the next few sections we shall describe how to resolve such contradictory and associated mathematical difficulties.

Assuming $\boldsymbol{x}^P$ is located on the fault plane Σ, we define a new plane $\Sigma_\epsilon = \Sigma - e_\epsilon + s_\epsilon$ (Fig. 2(b)) to replace the original fault plane Σ in equation (1), where, e_ϵ is a small ellipse with $\boldsymbol{x}^P$ on the center, and s_ϵ is a half rotary ellipsoid with the half major axis and the half minor axis being $\epsilon/\cos\delta$ and ϵ with $\epsilon \to 0$, respectively, such that the projection of e_ϵ onto the ground surface is a small circle with a radius ϵ. Now the boundary point $\boldsymbol{x}^P$ on original fault Σ becomes an inner point of region V with Σ_ϵ. Replacing $\boldsymbol{x}$ with $\boldsymbol{x}^P$, equation (1) gives

$$\tilde{\tau}_{kl}(\boldsymbol{x}^P, \omega) = c_{klnm}\left\{\lim_{\epsilon\to 0}\iint_{\Sigma - e_\epsilon} \Delta\tilde{u}_i(\boldsymbol{\xi}, \omega) c_{ijpq} v_j \partial_m \tilde{G}_{np,q}(\boldsymbol{x}, \boldsymbol{\xi}, \omega)\, \mathrm{d}S(\boldsymbol{\xi})\right\}$$
$$+ c_{klnm}\left\{\lim_{\epsilon\to 0}\iint_{s_\epsilon} \Delta\tilde{u}_i(\boldsymbol{\xi}, \omega) c_{ijpq} v_j \partial_m \tilde{G}_{np,q}(\boldsymbol{x}, \boldsymbol{\xi}, \omega)\, \mathrm{d}S(\boldsymbol{\xi})\right\}. \quad (2)$$

Since $\partial_m \tilde{G}_{np,q}(\boldsymbol{x}, \boldsymbol{\xi}, \omega) \sim 1/\epsilon^3$ as $\epsilon \to 0$, and $\mathrm{d}S \sim \epsilon^2$, the last term becomes

$$c_{klnm}\left\{\lim_{\epsilon\to 0}\iint_{s_\epsilon} \Delta\tilde{u}_i(\boldsymbol{\xi}, \omega) c_{ijpq} v_j \partial_m \tilde{G}_{np,q}(\boldsymbol{x}, \boldsymbol{\xi}, \omega)\mathrm{d}S(\boldsymbol{\xi})\right\} = \lim_{\epsilon\to 0}\frac{\mathcal{A}_{kl}}{\epsilon},$$

where, $\mathcal{A}_{kl}$ is a bounded quantity. Accordingly, the stress tensor field on the fault can be found as follows

$$\tilde{\tau}_{kl}(\boldsymbol{x}^P, \omega) = c_{klnm} c_{ijpq} v_j w_0 \lim_{\epsilon\to 0}\iint_{\Sigma' - e'_\epsilon} \left\{\Delta\tilde{u}_i(\boldsymbol{\xi}, \omega) \partial_m \tilde{G}_{np,q}(\boldsymbol{x}, \boldsymbol{\xi}, \omega)\right\}_{\xi_3 = p_\alpha \xi_\alpha} \mathrm{d}\xi_1\, \mathrm{d}\xi_2$$
$$+ \lim_{\epsilon\to 0}\frac{\mathcal{A}_{kl}}{\epsilon}. \quad (3)$$

Notice that in equation (3), the integral domain on the fault plane is now projected onto the ground surface. $\Sigma' - e'_\epsilon$ is the projection of $\Sigma - e_\epsilon$ on the ground surface, and e'_ϵ is a circle with a radius ϵ. $w_0 = \sqrt{1 + p_1^2 + p_2^2}$, in which p_α is the coefficient of the plane equation of the fault. The repeated Greek subscripts here or hereafter also follow the Einstein's summation convention but just running over 1 to 2. For the specific problem under consideration, $p_1 = 0$ and $p_2 = \tan\delta$.

As demonstrated above the two differentiations of the Green's function in equation (3) result in the singularities with order of $1/\epsilon^3$, i.e., the hypersingularities, which are generally non-removable. Fortunately, some of them can be regularized by simply integrating by parts (e.g., SLÁDEK and SLÁDEK, 1984; NISHIMURA and KOBAYASHI, 1989; FUKUYAMA and MADARIAGA, 1995, 1998). With the following relations

$$\tilde{G}_{ij,\alpha} = \frac{\mathrm{d}}{\mathrm{d}\xi_\alpha}\tilde{G}_{ij} - p_\alpha \tilde{G}_{ij,3}, \qquad \partial_\alpha \tilde{G}_{ij} = -\tilde{G}_{ij,\alpha},$$

and taking integration by parts, we obtain

$$\tilde{\tau}_{kl}(\boldsymbol{x}^P,\omega)=\lim_{\epsilon\to 0}\iint_{\Sigma'-e'_\epsilon}\{\mathcal{W}^{kl}_{i\alpha}(\boldsymbol{x}^P,\boldsymbol{\xi},\omega)\left[\Delta\tilde{u}_{i,\alpha}(\boldsymbol{\xi},\omega)+p_\alpha\Delta\tilde{u}_{i,3}(\boldsymbol{\xi},\omega)\right]$$
$$+\mathcal{U}^{kl}_i(\boldsymbol{x}^P,\boldsymbol{\xi},\omega)\Delta\tilde{u}_i(\boldsymbol{\xi},\omega)\}_{\xi_3=p_\beta\xi_\beta}\,\mathrm{d}\xi_1\,\mathrm{d}\xi_2+\lim_{\epsilon\to 0}\frac{\mathcal{A}_{kl}}{\epsilon}, \tag{4}$$

where,

$$\mathcal{W}^{kl}_{i\alpha}(\boldsymbol{x}^P,\boldsymbol{\xi},\omega)=w_0v_j\{c_{kln\alpha}c_{ijpq}\tilde{G}_{np,q}(\boldsymbol{x}^P,\boldsymbol{\xi},\omega)-c_{ijp\alpha}c_{kln\beta}p_\beta\tilde{G}_{np,3}(\boldsymbol{x}^P,\boldsymbol{\xi},\omega)$$
$$-c_{ijp\alpha}c_{kln3}\partial_3\tilde{G}_{np}(\boldsymbol{x}^P,\boldsymbol{\xi},\omega)\}, \tag{5a}$$

$$\mathcal{U}^{kl}_i(\boldsymbol{x}^P,\boldsymbol{\xi},\omega)=w_0v_j\{c_{kln\alpha}p_\alpha(c_{ijp3}-c_{ijp\beta}p_\beta)\tilde{G}_{np,33}(\boldsymbol{x}^P,\boldsymbol{\xi},\omega)$$
$$+c_{kln3}(c_{ijp3}-c_{kln3}c_{ijp\alpha}p_\alpha)\partial_3\tilde{G}_{np,3}(\boldsymbol{x}^P,\boldsymbol{\xi},\omega)\}. \tag{5b}$$

After taking the coordinate transform for tensor, stress in the fault coordinate system can be written as

$$\tilde{\tau}'_{mn}(\boldsymbol{x}^P,\omega)=M_{m'k}M_{n'l}\tilde{\tau}_{kl}(\boldsymbol{x}^P,\omega), \tag{6}$$

where, $M_{m'k}$ is the coefficient of the coordinate transform:

$$\begin{pmatrix} M_{1'1} & M_{1'2} & M_{1'3}\\ M_{2'1} & M_{2'2} & M_{2'3}\\ M_{3'1} & M_{3'2} & M_{3'3}\end{pmatrix}=\begin{pmatrix}1 & 0 & 0\\ 0 & \cos\delta & \sin\delta\\ 0 & -\sin\delta & \cos\delta\end{pmatrix}.$$

Equation (6), combining with equation (4), is the fundamental BIE which links the fault slip to the stress field on the fault plane. However, it is a singular integral equation consisting of an explicit singular term with order of $1/\epsilon$ and some implicit net singular terms buried in those hypersingular integrals for an arbitrary point ($\boldsymbol{x}^P$) on the fault. As demonstrated later, those implicit net singular terms have the same order singularity as the explicit singular term. On the other hand, the stress field in an arbitrary point on the fault plane should not be infinity. Therefore those singular terms, both explicit and implicit, should exactly cancel out each other. Therefore, the major task for regularizing such singular integral equations is to accurately extract those net singular terms from the implicit hypersingular integrals, simultaneously the remainders are non-singular terms and will be net contribution to the stress field on the fault.

In the following sections we shall briefly describe how to extract the net singular terms from the hypersingular integrals based on a particular discretization scheme.

2.2 Discretization of the Boundary Integral Equations

In order to numerically solve the fundamental BIE, we discretize it by using a simple box-like discretization scheme (e.g., COCHARD and MADARIAGA, 1994;

FUKUYAMA and MADARIAGA, 1998; AOCHI *et al.*, 2000), in which the slip rate is expanded by box function as follows,

$$\Delta \dot{u}'_\alpha(\xi', \tau) = \sum_{l,m,n} V'^{lmn}_{\alpha} \, B^l(\xi'_1) B^m(\xi'_2) \dot{S}^n(\tau), \tag{7}$$

where,

$$B^k(x) = H(x - k\Delta s) - H[x - (k+1)\Delta s], \tag{8a}$$

$$\dot{S}^n(\tau) = H[\tau - (n-1)\Delta t] - H(\tau - n\Delta t), \tag{8b}$$

and V'^{lmn}_{α} is the expansion coefficient of slip rate in element (l, m, n), Δs and Δt are spatial and temporal step sizes, respectively. $\dot{(\cdot)}$ denotes the partial derivative with respect to time. $H(\cdot)$ is the Heaviside step function. Throughout this paper, (l, m) and n denote the serial numbers of spatial and temporal grids of source point, respectively, while (i, j) and k are those of field points. Substituting equation (7) into equation (6) yields the following frequency-domain discrete BIE

$$\tilde{\tau}'^{ij}_{st}(\omega) = \sum_{l,m,n} \tilde{C}^{ij;lmn}_{st\eta}(\omega) V'^{lmn}_{\eta}. \tag{9}$$

After taking the inverse Fourier transformation on the above equation, we finally obtain the time-domain discrete BIE,

$$\tau'^{ijk}_{st} = \sum_{l,m,n} \mathcal{F}^{-1}\left\{\tilde{C}^{ij;lmn}_{st\eta}(\omega)\right\}\Big|_{t=k\Delta t} \cdot V'^{lmn}_{\eta}, \tag{10}$$

where, $\mathcal{F}^{-1}\{\cdot\}$ stands for inverse Fourier transform. Notice that although equation (10) is the same as that of AOCHI *et al.* (2000) in form, the kernels are completely different, because of the use of a different Green's function. The explicit expression of the frequency-domain kernel $\tilde{C}^{ij;lmn}_{st\eta}(\omega)$ in equation (10) is as follows,

$$\tilde{C}^{ij;lmn}_{st\eta}(\omega) = M_{sa}M_{tb}\left\{\mathcal{E}^{\mathrm{I}:ij;lm}_{ab\eta}(\omega) + \mathcal{E}^{\mathrm{II}:ij;lm}_{ab\eta}(\omega) + \mathcal{L}^{qp}_{ab\eta}\mathcal{F}^{\mathrm{I}:ij;lm}_{qp}(\omega) + \mathcal{K}^{qp}_{ab\eta}\mathcal{F}^{\mathrm{II}:ij;lm}_{qp}(\omega)\right\}\tilde{S}^n(\omega)$$
$$+ \lim_{\epsilon \to 0} \frac{\mathcal{A}^{ij;lmn}_{st\eta}(\omega)}{\epsilon}, \tag{11}$$

where,

$$\mathcal{L}^{qp}_{ab\eta} = w_0 v_g c_{abq\alpha} p_\alpha (c_{\eta gp3} - c_{\eta gp\beta} p_\beta), \tag{12a}$$

$$\mathcal{K}^{qp}_{ab\eta} = w_0 v_g c_{abq3} (c_{\eta gp3} - c_{\eta gp\alpha} p_\alpha), \tag{12b}$$

$$\mathcal{E}^{\mathrm{I}:ij;lm}_{ab\eta}(\omega) = \int_0^{\Delta s} \Big\{\mathcal{W}^{ab}_{\eta 1}[\boldsymbol{x}'^{(i,j)}; (l+1)\Delta s, m\Delta s + \xi, p_2(m\Delta s + \xi); \omega]$$
$$- \mathcal{W}^{ab}_{\eta 1}[\boldsymbol{x}'^{(i,j)}; l\Delta s, m\Delta s + \xi, p_2(m\Delta s + \xi); \omega]\Big\}\, \mathrm{d}\xi, \tag{12c}$$

$$\mathcal{E}^{\mathrm{II}:ij;lm}_{ab\eta}(\omega) = \int_0^{\Delta s} \Big\{ \mathcal{W}^{ab}_{\eta 2}[\boldsymbol{x}'^{(i,j)};\, l\Delta s + \xi, (m+1)\Delta s, p_2(m+1)\Delta s; \omega] \\ - \mathcal{W}^{ab}_{\eta 2}[\boldsymbol{x}'^{(i,j)};\, l\Delta s + \xi, m\Delta s, p_2 m\Delta s; \omega] \Big\}\, \mathrm{d}\xi, \tag{12d}$$

$$\mathcal{F}^{\mathrm{I}:ij;lm}_{qp}(\omega) = \lim_{\epsilon\to 0} \iint_{\Sigma' - e'_\epsilon} \tilde{G}_{qp,33}(\boldsymbol{x}'^{(i,j)}; \xi_1, \xi_2, p_2\xi_2; \omega) B^l(\xi_1) B^m(\xi_2)\, \mathrm{d}\xi_1\, \mathrm{d}\xi_2, \tag{12e}$$

$$\mathcal{F}^{\mathrm{II}:ij;lm}_{qp}(\omega) = \lim_{\epsilon\to 0} \iint_{\Sigma' - e'_\epsilon} \Big[\partial_3 \tilde{G}_{qp,3}(\boldsymbol{x}'^{(i,j)}; \xi_1, \xi_2, p_2\xi_2; \omega)\Big] B^l(\xi_1) B^m(\xi_2)\, \mathrm{d}\xi_1\, \mathrm{d}\xi_2, \tag{12f}$$

$$\left|\mathcal{A}^{ij;lmn}_{st\eta}(\omega)\right| < +\infty, \tag{12g}$$

and $\boldsymbol{x}'^{(i,j)}$ represents the $(i,j)^{\text{th}}$ spatial grid points on the fault plane. In the following, we called the elements with $(i,j) \neq (l,m)$ *non-collocation elements*, and those with $(i,j) = (l,m)$ *collocation elements*.

With the box-like discretization scheme, the fundamental BIE (6) has been derived to discrete BIE (10). Accordingly, our main task becomes to evaluate the kernel $\tilde{C}^{ij;lmn}_{st\eta}(\omega)$ which consists of four terms besides the explicit singular term. The equation which defined the kernel, equation (10), is obviously singular.

3. *Regularization and Computation of Integral Kernels*

To solve the singular discrete BIE, we shall first substitute the specific expression of the Green's function for an elastic half space into the singular equation (11) defined for the kernel, and then regularize the singularities by a two-step regularization approach.

3.1 Green's Function for Half Space

Since the pioneering work of LAMB (1904), numerous authors have presented a variety of effective methods to compute the Green's functions for elastic half space or flat multilayered half-space media (e.g., CAGNIARD, 1939; PEKERIS, 1955a,b; HELMBERGER, 1968; FUCHS and MÜLLER, 1971; JOHNSON, 1974; LUCO and APSEL, 1983; YAO and HARKRIDER, 1983; WANG and HERMANN, 1980; CHEN, 1999). In this study we adopt the Green's function given by the wavenumber integration method. The Green's function in frequency domain for an elastic half space can be expressed as follows (see e.g., CHEN, 1999)

$$\tilde{G}_{ij}(\boldsymbol{x}, \boldsymbol{\xi}, \omega) = \frac{1}{4\pi\mu} \sum_{p=1}^{5} g^{(p)}_{ij}(\gamma_1, \gamma_2) \int_0^{+\infty} A^{(1)}_p(x_3, \xi_3, k, \omega) J_{n_{(p)}}(kr)\, \mathrm{d}k, \tag{13}$$

where $\gamma_1 = (x_1 - \xi_1)/r$ and $\gamma_2 = (x_2 - \xi_2)/r$ are direction factors, $g^{(p)}_{ij}$ and $n_{(p)}$ are summarized in Table 1, and

Table 1

Expressions for $g_{ij}^{(p)}$ and values for $n_{(p)}$

p	1	2	3	4	5
$g_{11}^{(p)}$	$\gamma_1^2+\gamma_2^2$	$-\gamma_1^2+\gamma_2^2$	0	0	0
$g_{12}^{(p)}$	0	$-2\gamma_1\gamma_2$	0	0	0
$g_{13}^{(p)}$	0	0	$-\gamma_1$	0	0
$g_{21}^{(p)}$	0	$-2\gamma_1\gamma_2$	0	0	0
$g_{22}^{(p)}$	$\gamma_1^2+\gamma_2^2$	$\gamma_1^2-\gamma_2^2$	0	0	0
$g_{23}^{(p)}$	0	0	$-\gamma_2$	0	0
$g_{31}^{(p)}$	0	0	0	$-\gamma_1$	0
$g_{32}^{(p)}$	0	0	0	$-\gamma_2$	0
$g_{33}^{(p)}$	0	0	0	0	-1
$n_{(p)}$	0	2	1	1	0

$$A_1^{(1)}(x_3,\xi_3,k,\omega)=\frac{k\beta^2}{2\omega^2}\left[\left(\frac{k^2}{\gamma}e^{\mathrm{P}}-\nu e^{\mathrm{S}}\right)-\frac{\bar{\mathcal{R}}(k)}{\mathcal{R}(k)}\left(\frac{k^2}{\gamma}e^{\mathrm{PP}}+\nu e^{\mathrm{SS}}\right)+\frac{4k^2\nu\chi}{\mathcal{R}(k)}\left(e^{\mathrm{SP}}+e^{\mathrm{PS}}\right)\right]$$
$$+\frac{k}{2\nu}\left(e^{\mathrm{S}}+e^{\mathrm{SS}}\right), \tag{14a}$$
$$A_2^{(1)}(x_3,\xi_3,k,\omega)=A_1^{(1)}(x_3,\xi_3,k,\omega)-\frac{k}{\nu}\left(e^{\mathrm{S}}+e^{\mathrm{SS}}\right), \tag{14b}$$
$$A_3^{(1)}(x_3,\xi_3,k,\omega)=\frac{k^2\beta^2}{\omega^2}\Bigg[\mathrm{sgn}(x_3-\xi_3)\left(e^{\mathrm{P}}-e^{\mathrm{S}}\right)+\frac{\bar{\mathcal{R}}(k)}{\mathcal{R}(k)}\left(e^{\mathrm{PP}}+e^{\mathrm{SS}}\right)$$
$$-\frac{4\chi}{\mathcal{R}(k)}\left(k^2e^{\mathrm{SP}}+\nu\gamma e^{\mathrm{PS}}\right)\Bigg], \tag{14c}$$
$$A_4^{(1)}(x_3,\xi_3,k,\omega)=\frac{k^2\beta^2}{\omega^2}\Bigg[\mathrm{sgn}(x_3-\xi_3)\left(e^{\mathrm{P}}-e^{\mathrm{S}}\right)-\frac{\bar{\mathcal{R}}(k)}{\mathcal{R}(k)}\left(e^{\mathrm{PP}}+e^{\mathrm{SS}}\right)$$
$$+\frac{4\chi}{\mathcal{R}(k)}\left(\nu\gamma e^{\mathrm{SP}}+k^2e^{\mathrm{PS}}\right)\Bigg], \tag{14d}$$
$$A_5^{(1)}(x_3,\xi_3,k,\omega)=\frac{k\beta^2}{\omega^2}\left[\left(\gamma e^{\mathrm{P}}-\frac{k^2}{\nu}e^{\mathrm{S}}\right)+\frac{\bar{\mathcal{R}}(k)}{\mathcal{R}(k)}\left(\gamma e^{\mathrm{PP}}+\frac{k^2}{\nu}e^{\mathrm{SS}}\right)-\frac{4k^2\gamma\chi}{\mathcal{R}(k)}\left(e^{\mathrm{SP}}+e^{\mathrm{PS}}\right)\right], \tag{14e}$$

with

$$e^{\mathrm{P}}=e^{-\gamma|x_3-\xi_3|},\quad e^{\mathrm{S}}=e^{-\nu|x_3-\xi_3|},\quad e^{\mathrm{PP}}=e^{-\gamma(x_3+\xi_3)},\quad e^{\mathrm{SS}}=e^{-\nu(x_3+\xi_3)},$$
$$e^{\mathrm{SP}}=e^{-(\gamma x_3+\nu\xi_3)},\quad e^{\mathrm{PS}}=e^{-(\gamma\xi_3+\nu x_3)},$$

$\gamma=\sqrt{k^2-(\omega/\alpha)^2}$ $(\mathrm{Re}(\gamma)>0)$, $\nu=\sqrt{k^2-(\omega/\beta)^2}$ $(\mathrm{Re}(\nu)>0)$, $\chi=2k^2-(\omega/\beta)^2$, $\mathcal{R}(k)=\chi^2-4k^2\nu\gamma$ is the Rayleigh function, $\bar{\mathcal{R}}(k)=\chi^2+4k^2\nu\gamma$, α and β are the P and S velocity, respectively, k is the horizontal wavenumber, ω is the circular frequency, $r = \sqrt{(x_1 - \xi_1)^2 + (x_2 - \xi_2)^2}$, and $J_{n_{(m)}}(x)$ is the $n_{(m)}$-th order Bessel function. It is worth noticing that in ω-k domain, the contributions from the free surface are separated from those of direct waves which are the Green's functions for full space.

Substituting the explicit expressions of the half-space Green's function in equations (12c)–(12f), we can obtain the specific forms of the four subkernels, and find that all the elements of $\mathcal{E}^{\mathrm{I}:ij;lm}_{ab\eta}(\omega)$ and $\mathcal{E}^{\mathrm{II}:ij;lm}_{ab\eta}(\omega)$, and the non-collocation elements of $\mathcal{F}^{\mathrm{I}:ij;lm}_{pq}(\omega)$ and $\mathcal{F}^{\mathrm{II}:ij;lm}_{pq}(\omega)$ are nonsingular integrals which can be evaluated, in principle, by standard integration algorithms. The collocation elements of $\mathcal{F}^{\mathrm{I}:ij;lm}_{pq}(\omega)$ and $\mathcal{F}^{\mathrm{II}:ij;lm}_{pq}(\omega)$ are hypersingular integrals and need to be further regularized.

To end this section we derive the expressions for two particular second-order derivatives of half-space Green's function used in the singular integrals as follows,

$$\tilde{G}_{ij,33}(\boldsymbol{x}, \boldsymbol{\xi}, \omega) = \frac{1}{4\pi\mu}\sum_{p=1}^{5} g_{ij}^{(p)}(\gamma_1, \gamma_2)\int_0^{+\infty} A_p^{(2)}(x_3, \xi_3, k, \omega) J_{n_{(p)}}(kr)\,\mathrm{d}k, \tag{15a}$$

$$\partial_3\tilde{G}_{ij,3}(\boldsymbol{x}, \boldsymbol{\xi}, \omega) = \frac{1}{4\pi\mu}\sum_{p=1}^{5} g_{ij}^{(p)}(\gamma_1, \gamma_2)\int_0^{+\infty} A_p^{(3)}(x_3, \xi_3, k, \omega) J_{n_{(p)}}(kr)\,\mathrm{d}k, \tag{15b}$$

where,

$$A_p^{(2)}(x_3, \xi_3, k, \omega) = A_{p,33}^{(1)}(x_3, \xi_3, k, \omega), \tag{16a}$$

$$A_p^{(3)}(x_3, \xi_3, k, \omega) = \partial_3 A_{p,3}^{(1)}(x_3, \xi_3, k, \omega), \tag{16b}$$

and their explicit expressions are given in Appendix A.

3.2 Generalized Apsel-Luco Correction

Detailed investigation on equation (15) reveals that the integrands tend to be infinity at $k \to +\infty$ when $x_3 \to \xi_3$, which is difficult to compute numerically. We will adopt the following procedure to overcome this difficulty. By applying the Taylor's expansions at $k \to +\infty$, we separate the explicit asymptotic expressions of the integrands from the original ones. The asymptotic expressions are simple functions of k, such that the integrals of them can be carried out analytically, and the integrals of the remained parts of the original integrands become regular and can be carried out numerically. This idea was originally proposed by APSEL and LUCO (1983), where they used the static solution, rather than the Taylor's expansion, as asymptotic expression, thus we called the technique used here a *generalized Apsel-Luco correction*. Further investigations demonstrated that the usage of Taylor's expansion performs better in the accuracy, efficiency and the wide range applicability than the static solution (ZHANG, 2004; ZHANG and CHEN, 2005a).

In ω-k domain, as mentioned previously, the contributions from the free surface can be separated from those of full space. For example, $A_p^{(2)}$ and $A_p^{(3)}$ $(p = 1, 2, \ldots, 5)$ can be expressed as

$$A_p^{(l)}(x_3, \xi_3, k, \omega) = F_p^{(l)}(x_3, \xi_3, k, \omega) + S_p^{(l)}(x_3, \xi_3, k, \omega), (l = 2, 3;\ p = 1, 2, \ldots, 5), \quad (17)$$

where, $F_p^{(l)}$ and $S_p^{(l)}$ are terms due to full space and the free surface, respectively. The k-integrals in equation (15) can be rearranged in the following form

$$\int_0^{+\infty} A_p^{(l)}(x_3, \xi_3, k, \omega) J_{n_{(p)}}(kr)\,\mathrm{d}k = \int_0^{+\infty} \left\{\delta F_p^{(l)}(x_3, \xi_3, k, \omega) + S_p^{(l)}(x_3, \xi_3, k, \omega)\right\} \cdot J_{n_{(p)}}(kr)\,\mathrm{d}k + \int_0^{+\infty} F_{p,\mathrm{asy}}^{(l)}(x_3, \xi_3, k, \omega) J_{n_{(p)}}(kr)\,\mathrm{d}k, \quad (18)$$

with

$$\delta F_p^{(l)}(x_3, \xi_3, k, \omega) = F_p^{(l)}(x_3, \xi_3, k, \omega) - F_{p,\mathrm{asy}}^{(l)}(x_3, \xi_3, k, \omega), \quad (19a)$$

$$F_{p,\mathrm{asy}}^{(l)}(x_3, \xi_3, k, \omega) = \sum_{m=0}^{3} f_{p,m}^{(l)}(x_3 - \xi_3, \omega) k^m e^{-k|x_3 - \xi_3|}, \quad (19b)$$

where, $F_{p,\mathrm{asy}}^{(l)}$ is the asymptotic expression of $F_p^{(l)}$, which can be obtained by applying the Taylor's expansion on $F_p^{(l)}$ as $k \to +\infty$. Detailed expressions for $f_{p,m}^{(l)}$ are given in Appendix B. After subtracting $F_{p,\mathrm{asy}}^{(l)}$, the residual $\{\delta F_p^{(l)} + S_p^{(l)}\}$ has a rapid convergent property for any case, thus the associated integral in equation (18) is nonsingular and can be treated normally. The integral associated with $F_{p,\mathrm{asy}}^{(l)}$, however, can be evaluated analytically though it converges very slowly or does not converge at all as $\boldsymbol{x} \to \boldsymbol{\xi}$. We have

$$\int_0^{+\infty} F_{p,\mathrm{asy}}^{(l)}(x_3, \xi_3, k, \omega) J_{n_{(p)}}(kr)\,\mathrm{d}k = \sum_{m=0}^{3} f_{p,m}^{(l)}(x_3 - \xi_3, \omega) Q_{m,n_{(p)}}(r, x_3 - \xi_3). \quad (20)$$

where,

$$Q_{m,n_{(p)}}(r, x_3 - \xi_3) = \int_0^{+\infty} k^m J_{n_{(p)}}(kr) e^{-k|x_3 - \xi_3|}\,\mathrm{d}k, \quad (21)$$

and the detailed expression is listed in Appendix C. It is noticed that function $Q_{m,n_{(p)}}(r, x_3 - \xi_3)$ (see Appendix C) is singular as $\boldsymbol{x} \to \boldsymbol{\xi}$, and its singularity could be up to order three, i.e., hypersingualrity.

3.3 Regularization of Hypersingular Integrals

In the preceding section we have found that the hypersingularity of the subkernel integrals originate from those terms associated with $Q_{m,n_{(p)}}(r, x_3 - \xi_3)$, and the remaining integrals are nonsingular and can be treated normally. Next, we focus only on those singular terms associated with $Q_{m,n_{(p)}}(r, x_3 - \xi_3)$ in equations (12e) and (12f) for collocation elements, then apply the efficient method of KARAMI and DERAKHSHAN (1999) to extract the net hypersingular terms for finalizing the regularization.

Substituting equations (18) and (20) into equations (12e) and (12f) yields,

$$ {}^{S}\mathcal{F}^{\mathrm{I}:ij;ij}_{pq}(\omega)=\sum_{l=1}^{5}\sum_{m=0}^{3}\frac{f^{(2)}_{l,m}(x_3-\xi_3,\omega)}{4\pi\mu}\left\{\lim_{\epsilon\to 0}\iint_{\Delta^{ij}_{\epsilon}} g^{(l)}_{pq}(\gamma_1,\gamma_2)Q_{m,n_{(l)}}(r,x_3-\xi_3)d\xi_1 d\xi_2\right\}, \quad (22a)$$

$$ {}^{S}\mathcal{F}^{\mathrm{II}:ij;ij}_{pq}(\omega)=\sum_{l=1}^{5}\sum_{m=0}^{3}\frac{f^{(3)}_{l,m}(x_3-\xi_3,\omega)}{4\pi\mu}\left\{\lim_{\epsilon\to 0}\iint_{\Delta^{ij}_{\epsilon}} g^{(l)}_{pq}(\gamma_1,\gamma_2)Q_{m,n_{(l)}}(r,x_3-\xi_3)d\xi_1 d\xi_2\right\}, \quad (22b)$$

where ${}^{S}\mathcal{F}^{\mathrm{I}:ij;ij}_{pq}(\omega)$ and ${}^{S}\mathcal{F}^{\mathrm{II}:ij;ij}_{pq}(\omega)$ denote the singular parts of $\mathcal{F}^{\mathrm{I}:ij;ij}_{pq}(\omega)$ and $\mathcal{F}^{\mathrm{II}:ij;ij}_{pq}(\omega)$, respectively, Δ^{ij}_{ϵ} is the region of collocation element with the circle e'_{ϵ} in the center removed, as illustrated in Figure 3. It is now very clear that all hypersingularities originate from one general integral, i.e., the integral included in the bracket in the above two equations. With substitution of $g^{(l)}_{pq}$, we find that the general hypersingular integral is a linear combination of the following elementary hypersingular integrals,

$$P_{pqsmn} = \lim_{\epsilon\to 0}\iint_{\Delta^{\mathrm{col}}-e'_{\epsilon}} \frac{u_1^p u_2^q}{r^s} Q_{m,n}(u_1,u_2)\,\mathrm{d}u_1\mathrm{d}u_2, \quad (23)$$

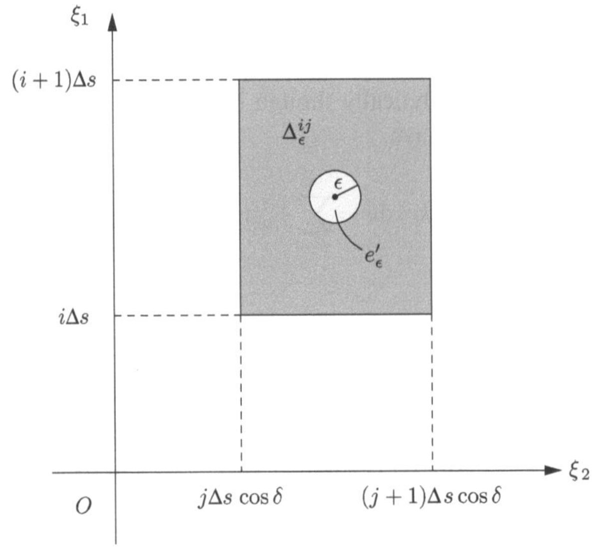

Figure 3

Sketch for the region Δ^{ij}_{ϵ} (in grey). $O\xi_1\xi_2$ is the coordinate system fixed on the free surface (c.f. Fig. 1). Rectangle is the projection of an element (i, j) on the fault to the free surface. e'_{ϵ} is a removed circle with a radius ϵ.

where, $u_1 = x_1 - \xi_1$, $u_2 = x_2 - \xi_2$, and Δ^{col} denotes the region of $\left[-\frac{1}{2}\Delta s, \frac{1}{2}\Delta s\right] \times \left[-\frac{1}{2}\Delta s \cos\delta, \frac{1}{2}\Delta s \cos\delta\right]$, which is the same for all collocation elements and independent of specific (i, j). For the problem we considered, (p, q, s) runs over (0,2,2), (2,0,2), (0,1,1) (1,0,1) and (0,0,0). $Q_{m,n}$ has an $(m+1)^{\text{th}}$-order singularity as $R \to 0$, here $R = \sqrt{(x_1 - \xi_1)^2 + (x_2 - \xi_2)^2 + (x_3 - \xi_3)^2}$. Therefore the singularity in equation (23) has an order of $m + s - p - q + 1 = 1$ (weak singularity) or 3 (hypersingularity).

For convenience, we transform the integral in equation (23) first into a local Cartesian coordinate, then into a local polar coordinate. After the first transform, the region $\Delta^{\text{col}} - e'_\epsilon$ becomes $\Delta^{\text{unit}} - e''_\epsilon$, in which Δ^{unit} denotes the region of $[-1, 1] \times [-1, 1]$ and e''_ϵ is an ellipse. Then P_{pqsmn} can be written in a local polar coordinate as

$$P_{pqsmn} = c \lim_{\epsilon \to 0} \int_0^{2\pi} \left[\int_{\rho_\epsilon(\phi)}^{\rho_M(\phi)} \rho^{p+q-s-m} \, d\rho \right] \frac{\cos^p\phi \sin^q\phi}{[M(\phi)]^s} Q'_{m,n}(\phi) \, d\phi, \tag{24}$$

where, $c = (\Delta s/2)^{p+q+1-s-m} (\cos\delta)^{q+1}$, $M(\phi) = \sqrt{\cos^2\phi + \cos^2\delta \sin^2\phi}$, $\rho_\epsilon(\phi)$ and $\rho_M(\phi)$ are the lower and upper bounds of $\rho(\phi)$, respectively, and expressions for $Q'_{m,n}(\phi)$ are listed in Appendix C.

According to equation (24), we can see that for integrals with weak singularity $(p + q - s - m = 0)$, the integrals become proper ones naturally. For integrals with hypersingularity $(p + q - s - m = -2)$, however, the integrals in polar coordinate are still with strong singularity (with factor of $1/\rho^2$). We can glean from equation (24) that

$$P_{pqsmn} = \lim_{\epsilon \to 0} \int_0^{2\pi} \frac{f(\phi)}{\rho_\epsilon(\phi)} \, d\phi - \int_0^{2\pi} \frac{f(\phi)}{\rho_M(\phi)} \, d\phi, \tag{25}$$

where

$$f(\phi) = \left(\frac{2}{\Delta s}\right) (\cos\delta)^{q+1} \frac{\cos^p\phi \sin^q\phi}{[M(\phi)]^s} Q'_{m,n}(\phi). \tag{26}$$

Substituting $r = \epsilon = \rho_\epsilon(\phi) M(\phi)$ (Karami and Derakhshan, 1999) into equation (25) yields

$$P_{pqsmn} = \lim_{\epsilon \to 0} \frac{1}{\epsilon} \int_0^{2\pi} f(\phi) M(\phi) \, d\phi - \int_0^{2\pi} \frac{f(\phi)}{\rho_M(\phi)} \, d\phi. \tag{27}$$

In equation (27), the term explicitly associated with factor $1/\epsilon$ is eventually extracted, while the remaining term is a normal integral. According to the principle of regularization mentioned previously, all such extracted pure singular terms together should be exactly cancelled with the explicit singular term (with same order singularity) in the integral kernels in equation (11), and most importantly the remaining terms' dose be an effective contribution to the kernel.

It is, however, not yet an easy task to compute the normal integrals in the kernels, because the kernels consist of a single k-integral and a double surface integral, which are both oscillatory multiple integrals. It is very time-consuming to accurately evaluate the kernel by using a standard integration algorithm. An efficient technique was proposed to speed up the computation of integral kernels (ZHANG, 2004; ZHANG and CHEN, 2005a). The most efficient algorithm to compute an oscillatory integral is to divide sub-intervals adaptively, and then apply standard numerical integral methods, such as the Gaussian integration method, to every sub-interval. The integral scheme we adopted is a combination of self-adaptive sampling based on proximate period estimation and the standard Gaussian integration method. Details of this technique can be found in ZHANG and CHEN (2005a).

4. Numerical Results of Dynamic Simulation

In a previous study (ZHANG and CHEN, 2005b), we have validated the BIEs and applied them to preliminarily investigate the effect of depth and dip angle of the fault, as well as barrier on the fault in half space by comparison with other existing available results. We found that the geometry of an dipping fault intersecting the free surface will strongly affect the rupture processes and slip distribution. Dip angle of a fault, however, has little effect on the overall pattern of rupture processes, though it will change the final slip distribution. Existence of a barrier on a fault may strongly influence the rupture process, but once the barrier is broken, final slip distribution is similar to that without the barrier. All these features are due to the interaction of the rupture and the free surface. Preliminary numerical simulation showed that, as can be expected, free surface plays an important role in the rupture behavior on a fault, especially that which intersects the free surface directly or is buried in a shallow

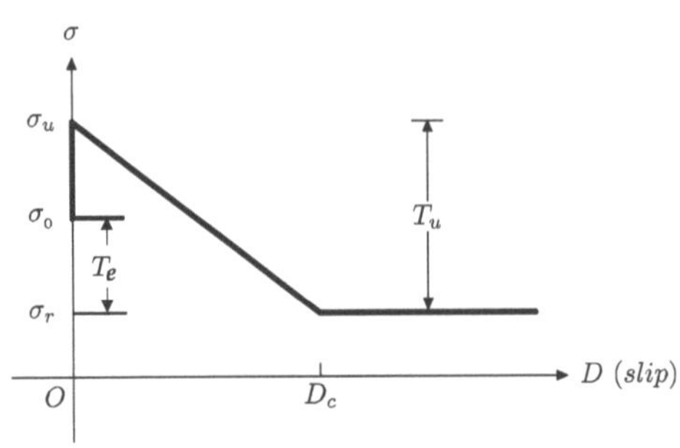

Figure 4
Sketch for the the slip-weakening law. $\sigma_u(=T_u+\sigma_r), \sigma_0(=T_e+\sigma_r)$ and σ_r are the peak strength, initial stress and residual stress, respectively, T_u is breakdown stress drop, and D_c is the critical slip-weakening distance. For simplicity and without loss of generality, σ_r is assumed to be zero throughout this paper.

depth. Further study, however, is required to understand the basic physics of the seismic rupture under the influence of free surface.

Once a rupture starts from a nucleation region, the development of the subsequent rupture is controlled by the friction law, as well as the initial stress field on the fault ($T_e + \sigma_r$) and the initial stress inside the region from which the rupture starts, i.e., asperity ($T_i + \sigma_r$). We assumed a simple friction law, i.e., the slip-weakening law as illustrated in Figure 4. Considering that only the relative stress level, i.e., stress drop, effectively influences the rupture dynamics, we set the residual stress σ_r to be zero in our simulation without loss of generality (e.g., FUKUYAMA and MADARIAGA, 1998; MADARIAGA *et al.*, 1998), and neglected all the heterogeneity to gain insight into the basic feature of the rupture. The effective controlling parameters of the slip-weakening law are the breakdown stress drop T_u and the slip-weakening distance D_c. In addition, behavior of the rupture is also influenced by geometrical parameters, such as the size of the asperity (R_{asp}), as well as depth (h) and dip angle (δ) of the fault (ZHANG and CHEN, 2005b).

In the following part, we shall investigate the effect of these physical and geometrical parameters on the rupture dynamics of a planar fault in 3-D half space by conducting a series of extensive sensitivity studies. For all the simulations below, the size of the fault is 100 km × 30 km with an element size $\Delta s = 1.0$ km. Resolution test showed that such element size is acceptable for simulating rupture dynamics of earthquake fault with a D_c of dozens of centimeters (see Appendix D). CFL ratio

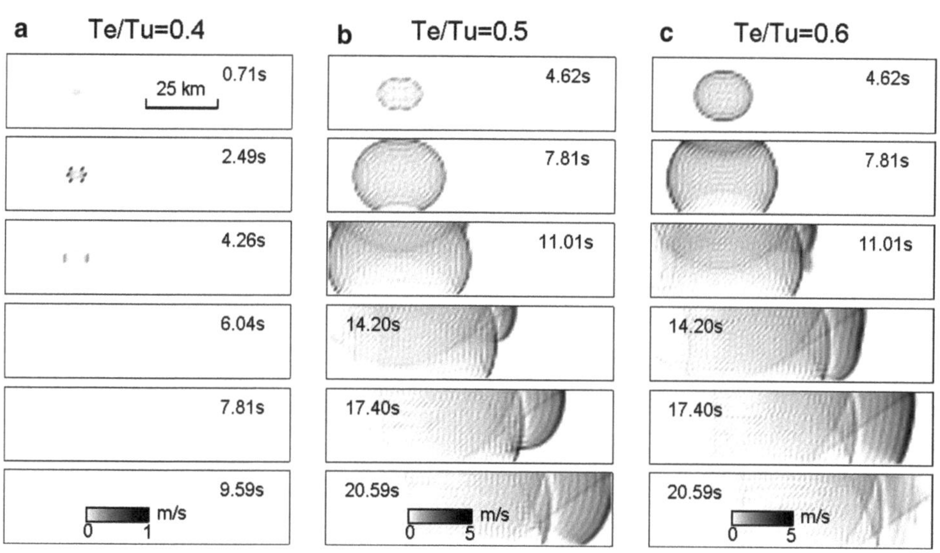

Figure 5
Comparisons of snapshots of slip rate (strike-slip) on fault in half space for (a) $T_e/T_u = 0.4$, (b) $T_e/T_u = 0.5$, and $T_e/T_u = 0.6$.

(e.g., MADARIAGA *et al.*, 1998) is 0.4. P- and S-wave velocities and density of the elastic half space are 5.60 km/s, 3.23 km/s and 3.30 g/cm^3, respectively. The breakdown stress drop T_u is fixed as 10 MPa, and T_i and T_e are given in (values the normalized values by T_u.)

4.1 *Effect of Relative Stress T_e/T_u*

Figures 5 and 6 show comparisons of snapshots of slip rate with different relative stress T_e/T_u (0.4, 0.5 and 0.6) for strike-slip and dip-slip, respectively. Other parameters are: $D_c = 29.0$ cm, $\delta = 60°$, $h = 0$ km, $T_i/T_u = 1.001$, $R_{asp} = 2.0$ km. In the strike-slip case (Fig. 5), the rupture elongates in a horizontal direction after nucleation and soon dies away for a low T_e/T_u value 0.4 (Fig. 5(a)). While for a moderate value 0.5 (Fig. 5(b)), a new rupture front propagating at a supershear speed occurs after the rupture hits the free surface. For a high value 0.6 (Fig. 5(c)), the rupture develops more rapidly at the first stage. After the appearance of a secondary front due to free surface, another new front occurs at the center of the original front at about $t = 11.01$ s, and the two secondary fronts soon coalesce and develop at a supershear speed.

In dip-slip case (Fig. 6), the rupture process for $T_e/T_u = 0.4$ is quite similar to that in the strike-slip case (Fig. 5(a)) except that the elongated direction is vertical. Unlike the corresponding results for the strike-slip case, no new secondary rupture front

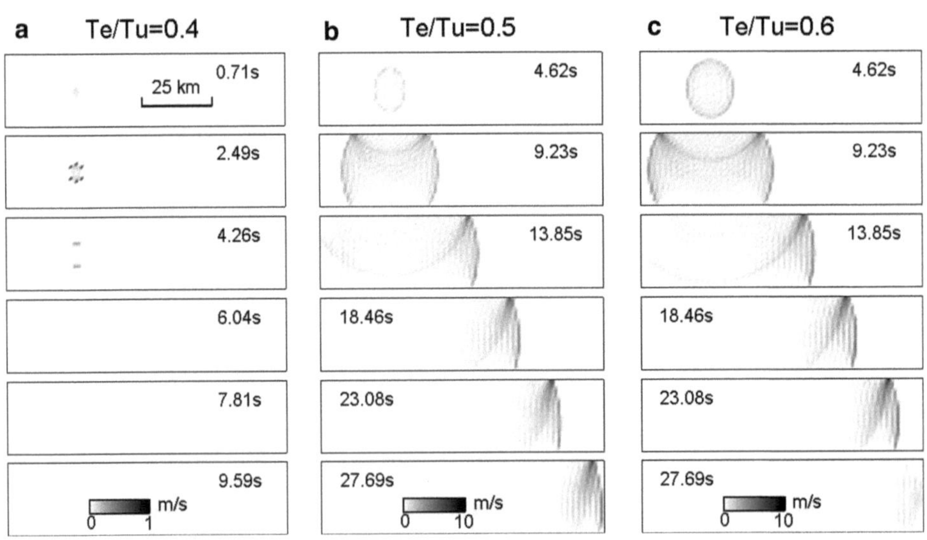

Figure 6
Comparisons of snapshots of slip rate (dip-slip) on fault in half space for (a) $T_e/T_u = 0.4$, (b) $T_e/T_u = 0.5$, and $T_e/T_u = 0.6$.

occurs for $T_e/T_u = 0.5$ and 0.6. However, the rupture process in Fig. 6(c) develops more rapidly than that in Figure 6(b) due to higher initial stress. Note that a reflected rupture front can be clearly recognized after the original front hits the free surface and is reflected. It is also interesting to note that compared with the strike-slip case, the maximum slip rate in the dip-slip case is considerably higher and concentrates near the free surface.

Figures 7(a) and (b) show the comparisons of the evolution of the rupture front along the horizontal center line of the fault for different values of T_e/T_u in the strike-slip and dip-slip cases, respectively. We can see that the major difference between the curves for $T_e/T_u = 0.5$ and $T_e/T_u = 0.6$ is that development of the rupture front for $T_e/T_u = 0.5$ at the first stage is apparently slower than that of $T_e/T_u = 0.6$. However, after some time ($\sim$10s), rupture fronts for both cases propagate at similar speed. For the strike-slip case, the occurrence of a secondary front propagating at a supershear speed for $T_e/T_u = 0.6$ is considerably earlier than that for $T_e/T_u = 0.5$. Besides, it can be clearly recognized that for $T_e/T_u = 0.6$, the new rupture front with supershear rupture speed develops more rapidly at first, and then slows down slightly. This is due to the occurrence of the new supershear front which is generated by the intrinsic nature of the fault plane rather than the influence of free surface.

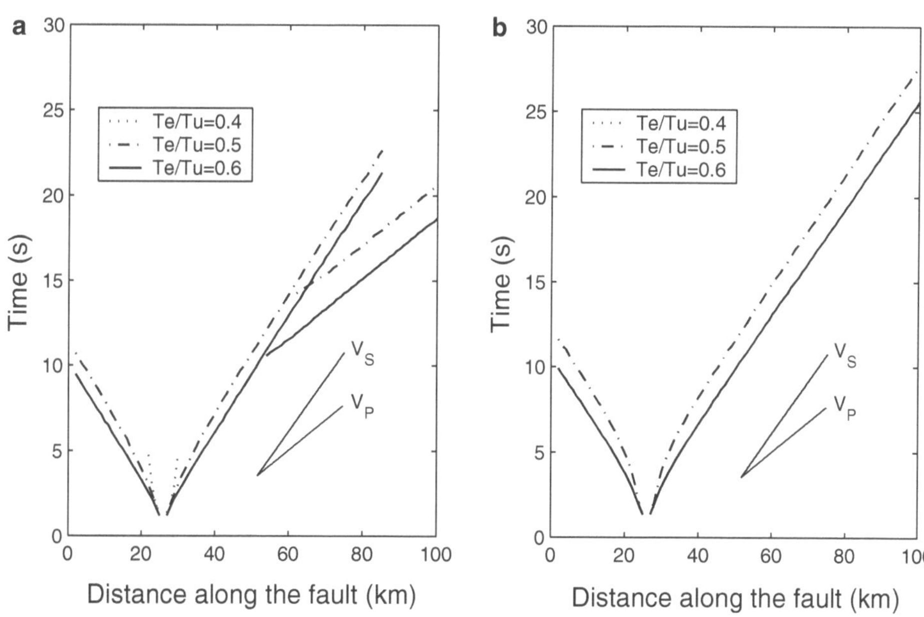

Figure 7
Comparisons of evolution history of points along the horizontal center of the fault with different T_e/T_u (0.4, 0.5 and 0.6) for (a) strike-slip and (b) dip-slip.

The results above also indicate that, on the one hand, initial stress distribution on the fault plane may determine whether the rupture can develop or not, which is well-known; on the other hand, once the rupture is able to propagate, different initial stresses may cause different rupture speed at the first stage, while only slightly influencing the later stage of development.

4.2 Effect of Slip-weakening Distance D_c

In slip-weakening law, the slip-weakening distance D_c is an important measurement of resistance on the fault plane. The larger D_c, the more difficult it is for the rupture against the friction with a given supplied energy. This is clearly demonstrated in the snapshots of slip rate for different values of D_c (11.6 cm, 29.0 cm and 46.5 cm) for both strike-slip and dip-slip (Figs. 8 and 9). In the computation, we fixed T_e/T_u as 0.5, and other parameters are the same as those in previous sections. Major features of these rupture processes are the same as those in Figures 5 and 6, and are not repeated here.

Figure 10 presents the comparisons of evolution of the rupture front along the horizontal center line of the fault for different values of D_c. Compared to the curves in Figure 7, the begin times of the development of rupture for different cases are significantly separated. The larger the D_c, the larger the resistance on the fault, and therefore the longer the time required for a rupture to propagate from the initial nucleation region.

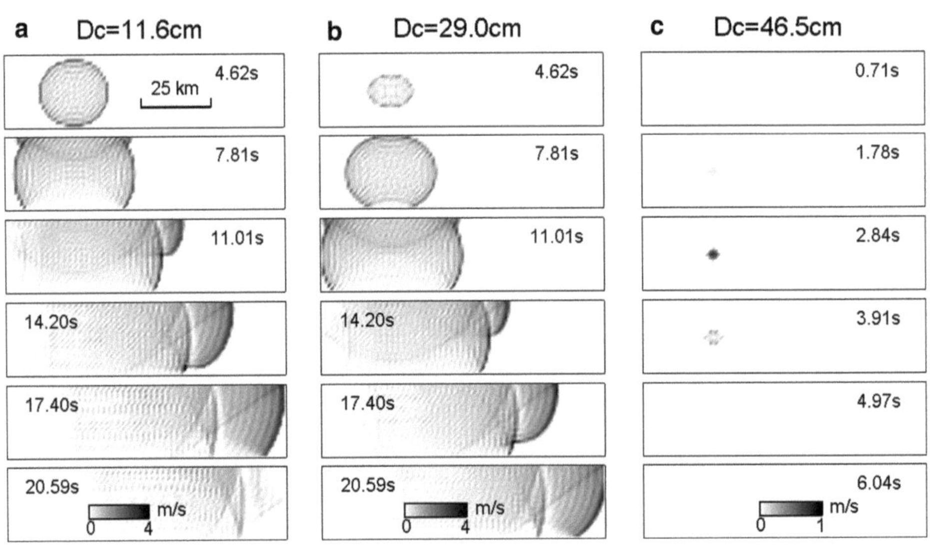

Figure 8
Comparisons of snapshots of slip rate (strike-slip) on fault in half space for (a) $D_c = 11.6$ cm, (b) $D_c = 29.0$ cm, and $D_c = 46.5$ cm.

Except T_e/T_u and D_c, other factors, such as the depth (h) and dip angle (δ) of the fault, the initial stress (T_i) inside the asperity and size (R_{asp}) of the asperity, may also influence the balance between the fracture energy and the available released strain energy, and hence the rupture process. In the following, we demonstrate the effect of these factors by comparing different evolution histories of rupture front with different values of parameter, and briefly discuss the results.

4.3 Effect of Dip Angle of the Fault δ

Figure 11 shows a comparison of evolution histories along the horizontal center line of the fault for different values of δ. We calculated results with $\delta = 60°$ and $90°$ for strike-slip (Fig. 11(a)) and $\delta = 30°$ and $60°$ for dip-slip (Fig. 11(b)). In the strike-slip case, the rupture front curves are consistent well at the beginning stage until the new secondary front occurs propagating at a supershear speed. Although the final rupture speed for $\delta = 60°$ and $90°$ is the same, the secondary front for $\delta = 60°$ occurs apparently earlier than that for $\delta = 90°$. As for the dip-slip case, the development of the rupture front for $\delta = 60°$ is slightly slower than that for $\delta = 30°$. These simulation results imply that the dip angle plays a less important role in the evolution history of the rupture process. However, on the one hand, if the rupture hits the free surface in a strike-slip, then the dip angle of the fault will strongly influence when the secondary front occurs; on the other hand, the final slip distribution depends on the

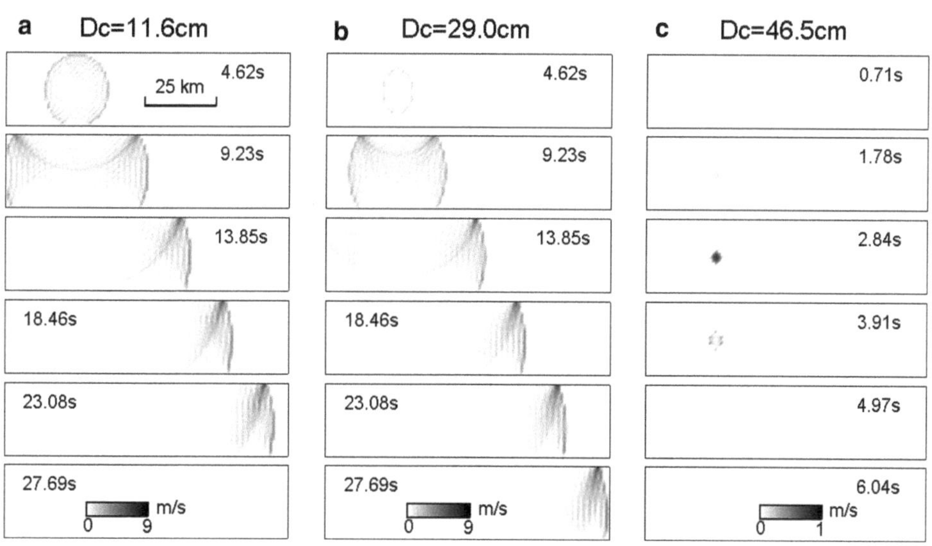

Figure 9
Comparisons of snapshots of slip rate (dip-slip) on fault in half space for (a) $D_c = 11.6$ cm, (b) $D_c = 29.0$ cm, and $D_c = 46.5$ cm.

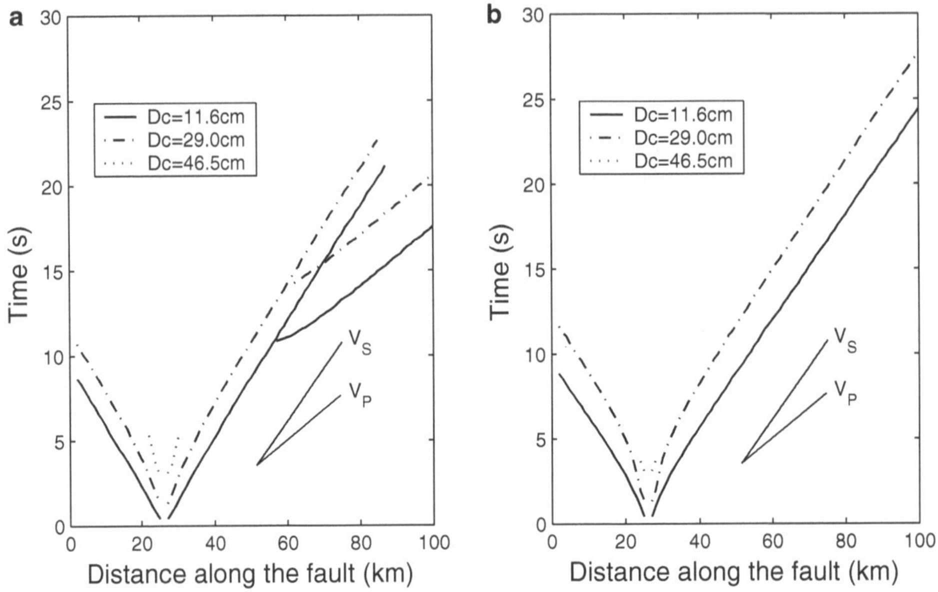

Figure 10
Comparisons of evolution history of points along the horizontal center of the fault with different D_c (11.6 cm, 29.0 cm and 46.5 cm) for (a) strike-slip and (b) dip-slip.

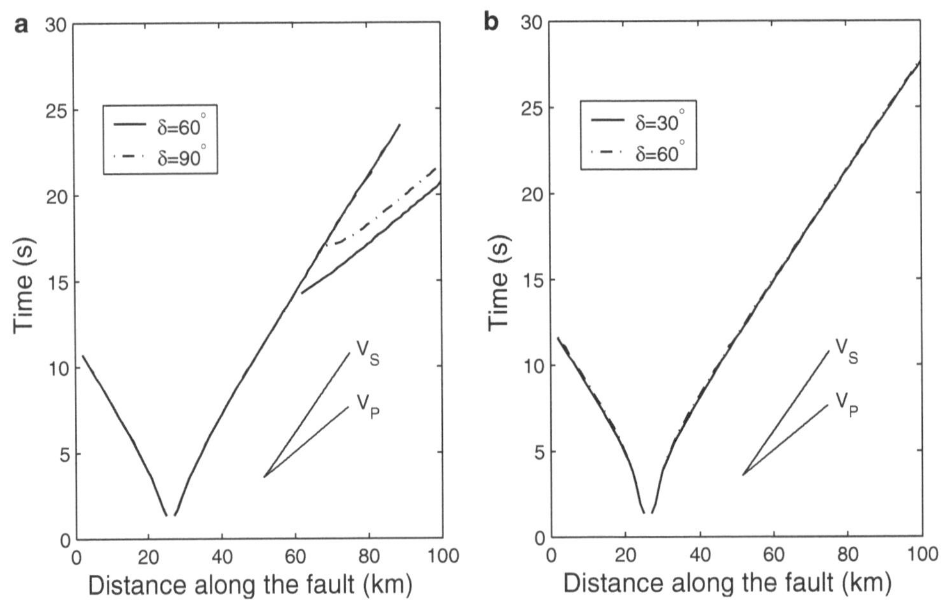

Figure 11
Comparisons of evolution history of points along the horizontal center of the fault with different δ for (a) strike-slip ($\delta = 60^\circ$ and 90°) and (b) dip-slip ($\delta = 30^\circ$ and 60°).

dip angle of the fault (ZHANG and CHEN, 2005b). Particularly, when the dip angle is very small, the final slip in the dip-slip case will be greatly enhanced (ZHANG, 2004).

4.4 Effect of Depth of the Fault h

We calculated rupture processes for various depths: $h = 0.0\,\text{km}$, $0.9\,\text{km}$ and $2.6\,\text{km}$ for both strike-slip and dip-slip, and obtained exactly the same evolution history of rupture front for $h = 0.9\,\text{km}$ and $2.6\,\text{km}$ (Fig. 12). Moreover, the curve for $h = 0.0\,\text{km}$ for dip-slip is also consistent with the one for the other two depths (Fig. 12(b)). For strike-slip, however, there is an essential difference between the result for $h = 0.0\,\text{km}$ and that for $h > 0.0\,\text{km}$, i.e., a new secondary rupture front appears (Fig. 12(a)). Such difference brings about a remarkable change in the distribution of final slip: not only is the amplitude of final slip considerably larger for $h = 0.0\,\text{km}$, but also the maximum final slip is concentrated near the free surface (ZHANG and CHEN, 2005b). It is worth mentioning that although the curves in Figure 12(b) are consistent, the final slip distribution is quite different for $h = 0.0\,\text{km}$ due to the reflected front (c.f. Figs. 6(b) and (c)). These results imply that whether the seismic fault rupture directly to the ground surface will strongly influence both the rupture process and the final slip distribution. For fault intersecting the free surface, a secondary rupture front propagating at a supershear speed may be generated for

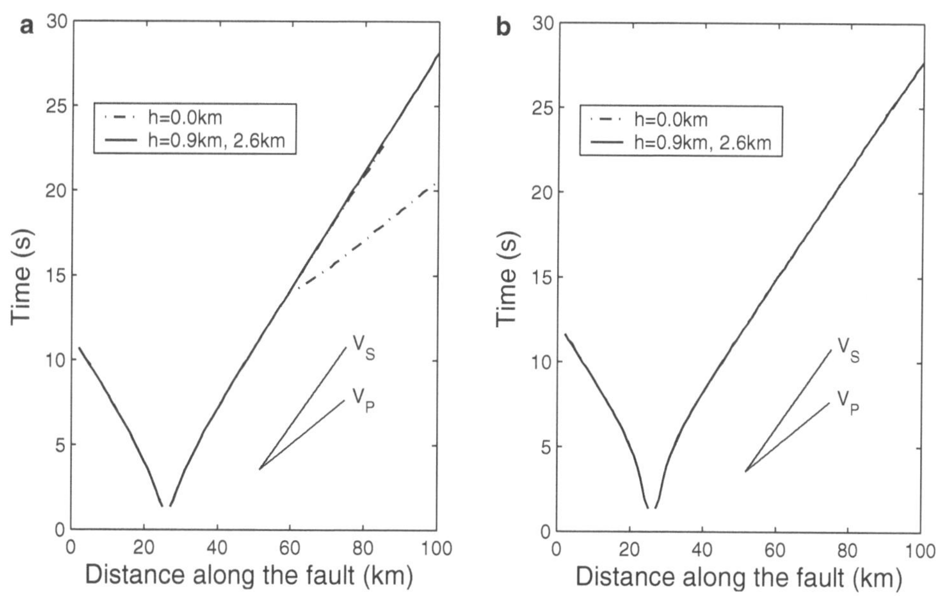

Figure 12
Comparisons of evolution history of points along the horizontal center of the fault with different h (0.0 km, 0.9 km and 2.6 km) for (a) strike-slip and (b) dip-slip.

strike-slip, and the maximum slip rate along the free surface will be greatly enhanced for dip-slip. Both these changes will play an important role in the radiation field, especially the near field.

4.5 Effect of the Stress inside the Asperity T_i

In the simulation throughout this study, we start the rupture by overstressing an asperity (e.g., FUKUYAMA and MADARIAGA, 1998). In the above models, we assumed $T_i/T_u = 1.001$, which means the initial stress inside the asperity T_i is just above the shear stress T_u. We believe this is the most common case in an real earthquake. Since T_u is assumed to be homogeneous on the entire fault, the value can be regarded as an average one. Shear stress on a real fault cannot be homogeneous; it is possible that in some patch the shear stress is larger and the rupture starts from this region first due to particular tectonic load.

We calculated rupture processes for various T_i: $T_i/T_u = 1.001$, $T_i/T_u = 1.2$ and $T_i/T_u = 1.5$ (Fig. 13). For both strike-slip and dip-slip, the evolution histories of rupture front for different T_i/T_u are similar. The difference lies in that the larger the T_i/T_u is, the shorter the acceleration time needed. As can be expected, since T_i is a measure of provided energy, larger T_i means that more energy is provided to break the fault.

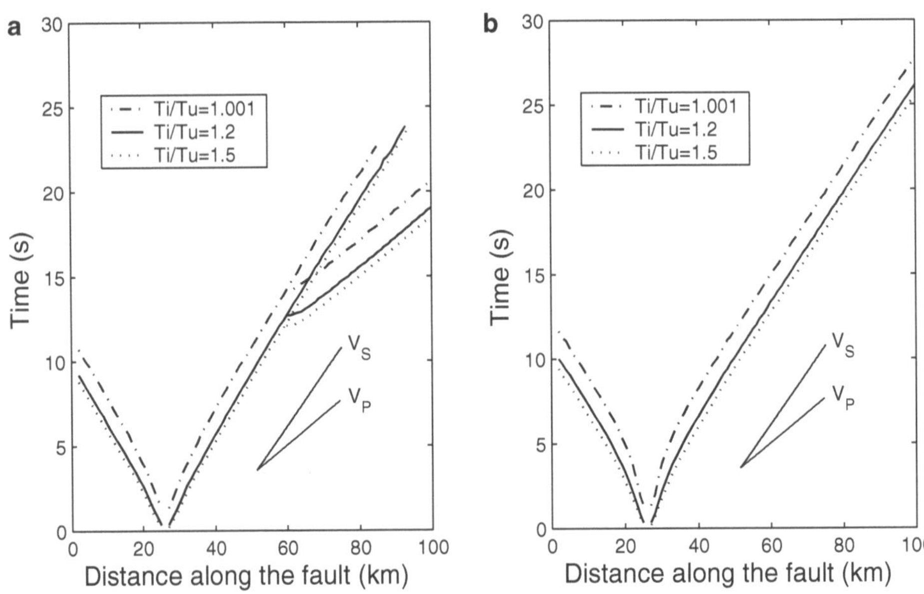

Figure 13
Comparisons of evolution history of points along the horizontal center of the fault with different T_i/T_u (1.001, 1.2 and 1.5) for (a) strike-slip and (b) dip-slip.

4.6 *Effect of the Radius of the Asperity* R_{asp}

Figure 14 shows a comparison of evolution histories along the horizontal center line of the fault for different values of R_{asp} (2 km, 3 km and 4 km). Notice that Figure 14 is almost the same as Figure 13, except the very beginning stage of evolution. This implies that after nucleation the rupture start to propagate is influenced by T_i not by R_{asp}. Similar to the feature revealed by Figure 13, the larger the R_{asp} is, the shorter the acceleration time needed. The final rupture speeds of fronts for $R_{asp} = 2$ km, 3 km and 4 km are almost the same. With a fixed T_i, the larger area of the asperity means more energy is provided, and therefore the rupture can propagate on the fault plane more easily.

5. *Conclusions and Discussions*

We have extended the BIEM, which is a powerful tool to study earthquake rupture dynamics on complicated fault systems but is limited to the full-space medium model to date, to the half-space model by incorporating the Green's function for 3-D half-space. Although the form of the final discrete BIEs are the same as those based on full-space Green's function (e.g., FUKUYAMA and MADARIAGA,

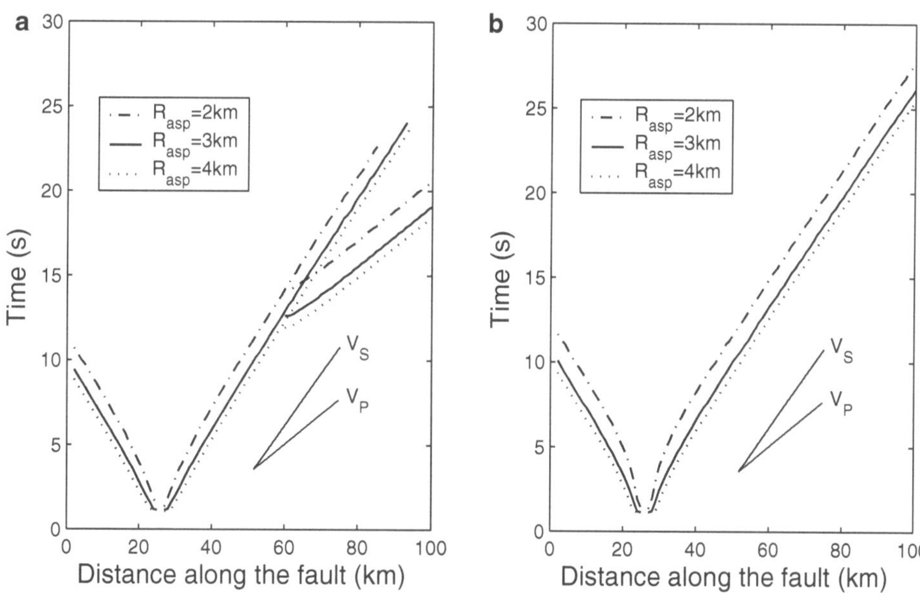

Figure 14
Comparisons of evolution history of points along the horizontal center of the fault with different R_{asp} (2 km, 3 km and 4 km) for (a) strike-slip and (b) dip-slip.

1998; Aochi *et al.*, 2000), the integral kernels, which are essential to the computation of the rupture process, are completely different. In order to compute the kernels accurately and effectively, we applied the generalized Apsel-Luco correction and the Karami-Derakhshan algorithm, to explicitly separate the singular part, which is imposed to be cancelled by the explicit singular terms in kernels. Normal integrals in the kernels are computed by using an adaptive integration scheme (Zhang and Chen, 2005b). The new BIEM provides a powerful tool for investigating the physics of earthquake dynamics.

We applied the new method to study several factors, including geometrical ones (h, δ and $R_{\rm asp}$) and physical ones (T_e, T_i and D_c). To investigate the effect of the above parameters on the fault rupture under the influence of the free surface, we compared snapshots of slip rate and evolution of rupture front for different values, and found that (1) overall pattern of the rupture depends on whether the fault runs up to the free surface or not, especially for strike-slip; (2) although final slip distribution is influenced by the dip angle of the fault, the dip angle plays a less important role in the major feature of the rupture progress; (3) different value of h, δ, $R_{\rm asp}$, T_e, T_i and D_c may influence the balance of energy and thus the acceleration time of the rupture, but the final rupture speed is not controlled by these parameters.

In the present scheme, computation of the integral kernels and the dynamic rupture process are fully separated, which is a great advantage. Since only geometry of the fault is required in the computation of kernels, it is possible to build a database for a certain fault, and the new BIEM is therefore very suitable for numerous of dynamic simulations. On the other hand, although oscillatory multiple integrals are involved and thus the computation of kernels is relatively time-consuming, the computation time required is fairly acceptable with the current computation condition based on the proposed numerical technique. For example, to compute all the kernels required for a fault divided into 80×20 elements with a maximum frequency 5.0 Hz in the computation of the Green's function, it will take about 4.8 hours on an ordinary PC-Cluster containing 16 Intel P4 microprocessors and each with 2.0 GHz and 1GB memory. In addition, there is still a large space to speed up the computation.

Acknowledgement

We dedicate this work in memory of Professor Keiiti Aki. One of us (XFC) vividly remembers the sunny and inspiring days with Kei in Southern California. We thank Prof. Yehuda Ben-Zion and Dr. Eric Dunham for critical review and constructive suggestions on this manuscript. This work was supported by the National Nature Science Foundation of China (Grant No. 40134010) and the National Basic Research Program of China (Grant No. 2004CB418404).

Appendix

A. Expressions for $A_p^{(2)}$ and $A_p^{(3)}$ $(p = 1, 2, \ldots, 5)$

$$
\begin{aligned}
A_1^{(2)}(x_3, \xi_3, k, \omega) = \frac{k\beta^2}{2\omega^2}\Big[&\big(k^2\gamma e^{\mathrm{P}} - v^3 e^{\mathrm{S}}\big) - \frac{\overline{\mathcal{R}}(k)}{\mathcal{R}(k)}\big(k^2\gamma e^{\mathrm{PP}} + v^3 e^{\mathrm{SS}}\big) \\
&+ \frac{4k^2 v\chi}{\mathcal{R}(k)}\big(v^2 e^{\mathrm{SP}} + \gamma^2 e^{\mathrm{PS}}\big)\Big] + \frac{kv}{2}\big(e^{\mathrm{S}} + e^{\mathrm{SS}}\big),
\end{aligned}
$$

$$
A_2^{(2)}(x_3, \xi_3, k, \omega) = A_1^{(2)}(x_3, \xi_3, k, \omega) - kv\big(e^{\mathrm{S}} + e^{\mathrm{SS}}\big),
$$

$$
\begin{aligned}
A_3^{(2)}(x_3, \xi_3, k, \omega) = \frac{k^2\beta^2}{\omega^2}\Big[&\mathrm{sgn}(x_3 - \xi_3)\big(\gamma^2 e^{\mathrm{P}} - v^2 e^{\mathrm{S}}\big) + \frac{\bar{\mathcal{R}}(k)}{\mathcal{R}(k)}\big(\gamma^2 e^{\mathrm{PP}} + v^2 e^{\mathrm{SS}}\big) \\
&- \frac{4v\chi}{\mathcal{R}(k)}\big(k^2 v e^{\mathrm{SP}} + \gamma^3 e^{\mathrm{PS}}\big)\Big],
\end{aligned}
$$

$$
\begin{aligned}
A_4^{(2)}(x_3, \xi_3, k, \omega) = \frac{k^2\beta^2}{\omega^2}\Big[&\mathrm{sgn}(x_3 - \xi_3)\big(\gamma^2 e^{\mathrm{P}} - v^2 e^{\mathrm{S}}\big) - \frac{\bar{\mathcal{R}}(k)}{\mathcal{R}(k)}\big(\gamma^2 e^{\mathrm{PP}} + v^2 e^{\mathrm{SS}}\big) \\
&+ \frac{4\gamma\chi}{\mathcal{R}(k)}\big(v^3 e^{\mathrm{SP}} + k^2\gamma e^{\mathrm{PS}}\big)\Big],
\end{aligned}
$$

$$
\begin{aligned}
A_5^{(2)}(x_3, \xi_3, k, \omega) = \frac{k\beta^2}{\omega^2}\Big[&\big(\gamma^3 e^{\mathrm{P}} - k^2 v e^{\mathrm{S}}\big) + \frac{\bar{\mathcal{R}}(k)}{\mathcal{R}(k)}\big(\gamma^3 e^{\mathrm{PP}} + k^2 v e^{\mathrm{SS}}\big) \\
&- \frac{4k^2\gamma\chi}{\mathcal{R}(k)}\big(v^2 e^{\mathrm{SP}} + \gamma^2 e^{\mathrm{PS}}\big)\Big],
\end{aligned}
$$

$$
\begin{aligned}
A_1^{(3)}(x_3, \xi_3, k, \omega) = \frac{k\beta^2}{2\omega^2}\Big[&-\big(k^2\gamma e^{\mathrm{P}} - v^3 e^{\mathrm{S}}\big) - \frac{\bar{\mathcal{R}}(k)}{\mathcal{R}(k)}\big(k^2\gamma e^{\mathrm{PP}} + v^3 e^{\mathrm{SS}}\big) \\
&+ \frac{4k^2 v^2\gamma\chi}{\mathcal{R}(k)}\big(e^{\mathrm{SP}} + e^{\mathrm{PS}}\big)\Big] - \frac{kv}{2}\big(e^{\mathrm{S}} - e^{\mathrm{SS}}\big),
\end{aligned}
$$

$$
A_2^{(3)}(x_3, \xi_3, k, \omega) = A_1^{(3)}(x_3, \xi_3, k, \omega) + kv\big(e^{\mathrm{S}} - e^{\mathrm{SS}}\big),
$$

$$
\begin{aligned}
A_3^{(3)}(x_3, \xi_3, k, \omega) = \frac{k^2\beta^2}{\omega^2}\Big[&-\mathrm{sgn}(x_3 - \xi_3)\big(\gamma^2 e^{\mathrm{P}} - v^2 e^{\mathrm{S}}\big) + \frac{\bar{\mathcal{R}}(k)}{\mathcal{R}(k)}\big(\gamma^2 e^{\mathrm{PP}} + v^2 e^{\mathrm{SS}}\big) \\
&- \frac{4v\gamma\chi}{\mathcal{R}(k)}\big(k^2 e^{\mathrm{SP}} + v\gamma e^{\mathrm{PS}}\big)\Big],
\end{aligned}
$$

$$
\begin{aligned}
A_4^{(3)}(x_3, \xi_3, k, \omega) = \frac{k^2\beta^2}{\omega^2}\Big[&-\mathrm{sgn}(x_3 - \xi_3)\big(\gamma^2 e^{\mathrm{P}} - v^2 e^{\mathrm{S}}\big) - \frac{\bar{\mathcal{R}}(k)}{\mathcal{R}(k)}\big(\gamma^2 e^{\mathrm{PP}} + v^2 e^{\mathrm{SS}}\big) \\
&+ \frac{4v\gamma\chi}{\mathcal{R}(k)}\big(v\gamma e^{\mathrm{SP}} + k^2 e^{\mathrm{PS}}\big)\Big],
\end{aligned}
$$

$$A_5^{(3)}(x_3,\xi_3,k,\omega) = \frac{k\beta^2}{\omega^2}\left[-(\gamma^3 e^{\mathrm{P}} - k^2 \nu e^{\mathrm{S}}) + \frac{\bar{\mathcal{R}}(k)}{\mathcal{R}(k)}(\gamma^3 e^{\mathrm{PP}} + k^2 \nu e^{\mathrm{SS}}) - \frac{4k^2\gamma^2\nu\chi}{\mathcal{R}(k)}(e^{\mathrm{SP}} + e^{\mathrm{PS}})\right].$$

B. Tables of Expressions for $f_{p,m}^{(l)}$

$f_{p,m}^{(3)} = -f_{p,m}^{(2)}$, $f_{4,m}^{(l)} = f_{3,m}^{(l)}$.

$l = 2$:

p	m			
	3	2	1	0
1	0	0	a	$b + \frac{3+p}{4}$
2	0	0	a	$\frac{b-1-p}{4}$
3	0	0	$2\zeta a$	$2\zeta b$
5	0	0	$2a$	$\frac{2b-1+p}{2}$

$l = 3$:

p	m			
	3	2	1	0
1	a	$b + \frac{5-p}{4}$	$c + f$	$-\frac{\omega^2}{16\beta^2}(7+p^2) + \frac{\omega^4}{8\beta^4}(x_3-\xi_3)^2 + g$
2	a	$b + \frac{1-p}{4}$	$d + f$	$\frac{\omega^2}{16\beta^2}(1-p^2) + g$
3	$2\zeta a$	$\zeta(2b+1-p)$	$2\zeta(3d+f)$	$\frac{\omega^4}{16\beta^4}\zeta(1-p^3)(x_3-\xi_3)^2 + 2\zeta g$
5	2a	$2b + \frac{1-3p}{2}$	$e + 2f$	$\frac{\omega^2}{8\beta^2}(1+3p^2) + \frac{\omega^4}{8\beta^4}p^3(x_3-\xi_3)^2 + 2g$

where, $p = \frac{\beta^2}{\alpha^2}$, $a = -\frac{1-p}{4}|x_3-\xi_3|$, $b = -\frac{\omega^2}{16\beta^2}(1-p^2)(x_3-\xi_3)^2$, $c = \frac{\omega^2}{16\beta^2}(9-p^2)|x_3-\xi_3|$, $d = \frac{\omega^2}{16\beta^2}(1-p^2)|x_3-\xi_3|$, $e = \frac{\omega^2}{8\beta^2}(1-5p^2)|x_3-\xi_3|$, $f = -\frac{\omega^4}{96\beta^3}(1-p^3)|x_3-\xi_3|^3$, $g = -\frac{\omega^6}{768\beta^4}(1-p^4)(x_3-\xi_3)^4$, and $\zeta = \mathrm{sgn}(x_3-\xi_3)$.

C. Tables of Expressions for $Q_{m,n_{(p)}}(r, x_3 - \xi_3)$ *and* $Q'_{m,n}(\phi)$

m	$n_{(p)}$		
	0	1	2
0	$\frac{1}{R}$	$\frac{R-\lvert x_3-\xi_3\rvert}{rR}$	$\frac{(R-\lvert x_3-\xi_3\rvert)^2}{r^2R}$
1	$\frac{\lvert x_3-\xi_3\rvert}{R^3}$	$\frac{r}{R^3}$	$\frac{2(R-\lvert x_3-\xi_3\rvert)}{r^2R} - \frac{\lvert x_3-\xi_3\rvert}{R^3}$
2	$\frac{2(x_3-\xi_3)^2-r^2}{R^5}$	$\frac{3r\lvert x_3-\xi_3\rvert}{R^5}$	$\frac{3r^2}{R^5}$
3	$\frac{3\lvert x_3-\xi_3\rvert\left[2(x_3-\xi_3)^2-3r^2\right]}{R^7}$	$\frac{3r\left[4(x_3-\xi_3)^2-r^2\right]}{R^7}$	$\frac{15r^2\lvert x_3-\xi_3\rvert}{R^7}$

$$Q'_{0,0}(\phi) = 1,$$

$$Q'_{1,0}(\phi) = \cos\delta\sin\phi,$$

$$Q'_{2,0}(\phi) = \left(2\sin^2\delta - \cos^2\delta\right)\sin^2\phi - \cos^2\phi,$$

$$Q'_{3,0}(\phi) = 3\sin\delta\sin\phi\left[\left(2\sin^2\delta - 3\cos^2\delta\right)\sin^2\phi - 3\cos^2\phi\right],$$

$$Q'_{0,1}(\phi) = [1 - \sin\delta\sin\phi]/M(\phi),$$

$$Q'_{1,1}(\phi) = M(\phi),$$

$$Q'_{2,1}(\phi) = 3\sin\phi\sin\phi M(\phi),$$

$$Q'_{3,1}(\phi) = 3M(\phi)\left[\left(4\sin^2\delta - \cos^2\delta\right)\sin^2\phi - \cos^2\phi\right],$$

$$Q'_{0,2}(\phi) = \left[Q'_{0,1}(\phi)\right]^2,$$

$$Q'_{1,2}(\phi) = 2Q'_{0,1}(\phi)/M(\phi) - Q'_{1,0}(\phi),$$

$$Q'_{2,2}(\phi) = 3[M(\phi)]^2,$$

$$Q'_{3,2}(\phi) = 15\sin\delta\sin\phi[M(\phi)]^2.$$

D. Resolution Test

Since the effect of grid size is independent of the medium model, we used the BIEM based on the full-space Green's function (AOCHI *et al.*, 2000), in which the kernels can be obtained analytically and is hence considerably more efficient, to perform the resolution test.

The fault model is shown in Figure D1. The fault is 38 km × 38 km with different grid sizes $\Delta s = 2$ km, 1 km, 0.5 km and 0.25 km (Figs. D1(a) to (d)). The grey region in the center of the fault is assumed to be the start region to approximate a "circular asperity" (circle in Fig. D1(a)).

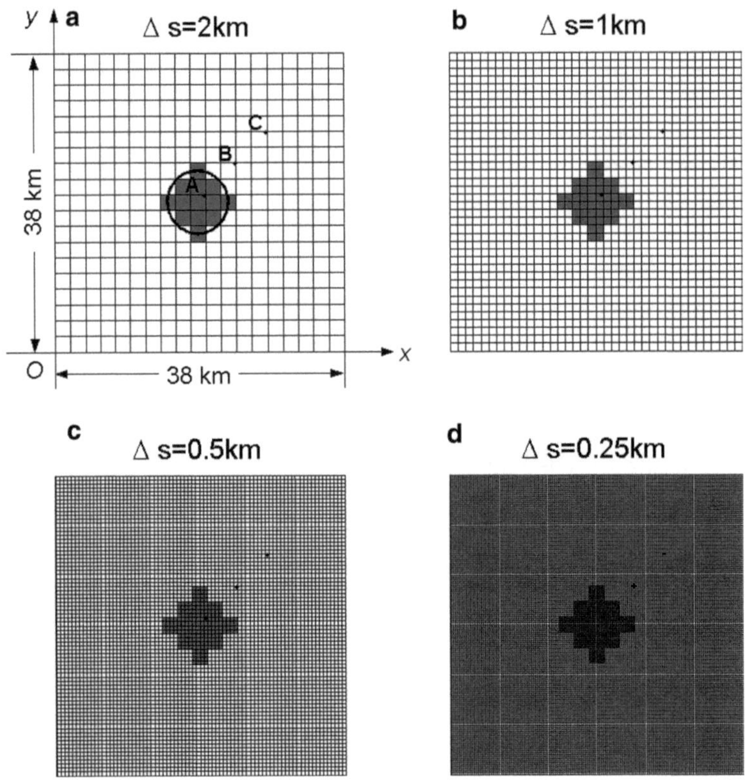

Figure D1
Fault models with different grid sizes, (a) $\Delta s = 2$ km, (b) $\Delta s = 1$ km, (c) $\Delta s = 0.5$ km and (d) $\Delta s = 0.25$ km. Grey region surrounded by a circle in the center of the fault is the start region. Points A to C denote the positions we showed the time history of slip and stress (see Fig. D3).

Although we used nondimensional variables followed MADARIAGA *et al.* (1998) and hence the absolute values of T_u, T_i and T_e are not important in our simulation, we set the following values to make a quantitative estimation: yield stress $T_u = 10$ MPa, initial stress inside the asperity $T_i = 1.01\ T_u$, initial stress outside the asperity $T_e = 0.45\ T_u$, critical slip-weakening distance $D_c = 13.9$ cm. Other parameters are the same as in the text.

Figure D2 shows the snapshots of slip rate with different grid sizes. As can be expected, the finer the grid is, the more details of the rupture history can be revealed. For $\Delta s = 2$ km, the overall feature of the rupture can be roughly recognized, however, values of slip rate, especially on/near the rupture front, are badly discontinuous. This is obviously due to an inadequate division of elements. For $\Delta s = 1$ km, although the result still seems rough, more details can be observed, and the overall pattern of rupture can be well resolved, at least in a qualitative sense.

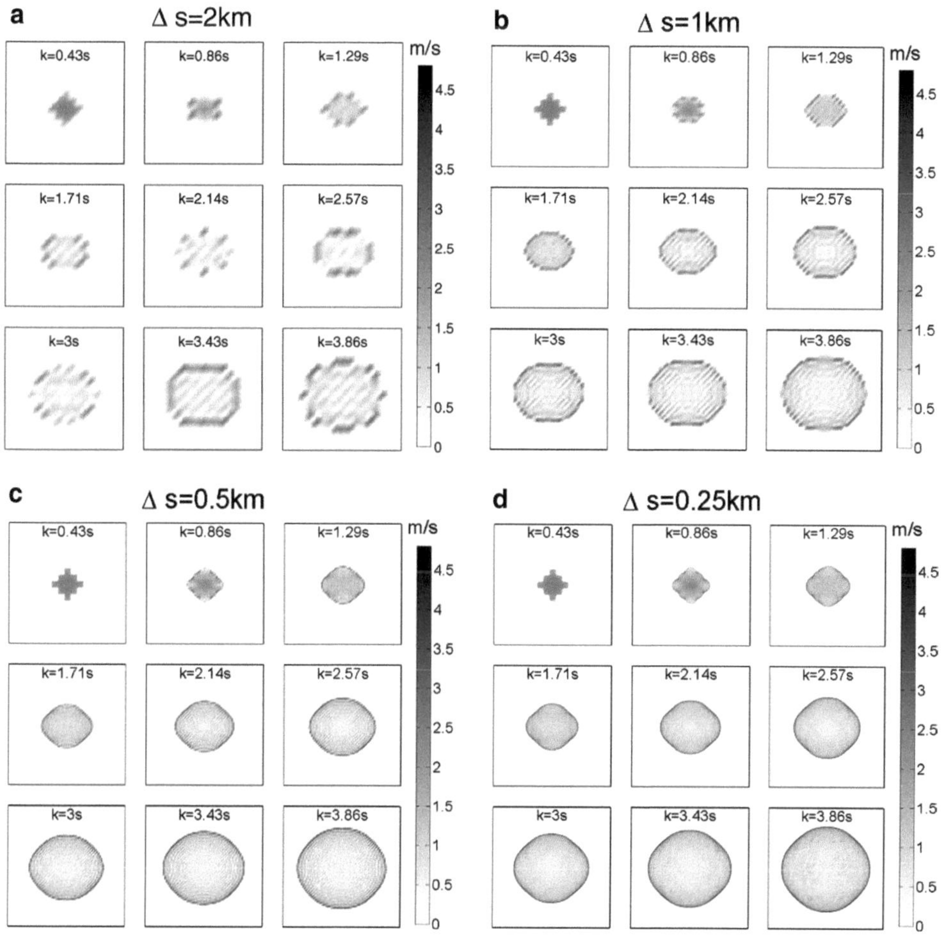

Figure D2
Snapshots of slip rate for (a) $\Delta s = 2\,\text{km}$, (b) $\Delta s = 1\,\text{km}$, (c) $\Delta s = 0.5\,\text{km}$, and (d) $\Delta s = 0.25\,\text{km}$.

Results for $\Delta s = 0.5$ km and $\Delta s = 0.25$ km are quite similar, indicating the results have already converged to a stable level.

To better see the resolution, we show comparisons of slip and stress as a function of time at (19.875, 19.875)(km) (Figs. D3 (a) and (b), see point A in Fig. D1 (a) for reference), (23.875, 23.875)(km) (Figs. D3 (c) and (d), see point B in Fig. D1 (a) for reference), and (27.875, 27.875)(km) (Figs. D3 (e) and (f), see point C in Fig. D1 (c) for reference). For points inside/near the asperity (points A and B), the slip and stress for $\Delta s = 1\text{km}$, $\Delta s = 0.5\,\text{km}$ and $\Delta s = 0.25\,\text{km}$ are favouarably consistent. At point B, slip for $\Delta s = 1\,\text{km}$ increases stepwise, taking the curve for $\Delta s = 0.25\,\text{km}$ as an envelope. The same pattern can be recognized for point C, although the difference is

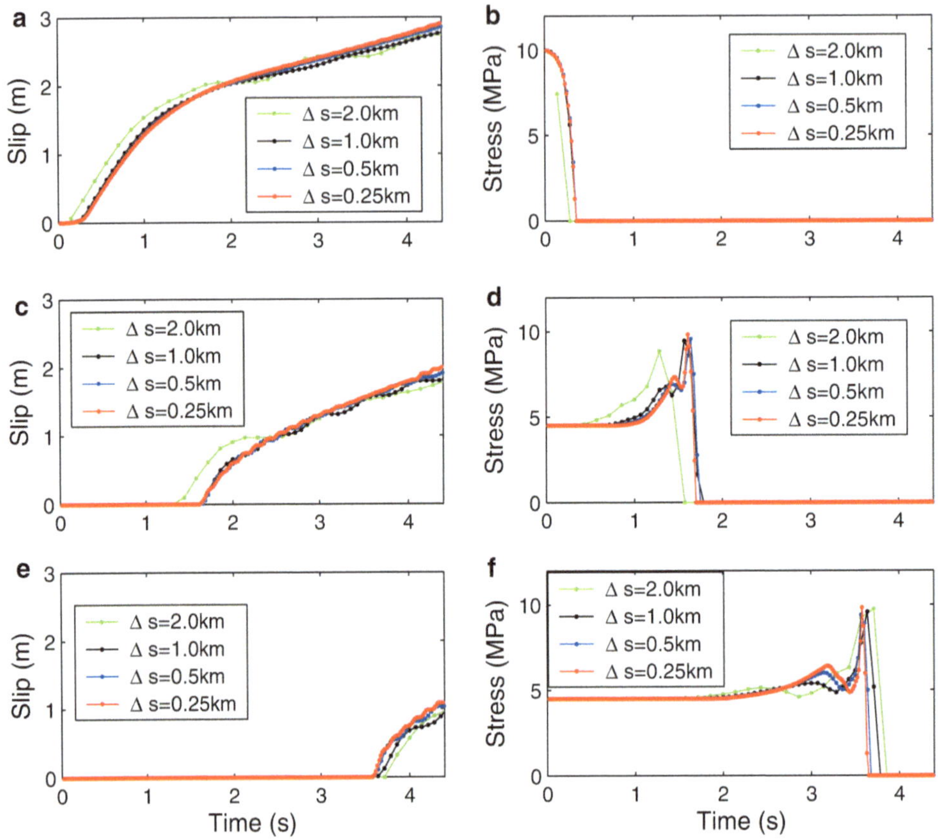

Figure D3
Comparisons of slip and stress as a function of time at points A ((a) and (b)), B ((c) and (d)) and C ((e) and (f)) (See Fig. D1 for reference).

relatively large. Results for $\Delta s = 2$ km are apparently different from others, especially at the first stage of the rupture.

The comparisons indicate that for D_c around a dozen centimeters, rupture timing has converged to an acceptable level at 1 km grid spacing, though the finer the better. Compromising the accuracy and the cost of computation of kernels for half space, we set the grid size $\Delta s = 1$ km in this study.

References

AAGAARD, B.T., HEATON, T.H., and HALL, J.F. (2001), *Dynamic earthquake ruptures in the presence of lithostatic normal stresses: Implications for friction models and heat production*, Bull. Seismol. Soc. Am. *91*, 1765–1796.

AKI, K. and RICHARDS, P.G., *Quantitative Seismology*, 2nd ed. (University Science Books, Sausality, California, 2002).

ANDREWS, D.J. (1976), *Rupture propagation with finite stress in antiplane strain*, J. Geophys. Res. *81*, 3575–3582.

AOCHI, H. and FUKUYAMA, E. (2002), *Three-dimensional nonplanar simulation of the 1992 Landers earthquake*, J. Geophys. Res. *107*, 10.1029/2000JB000061.

AOCHI, H., FUKUYAMA, E., and MATSU'URA, M. (2000), *Spontaneous rupture propagation on a non-planar fault in 3-D elastic medium*, Pure Appl. Geophys. *157*, 2003–2027.

APSEL, R.J. and LUCO, J.E. (1983), *On the Green's functions for a layered halfspace. Part II*, Bull. Seismol. Soc. Am. *73*, 931–951.

CAGNIARD, L., *Réflexion et réfraction des ondes séismiques progressives* (Gauthier-Villars, Paris, 1939).

CHEN, X.F. (1999), *Seismogram synthesis in multi-layered half space Part I. Theoretical formulations*, Earth. Res. China *13*, 149–174.

CHEN, X.F. and AKI, K. (1996), *An effective approach to determine the dynamic source parameters*, Pure Appl. Geophys. *146*, 689–696.

CHEN, X.F. and ZHANG, H.M. (2004), *Spontaneous rupture dynamics of a planar fault with arbitrary dipping orientation embedded in 3-D elastic half-space*, EOS-Trans. AGU, vol. 85, no. 47.

CHEN, Y.T., CHEN, X.F., and KNOPOFF, L. (1987), *Spontaneous growth and autonomous contraction of a two-dimensional earthquake fault*, Tectonophysics *144*, 5–17.

COCHARD, A. and MADARIAGA, R. (1994), *Dynamic faulting under rate-dependent friction*, Pure Appl. Geophys. *142*, 419–445.

CRUZ-ATIENZA, V.M. and VIRIEUX, J. (2004), *Dynamic rupture simulation of non-planar faults with a finite-difference approach*, Geophys. J. Int. *158*, 939–954.

DAS, S. (1980), *A numerical method for determination of source time functions for general three-dimensional rupture propagation*, Geophys. J. R. astr. Soc. *62*, 591–604.

DAS, S. (1981), *Three-dimensional rupture propagation and implications for the earthquake source mechanism*, Geophys. J. R. astr. Soc. *67*, 375–393.

DAS, S. and AKI, K. (1977), *A numerical study of two-dimensional spontaneous rupture propagation*, Geophys. J. R. astr. Soc. *50*, 643–668.

DAS, S. and KOSTROV, B.V. (1987), *On the numerical boundary integral equation method for 3-D shear-crack problems*, J. Appl. Mech. *54*, 99–104.

DAY, S.M. (1982a), *Three-dimensional finite-difference simulation of fault dynamics: Rectangular faults with fixed rupture velocity*, Bull. Seismol. Soc. Am. *72*, 705–727.

DAY, S.M. (1982b), *Three-dimensional simulation of spontaneous rupture: The effect of nonuniform prestress*, Bull. Seismol. Soc. Am. *72*, 1881–1902.

DUNHAM, E.M., FAVREAU, P., and CARLSON, J.M. (2003), *A supershear transition mechanism for cracks*, Science *299*, 1557–1559.

FUCHS, K. and MÜLLER, G. (1971), *Computaion of synthetic seismograms with the reflectivity method and comparison of observations*, Geophys. J. R. astr. Soc. *23*, 417–433.

FUKUYAMA, E. and MADARIAGA, R. (1995), *Integral equation method for plane crack with arbitrary shape in 3-D elastic medium*, Bull. Seismol. Soc. Am. *85*, 614–628.

FUKUYAMA, E. and MADARIAGA, R. (1998), *Rupture dynamics of a plannar fault in a 3-D elastic medium: Rate- and slip-weakening friction*, Bull. Seismol. Soc. Am. *88*, 1–17.

HARRIS, R.A. and DAY, S.M. (1993), *Dynamics of fault interaction: Parallel strike-slip faults*, J. Geophys. Res. *98*, 4461–4472.

HARTZELL, S.H. and HEATON, T.H. (1983), *Inversion of strong ground motion and teleseismic waveform data for the fault rupture history of the 1979 Imperial Valley, California, earthquake*, Bull. Seismol. Soc. Am. *73*, 1553–1583.

HELMBERGER, D.V. (1968), *The crust-mantle transition in the Bering Sea*, Bull. Seismol. Soc. Am. *58*, 179–214.

IWATA, T., SEKIGUCHI, H., and PITARKA, A. (2000), *Source and site effects on strong ground motions in near-source area during the 1999 Chi-Chi, Taiwan, earthquake*, Eos Trans. AGU *81*(48), Fall Meet. Suppl., Abstract S72B07.

JOHNSON, L. (1974), *Green's function for Lamb's problem*, Geophys. J. R. astr. Soc. *37*, 99–131.

KARAMI, G. and DERAKHSHAN, D. (1999), *An efficient method to evaluate hypersingular and supersingular integrals in boundary integrals equations analysis*, Eng. Anal. Boundary Elements *23*, 317–326.

KOLLER, M.G., BONNET, M., and MADARIAGA, R. (1992), *Modelling of dynamical crack propagation using time-domain boundary integral equations*, Wave Motion *16*, 339–366.

KOSTROV, B.V. (1964), *Self-similar problems of propagation of shear cracks*, J. Appl. Math. Mech. *28*, 1077–1087.

KOSTROV, B.V. (1966), *Unsteady propagation of longitudinal shear cracks*, J. Appl. Math. Mech. *30*, 1241–1248.

LAMB, H. (1904), *On the propagation of tremors over the surface of an elastic solid*, Phil. Trans. R. Soc. London A, *203*, 1–42.

LUCO, J.E. and APSEL, R.J. (1983), *On the Green's functions for a layered half-space. Part I*, Bull. Seismol. Soc. Am. *73*, 909–929.

MADARIAGA, R. (1976), *Dynamic of an expanding circular fault*, Bull. Seismol. Soc. Am. *66*, 639–667.

MADARIAGA, R. and COCHARD, A. (1996), *Dynamic friction and the origin of the complexity of earthquake sources*, Proc. Natl. Acad. Sci. USA *93*, 3819–3824.

MADARIAGA, R. and OSLEN, K.B. (2000), *Criticality of rupture dynamics in 3-D*, Pure Appl. Geophys. *157*, 1981–2001.

MADARIAGA, R., OLSEN, K.B., and ARCHULETA, R.J. (1998), *Modeling dynamic rupture in a 3D earthquake fault model*, Bull. Seismol. Soc. Am. *88*, 1182–1197.

MIKUMO, T. and MIYATAKE, T. (1978), *Dynamic rupture process on a three-dimensional fault with non-uniform frictions and near-field seismic waves*, Geophys. J. R. astr. Soc. *54*, 417–438.

MIKUMO, T. and MIYATAKE, T. (1995), *Heterogeneous distribution of dynamic stress drop and relative fault strength recovered from the results of waveform inversion*, Bull. Seismol. Soc. Am. *85*, 178–193.

MIYATAKE, T. (1980a), *Numerical simulations of earthquake source process by a three-dimensional crack model. Part I. Rupture process*, J. Phys. Earth *28*, 565–598.

MIYATAKE, T. (1980b), *Numerical simulations of earthquake source process by a three-dimensional crack model. Part II. Seismic wave and spectrum*, J. Phys. Earth *28*, 599–616.

MIYATAKE, T. (1992), *Reconstruction of dynamic rupture process of an earthquake with constrains of kinematic parameters*, Geophys. Res. Lett. *19*, 349–352.

NISHIMURA, N. and KOBAYASHI, S. (1989), *A regularized boundary integral equation method for elastodynamic crack problems*, Comput. Mech. *4*, 319–328.

OGLESBY, D.D., ARCHULETA, R.J. and NIELSON, S.B. (1998), *Earthquakes on dipping faults: the effects of broken symmetry*, Science *280*, 1055–1059.

OGLESBY, D.D., ARCHULETA, R.J., and NIELSON, S.B. (2000a), *The three-dimensional dynamics of dipping faults*, Bull. Seismol. Soc. Am. *90*, 616–628.

OGLESBY, D.D., ARCHULETA, R.J., and NIELSON, S.B. (2000b), *The dynamics of dipping faulting: Explorations in two dimensions*, J. Geophys. Res. *105*, 13643–13653.

OGLESBY, D.D. and DAY, S.M. (2001a), *The effect of fault geometry on the 1999 Chi-Chi (Taiwan) earthquake*, Geophys. Res. Lett. *28*, 1831–1834.

OGLESBY, D.D. and DAY, S.M. (2001b), *Fault geometry and the dynamics of the 1999 Chi-Chi (Taiwan) earthquake*, Bull. Seismol. Soc. Am. *91*, 1099–1111.

OLSEN, K.B., MADARIAGA, R., and ARCHULETA, R.J. (1997), *Three-dimensional dynamic simulations of the 1992 Landers earthquake*, Science *278*, 834–838.

PEKERIS, C. (1955a), *The seismic surface pulse*, Proc. Nat. Acad. Sci. *41*, 469–480.

PEKERIS, C. (1955b), *The seismic buried pulse*, Proc. Nat. Acad. Sci. *41*, 629–639.

SLÁDEK, V. and SLÁDEK, J. (1984), *Transient elastodynamic three-dimensional problems in crack bodies*, Appl. Math. Modelling *8*, 2–10.

TADA, T., FUKUYAMA, E., and MADARIAGA, R. (2000), *Non-hypersingular boundary integral equations for 3-D non-planar crack dynamics*, Comput. Mech. *25*, 613–626.

WANG, C.Y. and HERRMANN, R.B. (1980), *A numerical study of P-, SV-, and SH-wave generation in plane-layered medium*, Bull. Seismol. Soc. Am. *70*, 1015–1036.

YAMASHITA (1976)

YAO, Z.X. and HARKRIDER, D.G. (1983), *A generalized reflection-transmission coefficient matrix and discrete wavenumber method for synthetic seismograms*, Bull. Seismol. Soc. Am. *73*, 1685–1699.

ZHANG, H.M. (2004), *Theoretical study on the spontaneous propagation of 3-D seismic rupture on a planar fault in half space* (in Chinese), Ph.D. Thesis, Peking University, Beijing.

ZHANG, H.M. and CHEN, X.F. (2005a), *Dynamic rupture on a planar fault in 3-D half space. I. Theory*, submitted to Geophys. J. Int.

ZHANG, H.M. and CHEN, X.F. (2005b), *Dynamic rupture on a planar fault in 3-D half space. II. Validations and numerical experiments,* submitted to Geophys. J. Int.

ZHANG, W.B., IWATA, T., IRIKURA, K., PITARKA, A., and SEKIGUCHI, H. (2004), *Dynamic rupture process of the 1999 Chi-Chi, Taiwan, earthquake*, Geophys. Res. Lett. *31*, L10605, doi:10.1029/2004GL109827.

(Received May 1, 2005; accepted October 14, 2005)
Published Online First: February 8, 2006

Pure appl. geophys. 163 (2006) 301–349
0033–4553/06/030301–49
DOI 10.1007/s00024-005-0023-9

Pure and Applied Geophysics

Geological Observations of Damage Asymmetry in the Structure of the San Jacinto, San Andreas and Punchbowl Faults in Southern California: A Possible Indicator for Preferred Rupture Propagation Direction

Ory Dor,[1] Thomas K. Rockwell,[2] and Yehuda Ben-Zion[1]

Abstract—We present new *in situ* observations of systematic asymmetry in the pattern of damage expressed by fault zone rocks along sections of the San Andreas, San Jacinto, and Punchbowl faults in southern California. The observed structural asymmetry has consistent manifestations at a fault core scale of millimeters to meters, a fault zone scale of meters to tens of meters and related geomorphologic features. The observed asymmetric signals are in agreement with other geological and geophysical observations of structural asymmetry in a damage zone scale of tens to hundreds of meters. In all of those scales, more damage is found on the side of the fault with faster seismic velocities at seismogenic depths. The observed correlation between the damage asymmetry and local seismic velocity structure is compatible with theoretical predictions associated with preferred propagation direction of earthquake ruptures along faults that separate different crustal blocks. The data are consistent with a preferred northwestward propagation direction for ruptures on all three faults. If our results are supported by additional observations, asymmetry of structural properties determined in field studies can be utilized to infer preferred propagation direction of large earthquake ruptures along a given fault section. The property of a preferred rupture direction can explain anomalous behavior of historic rupture events, and may have profound implications for many aspects of earthquake physics on large faults.

Key words: Earthquake physics, fault zone structure, rock damage, material interfaces, geologic mapping, dynamic rupture.

1. *Introduction*

Small earthquakes expanding in two directions on a fault are in general a mixture of modes II and III shear ruptures. However, moderate and large earthquakes on strike-slip faults (e.g., events with magnitude larger than about M6.5) become, once they saturate the seismogenic zone, predominantly mode II ruptures (Fig. 1). Theoretical works indicate that mode II ruptures on a fault that separates different media (e.g., Weertman, 1980; Adams, 1995; Andrews and Ben-Zion, 1997;

[1]Department of Earth Sciences, University of Southern California, Los Angeles, CA 90089-0740, USA
[2]Department of Geological Sciences, San Diego State University, San Diego, CA 92182-1020, USA

Cochard and Rice, 2000; Ben-Zion and Huang, 2002; Shi and Ben-Zion, 2006) tend to evolve with continued propagation to a narrow "wrinkle-like" pulse that propagates preferentially in the direction of slip on the compliant side of the fault (Fig. 2). Characteristic features of the wrinkle-like pulse include: (1) strong dynamic reduction of normal stress at the propagating tip, (2) asymmetric particle motion on the different sides of the fault, (3) self-sharpening with propagation distance, and (4) preferred direction of rupture propagation and associated directivity effects. Properties (1)–(3) can produce tensile components of particle motion, leading in some conditions to "opening modes" of rupture. Property (4) can lead to strong directivity effects and associated asymmetric ground motion. Anooshehpoor and Brune (1999) observed features (1)–(4) of the wrinkle-like rupture in laboratory experiments of sliding on an interface between two different foam rubber blocks. If these properties characterize ruptures on natural faults, they would have fundamental consequences for many aspects of earthquake dynamics, including effective constitutive laws, suppression of branching, the heat flow paradox, short rise-time of earthquake slip, and expected seismic shaking hazard (Ben-Zion and Andrews, 1998; Ben-Zion, 2001). It is thus important to test the above theoretical predictions with detailed *in situ* observations.

Large fault zones that accommodated significant slip have interfaces that separate different media (Ben-Zion and Sammis, 2003, and references therein) due to the production of damaged fault zone material and the juxtaposition of different rock bodies across the fault. Such material interfaces are seen directly in exhumed fault

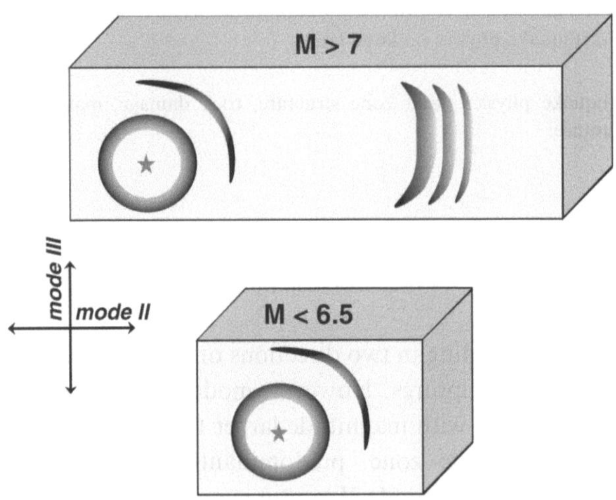

Figure 1
Schematic illustration of rupture modes of small to moderate (M ≤ 6.5) and large (M ≥ 7) earthquakes on a strike-slip fault. Earthquakes smaller than about M6.5 propagate as mixed modes II and III rupture (bottom), while ruptures of M > 7 earthquakes saturate the seismogenic zone and then propagate predominantly as mode II rupture (top).

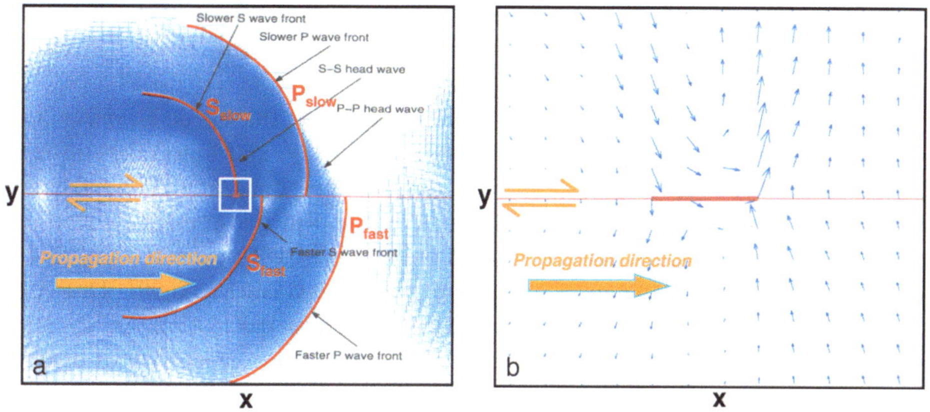

Figure 2

(a) Particle velocities at a given time generated by a wrinkle-like rupture pulse (small red bar within the white box) propagating to the right along a right-lateral strike-slip fault between different elastic solids (thin red horizontal line). The slower velocity block is above the fault as manifested by the fronts of the radiated seismic waves. (b) Enlarged view of the white box in (a) showing asymmetric particle velocities around the rupture pulse. After BEN-ZION (2001).

zones (e.g., CHESTER and LOGAN, 1986; EVANS and CHESTER, 1995; SIBSON, 2003) and were imaged in the structure of large strike-slip faults using direct- and coda-wave tomography (e.g., EBERHART-PHILLIPS and MICHAEL, 1993; SCOTT *et al.*, 1994; SHAPIRO *et al.*, 2005), reflection/refraction seismology (e.g., FUIS *et al.*, 2001; LUTTER *et al.*, 2004), seismic fault zone head waves (BEN-ZION and MALIN, 1991; BEN-ZION *et al.*, 1992; MCGUIRE and BEN-ZION, 2005) and geodetic data (LE PICHON *et al.*, 2005).

Testing whether mode II ruptures on faults that separate different solids have a preferred propagation direction, which correlates with the velocity structure as predicted by theory, should ideally be done with data associated with a large population (for high statistical significance) of earthquakes that are large enough to saturate the seismogenic zone (of strike-slip faults) and propagate predominantly as mode II ruptures. Tests associated with smaller events should especially employ large data sets, as the bimaterial effect is not fully developed for small to moderate mixed-mode events. High resolution locations of several thousand small events on the San Andreas Fault (SAF) show directional asymmetry that is compatible with a preferred propagation direction associated with the local velocity structure (RUBIN and GILLARD, 2000; RUBIN, 2002). However, the employed events are small and it is important to obtain additional evidence associated with large earthquakes behavior. MCGUIRE *et al.* (2002) and HENRY and DAS (2001) analyzed rupture properties of over 100 large global earthquakes and found that most are predominantly unidirectional. However, the examined earthquakes are associated with different faults and therefore do not necessarily indicate a preferred propagation direction of large events on a given fault section. Testing the latter can, at present, be done only with geological studies that search for predicted signals that might reflect the cumulative effect of many large ruptures.

POLIAKOV *et al.* (2002) examined the stress field near the tip of a semi-infinite crack propagating in one direction with constant velocity, and found that off-fault damage is predicted to occur primarily in the tensional quadrant of the assumed propagation direction, in a zone that depends on the principal prestress directions, rupture velocity and frictional parameters. RICE *et al.* (2005) extended the results for the case of a steady unidirectional slip pulse. YAMASHITA (2000), DALGUER *et al.* (2003) and ANDREWS (2005) performed numerical calculations of dynamic ruptures on a fault in a homogenous solid with spontaneous generation of damage in the material off the fault. In such cases, ruptures propagate as symmetric bilateral cracks and damage in a single event is produced primarily in anti-symmetric triangular zones, with width proportional to rupture length, in the two tensional quadrants (Fig. 3a). If earthquakes on a given fault section propagate predominately as such bilateral ruptures, or as unilateral ruptures without a preferred propagation direction, the cumulative pattern of rock damage generated by many events will be approximately symmetric across the fault. On the other hand, if earthquakes on a given fault section have a preferred propagation direction, due to any mechanism, the cumulative damage pattern will be asymmetric, with more damage in the tensional quadrant associated with the preferred propagation direction.

BEN-ZION and SHI (2005) performed simulations of ruptures along a material interface with spontaneous generation of damage in the bulk and showed that in such cases damage is generated primarily on the stiffer side of the fault, which for the preferred propagation direction is persistently in the tensional quadrant of the radiated seismic field. The damage is generated in a strip of approximately constant width that is related to the pulse width (Fig. 3b). The cumulative damage generated by many such ruptures is expected to be similar to the asymmetric pattern of Figure 3b, with significantly more damage on the side of the fault with higher seismic velocity. The above results provide clear predictions that can be tested in the field by performing geologic observations of symmetry properties of rock damage across large strike-slip faults.

In the following sections, we present systematic *in situ* observations of symmetry properties of rock damage in the structure of large strike-slip faults of the San Andreas system in southern California. The observations indicate overall a consistent damage asymmetry across the Principal Slip Surface of a given fault section at several scales. The observed sense of damage asymmetry at these scales correlates with available information on the seismic velocity structure as predicted by the theory of rupture along a material interface. The results open up the possibility of inferring preferred directions of rupture propagation of large earthquakes on large fault sections from geologic field observations.

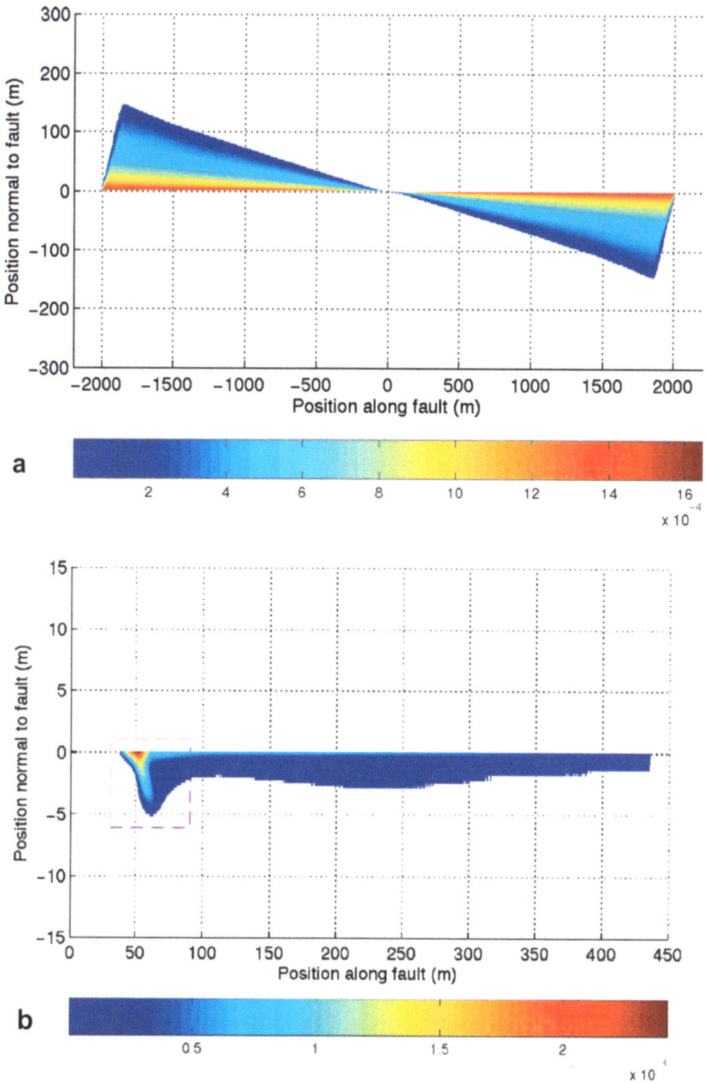

Figure 3
(a) Plastic strain (color scale) generated by a bilateral crack-like rupture on a fault at $y = 0$ in a homogeneous solid. (b) Plastic strain (color scale) generated by a wrinkle-like rupture pulse propagating to the right on a fault separating a compliant material at $y > 0$ from stiffer material at $y < 0$. The stiffer side of the fault is in the tensional quadrant of the radiation pattern for the preferred propagation direction of the wrinkle-like pulse. The dashed box marks the nucleation zone. From BEN-ZION and SHI (2005).

2. Approach and Methodology

We study *in-situ* structural properties of fault zones that may reflect the long-term signature of a large population of earthquake ruptures. The examined properties include the intensity and style of wall-rock and gouge damage, the orientation and density of fractures, the presence and orientation of subsidiary faults, and geomorphologic features. Because these properties can be affected by local fault geometry and complexity, the recognition of persistent asymmetry that is independent of local site effects requires documentation of multiple signals at multiple scales at several locations along a relatively straight fault segment.

We utilized both natural and excavated exposures for our studies. In general, active faults generate detritus that tends to obscure or bury the fault and immediate damage zone, and we found this to be the case along the San Andreas and San Jacinto faults. In contrast, the Punchbowl fault, an older inactive strand of the San Andreas system, is incised and easy to study in natural exposures.

Scales of Observations: We discuss primarily two types of observations associated with different scales. The first type includes observations on the near-field damage of millimeters to meter fault core scale within gouge zones and their immediately adjacent wall-rocks. We refer to gouge as crushed and ground-up rock produced by shear. Increasing displacement tends to reduce the grain size (ENGELDER, 1974; BIEGEL *et al.*, 1989) and may enhance the development of shear fabric, possibly due to the competency contrast between the wall-rocks and the gouge material (GOODWIN and TIKOFF, 2002). Well-developed gouge zones, like most of those described in this paper, do not display primary structures, probably due to the accommodation of large shear strain. The composition of a gouge zone is derived from the adjacent wall-rocks but it often also contains translocated (pedogenic) clay and other material that migrate from the surface environment into the fault, especially if it is close to the ground surface. Such alien material often has blackish organic appearance. The gouge zones of the active faults studied here are typically dominated by claylike unlithified material, which behaves plastically when moistened and contains very small amounts of visible porphyroclasts (unless otherwise stated). Hence, some of the described gouge elements match the definition of SCHOLZ (2002) for foliated gouge, modified after SIBSON (1977), although we also commonly observe cohesive and stiff gouge elements that do not fall into any of Scholz's categories. The examined gouge layers of the active faults are the zones that accommodate most of the displacement along the fault, and are therefore the surface analogue of the fault core, with concentrated slip, in the structure of several exhumed faults (e.g., CHESTER *et al.*, 1993). We use the term 'proto-gouge' to describe a fault zone layer within or close to the fault gouge or core in which the fabric and the composition are in an intermediate stage between a typical gouge and the local wall-rock.

The second type of observations is made on the distribution of secondary faulting and associated features in the meters to tens of meters fault zone scale. A fault zone includes mesoscopic fractures and small secondary faults, typically extending to a distance of about 50 to 70 m from the fault core (e.g., SCHULZ and EVANS, 2000). We discuss also a third type of observations, made on the distribution of damage to the bounding rocks in the tens to hundreds of meters damage zone scale. This type of observation includes zones of pulverized rocks and seismic fault zone trapping structures, with considerably more damage than the background damage of the country rocks.

Frame of Reference for Symmetry Properties: Our frame of reference for descriptions and measurements of symmetry properties is the currently active or most recently active Principal Slip Surface (PSS). In the fault core scale, it is the major through-going slip surface within the fault core that appears to accommodate most of the displacement during the current phase of faulting, or the life time of the evolving gouge. In the fault zone scale, the PSS is the most dominant and most recently active strand among the secondary faults that compose the fault zone (indicated mostly by the geomorphology). The PSS usually, but not necessarily, coincides with the major lithological boundary. In the damage zone scale, symmetry is examined with respect to the zone that accommodates the majority of the long-term displacement. We use the active (or most recently active) slip zone because the composition and structure of faults generally evolve with time, and we are interested in structural features that were produced in the current or most recent phase of faulting, which may be correlated with available information on the current velocity structure. We cannot always distinguish between different generations of damage and this is another reason to perform observations on several scales and establish a consistency along a given fault section.

The symmetry of structural properties is evaluated by comparing intensity and style of deformation between the two sides of the PSS. In most of the gouge exposures, there is usually one continuous surface that becomes clearly visible when pieces and fragments of gouge are peeled away. When more than a single significant slip surface is present within the fault core, we describe symmetry properties with respect to the PSS that is the most continuous and dominant. We assume that every faulting event includes motion on the currently-active coherent slip surfaces, perhaps with ongoing destruction or overprinting of older, previously-active slip surfaces. Evidence for such behavior can be frequently seen in trench logs of paleoseismic studies, where the relative timing of events is known and one can see the destruction or overprinting of an older slip surface by a younger trace (e.g., ROCKWELL *et al.*, 1990).

Overview of Investigation Sites: For our first set of sites along the San Jacinto fault, we worked along a stretch of fault with deeply incised geomorphology resulting from a recent stream capture. In that place the fault is poorly exposed but could be accessed with minimal work. We mapped the damage in a road-cut and in backhoe and hand-excavated pits, focusing on the mm to m fault core scale damage. For the San Andreas fault zone, we studied several exposures at the core scale along recently active traces, and also opened trenches in bedrock at a number of localities to explore the tens of meters wide fault zone scale damage. In a parallel study, we examined the broader damage zone scale that may extend hundreds of meters from the principal slip surface and includes such characteristics as pulverization of rocks (DOR *et al.*, 2006; BRUNE, 2001; WILSON *et al.*, 2005) and chemical alterations. In the Punchbowl fault within and in the vicinity of the Devil's Punchbowl County Park, we studied symmetry properties of the fault core with respect to the ultracataclasite layer (CHESTER and CHESTER, 1998) as well as fault zone scale distribution of damage. Observations of symmetry properties in the microscale are also a natural component of a study of this kind, but are beyond the scope of this paper and are left for future studies.

Methods: The exposures were first cleaned and flattened, then moistened and finally scraped gently so as not to disturb the natural fabric. We found the natural gouge fabric to be better defined after several hours, when the exposures had sufficiently dried. In some cases, additional cleaning with a soft brush and compressed air were used to better express the gouge fabric. Using optimal natural light conditions, high resolution digital photos were taken. For most of the smaller scale study sites, we mapped the fractures on the images and digitized them for quantification of the fracture density in the different gouge and wall-rock layers on both sides of the PSS. For some of the other sites we described the overall pattern of structural features without quantifying them. Fault and damage zone rocks in the exhumed Punchbowl fault are well exposed and composed of hard clastic and metamorphic rocks that require just a light brushing and dusting before documentation. Study of fabric in soft fault rocks was done only in exposures that are at a minimum distance of 1–1.5 m below the original ground surface (mainly excavated by us), to avoid bias of the results by fabric that results from thermal cycling, wet-dry cycles, or other surface phenomena.

The intensity of deformation was evaluated in several ways. First, we measured Fracture Density (FD) across the fault-core or the fault-zone on digitized fracture maps in rectangular 'digital samples' employing a "box-count" method, which was similarly used by DERSHOWITZ and HERDA (1992) and others, involving counting the cumulative length of fractures and dividing by the sample area. Alternatively, we applied those measurements to an entire digitized gouge or wall-rock layer. We also describe the mode of fractures, i.e., joints (mode I) or shear fractures, when such can be typified, along with their length and orientation. A second method to evaluate damage was to count fractures along scan-lines perpendicular to the fault strike, similar to the linear fracture density method used by WILSON *et al.* (2003) and PRIEST (1993). For some exposures, we compared relative strike-slip activity across the PSS either qualitatively or by counting secondary strike-slip faults that were identified using kinematic slip indicators. Finally, we describe symmetry properties of geomorphic features that characterize the various sites.

Distinguishing Fabrics: For several cases, we discuss relative activity time for the different gouge elements within a gouge zone. The discussion is not based on an absolute dating method, but rather on qualitative characteristics of the shear fabric.

The most recently active principal shear zone, as defined by the surface geomorphology, tends to break out and can be expressed with a scaly fabric (AGAR *et al.*, 1989), with highly sheared and polished anastomosing facoidal lenses of gouge, and abundant evidence of shear in the form of striae and mullions on curviplanar surfaces. In contrast, what we interpret as older shears tend to be considerably more cohesive and either do not have as easily recognized parting surfaces or the surfaces are compressed and welded. We have found that gouge with high moisture content appears to behave plastically, whereas lower moisture content gouges tend to have more brittle characteristics (undoubtedly fluctuates with the seasons). Nevertheless, the PSS along the active fault core expresses highly sheared parting surfaces even when the moisture content is high, while older shears tend to develop good parting surfaces only when relatively dry. Overall, the more cohesive fractures were found away from the active zone of the gouge when the active zone could be identified independently. In contrast, the less cohesive shear fractures were found within the more moistened and more plastic gouge elements. In the presence of several gouge layers embedded within the fault zone, the one that can be correlated with the geomorphically active trace of the fault is typically the more moistened/plastic, which presents relatively incohesive fractures. We use these criteria when the active or the most recently active zone of the gouge cannot be identified independently.

3. Field Observations

We discuss our field observations by fault zones, starting with the exposures along the San Jacinto fault near Anza. We then describe the several sites we studied along the San Andreas and the Punchbowl faults, an inactive ancestral strand of the San Andreas fault system.

3.1 The San Jacinto Fault

The San Jacinto fault (SJF) is a primary strand of the San Andreas fault south of the Transverse Ranges (Fig. 4) and is currently the most active fault zone in southern California (CROWELL, 1962; ALLEN *et al.*, 1965; SANDERS and KANAMORI, 1984). The 20-km long Anza section of the SJF is known as the Anza Seismicity gap (BRUNE, 1968; SANDERS and KANAMORI, 1984) and includes the study area. This section of the fault is well expressed by geomorphic features and the fault is relatively simple with only one major active strand, the Clark fault (ROCKWELL *et al.*, 1990), that separates surficial tonalite and other leucocratic plutonic rocks on the northeast against gabbroic and more melanocratic rocks to the southwest (SHARP, 1967). These lithologies probably do not extend very deep below the surface, as the local velocity structure indicates faster basement velocities at seismogenic depths on the northeastern side of the fault (SCOTT *et al.*, 1994). The total right-lateral displacement across all strands of the SJF southeast of Hemet is estimated to be 29 km (SHARP, 1967). The surface slip rate was estimated by trench studies in the Anza area, a few km southeast of our study area, to be on the order of 12–18 mm/y (ROCKWELL *et al.*, 1990; MERIFIELD *et al.*, 1991). More recently, in a new set of trenches across the SJF in Hog Lake, ROCKWELL *et al.* (2005) documented the occurrence of 15 surface ruptures during the past 3500 years, with the most recent event at approximately A.D. 1760.

The study area is on the edge of the Ramona Indian Reservation, northeast of the Anza Valley. The surface expression of the fault within the study area indicates that the fault is highly linear and localized, with one narrow principal strand; in places it has a well-developed meter-wide gouge zone that juxtaposes surficial middle Pleistocene alluvial deposits on the southwest against late Pleistocene alluvial deposits on the northeast. The following sections describe the structure and symmetry properties of three fault core scale exposures that we excavated and utilized along a 140 m section of the fault (Fig. 5). The exposures are between 1.5 to 5 m from the original ground surface, but during the last earthquake (approximately

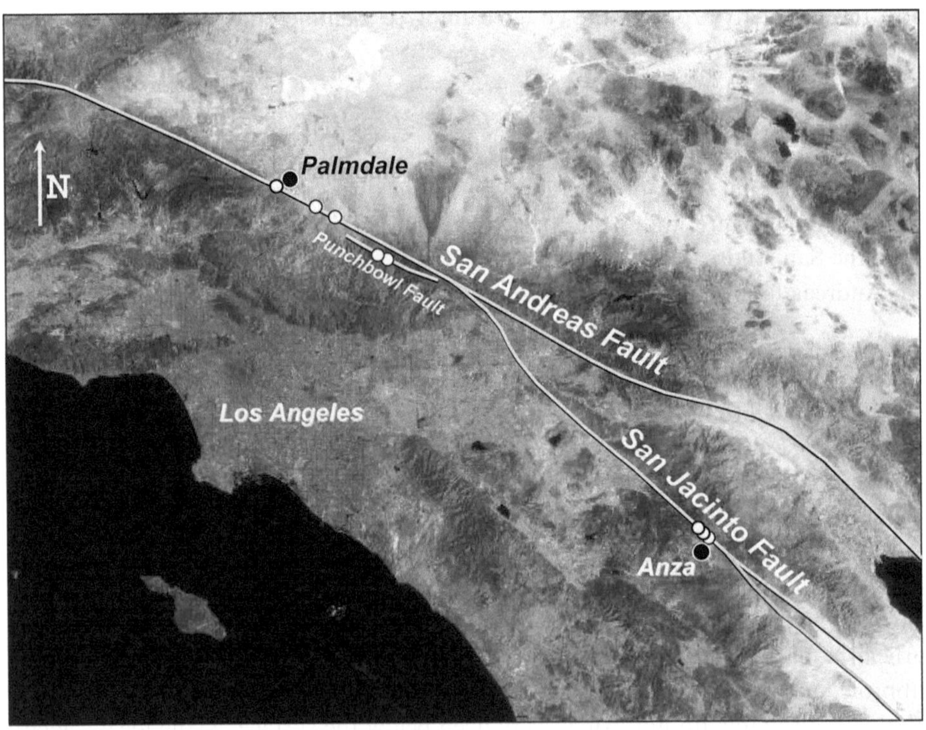

Figure 4
Faults of the southern San Andreas system included in this study and location of investigation sites (white circles). The town of Anza and city of Palmdale are indicated by black circles.

245 years ago) their depth was about twice the current one based on estimates of local erosion rates.

The Road Cut and Trench Exposure: We utilized a steep road cut, facing towards the southeast, where we exposed the meter-wide gouge zone and adjacent damaged wall-rocks with a backhoe. The exposed fault can be divided into five structural domains (Fig. 5b). To the west, the southwest wall-rock is a well-cemented, sand-rich middle-Pleistocene conglomerate with sparse cm-sized pebbles. The conglomerate appears to be macroscopically intact even in the immediate vicinity of its sharp contact with the gouge zone. The westernmost gouge element is a 15 to 20 cm wide, highly cohesive, stiff and massive, gray-black and clayey, with a relatively large amount (compare to other gouge elements in this exposure) of sandy porphyroclasts. There are very few macroscopic fractures in this layer and the lack of foliation or any other evidence for shear fabric suggests that the layer is inactive. The contact with the southwest wall-rock is well defined. The "massive" gouge is bound on the northeast by the principal slip zone and a PSS.

There is a sharp transition in the nature of the gouge that takes place across the PSS. The 40 cm wide gouge layer northeast of the PSS has a dark brownish color, is highly sheared, and includes numerous small slip surfaces (shear fractures). These curviplanar, facoidal slip surfaces are polished, shiny and contain low-rake slip striations. Most of the surfaces are oriented parallel to the fault. The cohesion on the slip surfaces is low as they are easy to separate. The slabs of the material between the fractures are claylike, flexible and lack significant amount of visible porphyroclasts. Farther northeast, the next gouge layer is about 40 to 50 cm wide and provides a gradual transition from the sheared gouge on the southwest to the northeastern wall-rock. This gouge layer is also rich in clay over porphyroclasts but slightly more stiff and cohesive with respect to the adjacent sheared gouge layer. It has a mixed population of fracture modes, dominated by joints (mode I fractures) with just minor evidence of shearing; it therefore appears to be mostly shattered. The fracture orientation is more diverse and exhibits an overall higher angle to the PSS compared to the fractures in the sheared gouge layer. Finally, the northeastern wall-rock is a late Pleistocene conglomerate that is similar in texture to the SW wall-rock but is more damaged with distributed macroscale fractures. It is important to note that these cemented sediments are considerably younger than the less damaged middle Pleistocene sediments on the southwest side of the fault.

We created a highly detailed digital image framing a 180 cm × 50 cm rectangle across the fault. Figure 5c shows 900 fractures that were mapped on the image. The gray spots mark zones where visibility on the image was not good enough for accurate mapping. The fracture pattern in the digitized image delineates the structural divisions. In order to quantify the fracture density gradient, we defined 29 rectangular 4 cm × 12 cm frames in a continuous array across the image. The distribution of the rectangles on the image was dictated by the zones of low visibility. The Cumulative Fracture Length (CFL, in cm) within each of these frames was calculated, and the results are shown by the red curve against distance across the fault in Figure 5b.

The CFL in the southwest wall-rock is practically zero at this mapping resolution. In the massive gouge, the CFL rises to several cm/frame, whereas in the sheared gouge it jumps across the PSS to between 20 and 50 cm/frame. The CFL stays at a level of 30–40 cm/frame in the shattered gouge and drops to between 0 to 20 cm/frame in the NE wall-rock.

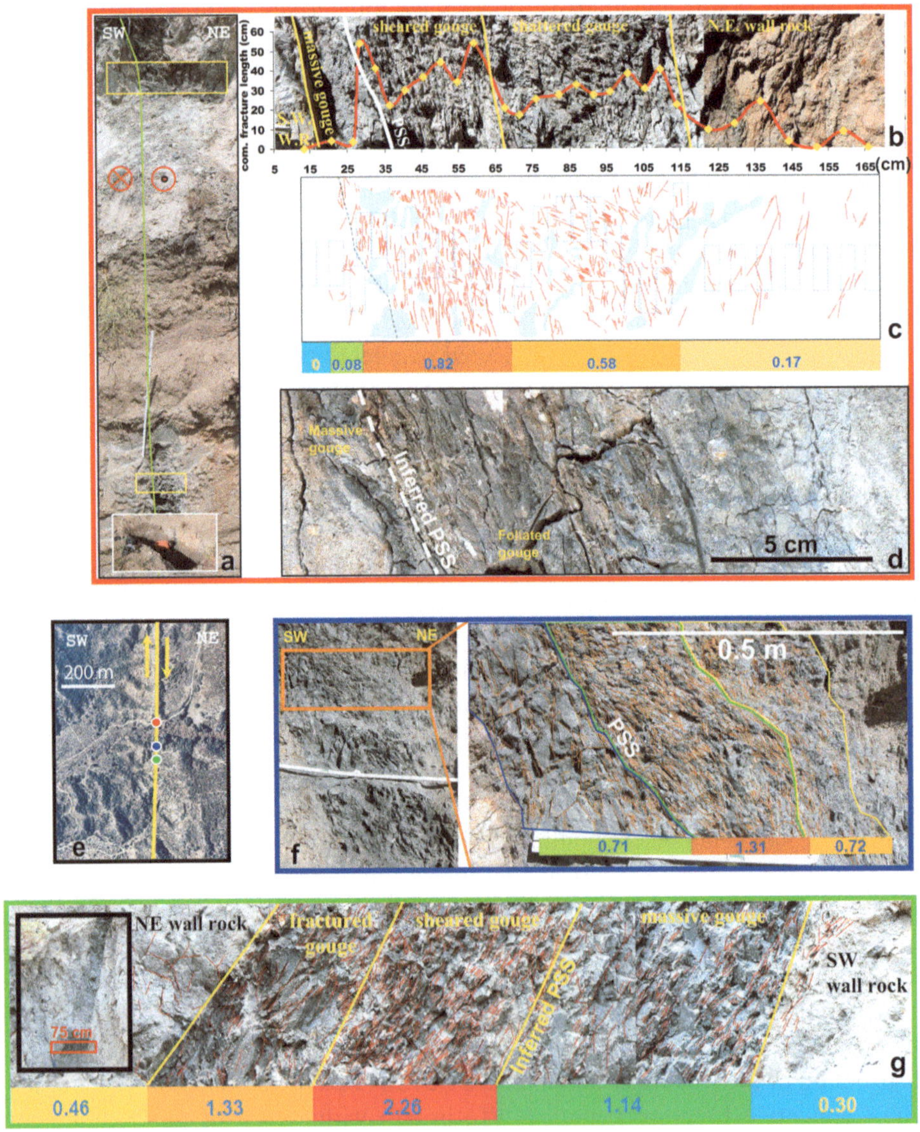

To correspond with the structure of the fault zone, the CFL curve was divided along the length axis into domains corresponding to the above structural divisions. Each structural domain was assigned a FD index in units of cm/cm^2 based on the sum of the cumulative fracture length from all the frames within a domain, divided by the cumulative area of all the frames within the domain. The FD index values for the structural domains are as follows: SW wall-rock: 0; massive gouge: 0.08; sheared gouge: 0.82; shattered gouge: 0.58; and NE wall-rock: 0.17. The results indicate that the structural division can be reasonably considered as fracture density division. It is especially interesting that the FD in the northeast wall-rock is higher than the FD in the massive gouge.

We extended the road cut exposure downward by digging a trench on the side of the road three meters below the location of the exposure described above (Figure 5a). We estimated the bottom of the trench to be about five meters below the original ground surface prior to the excavation of the road. The gouge width shrinks from one meter in the road cut to only 15 cm at the base of the trench. The structural division of the gouge in the road cut can in general be applied to the gouge in the trench, but it is substantially more compressed and the PSS cannot be identified macroscopically (Fig. 5d). The visible cracks in Figure 5d were developed upon dehydration of the exposure face and it is not clear to what extent their pattern is controlled by the gouge fabric. The relatively cohesive and massive gray gouge layer

◄

Figure 5
Structural units and fracture density in the gouge of the San Jacinto fault zone south of the Ramona Indian Reservation. The colors of the different frames orrespond to the colors of the circles in the small location map (panel e). **Red frame**: (a) Panoramic view showing the upper and lower road cut exposures with the inferred location of the PSS (green) and the analyzed frames (yellow boxes) presented to the right (panels b and c). In the center of the panel the fault is covered with debris. (b) A cross fault view of the gouge and damage zone, with a division to 5 structural domains. The red graph shows the trend of fracturing intensity across the fault. Each yellow dot indicates the cumulative fracture length (CFL) in cm measured in the rectangle frame right below it in the middle panel. (c) A digitized fracture map of the upper panel. Light blue spots mark zones with insufficient visibility for detailed mapping. Dashed line marks the PSS. The rectangles are samples of CFL. The numbers on the lower horizontal bar give fracture density (FD) calculated for each of the structural domains from the upper panel by dividing the CFL from all the samples in the domain with the cumulative area of the used samples. A sharp contrast exists between the FD of the sheared and massive gouge. Note also that the FD of the NE wall-rock is higher than the FD of the massive gouge. (d) The gouge 3 m below the upper road cut narrows down to 15 cm but maintains its structural division. Massive, gray, 1 cm wide gouge appears on the SW and highly foliated (sheared) gouge appears on the NE. **Blue frame**: (f) The "Cliff" exposure, 100 m southeast of the road cut. The picture on the left shows the artificial exposure with difference between the dark cohesive gouge on the left to the fractured gouge on the right. The framed zone is 70 cm wide. The structural division in the digitized map on the right is consistent with the structural division shown for the road cut (red frame), with similar trend of the FD indicated by the numbers at the colored bar at the bottom of the panel. **Green frame**: (g) Digital fracture map in artificial exposure (inset) 140 m SE of the road cut. The gouge and adjacent wall-rocks were divided into 5 structural domains that can be correlated to the structural division of the road cut exposure (red frame). The PSS is not clear here but based on the other two exposures with similar structure, the PSS should be between the massive and the sheared gouge. The FD index is shown for each domain on the lower colored bar. Higher damage is on the NE side of the fault.

on the southwest side of the zone is only 1 cm wide and shows no original visible fractures. The next gouge layer is more brownish and exhibits dense foliation with fracture spacing in the sub-millimeter scale, and we correlate this layer upward to the sheared gouge in the road cut. The FD of these gouge layers can be studied and quantified only on the microscale, which is beyond the scope of this paper.

The Cliff Exposure: The "Cliff" exposure is 100 m southeast of the road cut exposure on a natural steep slope facing to the southeast. We manually excavated the fault to a depth of 1.5 m below and away from the slope; this area was not accessible to a backhoe. The exposure was cleaned following the procedure described above, and a sequence of digital photos was taken across it upon which the fractures were mapped and digitized. The 70-cm-wide gouge zone at this exposure can be divided into three distinct structural domains, and each domain can be correlated to the gouge domains in the road cut. The structural domains also coincide with the divisions in fracture density domain (Fig. 5f). This division is sharp and continuous enough, and the quality of the digital image is sufficiently good, to enable us to directly calculate the FD of the structural domains without using a continuous array of digital samples.

The gouge layer on the southwest side of the PSS is massive and highly cohesive, gray dark and rich in clasts. It is especially stiff and brittle. The layer contains fractures that are relatively large, showing inconsistent orientations, but no macroscopic foliation and shear fabric. We measured a FD of 0.71 for this unit. Many of the fractures are incohesive and they seem to have fresh faces of recently broken material; therefore, they may have been created when the outcrop was excavated, possibly due to the material stiffness. We therefore conjecture that the actual *in situ* FD is less than the measured FD. The gouge layer on the northeast side of the PSS is highly sheared and characterized by a higher value of FD of 1.31. It is much less cohesive, sparse with porphyroclasts and contains many anastomosing shear fractures with curvy faces and stria; the fractures are relatively small, giving to the gouge a flaky texture. The majority of the fracture population is parallel to the fault. Farther to the northeast, the next gouge layer is a transition from the sheared gouge to the northeast wall-rock, with a FD of 0.72. The fractures orientation in this gouge layer are less uniform with respect to orientation of fractures in the sheared gouge immediately to the southwest side but more internally consistent with respect to those of the massive gouge from the southwest side of the PSS. Their surfaces are also more cohesive and do not have a fresh appearance. Thus, unlike fractures in the massive gouge, the fractures in this gouge layer are believed to be part of the original rock fabric prior to excavation, and therefore more related to faulting. We suspect that the FD contrast between this gouge layer and the massive gouge is higher than we measured.

The SE Exposure: The southeast exposure is an additional 40 m to the southeast from the "Cliff" exposure. The outcrop faces to the northwest, opposite the direction of the two previously described exposures. We manually excavated an exposure to a depth of about 1.5 m away from the current slope surface (this site was also not accessible by backhoe) and performed the same analysis as in the other exposures. Here the fault zone is similar in its structural division, in its FD distribution and in the sense of asymmetry to those presented for the road cut, trench and Cliff exposures (Fig. 5g). The FD indexes on the southwest side of the PSS are 0.3 for the southwest wall-rock and 1.14 for the massive gouge. Northeast of the PSS, the FD indices are 2.26 for the sheared gouge, 1.33 for the shattered gouge, and 0.46 for the northeast wall-rock.

Summary: All the three exposures along a 140 m long section of the San Jacinto fault near Anza exhibit consistent and distinct structural division and sense of asymmetry, with the northeast side of the fault always more damaged, regardless of the slope aspect or depth of the exposure. In each of the three studied exposures, the gouge on the northeast side of the PSS is highly sheared, followed by a more shattered gouge layer farther to the northeast. The gouge layer on the southwest side of the PSS is always cohesive and massive and appears to be inactive. Fracture density analysis highlights and confirms the apparent structural division and the sense of asymmetry.

3.2 The Mojave Section of the San Andreas Fault

The San Andreas fault system, in a cross section that includes our study area in the Mojave, contains in its broader sense the San Gabriel fault on the south side of the San Gabriel Mountains and several parallel strands on the north side of the San Gabriel Mountains. The latter include the currently active SAF and the Little Rock, Punchbowl, southern and northern Nadeau faults, as well as several other smaller faults. These faults all exhibit a zone with faulting-related structural features that measures tens of meters in width (CHESTER and LOGAN, 1986; CHESTER *et al.*, 2004), and they can potentially have damage zones at a scale of ~100 m that correspond to the observed low velocity trapping structures for several faults (e.g., BEN-ZION *et al.*, 2003; PENG *et al.*, 2003; LEWIS *et al.*, 2005), and to the ~100 m wide zone of pulverized fault zone rocks observed for the SAF (WILSON *et al.*, 2005; DOR *et al.*, 2006). A broader damage zone that reflects the entire deformation history of a large fault can consist of a few km wide region with enhanced anisotropy and reduced density and elastic moduli (BEN-ZION and SAMMIS, 2003; FIALKO, 2004; PENG and BEN-ZION, 2004).

The San Andreas fault system in southern California has accommodated altogether about 240 km of displacement, based on the separation of the correlative Neenach and Pinnacles volcanic rocks (MATTHEWS, 1976), but this may not include 50–60 km of slip on the San Gabriel fault. The displacement may be even larger, in excess of 300 km, based on the separation of Pelona schist and other basement units (DIBBLEE, 1989). The displacement was accommodated on the ancestral faults of the system, and subsequently on the modern elements of the SAF as they developed

(Powell and Weldon, 1992). Each fault remains a narrow zone of discontinuity along which local dip-slip adjustments have occurred, sometimes into Holocene times (Barrows *et al.*, 1985).

The individual displacement values on each of the primary strands of the San Andreas fault system in the central and southeastern Mojave are only partially constrained in most cases. The San Gabriel fault accounts for about 50 to 60 km of the total slip, whereas the Punchbowl and Nadeau group of faults accounts for 44 and 16 km, respectively. The Little Rock fault has only a minimum value of 21 km of displacement, and the currently active trace of the SAF appears to have only 21 km of slip based on the displacement of the late Pliocene Juniper Hills formation. The discrepancy between the sum of those displacements ($\sim$160 km) to the assumed total $\sim$300 km could be accounted for by additional displacement on the Little Rock fault, displacement on the Hitchbrook fault northwest of our study area, and larger earlier displacement on the current SAF (Barrows *et al.*, 1985) or on the Clemens Well, Fenner and San Francisquito faults (Powell and Weldon, 1992).

The broad SAF system in the study area is therefore a 7 km wide zone of extended deformation, and with the San Gabriel fault the zone is $>$30 km wide. This broad zone experienced several generations of deformation after the middle Miocene, with complex partitioning of displacement between its components; each deformation phase could have overprinted its signature on rocks within the system. Individual rock bodies are therefore likely to have recorded a complex displacement and deformation history. The existence of numerous geometrical perturbations at many scales, from the big bend to smaller step-overs, and interaction between the main trace to subsidiary faults all influence the faulting pattern. Even rocks along a relatively straight portion of the fault may carry in their 'geological memory' an imprint from a zone of structural complexity farther along the fault. In such a complex system with very large displacement, the velocity structure across the fault may have changed several times. A single rock body juxtaposed along the fault could have experienced a contrasting velocity structures over time, and maybe even have experienced an opposite preferred rupture propagation direction, leaving contradicting long-term signals in its fabric.

The task of finding a coherent geological signal along such a fault that records the long-term behavior is rather challenging. It is particularly complex in the active environment of the SAF zone with sparse bedrock exposures, even in the arid portions of the Mojave Desert. As we anticipated the signal to be noisy and possibly confusing, we searched for evidence of several types and at several scales.

We present here observations from three sites in the Little Rock to Palmdale area, spanning a 20 km long section of the fault. At two of the sites near Little Rock, observations are on the fault core scale, and at one site near Palmdale the observations are at the fault zone scale. We also refer to observations in the damage zone scale made along the entire 140 km fault section of the Mojave in a parallel study (Dor *et al.*, 2006). The sites are not on the geometrically simplest section of the fault, nor are all of the sites on the currently active strand of the SAF, but the region has a semi-arid climate that provides opportunities for examination of the fault in bedrock exposures with minimal excavation.

The Little Rock Paleoseismic Site: The Little Rock site, in the vicinity of E. 96th St. south of the town of Little Rock (Fig. 4), hosted an investigation of the Holocene slip rate of the SAF (SCHWARTZ and WELDON, 1986; WELDON and FUMAL, 2005). They excavated a system of trenches and exposed several parallel strands of the fault, including the one that accommodated the 1857 rupture. According to their unpublished mapping, only one strand that separates bedrock units was found at the bottom of the trench system. This strand of the SAF is about 5 to 8 meters north of the 1857 strand and below an active channel (Fig. 6a). We re-trenched and exposed the upper 60 cm of this bed rock fault zone, which is cut across by middle Holocene (3.5 ka) unconsolidated alluvial deposits (Tom Fumal, pers. comm., 2004).

This fault strand juxtaposes granite on the northeast against the conglomeratic late Pliocene Juniper Hills formation. These two units are separated by a 5 cm wide soft and claylike dark brown gouge zone with distributed shear fabric, and a through-going PSS inclined about 60° to the southwest (Fig. 6b). This strand probably represents the post-Pliocene long-term SAF in this area. Individual earthquakes may still occupy shallow branches close to the ground surface, as indicated by nearby faults within young sediments.

The Juniper Hills conglomerate of the southwestern block is essentially intact. A 10 cm long pebble located immediately south of the gouge was reoriented parallel to the PSS and shows no macroscopic fractures. Damage in the northeastern block is substantially more intense. The granite is pulverized, with the original grain-scale fabric preserved. The rock has a powder-like texture similar to the one that characterizes the outcrops of Tejon Lookout granite in Tejon Pass (WILSON *et al.*, 2005) and in other localities along the Mojave section of the SAF (DOR *et al.*, 2006). In addition, cm-to-meter long fractures cut sporadically through the rock mass. This texture and fabric disappear within 30 to 40 m from the contact of the northeastern block with the gouge. The rock belt immediately northeast of the gouge is gray, have powdery-plastic texture, and seems to include a mixture of siliceous material from the granite on the northeast and claylike material from the gouge on the southwest. This belt exhibits a diffuse contact with the granite and a sharp contact with the gouge zone. We interpret this as a proto-gouge layer, representing an intermediate development stage between the granitic protolith and the clay-rich gouge.

The history of deformation in the northeastern block of this fault is unknown and the granite could have been pulverized and fractured during a previous episode of faulting. Some of this deformation could therefore have occurred against other rock bodies and along other fault strands. Nevertheless, the fault core, which is the zone that accommodated the recent displacement and experienced mineralogical and textural changes associated with slip along this fault strand, includes in addition to the PSS and gouge, a proto-gouge layer of reworked granite within the northeastern

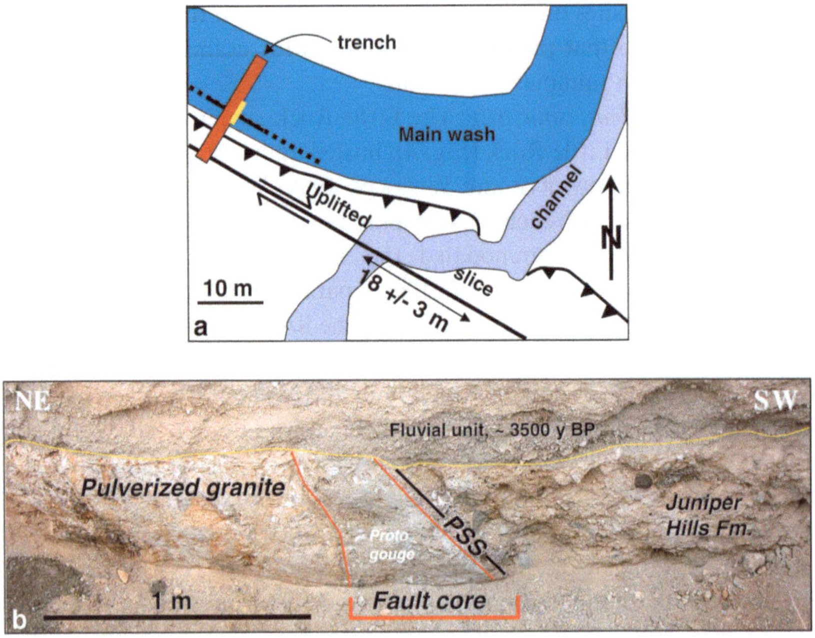

Figure 6

(a) Paleoseismicity site south of the town of Little Rock (after WELDON and FUMAL, 2005). The active strand of the SAF is marked by the line with the slip sense marks; the amount of channel deflection is indicated. The red heavy line represents the re-trenched strip and the yellow box marks the location of the trench exposure shown in (b). The line with triangles is a thrust fault, footwall to the north. (b). The trench exposure in the yellow box of (a). The PSS is within a 5 cm wide gouge zone separating the Juniper Hill Fm. from pulverized granite. The intense damage zone (fault core) related to displacement of this fault outside the gouge is exclusively within the granite. The damage is asymmetrically located on the NE side of the fault.

block; no similar changes or damage to the rock can be identified within the Juniper Hills conglomerate. The fault core here is the result of a limited history of faulting, as indicated by a narrow gouge, and was not likely developed within the granite during a previous faulting phase. The smooth transition from pulverized granite to the proto-gouge layer immediately adjacent to the PSS indicates that this layer and the PSS are genetically related. Hence, the fault core that is associated with slip along this fault strand, during which the fault juxtaposed the Juniper Hills conglomerate and the granite, was developed entirely on the northeast side of the PSS. The asymmetric structure of the fault core with respect to the PSS indicates that the creation of 'fault core scale' damage during SAF earthquakes favors the northeastern side of the fault in this location.

Little Rock Creek Site: The active strand of the SAF southwest of the town of Little Rock and north of the Little Rock reservoir (Fig. 7) has strong geomorphic expression with a right-lateral deflection of 600 m in the channel of Little Rock

Creek. Bedrock outcrops with expression of fault zone features are absent from this active fault environment due to burial by young sediments. However, 300 m south of the currently active strand, a continuous exposure of a parallel, presumably inactive strand of the SAF can be traced for several hundred meters. This fault is especially prominent on the southeast wall of the Little Rock channel, where the gouge zone is more than 6 meters wide. It continues in a relatively straight manner toward the southeast, branching locally into two parallel strands. Along most of its exposed length, the fault expresses a 1–2 m wide gouge zone, separating slightly metamorphous granodiorite on the northeast from fine sandstones interbedded with shale on the southwest. The sandstone-shale sequence is plastically deformed with fold limbs at the meter scale. The time of activity of this ancestral fault strand is unknown but there are no obvious geomorphic features to indicate significant Holocene slip.

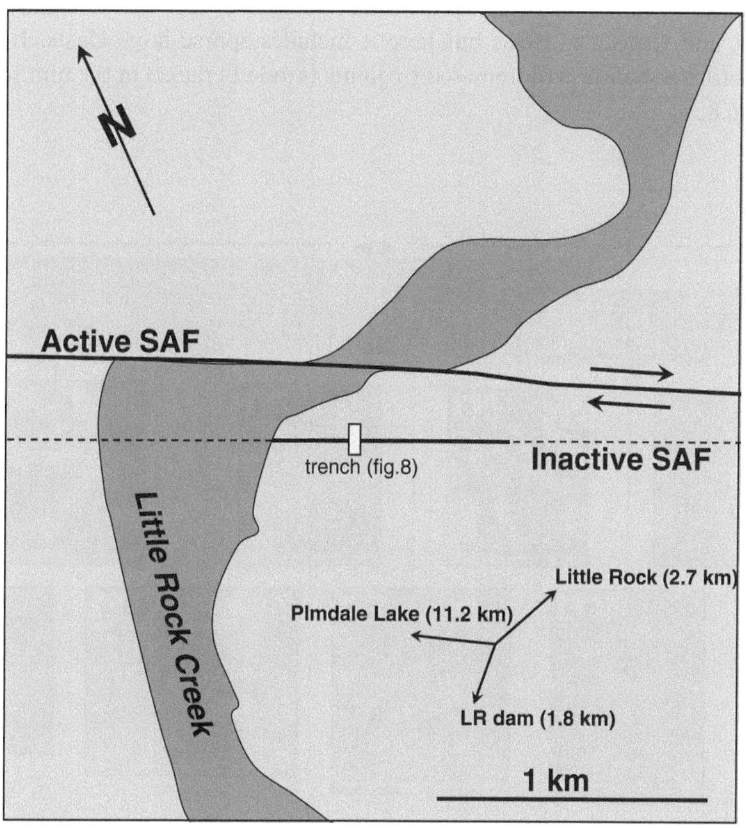

Figure 7
Locations of the Little Rock Creek study site and features of the SAF near by. The location of the trench is marked by a vertical white box. Note the ~600 m deflection of the Little Rock Creek channel by the active strand of the SAF. The trench was excavated 300 meters to the south, across an inactive strand of the fault.

We manually excavated a 1.2-m-deep and 4-m-long trench (Fig. 8) covering the span of the fault core, across the gouge zone and the adjacent wall-rocks where the fault is expressed as a single straight lineament. The fault core here has a composite structure with several fault rock layers; each of them has a distinct lithology, texture and color. We applied the box-count methods for each of the layers within the fault core, using rectangular zones with homogeneous texture and minimum irregularities for FD analysis.

On the northeast, the granodiorite is pulverized, exhibiting textures similar to those found in the pulverized Tejon Lookout granite at Tejon Pass (WILSON *et al.*, 2005). However, the width of the pulverized zone is substantially narrower and extends to between several tens of cm to several meters from the gouge. The fractures here are in the submicron scale and therefore mesoscale fracture mapping is not effective for reliable FD analysis. Nevertheless, the FD must be high at the fine scale.

Adjacent to the northeast wall-rock, there is a 90-cm-wide, fine cataclasite that is macroscopically alike the ultracataclasite described for the Punchbowl fault (CHESTER and CHESTER, 1998) but here it includes sparse large clasts. It contains dense fracture population, dominated by joints (mode I cracks) in the mm scale, with a FD of 6.8.

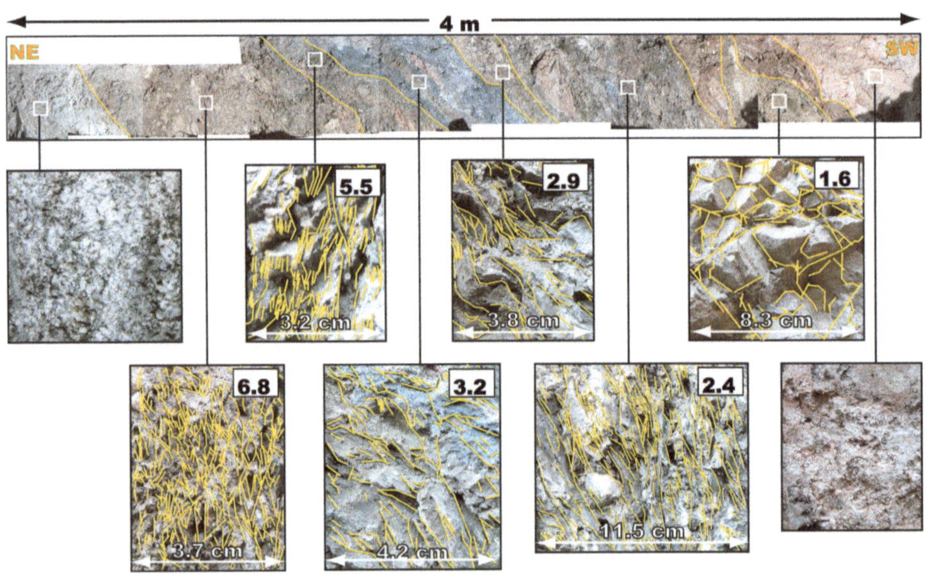

Figure 8
Trench log from the Little Rock Creek site and digitized fracture maps of framed zones from the different gouge layers within the gouge zone. The numbers in the rectangles indicate the fracture density (cm/cm^2). There is a clear trend of decreasing FD from NE to SW. The fractures in the NE pulverized granodiorite are in the microscale.

The next three gouge layers to the southwest, with a cumulative width of one meter, differ in their color (from northeast to southwest: gray, blue and brown) but share the same general texture and overall clayey composition. The gouge layers are stiff and cohesive but contain a curvy and apparently bended shear fractures. The fractures also appear to be well-welded (cohesive) and are therefore probably recently inactive. They are on average parallel to the orientation of the fault. The FD values of the three layers from northeast to southwest are 5.5, 3.2 and 2.9, respectively.

Farther southwest, a 75-cm-wide gouge layer is probably the most recently active shear zone based on its similarity to the active gouge zone in other places. The fractures are small to medium in the cm scale (1–7 cm long); they are curved, shiny and have slip striations with low rakes. The fracture surfaces are less cohesive with respect to fracture surfaces in the other gouge layers and the clayey flakes between them are flexible. Based on this texture and the above similarity, it is possible that this gouge layer was the main zone to participate in slip events during earthquakes on the modern SAF until fairly recently. This gouge layer has a through-going slip surface, which we interpret as the most recent PSS within the fault zone, dipping 65° and located slightly to the northeast from the center of the layer. Nevertheless, its FD of 2.4 is somewhat less than those of the gouge layers to the northeast. The sheared gouge layer is separated from the southwest host rock by a 20-cm-wide layer of slightly more massive gouge. Finally on the southwest, there is a broken sandstone unit with FD of 1.6. Fractures are relatively large and have no preferred orientation.

Overall, the fracture density within the gouge decreases systematically from the northeast to the southwest. Although we are not sure whether the entire structure and damage pattern was developed during the activity period of the current PSS, it appears that throughout the history of this fault strand, damage accumulated preferentially on the northeast side of the fault.

West Palmdale Trench Site: Between Highway 14 and Tierra Subida road in Palmdale, the SAF juxtaposes surficial Quaternary alluvial deposits on the southwest against similar deposits with slivers of the sandstone member of the late Pliocene Anaverde formation on the northeast (BARROWS *et al.*, 1985). A series of deflected channels, benches and linear ridges attest to the localized nature of the fault in this area, with only minor indications for small-scale step-overs and active secondary parallel strands.

We excavated a 45-m-long trench with a general trend of 200°, perpendicular to the strike of the fault, in order to expose the structure of the fault zone. The trench crosses the fault where it deflects an active channel by about 15 m. Fortunately, the sedimentary and crystalline bedrock that host the fault are either exposed on the surface or just shallowly buried. Due to extensive damage and pulverization, we found excavation through those rocks to be remarkably easy, except in places where the backhoe had to penetrate old and tremendously cohesive gouge zones associated with inactive fault strands.

Two fault structures dominate the trench exposure. One is a massive dip-slip fault zone with a minimum horizontal width of 21 m. The second structure, which occupies much of the rest of the trench, manifests strike-slip faulting related to the active SAF. A detailed description of structural features, their relations within the trench exposure, and their general geological context is presented below and corresponds to Figure 9.

A primary interpreted aspect of these two structures is that the dip-slip structure is inactive, as indicated by the fabric of its gouge, by the lack of geomorphic evidence for its activity and by a cluster of relatively young strike-slip faults that cross-cut its gouge fabric and juxtaposes it against young sediments. In contrast, the strike-slip structure has several active or recently active components, as indicated by their correlation with surface geomorphic features and by the nature of their fabric at several scales (consistent with the discussed indicative fabrics in section 2). Structural relationships indicate that the strike-slip structure is superimposed on the dip-slip structure.

The distribution of bedrock and other lithologic units exposed in the trench is shown in Figure 9. Sandstone of the Anaverde formation (light brown and orange) appears in the northeastern half of the trench. Several slivers of granite and gabbro (dark purple) appear within the sandstone, separated from it by strike-slip faults. Finally, an assemblage of crystalline rocks is present on the southwest side of the main fault as slivers within the dip-slip fault zone (pink). All of the sedimentary and crystalline rocks yield powdery substance upon light pressure suggesting that they were pervasively pulverized (Dor et al., 2006). Pleistocene (and younger?) alluvial material crops out at the southern end of the trench (yellow). Shallow soil and recent alluvium were not included in the trench log except in the active channel area (dark brown). Structural features include faults (red and black lines), shear fabric and its orientation (red and black small bars), gouge zones (color coded according to type) and slip striations with a separate symbol for strike-slip, dip-slip and oblique-slip, marked where kinematic indicators were observed.

Features related to the dip-slip fault appear mainly in the southwest and central parts of the trench. They comprise a wide dip-slip fault zone parallel to the strike of the SAF and dipping 35° to the southwest. Dark blue zones in Figure 9 represent massive and extremely cohesive, clayey and heavily welded dark brown to black gouge, having large planar slip surfaces that are parallel to the overall structure. The more pronounced, dominant and continuous surfaces appear as black heavy lines that in most cases separate the massive gouge from other units. Smaller shear surfaces (short black bars) within the gouge and along the major slip surfaces show a consistent set of pure dip-slip striations. The striations are clearly printed as long, straight and parallel mini troughs and ridges into the gouge surfaces (mullions), and are often associated with linear gypsum crystal growth. Zones in light blue are gouge layers of the same nature as above, also having a substantial amount of silty material. In its main part (south of the central part of Figure 9), this gouge is shown with

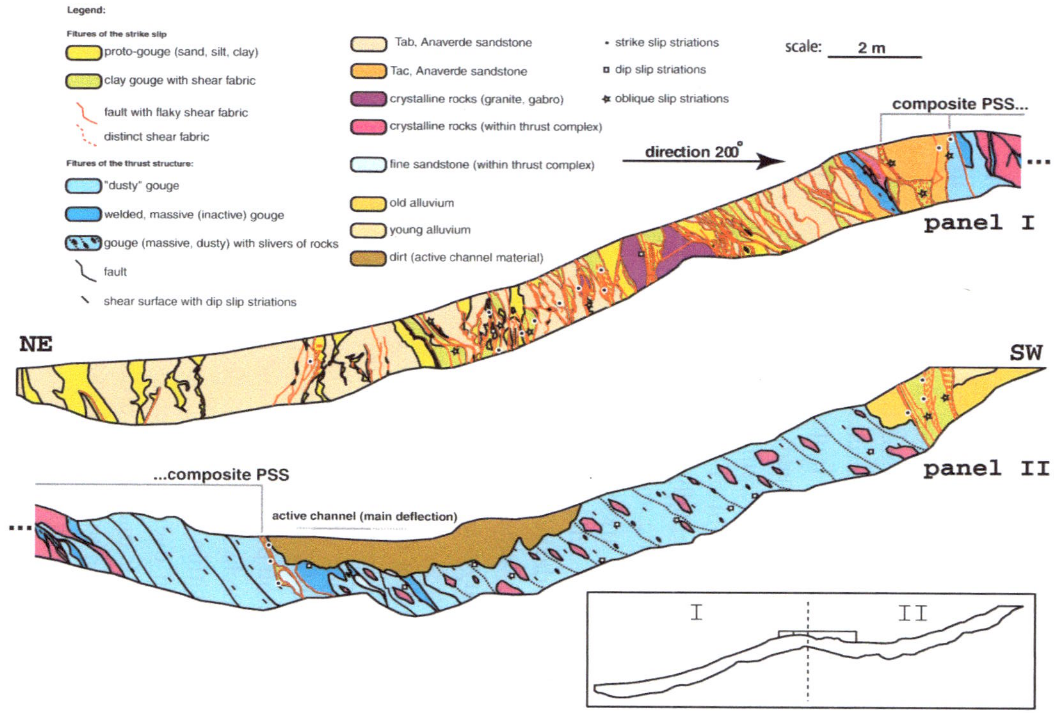

Figure 9
A trench log displaying a 45 m long cross section of the San Andreas fault zone near Palmdale with mapping of structural features and lithology. For displaying purposes, the trench profile is divided (bottom inset) into panels I and II showing the northeastern and southwestern parts, respectively. Features related to the strike-slip structure are shown with red, light-green and dark-green colors. The active strike-slip structure is superimposed on an inactive exhumed dip-slip structure. Elements of the composite PSS appear in the middle of the trench (shown in both panels). Most of the strike-slip activity is concentrated in panel I northeast of the inferred PSS. See details in the legend and in the text.

dotted lines and elongated pink patches, representing and generalizing numerous slip surfaces and slivers of crystalline rock inside the gouge, respectively.

Strike-slip faulting dominates in the northeastern and central part of the trench and is manifested by several tens of strike-slip faults and fault segments (red lines) with various inclinations from 30° to vertical, dipping both to the north and to the south. These faults are typically narrow and localized (0.5 to 4 cm wide), defined by brown-red to brown-dark, incohesive gouge layers with flaky fabric and small (mm to cm scale) shear surfaces that display mostly horizontal to oblique slip striations. The material itself is soft, claylike and appears to lack visible porphyroclasts. The strike-slip faults are surrounded or bounded by additional two types of gouge zones: the first type (green) is brown to dark-brown and has clayey soft content with some silty material. In places, it exhibits shear fabric parallel to the orientation of the nearest fault (dashed red lines). The second type appears to be a "proto-gouge" (light green). Those gouge elements are normally gray or light brown and seem to have smaller amounts of clayey material with respect to silt; their visible clasts content is higher than that of the other gouge elements and may bring them marginally under the definition of breccia (SIBSON, 1977). They lack distinct shear fabric and preserve some of the grain fabric of the host rock (in most places, the arkosic member of the Anaverde Fm.). The relationship of the 'proto-gouge' units to strike-slip faulting is indicated by their close association with the other darker and sheared gouge units, with the strike-slip faults themselves, and by their subvertical orientation. These three gouge types seem to reflect different maturity stages of gouge zones and they all appear to be associated with recent faulting.

We attribute the PSS to one of the following two candidates (Fig. 9). The first is the isolated strike-slip fault immediately below the northern end of the active channel; this fault was the major carrier of displacement during at least the last several earthquakes as indicated by its strong geomorphic signature. Associated with this fault is a 15 m large deflection of the channel and additional geomorphic features along strike, such as aligned notches and other deflected channels. It reactivates two or more ancient slip surfaces within the dip-slip structure, while stepping from one surface to the other, creating an overall steeper inclination than the host surfaces. The other candidate for the PSS is one of the vertical faults that bound the sliver of fine-grained sandstone (orange) on the top of the hill approximately at the center of the trench. Those faults juxtapose distinctive lithological units (the coarse against the fine-grained members of the Anaverde formation and the entire northern sequence of rocks against the major part of the dip-slip structure). They also separate the same set of units in another trench, located ~100 m to the east, where they also correlate with the dominant geomorphic expression for recent faulting. Due to this uncertainty with regard to the definition of the PSS, we refer to the zone that includes both candidates as the composite PSS zone.

The distribution of strike-slip features across the fault zone in Figure 9 shows that strike-slip activity is drastically stronger and more intense on the northeast side of the composite PSS with respect to strike-slip activity on its southwest side. Only three vertical strike-slip faults clustered in a one-meter-wide belt appear southwest of the composite PSS. They most probably slipped during young SAF activity as they separate the dip-slip complex from young (late Quaternary?) alluvial material. However, they do not have a clear geomorphic signature that indicates recent activity. The ground surface outside of the trench southwest of this belt is covered with old alluvium and soil, and shows no indication of recent fault activity as well. From the northeastern side of the composite PSS, several tens of strike-slip faults cut through the exposed rocks. They have most probably been active during recent SAF events and may have accommodated a fraction of the displacement. This is suggested by the fresh appearance of their shear fabric, which is alike to the fabric found in the faults of the composite PSS and in other active gouge zones (such as the gouge of the road cut exposure along the SJF near Anza, as described above). The association of this large group of faults with the active faults of the composite PSS is supported not only by their textural similarity, but also by an overall decrease of their local density as a function of distance from the major active area of the fault zone. A dense group of faults is clustered around slivers of crystalline rocks north of the PSS and their density decreases for the adjacent five meters to the north, after which the density further decreases over the next few meters. They become more diffuse with larger volumes of the host rock between them, and the last seven northern meters or so of the trench exhibit sparse direct evidence for shear, although there are several proto-gouge zones that may indicate minor strike-slip activity. The overall trend is therefore of gradual decrease in strike-slip activity from the composite PSS toward the northeast.

In addition, a 'fault zone valley' is correlated with the zone of intense strike-slip activity and can be partially inferred from the profile of the trench in Figure 9. There is no lithological reason for the development of such a valley on this side of the fault zone, as rocks on the valley side of the fault are sandstones and crystalline rocks that are likely more resistant to erosion than the alluvial material on the other side of the fault. Therefore, we attribute this effect to the increase in damage, and in particular, the partial pulverization of the rock and the weakness of the active gouge zones on the northeastern side of the fault zone, thereby enhancing the erosion.

To eliminate the possibility that this pattern reflects a local step-over or another site-related complexity, we opened another trench 100 m to the southeast. We observed there similar distribution of structural properties and geomorphic features, confirming the general asymmetric nature of the fault zone structure in the area.

Summary: We observe asymmetry of structural properties in three different exposures along the Mojave section of the SAF. Two of the observations (Little Rock Paleoseismicity site and in Little Rock Creek) are at the scale of cm to meters (fault core scale), whereas the site west of Palmdale is at the scale of meters to 10s of meters (fault zone scale). Asymmetry has a unique form in each of the exposures, but they all share the same sense: the northeastern side of the PSS is systematically more damaged.

3.3 The Punchbowl Fault

The Punchbowl fault is an inactive, exhumed strand of the San Andreas system in the central Transverse ranges of Southern California (Fig. 4). The fault has a long wavelength sinuous trace, dips steeply to the southwest, and was primarily a right-lateral strike-slip fault with a minor reverse component (CHESTER and LOGAN, 1986). The Punchbowl fault is parallel to the SAF and it is truncated to the northwest and to the southeast by the active SAF. Most of the Punchbowl trace is positioned about 5 km southwest of the SAF. The Punchbowl fault is therefore considered to be part of the SAF system (DIBBLEE, 1968, 1987). It juxtaposes along much of its length Precambrian to Cretaceous rocks of the San Gabriel basement complex against arkosic sandstones and conglomerates of the Miocene-Pliocene Punchbowl formation. The Punchbowl fault and the SAF form the boundaries of the Punchbowl basin, which was most likely created as a pull-apart basin in which the Punchbowl formation accumulated to a thickness of more than 1 km (NOBLE, 1954; WOODBURNE, 1975). Subsidiary faults within the formation suggest that deposition of this unit was contemporaneous with the movement on the Punchbowl fault (Chester, unpublished mapping, 1995). According to WELDON *et al.* (1993), at least half of the displacement on the fault occurred during the Pliocene and Pleistocene, during and after the deposition of the Punchbowl formation. The overall right lateral displacement on the Punchbowl fault complex is assumed to be of the order of 40 to 50 km, with 44 km proposed by SCHULZ and EVANS (2000) and CHESTER *et al.* (2004) based on the separation of the San Francisquito formation and the Fenner faults (DIBBLEE, 1967, 1968), and in agreement with the offset of the Punchbowl basin from its inferred sediment source (WELDON *et al.*, 1993).

The Punchbowl fault zone has a well-defined structure with damage intensity that increases progressively from the outer fault zone toward the fault core and the ultracataclasite layer (CHESTER *et al.*, 1993). Several workers have shown that shear is restricted to the fault core, the zone that includes the ultracataclasite layer and a zone, up to a few meters wide, of foliated cataclasite on both sides of the ultracataclasite layer (CHESTER *et al.*, 1993; SCHULZ and EVANS, 1998; CHESTER and CHESTER, 1998). An argument in favor of localization is the continuous lithology across each block adjacent to the fault compared to the sharp contrast in lithology on both sides of the narrow contact zone (CHESTER *et al.*, 2004). Additional support comes from fabric analyses that show sharp mechanical and mineralogical boundaries between the ultracataclasite and the host rock (CHESTER *et al.*, 1993). Detailed ultracataclasite studies have shown that slip was further localized at the last stages of faulting, macroscopically and microscopically, on a prominent slip surface within the ultracataclasite layer (CHESTER and CHESTER, 1998). For this reason the ultracataclasite layer is our frame of reference for the purpose of the study of symmetry properties in the Punchbowl fault. Effective study of symmetry properties within the ultracataclasite layer can be done only on the microscale.

The present-day exhumation is the result of a post-Pliocene uplift and erosion of the San Gabriel Mountains. According to the inferred uplift and erosion rate (OAKESHOTT, 1971; MORTON and MATTI, 1987), the thickness of sedimentary sequence in the Devil's Punchbowl basin cut by the Punchbowl fault, and the mineral assemblage and microstructures of fault rocks (ANDERSON *et al.*, 1983; CHESTER and LOGAN, 1986; EVANS and CHESTER, 1995), 2 to 4 kilometers of rock sequence has been eroded since the Pliocene, exposing a depth that is comparable to the top of the seismogenic zone of the modern San Andreas fault (SCHULZ and EVANS, 2000). Our study area includes about 2 km of the fault length within the Devil's Punchbowl County Park area, and in the Forest Service lands east of the park down to the South Fork campground (Fig. 10). We first discuss observations from the South Fork area and then from the vicinity of Devil's Chair. Both areas have been studied in various contexts in previous studies (CHESTER, unpublished mapping, 1995; CHESTER *et al.*, 2004 and references therein).

South Fork Area: With a simple geometry and an accessible 100-m-long continuous exposure of the fault and the wall-rocks, the South Fork area provides an excellent investigation site for comparison of structural properties between the two sides of the fault (Fig. 10). The basement complex of the southwest block and the Punchbowl sandstone of the northeast block are separated by an ultracataclasite layer that is a few tens of centimeters wide and dips 75 degrees to the SSW.

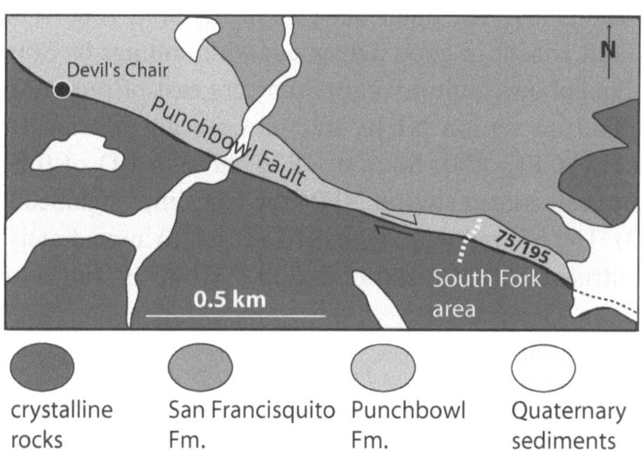

Figure 10
Geologic map of the Punchbowl fault in the South Fork and Devil's Chair study areas (after DIBBLEE and MINCH, 2002). The amount and direction of dip are indicated in the South Fork area. The array of white dots marks locations of fracture density measurements shown in Figure 11.

We evaluated the fault zone properties across the fault with an 85-m-long fault-perpendicular traverse (Figs. 10, 11). Nine stations, five of them in the Punchbowl sandstone and the other four in the basement complex, were chosen to represent the fracture density at a given distance from the fault. They are designated as NE55, NE40, NE25, NE04 and NE00.7 for stations on the northeast side of the fault, and SW30, SW12, SW02 and SW00.1 for stations on the southwest side of the fault, with the numbers indicating the distance in meters from the ultracataclasite layer. The choice of each station took into consideration the desired spacing that enables a reliable representation of FD gradient with the highest possible exposure quality. For quantification of the FD we used the box-count method with a 25 cm × 25 cm frame of exposed rock in each station. The frames were photographed and the fractures were mapped in the field on the photos and were later digitized. Figure 11a shows pictures of these nine frames with the corresponding digitized fractures maps; the assigned numbers are the Cumulative Fracture Length (CFL) for each frame in cm.

Several tens of meters northeast of the fault in station NE55, we measured 105 cm of CFL which corresponds to a FD of 0.168. Most (if not all) of the fractures are tensile. Fractures here are limited to pebbles and are believed to be the reaction of the tectonic stress field applied to rigid inclusions within a compliant material (EIDELMAN and RECHES, 1992). In the current state of the rock, the difference in mechanical properties between the host rock and the pebbles is negligible and therefore fracturing should not be selective with respect to the different lithological domains. The fractures could "select" the pebbles only when the pebbles were much stiffer than the matrix; this could happen only before the burial and litification of the conglomerate. Therefore, the faulting-related FD is very low and believed close to the background damage level.

The decrease in damage level around station NE55 with respect to stations closer to the fault probably marks the outer margins of the fault zone. This constraint on the width of the fault zone is in good agreement with findings by SCHULZ and EVANS (2000) from the Punchbowl fault several kilometers east of our study area.

Closer to the fault at station NE40, fractures are not restricted to only pebbles and the FD is 0.118 (CFL = 74). Station NE25 shows a FD of 0.194 (CFL = 121), whereas within the fault core at station NE04 the FD jumps significantly to a value of 0.885 (CFL = 553). The FD reduces to 0.698 (CFL = 436) immediately adjacent to the contact with the ultracataclasite at station NE00.7. However, the nature of fracturing

▶

Figure 11
(a) Fracturing intensity measurements along an 85-m-long traverse across the Punchbowl fault in the South Fork area (Fig. 10). The upper orange panel includes measurement stations on the NE side of the fault and the lower blue panel includes stations on the SW side. The pictures frame 25 × 25 cm of rock exposures. Titles of the pictures indicate the relative location from the fault (e.g., NE25 is 25 m NE of the fault). The fracture maps are shown both on the pictures and below them with numbers indicating cumulative fracture length cm/frame. (b) Close-up view of fracture map of station SW00.1 illustrating the inhomogeneous nature of the fractures orientation and intensity.

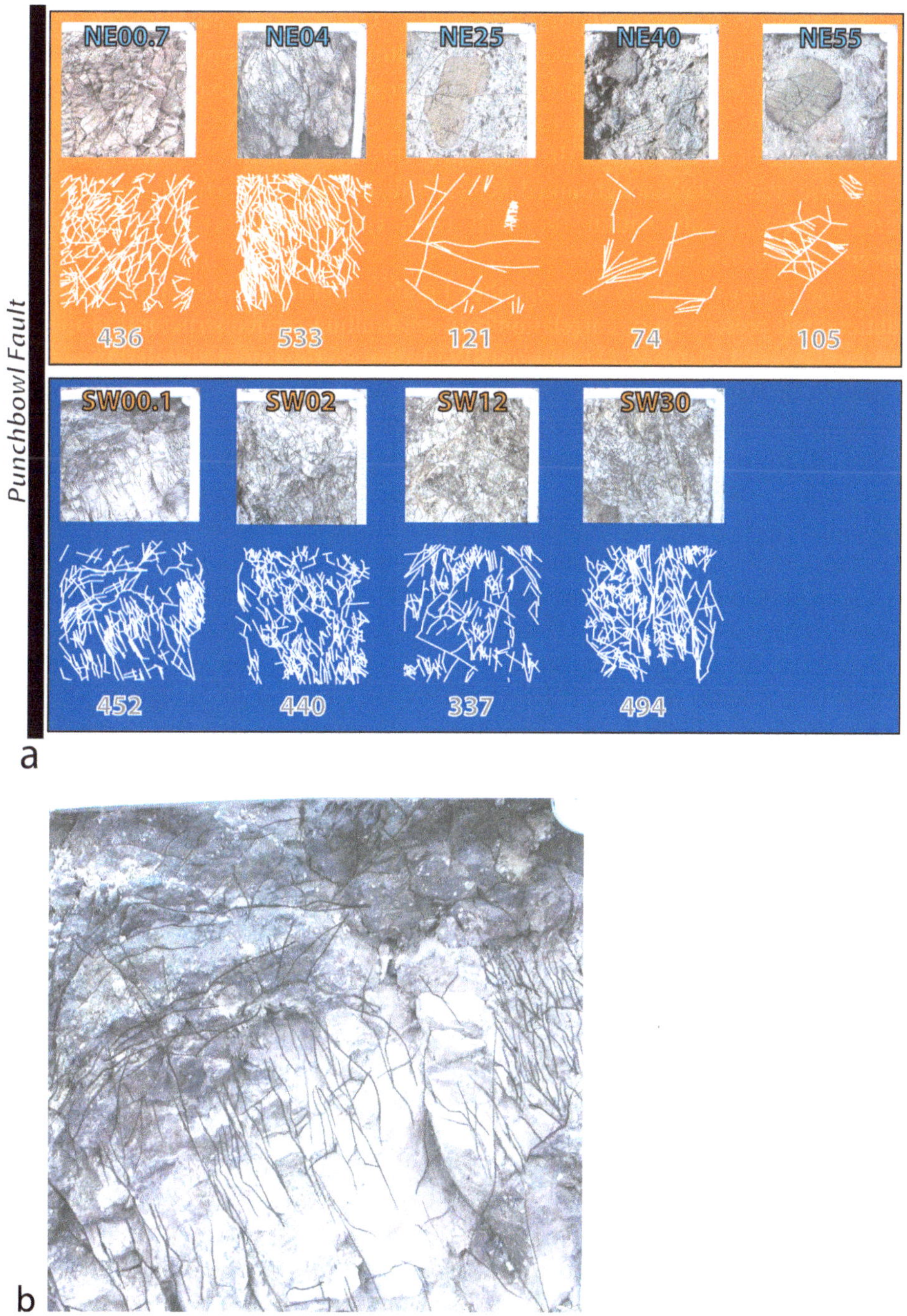
Punchbowl Fault
NE00.7
NE04
NE25
NE40
NE55
436
533
121
74
105
SW00.1
SW02
SW12
SW30
452
440
337
494
a
b

in samples from these two stations is similar: all macroscopic surfaces show slickensides population with internally consistent rake and the rock is partially pulverized. Therefore, we expect that a large portion of the fracture surface area is in the microscale below our mapping resolution. The ~20 cm wide sandstone belt immediately northeast of the contact with the ultracataclasite appears to be partially pulverized and fracture mapping in the macroscale cannot express the effective fracture density. These results express the overall qualitative observed damage pattern on the northeast side of the fault along this section: damage to the rock starts to increase above the background level several tens of meters from the fault and increases dramatically within the fault core, causing fragmentation and some pulverization of the rock.

On the southwest side of the Punchbowl fault, the granodiorite complex is highly fractured but presents poor correlation between faulting-related structural properties and fault-normal distance. At station SW30, the southwestern distal station with respect to the fault, the FD is measured as 0.790 (CFL = 494), the highest among the mapped stations on this side of the fault. The true FD is much higher as most of the fractures are in the mm scale, below the mapping resolution limit. Nevertheless, the rock is not pulverized and no macroscopic surfaces with slickensides were found. A similar level of fracturing intensity continues farther to the southwest for a distance of about 100 m along the accessible portion of the wall-rock (i.e., no qualitative FD gradient is observed). At stations SW12 and SW02, the FD is 0.539 and 0.704 (CFL = 337 and 440), respectively. The fractures at these two stations have characteristics similar to those described for station SW30. The fracture density rises slightly to 0.723 (CFL = 452) close to the fault at station SW00.1, within a zone of shear belts secondary to the ultracataclasite. However, even in this fine mapping scale, the distribution of fracture density and fracture orientation is not homogeneous. A careful analysis of the photograph and map at this station (Fig. 11b) shows that the chloritized foliated rock at the upper left corner presents a medium FD with subhorizontal long fractures that might be related to an inherited metamorphic fabric or to a fault parallel shear. The white rock shows a very high FD with relatively short subvertical fractures at its upper right part, and medium to low FD at its upper and left parts with essentially no fractures at its lower right part. The clear inhomogeneity of the damage in such a proximity to the fault, without significant rheological differences between the different fracture domains, suggests that most of these fractures are not fault-derived. The general appearance of the crystalline rocks immediately southwest of the ultracataclasite is almost intact in many places and presents very minor damage, in a sharp contrast to the shattered appearance of the Punchbowl sandstone (Fig. 12). Rocks in such a proximity to the fault are expected to experience pervasive, intense and relatively homogeneous fracturing during rupture, at least like the Punchbowl sandstone on the other side of the fault, if no symmetry considerations are involved. From a mechanical point of view, the crystalline rocks are expected to experience more fracturing, as the

Figure 12
Geomorphologic expression of the damage pattern across the Punchbowl fault, view to the southeast. Slip is localized in this area in the southwest side of the ultracataclasite (bottom of the trough, mostly covered with debris), separating the slightly fractured (almost intact) basement rocks on the right from the heavily fragmented Punchbowl conglomerate on the left. The conglomerate is significantly more degraded than the basement rocks within the fault core. The asymmetry in the topographic profile of the trough is clearly related to the asymmetry in fragmentation intensity across the fault.

sandstone has higher porosity and can more easily diffuse sharp stress concentrations. Therefore, the zone of minimum damage in Figure 11b may represent the fault-related FD.

Overall in this area, we did not observe a consistent gradient of damage level on the southwest side as a function of distance from the fault. We also noticed the absence of slip striae and fracture sets with preferred orientation, which are structural features that can be associated with faulting on the Punchbowl fault and were found on the northeast side. Despite the difficulties in comparing damage level between the two types of rocks, the details of the fracture density traverse and the overall intensity pattern of damage show that the damage pattern associated with slip on the Punchbowl fault is asymmetric with more damage on the northeast side of the fault. This asymmetry is especially pronounced by the contrast of damage intensity within the fault core.

An independent measure for the contrast in damage content is expressed by the morphology of the fault core (Fig. 12). While the basement metamorphic rocks on the southwest side are bounded by a high and continuous vertical cliff immediately along the contact with the ultracataclasite, the sandstone on the northeast side has developed a meter wide and a meter deep trough along the fault. The two sides of the fault are clearly subjected to significantly different rates of degradation. The trough is correlated with the ultracataclasite and the zone of maximum fragmentation and pulverization on the sandstone side. It is likely that weathering and fragmentation enhanced each other in a positive feedback mechanism, causing the development of this trough. The steep topography on the southwest side of the fault indicates the low rate of degradation that is possible when the damage level is low. The possibility that rocks on the southwest side gained strength to an almost intact state by healing is unlikely because re-strengthening by healing should have operated at least partially also on the northeast side, yet no significant healing occurred within the sandstone. Also, we only observed a few healed fractures in the cliff on the southwest side of the core zone, supporting the geomorphic observation.

Devil's Chair Area: CHESTER and CHESTER (1998) studied the fine structure of the fault in the Devil's Chair area and determined that slip was localized along the ultracataclasite layer. Their study further suggests that a single prominent slip surface accommodated most of the fault's displacement.

We compared the intensity of damage between the igneous rocks on the southwest and the Punchbowl sandstone on the northeast within the fault core by measuring linear fracture density along a meter-long scan line perpendicular to the ultracataclasite on both of its sides. Three stations along 250 m of the fault were chosen for this measurement (Fig. 13) and in all of them, the density of fractures is larger within the Punchbowl sandstone compared to the amount of fractures in the

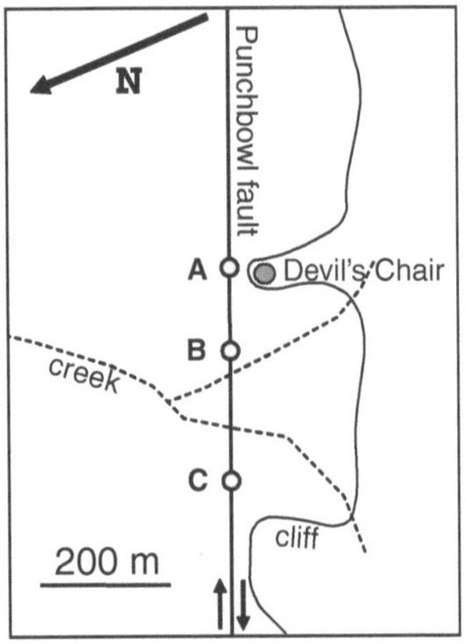

Figure 13
Major geographic features in the Devil's Chair area and locations of the FD measurement sites marked with white dots (see Fig. 10 for general location). Station C is in the vicinity of the study sites of CHESTER and CHESTER (1998).

Table 1

Number of fractures per meter on each side of the Punchbowl fault, as measured in the Devil's Chair area. The overall asymmetry is 71%

Site	Punchbowl (fractures/m)	Basement (fractures/m)
A	56	33
B	73	19
C	43	17
Mean	**57**	**23**

monzogranite and other rocks southwest of the fault (Table 1). The overall asymmetry in fracture count is about 71%.

Summary: Fracture density measurements done at two different scales at two different sites 1.5 km apart along the Punchbowl fault indicate that the Punchbowl sandstone on the northeast side of the fault is more damaged than the igneous complex on the southwest side of the fault. The FD traverse near South Fork shows that the intensity and type of damage in the Punchbowl sandstone is clearly related to faulting on the Punchbowl fault as damage intensity is proportional to the distance from the fault, and the fractures are organized with a preferred orientation and exhibit internally consistent kinematic indicators. In contrast, damage in the granodiorite shows no clear dependency on distance from the fault and the overall damage style may be related to other phases of deformation or to inherited metamorphic fabric. The morphology of the fault core at the South Fork site independently shows that the northeast side is significantly more damaged, leading to the development of a distinct geomorphic trough within the sandstone, as opposed to a cliff of nearly intact rocks on the southwest.

4. *Discussion*

We have established that at all examined localities there is an asymmetric pattern of fault zone damage with respect to the current principal slip surface and that the sense of damage asymmetry is consistent along each studied fault section. Those patterns are manifested at different scales, in different forms, and we expect to add to our understanding of the breadth of expression of damage and symmetry properties as we continue to explore new sites. Nevertheless, our observations demonstrate that macro-scale fault zone damage from the gouge zone outward to at least tens of meters expresses distinct asymmetry in form and extent. Those sets of observations can be interpreted as the result, and hence also as indicators, of preferred rupture direction of earthquakes. Below we assess the possibility that the seismogenic depth structure of a fault is controlling the geologic damage pattern by dictating the preferred direction of ruptures. This can be done by correlating the sense of observed damage asymmetry to the local velocity structure at depth, since this correlation is an expected outcome of rupture along a material interface (BEN-ZION and SHI, 2005).

4.1 Asymmetry of Structural Properties in Light of Velocity Structure

The San Jacinto Fault, Anza: The gouge fabric found in the three studied exposures on the SJF near Anza shows that the northeastern side of the current PSS is more damaged. This sense of asymmetry is consistent with high resolution seismic imaging by LEWIS *et al.* (2005), based on fault zone trapped waves recorded across the three branches of the SJF south of Anza. Their inversion results indicate the existence of ~100 wide fault zone trapping structures at the different branches that extend to a depth of about 3.5 km. The seismic imaging shows further that the trapping structure at each of the branches is not centered on the surface trace of the fault, but is offset 50–100 m to the northeast. The results imply that more damaged fault zone material is present on the northeast side of each fault. The damage asymmetry in our geological mapping and the seismic trapped waves study is compatible with northwestward rupture propagation direction, with the primary tensional quadrant of most rupture events on the northeast side of the fault.

The local velocity structure of the Anza seismic gap was imaged by SCOTT *et al.* (1994), who showed that at depths of 3 km, 6 km and to some extent also 9 km, the northeast side of the fault has faster seismic velocities. This is the side that has more damaged rock in our geological studies and in the seismic imaging of LEWIS *et al.* (2005). These results are in agreement with regional imaging of MAGISTRALE and SANDERS (1995) and SHAPIRO *et al.* (2005), who show the same sense of velocity contrast in the Anza area and farther to the southeast. The existence of more rock damage on the side of the fault that has faster seismic velocity at seismogenic depth is expected if a material interface controls the preferred propagation direction on the fault.

The San Andreas Fault in the Mojave: The consistency of our gouge and fault zone scale observations from the Palmdale-Little Rock area of the SAF most probably reflects the asymmetric damage pattern across the fault in this area with the northeastern side more damaged. This conclusion probably also applies for the entire Mojave section of the fault, based on the compatibility of the sense of asymmetry discussed in this work with results associated with pulverized fault zone rocks. DOR *et al.* (2006) mapped the distribution of pulverized rocks along a 140-km-long section of the SAF in the Mojave, overlapping with our study area. They found that the pulverized rock is a systematic structural feature of the fault that can be found, with just a few exceptions, wherever crystalline rocks crop out within up to 200 m from the active trace of the fault. The distribution of the outcrops with respect to the fault was found to be asymmetric, with most of the pulverized rock exposures being on the northeastern side of the fault. This pattern may reflect, at least partially, the current distribution of available exposures. Nevertheless, in several places where comparison is possible between the two sides of the fault, pulverization is more intense and the zone of pulverization is wider on the NE side of the fault, supporting the near-field macroscopic observations presented here. This combined multi-scale set of observations suggests that the northeastern side of the SAF in the Mojave has accumulated more damage at various scales and forms during repeated ruptures. A possible explanation for this damage distribution pattern is that most of the ruptures during the recent geological history of the fault propagated from southeast to northwest in the Mojave area, having their tensional quadrant on the northeastern side of the fault.

The velocity structure along the Mojave section of the SAF has been studied to some extent. SHAPIRO *et al.* (2005) cross-correlated ambient seismic noise at USArray stations and constructed tomographic images of principal geologic units in California. Their 7.5 and 15 s period images based on surface waves indicate that overall, the Mojave block shows higher seismic velocities with respect to the southwest side of the fault. This velocity contrast continues farther to the southeast and is of opposite sense to the velocity contrast across the SAF northwest of the big bend. As part of the Los Angeles Region Seismic Experiment (LARSE), FUIS *et al.* (2001) report for line 1 (crossing the SAF slightly southeast of our study areas) that in contrast to velocities below the San Gabriel Mountains, velocities to 10 km depth below the Mojave are consistently higher than average laboratory velocities for the Pelona schist, and therefore the Pelona schist does not appear to be present in the Mojave desert beneath line 1. This agrees with aeromagnetic data along this transect. Line 2 of the LARSE experiment crossed the SAF in the Elizabeth Lake area, northwest of our west Palmdale site. LUTTER *et al.* (2004) report for line 2 that although poorly resolved, basement velocities at depth are higher north of the SAF. That inference is also supported by gravity data along the transect. FUIS *et al.* (2003) show generalized velocity cross sections based on lines 1 and 2, with the Mojave side having 8–10% faster seismic velocities at depth. If these two studies are representative of the entire Mojave segment, the faster velocity side of the fault at seismogenic depth correlates with the side of the fault that expresses more damage, as expected for rupture along a material interface (BEN-ZION and SHI, 2005).

In both studied sections of the SAF and the SJF, the asymmetry of structural damage may reflect a preferred propagation direction associated with contrasting elastic properties at depth. Across both the SAF and the SJF, the velocity contrast continues farther southeast of the study areas, where ruptures could have nucleated in order to propagate to the northwest through our investigation sites.

The Punchbowl Fault: The excess in fault–related damage on the northeast side of the Punchbowl fault compared to its southwest side in the studied area suggests that more paleoearthquakes propagated to the northwest than to the southeast when the fault was active. The velocity structure of the Punchbowl fault at seismogenic depth is unknown. Based on our observed damage pattern we estimate that the southwest side of the fault has slower seismic velocity at depth with respect to the Mojave side.

Our observations are compatible with persistent occurrence of earthquakes along the examined fault sections in the form of wrinkle-like ruptures on material interfaces with preferred propagation direction. We are aware that the observations are subjected to various uncertainties, and of the general nonunique nature of any interpretation (see details in section 4.4). Nevertheless, the observed correlation of multiple manifestations of rock damage with the velocity contrast across a fault, at several scales along different sites of geometrically simple segments, supports our interpretation. We also note that wrinkle-like mode of rupture provides a possible explanation for a variety of outstanding geophysical issues, including the lack of frictional melting products, short rise time of earthquake slip and suppression of branching in the structures of large faults (e.g., BEN-ZION, 2001).

4.2 *Historic Earthquake Behavior that Might be Explained by Preferred Rupture Propagation Direction*

While a few earthquakes are statistically insignificant and cannot indicate the long-term preferred rupture direction associated with a fault section (which is why we study the geologic record, as mentioned in the introduction), the timing and spatial distribution of historical earthquakes on major strike-slip faults may be explained, in part, by the existence (or lack of) a preferred rupture direction.

The 19th Century Sequence of Earthquakes on the SAF: The elapsed time of about 300 years since the penultimate earthquake prior to 1857 (early 1500 A.D., LINDVALL *et al.*, 2002) and the sequence of the 1812 and 1857 earthquakes suggest that in the early 19th century, the entire south-central part of the SAF was ready to fail. It is thus interesting to ask why the 1812 rupture did not extend over this entire section of the fault. Moreover, why did the 1857 rupture overlap about 60 to 100 km with the 1812 rupture

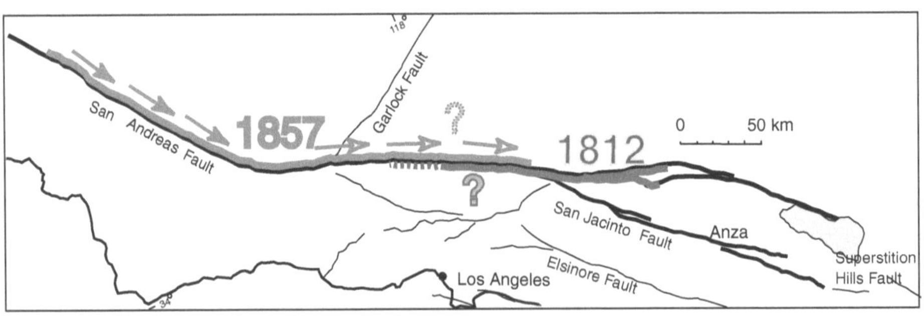

Figure 14
The south-central San Andreas Fault system in California and the known extent of the partially overlapping 1812 and 1857 ruptures (heavy and light gray, respectively). The propagation direction of the 1812 rupture is unknown while that of the 1857 rupture was inferred to be toward the southeast, based on locations of candidate foreshocks. Even if correct, the propagation direction in the Mojave may have been different if the earthquake was a multi-shock event.

(Fig. 14), overshooting several meters of slip into the rupture zone of the earlier earthquake only 45 years later (WELDON *et al.*, 2002; FUMAL *et al.*, 1993)? And why is the inferred rupture direction for the 1857 Ft. Tejon earthquake (SIEH, 1978b; AGNEW and SIEH, 1978) the opposite of our inferred preferred rupture direction for the Mojave?

To address these questions we need to examine the available information. Seismic imaging of the SAF near Parkfield shows 5–20% lower seismic velocity in the seismogenic zone on the northeast side of the fault (e.g., BEN-ZION *et al.*, 1992; EBERHART-PHILLIPS and MICHAEL, 1993). Therefore, a southeastward propagation direction of large ($M > 7$) earthquakes is expected in that area according to the model of rupture along a material interface. Nucleation in the Carrizo plain and southeastward propagation was inferred by SIEH (1978b) for the 1857 earthquake. In the last 40% or so of its final length, the rupture propagated in the Mojave, against our assumed long-term preferred rupture direction inferred from the observed rock damage asymmetry. High resolution seismic imaging of the SAF south of the Parkfield area is lacking, but as mentioned earlier the regional imaging by SHAPIRO *et al.* (2005) and traverses of the LARSE experiment (FUIS *et al.*, 2003) suggest that the sense of velocity contrast observed for the Parkfield area flips somewhere toward the Mojave. It is thus plausible that the last part of the 1857 event propagated into a zone of opposite sense of velocity contrast in the Mojave area, and was finally arrested toward Cajon Pass. A support for this comes from the slip function of this earthquake (SIEH, 1978a). The displacement peaked at ~8–9 m around Wallace Creek, then subsided to ~6–7 m along the southern Carrizo plain and dropped to ~3 m passing Tejon Pass into the Mojave with values as low as 1–1.5 m in Wrightwood. The 1812 earthquake has an unknown propagation direction, but from the correlation between the sense of damage asymmetry and the apparent velocity contrast we speculate that this earthquake propagated to the northwest. The strong velocity contrast in the vicinity of its southeastern most known part (SHAPIRO *et al.*, 2005), is compatible with northwestward propagation direction. If this is correct, the 1812 rupture was presumably arrested in the northwestern edge of the Mojave due to a change in the velocity structure, and thus did not leave evidence in the paleoseismic record of Frazier Mountain (LINDVALL *et al.*, 2002). More details about the velocity structure of the SAF in this area and additional paleoseismic data about the 1812 rupture are needed in order to clarify more aspects of this problem. We also note that with the limited available information on the rupture history of the 1857 earthquake, it is possible that this earthquake was a compound event with two (or more) shocks. The first and largest one nucleating in the northwestern Carrizo plain, propagating southeastward and arresting somewhere in the northwestern Mojave (maybe due to its entrance into a zone of decrease and then a change in the velocity contrast); the other shock, responding to the large stress transfer of the first one and overriding part of the previously yielded 1812 rupture zone, nucleating in the southeastern Mojave and propagating toward the northwest. A two-shock scenario as above is compatible with the known velocity structure along the rupture length and with the geological signal for preferred rupture direction in the Mojave.

The 20th Century Sequence of Earthquakes on the North Anatolian Fault (NAF): During the past century, the NAF experienced a remarkable sequence of ruptures that began with the great 1939 (M7.9) earthquake near Erzincan and proceeded westward with earthquakes in 1942, 1943, 1944, 1951, 1957, 1967, 1999a, and 1999b. STEIN *et al.* (1997) attributed this rupture sequence to successive failures of fault segments due to stress transfer from failed neighboring segments. However, the 1943 earthquake did not nucleate in the region of stress increase but rather at the opposite end of the final rupture, far to the west. Similarly, the 1999 August earthquake nucleated near the head of Izmit Bay and ruptured bilaterally with the main rupture directed eastward toward the earlier failures. This was followed by the continuation of rupture to the east in the November 1999 Düzce earthquake.

The general westward migration of the earthquakes is an expected outcome of rupture along a material interface at the NAF, if the southern block is generally the side with lower seismic velocity (e.g., SENGOR *et al.*, 2005). However, three events of the 1939–1999 sequence ruptured eastward against the overall unzipping direction of the fault. The 1943 and the 1999a earthquakes also nucleated away from the area of largest inferred stress-transfer loading. The loading, nucleation and rupture scenarios of these events are similar to the experimental setting and results of ANOOSHEHPOOR and BRUNE (1999) with two different foam rubber blocks. This suggests that the rupture directions of those events are controlled by a reversed local velocity structure (i.e., lower seismic velocity north of the fault). If correct, we expect in those fault sections that more damage is present in the southern block with the assumed faster seismic velocity, and opposite sense of damage asymmetry at the other rupture zones along the fault. These expectations can be tested with detailed field observations of the type discussed in Section 3.

4.3 Possible Related Observations

In the European Alps, study of the exhumed Gole Larghe fault showed a distinct asymmetry that was interpreted as the result of preferred direction of rupture propagation (DI TORO *et al.*, 2005). The Gole Larghe fault branches to the east from the large right lateral Tonale fault into a tonalitic intrusion. Right-lateral slip along the Gole Larghe fault took place at a depth of 9–11 km about 30 Ma ago (DI TORO and PENNACCHIONI, 2004) and was at least in part seismic as indicated by numerous veins of pseudotachylytes. DI TORO *et al.* (2005) studied the orientation of over 600 injected veins filled with pseudotachylytes, branching from several secondary faults of the Gole Larghe fault zone and exposed in the glaciated outcrops of upper Val di Genova. They found that veins are asymmetrically distributed with respect to the fault, with 67.7% of the veins intruding the southern bounding block. They concluded that those veins were formed in the tensional quadrant of propagating mode II ruptures and that the asymmetric fracturing pattern indicates that most of the paleoearthquakes on the Gole Larghe fault propagated from west to east, either nucleating at the junction of the Gole Larghe fault with the larger Tonale fault or branching during large Tonale earthquakes. The Tonale fault is a segment of the major tectonic lineament of the Alps and one of the main lithological boundaries (STECK and HUNZIKER, 1994). It is thus likely that the large Tonale paleoearthquakes were affected by material contrast which may have dictated the direction of ruptures.

The Nojima fault in Japan, along which the 1995 Kobe earthquake occurred, displays structural asymmetry at multiple scales and different types of signals. The Nojima fault is a right-lateral strike-slip fault with a minor reverse component, striking SW-NE along the northwestern margins of Awaji Island, Japan (MIZUNO *et al.*, 1990). The fault juxtaposes the Miocene Iwaya Fm. and the Plio-Pleistocene Osaka group on the northwest side of the fault against Cretaceous granitoids on the southeastern side of the fault. The geological survey of Japan recovered a drilling core from the Nojima fault about a year after the Kobe earthquake. The inclination of the fault near the drilling site was estimated to be 83° and the main fault crossed the drill hole at depth of 625 m. TANAKA *et al.* (2001) and OHTANI *et al.* (2000) identified one of the seven shear zones found within a 55-m-wide fault zone as the primary seismically-active fault core. Within the fault core, the breccia and ultracataclasite layers are located almost exclusively on the southeastern side of the principal slip surface of the fault. Except locally, the northwest side of the fault is weakly deformed (OHTANI *et al.*, 2000). TANAKA *et al.* (2001) defined the principal Nojima fault surface as the northwestern boundary of the ultracataclasite layer at a depth of 625.27 m, where it separates the fault core (ultracataclasite + breccia) on the southeast from damaged fault zone rocks on the northwest.

The structure of the Nojima Fault zone as it crops out southwest of the drill site (MIZOGUCHI *et al.*, 2000) can be correlated to the structure of the fault core at depth, with the same sense of asymmetry and similar contrast between the two sides of the fault. The PSS, represented here by a 0.1–0.15-m-wide layer of fault gouge, separates the Plio-Pleistocene Osaka group on the northwest from a two-meter-wide breccia layer on the southeast, with gradual transition to a 10-meter-wide zone of fractured granite. Additional manifestation of the asymmetry at the fault zone scale is given by laboratory analysis of rock strength of recovered core samples (LOCKNER *et al.*, 1999). The results from that study suggest that within a fault-zone width of about 40 m, rocks on the northwest side have about 40% higher peak shear strength with respect to rocks on the southeast side (although based on a small population of samples; D. Lockner, pers. comm.). Thus, both a careful structural study at the fault core scale and initial results of lab analysis of rock samples at the fault zone scale of the drilling core of the Nojima fault correlate well with the fault zone structure in a surface exposure. All those observations indicate that the fault zone structure is highly asymmetric with the southeast side being more damaged. These observations are consistent with a statistical preference of rupture direction to the northeast. The velocity structure associated with the surface geology at the Awaji Island region has rocks with faster seismic velocity to the southeast. If this velocity structure extends to depth, the observed damage asymmetry is compatible with the theoretical predictions for ruptures along a material interface.

4.4 Possible Interpretation Problems

Many processes can contribute to the fabric and composition of fault-zone rocks. First, a mature fault zone reflects the cumulative effect of many thousands of earthquakes, and the mechanical nature of those earthquakes may have changed over the fault's history, printing different and sometimes contrasting signals in the fault-zone rocks. Where migration of the active strand of the fault occurred, the apparent observed symmetry properties may not reflect the current distribution of damage around the currently active fault. Second, a single rupture can alter the fabric in several ways and have diverse imprints on the rocks (e.g., fracturing due to frictional instability vs. fracturing due to reduction of normal stress). Third, there are many interseismic processes acting in a fault zone that can influence not only the composition but also the internal structure, mainly the microstructure of fault rocks. Finally, all the above are influenced by both the overall geological setting and local site effects. In addition, we have to justify our ability to infer properties of dynamic earthquake ruptures from surface observations. These issues are discussed briefly below.

Deducing Dynamic Rupture Behavior at Seismogenic Depth from Surface Observations: Earthquake-related observations would be more reliable if done at seismogenic depth where the bulk of seismic slip occurs. However, the focal depths of earthquakes are generally not accessible, other than in deep mines (e.g., DOR *et al.*, 2001; RECHES and DEWERS, 2004; WILSON *et al.*, 2005). Many valuable studies have been conducted on inactive exhumed faults (SIBSON, 1989; CHESTER *et al.*, 1993; SCHULZ and EVANS, 2000, and references therein), but these studies are limited in the sense that they lack knowledge on the seismogenic behavior of the faults (and alterations during the exhumation process). In this study we use a combined approach in which we utilize and create exposures on exhumed (Punchbowl) and on currently active faults (SAF, SJF). While the bulk of seismic slip occurs at seismogenic depth, simulations of damage generation indicate that conditions in favor of fault damage creation are limited to the upper three km or so of the crust (BEN-ZION and SHI, 2005). This is similar to the depth extent of imaged low velocity trapping structures (e.g., BEN-ZION *et al.*, 2003; PENG *et al.*, 2003; LEWIS *et al.*, 2005) and is supported by our related geologic observations of damaged fault zone rocks that were never buried deep (DOR *et al.*, 2006). The behavior at the surface is assumed to be generally dictated by the behavior of dynamic rupture at depth. This is especially so in the case of symmetry properties that are expected to result from a wrinkle-like rupture mode, because although the normal stress decreases with decreasing depth, the material contrast is generally larger in the top few km of the crust (e.g., BEN-ZION *et al.*, 1992; EBERHART-PHILLIPS and MICHAEL, 1993), thus promoting strong signals at shallow depths. The rupture is expected to carry its asymmetric dynamic properties to the surface, where damage generation may be amplified due to the increase in contrast of the velocity structure. Creation of damage that is concentrated in the top few km will not affect substantially the elastic properties of the rocks at depth and therefore will not change significantly the velocity contrast where it has a control on dynamic effects. Hence, the process is controlled by stable conditions in the seismogenic depth and has increased manifestations as the depth becomes shallower.

Contrast in Surface Lithology and its Influence on Damage Pattern: Faults commonly juxtapose on the ground surface rocks with contrasts in their strength, competency, isotropy, flow density and other factors that, to some extent, might influence the damage distribution across them. Moreover, if this is a dominant factor, we expect to see some correlation between the strength of the rocks and the intensity of damage they present. However, observations made in this study and in a parallel work (DOR *et al.*, 2006) indicate that damage favors the northeast side of the SAF, regardless of the rock type. In addition, we show here that in the Punchbowl fault, even though pulverization should favor the denser igneous rocks of the San Gabriel complex, the more porous sandstone on the northeast side of the fault is more damaged and perhaps even pulverized.

The Effect of Fault Dip on Asymmetry: This study focuses on strike-slip faults because the dip of the fault alone can break the symmetry. BRUNE *et al.* (1999) showed that more damage is expected in the hanging wall of a thrust fault and especially in the near-surface region due to waves trapped in the wedge-shaped hanging wall of the fault. Not all the fault strands that we studied are perfectly vertical and their inclination close to the ground surface ranges between 60 to 90 degrees, with a common secondary component of dip-slip. If so, how do we eliminate the inclination factor? If the dip of the fault was a governing factor in the distribution of damage across the faults, we would expect the damage pattern to be correlated with the dip. However, the fault core scale damage in the SJF is concentrated in the hanging wall side, whereas fault core scale damage in exposures with dipping structures of the SAF and overall damage in the Punchbowl fault is more pronounced on the footwall side.

Complex History of the Fault, Temporal Changes in Velocity Contrast at One Site: Faults with large cumulative displacement may juxtapose different rock bodies against each other during the faulting history, and any segment may have experienced changes in the sense of velocity contrast. The geology of one locality can potentially represent the cumulative effects of several faulting phases in which the preferred rupture direction has flipped. A migration of the active strand of the fault over time can also bias the apparent symmetry properties of the fault. In the clay-rich active or recently active fault gouge, there is less danger of confusing former faulting phases with the current one, because the material itself evolves with time. In addition to the accumulated primary (detrital) material, newly-grown (authigenic), fine-grained phyllosilicates are constantly added (VAN DER PLUIJM *et al.*, 2001). Additional detrital material is also added, as evident in the translocation of dark, organic-rich clay and the addition of detrital sand and silt downward into open fractures that are produced by surface ruptures. The entire gouge body is therefore evolving fast enough so that it is unlikely that ancient fabrics will be preserved for any length of time. This is confirmed by the relatively incohesive nature of the fracture surfaces in an active gouge. If these fractures are inactive long enough, they become more cohesive and massive (like the gouge on the southwest side of the SJF and in the dip-slip structure in the trench exposure near Palmdale).

The wall-rock within the fault zone may preserve fabric from ancient faulting phases, but this fabric will most likely be overprinted by damage from the current or most recent faulting phase. In some cases it is possible to separate different generations of fractures using cross-cutting relations (e.g., VERMILYE and SCHULZ, 1998, 1999). When uncertainty still exists, it is essential to study multiple signals at multiple scales as done here, and to correlate the sense of asymmetry with additional independent observations such as seismic trapping structures. The correlation of several expressions of the signal from several sites along the fault can reduce the non-uniqueness that is naturally associated with observations at one site. In this study we have documented such a correlation of a consistent excess of damage on one side of the fault expressed in different ways and at several scales. The systematic damage asymmetry is correlated with the local velocity structure as predicted for rupture along a material interface (BEN-ZION and SHI, 2005).

Deviation of a Single Rupture or a Temporal Cluster of Ruptures from the Statistical Preferred Direction: A fault-zone is a complex system with many possible influencing factors; the interaction of the fault with other faults can make its behavior even more complex. Therefore, the preference for rupture direction is only expected to be statistical over the long term, and single ruptures might propagate against the preferred direction. Segments of a fault can also be in the spatial or the temporal transition from one preferred direction to the other, or may not have a preferred rupture direction at all. Various factors can reduce the statistical preference related to velocity contrast as a prime factor and hence obscure the intensity of the geological signals discussed in this work.

The Influence of Geometrical and Compositional Complexities: Kinks, bends, step-overs, branches and secondary faults can locally cause compression or tension, and possibly bias the signature of the fundamental dynamic effects associated with large-scale ruptures. The scale of geometrical irregularities that may interfere with the signal is not very clear. For example, it is unknown to what extent the regional big bend of the SAF around Tejon Pass may affect fault core scale damage pattern in that area. Significant compositional heterogeneities of the rocks at seismogenic depths may also influence the faulting pattern by locally changing the sense of the velocity contrast (i.e., a maffic, dense, regional scale intrusion in a leucocratic lighter rock). To overcome difficulties associated with geometrical perturbations, we selected sites with minimum structural complexities in sections of the faults that are relatively straight, simple and vertical. Nevertheless geometrical irregularities can occur at all scales. It is therefore important to show consistency in multi-signal multi-scale observations along a fault section. Our results from the examined exposures show an overall consistent pattern of damage asymmetry that is independent of known or unknown structural and compositional complexities in the vicinity of a particular exposure.

Finally, we emphasize that while single ruptures can produce asymmetric damage pattern at various scales (e.g., JOHNSON *et al.*, 1997; VERMILYE and SCHULZ, 1998, 1999; RECHES and LOCKNER, 1994), the observations discussed in this work reflect clearly the imprint of a large population of earthquakes. This is attested by the intensity, multi-signal and multi-scale nature of the observations (from fault core to geomorphic signatures), and by the observed fabrics and field relations which indicate that certain fault zone components were active at different time periods.

5. *Conclusions*

We have made numerous *in situ* observations of damage asymmetry at various scales along three major faults in southern California. We attempted to develop a methodology to infer systematic structural asymmetry at different scales (recognizing various potential interpretation problems), and use the observations to test the hypothesis that large earthquake ruptures on strike-slip faults that separate different crustal blocks have a preferred propagation direction (e.g., WEERTMAN, 1980; ANDREWS and BEN-ZION, 1997; RANJITH and RICE, 2001; BEN-ZION and HUANG, 2002; SHI and BEN-ZION, 2006). The results indicate that the observed sense of asymmetry is correlated with the local velocity structure, where known, and suggests that the asymmetry is the long-term result from a preferred dynamic rupture direction, in agreement with the theoretical predictions. In addition, we pointed out with examples of historic rupture events from the San Andreas and North Anatolian faults, that the property of a preferred rupture direction may partially explain the distribution in time and space of large earthquake ruptures and their nucleation points. The inferences that were presented here could possibly apply to other large strike-slip faults, where geological studies of symmetry properties may be utilized to infer on possible preferred propagation directions of earthquake ruptures.

The propagation of earthquakes as wrinkle-like ruptures in a preferred direction can have fundamental implications for many aspects of earthquake physics and hazard. The interaction between slip and normal stress along a material interface can dynamically reduce the frictional strength, potentially to zero, making material interfaces mechanically favored surfaces for rupture propagation (e.g., BEN-ZION, 2001; BRIETZKE and BEN-ZION, 2005). This may affect strongly the effective constitutive laws, short rise-time of earthquake slip, the generated frictional heat, suppression of branching, the evolution of fault zones, and expected seismic shaking hazard.

The geological signals for preferred rupture direction are highly diverse. In this paper, we presented several such signals based on observations in the gouge and fault zone scales of large strike-slip faults in southern California. In a parallel study by DOR *et al.* (2006), we discuss additional observations associated with pulverized fault zone rocks. Further progress in theory can provide more detailed predictions on expected properties of the damage that may be tested further by additional multi-disciplinary multi-scale observations. A continuing feedback between theoretical predictions and field observations can lead to the recognition of new types of signals that may be used to deduce fundamental properties of earthquake behavior from *in situ* observations of fault zone structure.

Acknowledgements

We thank Jim Brune for useful discussions that sharpened our description of the results, and Joe Ibarzabal for allowing us to work in his property near Tierra Subida Road in Palmdale. The paper benefited from useful comments by Judi Chester, Diane Moore, Rasool Anooshehpoor and Nathan Benesh. The work was funded by the National Science Foundation (grant EAR-0409048) and the Southern California Earthquake Center (based on NSF cooperative agreement EAR-8920136 and United States Geological Survey cooperative agreement 14-08-0001-A0899).

REFERENCES

ADAMS, G.G. (1995), *Self-excited oscillations of two elastic half spaces sliding with constant coefficient of friction*, J. Appl. Mech. *62*, 867–872.

AGAR, S.M., PRIOR, D.J., and BEHRMANN, J.H. (1989), *Back-scattered electron imagery of the tectonic fabrics of some fine-grained sediments: Implications for fabric nomenclature and deformation processes.* Geology *17*, 901–904.

AGNEW, D.C. and SIEH, K.E. (1978), *A documentary study of the felt effects of the great California earthquake of 1857*. Bull. Seismol. Soc. Am. *68*, 6, 1717–1729.

ALLEN, C.R., AMAND, P.St., RICHTER, C.F., and NORDQUIST, J.M. (1965), *Relationship between seismicity and geological structure in the Southern California region*, Bull. Seismol. Soc. Am. *55*, 753–797.

ANDERSON, J.L., OSBOURNE, R.H., and PALMER, D.F. (1983), *Cataclastic rocks of the San Gabriel fault: An expression of deformation at deeper level in the San Andreas fault zone*, Tectonophysics *98*, 209–251.

ANDREWS, D.J. (2005), *Rupture dynamics with energy loss outside the slip zone*, J. Geophys. Res. *110*, B01307, doi:10.1029/2004JB003191.

ANDREWS, D.J. and BEN-ZION, Y. (1997), *Wrinkle-like slip pulse on a fault between different materials*, J. Geophys. Res. *102*, 553–571.

ANOOSHEHPOOR, A. and BRUNE, J.N. (1999), *Wrinkle-like Weertman pulse at the interface between two blocks of foam rubber with different velocities*, Geophys. Res. Lett. *23*, 2025–2028.

BARROWS, A.G., KAHLE, J.E., and BEEBY, D.J. (1985), *Earthquake hazard and tectonic history of the San Andreas fault zone, Los Angeles County, California*, Open File Report 85–10 LA., California Department of Conservation, Division of Mines and Geology.

BEN-ZION, Y. (2001), *Dynamic rupture in recent models of earthquake faults*, J. Mech. Phys. Solids. *49*, 2209–2244.

BEN-ZION, Y. and ANDREWS, D.J. (1998), *Properties and implications of dynamic rupture along a material interface*, Bull. Seismol. Soc. Am. *88*, 1085–1094.

BEN-ZION, Y. and HUANG, Y. (2002), *Dynamic rupture on an interface between a compliant fault zone layer and a stiffer surrounding solid*, J. Geophys. Res. *107*, 2042, doi: 10.1029/2001JB000254.

BEN-ZION, Y., KATZ, S., and LEARY, P. (1992), *Joint inversion of fault zone head waves and direct P arrivals for crustal structure near major faults*, J. Geophys. Res. *97*, 1943–1951.

BEN-ZION, Y, and MALIN, P. (1991), *San Andreas fault zone head waves near Parkfield, California*, Science *251*, 1592–1594.

BEN-ZION, Y., PENG, Z., OKAYA, D., SEEBER, L., ARMBRUSTER, J.G., OZER, N., MICHAEL, A.J., BARIS, S., and AKTAR, M. (2003), *A shallow fault zone structure illuminated by trapped waves in the Karadere-Dusce branch of the north Anatolian fault, Western Turkey*, Geophys. J. Int. *152*, 699–717.

BEN-ZION, Y. and SAMMIS, C.G. (2003), *Characterization of fault zones*, Pure Appl. Geophys. *160*, 677–715.

BEN-ZION, Y. and SHI, Z. (2005), *Dynamic rupture on a material interface with spontaneous generation of plastic strain in the bulk*, Earth. Plan. Sci. Lett. *236*, 486–496, doi: 10.1016/j.epsl.2005.03.025.

BIEGEL, R.L., SAMMIS, C.G., and DIETERICH, J.H. (1989), *The frictional properties of a simulated gouge having a fractal particle distribution*, J. Structural Geol. *11*, 827–846.

BRIETZKE, G.B. and BEN-ZION, Y. (2005), *Examining tendencies of in-plane rupture to migrate to material interfaces*, submitted to Geophys. J. Int.

BRUNE, J.N., ANOOSHEHPOOR, R., and ZENG, Y. (1999), *Ground motions for shallow angle thrust faults. Final technical report*, University of Nevada at Reno, Seismological Laboratory, United States.

BRUNE, J.N. (1968), *Seismic moment, seismicity, and rate of slip along major fault zones*, J. Geophys. Res. *73*, 777–784.

BRUNE, J.N. (2001), *Fault normal dynamic loading and unloading: an explanation for "non-gouge" rock powder and lack of fault-parallel shear bands along the San Andreas fault*, EOS Trans. Am. Geophys. Union *82*, Abstract.

CHESTER, F.M. (1995), unpublished mapping.

CHESTER, F.M. and LOGAN, J.M. (1986), *Implications for mechanical properties of brittle faults from observation of the Punchbowl fault zone. California*, Pure. Appl. Geophys. *124*, 79–106.

CHESTER, F.M. and CHESTER, J.S. (1998), *Ultracataclasite structure and friction processes of the Punchbowl fault, San Andreas system, California*, Tectonophysics *295*, 199–221.

CHESTER, F.M., EVANS J.P., and BIEGEL, R.L. (1993), *Internal structure and weakening mechanisms of the San Andreas fault*, J. Geophys. Res. *98*, 771–786.

CHESTER, F.M., CHESTER, J.S., Kirschner, D.L., SCHULZ, S.E., and EVANS, J.P. (2004), *Structure of large-displacement strike-slip fault zones in the brittle continental crust*. In (Karner, Gary D., Taylor, Brian, Driscoll, Neal W., Kohlstedt, David L., eds.), *Rheology and Deformation in the Lithosphere at Continental Margins*, (Columbia University Press, New York, MARGINS Theoretical and Experimental Earth Science Series 1, 223–260).

COCHARD, A. and RICE, J.R. (2000), *Fault rupture between dissimilar materials: Ill-posedness, regularization, and slip-pulse response*, J. Geophys. Res. *101*, 25,321–25,336.

CROWELL, J.C. (1962), *Displacement along the San Andreas fault, California*, Spec. Pap. Geol. Soc. Am. *71*, 61 pp.

DALGUER, L.A., IRIKURA, K., and RIERA, J. (2003), *Simulation of tensile crack generation by 3-D dynamic shear rupture propagation during an earthquake*, J. Geophys. Res. *108*(B3), 2144, doi:10.1029/2001JB001738.

DERSHOWITZ, W.S. and HERDA, H.H. (1992), *Interpretation of fracture spacing and intensity*, Proc. U.S. Symp. Rock Mech., 33rd, 757–766.

DIBBLEE, T.W. Jr. (1967), *Displacements on the San Andreas fault system in the San Gabriel, San Bernardino, and San Jacinto Mountains, southern California. (in Conference on geologic problems of San Andreas fault system, Stanford, Calif.*, Proc. Stanford Univ. Pubs. Geol. Sci. v. 11, p. 260–276, illus).

DIBBLEE, T.W. (1968), *Displacement on the San Andreas fault system in the San Gabriel, San Bernardino and San Jacinto Mountains, Southern California*. In (Dickinson, W.R. and A. Grantz, eds.), *Proc. Conf.*

Geologic Problems of San Andreas fault system, Geological Science, 11 (Stanford University Publications, 1968), 260–278.

DIBBLEE, T.W., Jr. (1987), *Geology of the Devil's Punchbowl, Los Angeles County, California*, Geological Soc. Am. Centennial Field Guide Cordilleran Section, 207–210.

DIBBLEE, T.W., Jr. (1989), *The San Andreas Fault and major rock terranes of California displaced by it and its tectonics.* In (Baldwin, E. Joan, *et al.*), Annual Field Trip Guidebook. South Coast Geological Society, Vol. 17, Vol. 1, 223–275.

DIBBLEE, T.W. and MINCH, J.A. (2002), *Geologic map of the Valyermo quadrangle, Los Angeles County, California*: DIBBLEE Geological Foundation, Map DF-80, scale 1:24000.

DI TORO, G., NIELSEN, S., and PENNACCHIONI1, G. (2005), *Earthquake rupture dynamics frozen in exhumed ancient fault*, submitted to *Nature*.

DI TORO, G. and PENNACCHIONI, G. (2004), *Superheated friction-induced melts in zoned pseudotachylytes within the Adamello tonalites (Italian Southern Alps)*, J. Struct. Geol. *26*, 1783–1801.

DOR, O., RECHES, Z., VAN ASWEGEN, G., and BOSMAN, K. (2001), *Earthquake Rupture at the Focal Depth of M = 5.1 and M = 3.7 Earthquakes in Gold Mines, South Africa*, EOS Trans. Amer. Geophys. Union. *81*, Abstract.

DOR, O., BEN-ZION, Y., ROCKWELL, T.K., and BRUNE J.N. (2006), *Pulverized rocks in the Mojave section of the San Andreas Fault Zone*, Earth Planet. Sci. Lett., in review.

EBERHART-PHILLIPS, D. and MICHAEL, A.J. (1993), *Three-dimensional velocity structure, seismicity and fault structure in Parkfield region, central California*. J. Geophys. Res. *98*, 9, 15,737–15,758.

EIDELMAN, A. and RECHES, Z. (1992), *Fratured pebbles: a new stress indicator*. Geology., *20*, 4, 307–310.

ENGELDER, J.T. (1974), *Cataclasis and the generation of fault gouge*, Geol. Soc. Am. Bull. *85*, 10, 1515–1522.

EVANS, J.P. and CHESTER, F.M. (1995), *Fluid-rock interaction in faults of the San Andreas system—inferences from San-Gabriel fault rock geochemistry and microstructures*. J. Geophys. Res. *100*, 13007–13020

FIALKO, Y. (2004), *Probing the mechanical properties of seismically active crust with space geodesy: Study of the co-seismic deformation due to the 1992 Mw7.3 Landers (southern California) earthquake*, J. Geophys. Res. 109, doi:10.1029/2003JB002756.

FUIS, G.S., RYBERG, T., GODFREY, N., OKAYA, D.A., and MURPHY, J.M. (2001), *Crustal structure and tectonics from the Los Angeles basin to the Mojave Desert, southern CA*, Geology *29*, 15–18.

FUIS, G.S., CLAYTON, R.W., DAVIS, P.M., RYBERG, T., LUTTER, W.J., OKAYA, D.A., HAUKSSON, E., PRODEHL, C., MURPHY, J.M., BENTHIEN, M.L., BAHER, S.A., KOHLER, M.D., THYGESEN, K., SIMILA, G., and KELLER, G.R. (2003), *Fault systems of the 1971 San Fernando and 1994 Northridge earthquakes, southern California: Relocated aftershocks and seismic images from LARSE II*, Geology *31*, 171–174.

FUMAL, T.E., PEZZOPANE, S.K., WELDON, R.J.II, and SCHWARTZ, D.P. (1993), *A 100-year average recurrence interval for the San Andreas fault at Wrightwood, California*, Science *259*, 199–203.

GOODWIN, L.B. and TIKOFF, B. (2002), *Competency contrast, kinematics, and the development of foliations and liniations in the crust*, J. Struc. Geol. *24*, 1065–1085.

HENRY, C. and DAS, S. (2001), *Aftershock zones of large shallow earthquakes: Fault dimensions, aftershock area expansion, and scaling relations*, Geophys. J. Int. *147*, 272–293.

JOHNSON, A.V., FLEMING, R.W., CRUIKSHANK, K.M., MARTOSUDARMO, S.Y., JOHNSON, N.A., JOHNSON, K.M., and WEI, W. (1997) *Analecta of structures formed during the 28 June 1992 Landers, Big Bear, California earthquake sequence*, USGS Open File Report 97–94.

LE PICHON, X., KREEMER, C., and CHAMOT-ROOKE, N. (2005), *Asymmetry in elastic properties and the evolution of large continental strike-slip faults*, J. Geophys Res. *110*, xB03405 MAR 19.

LEWIS, M., PENG, Z., BEN-ZION, Y., and VERNON, F. (2005), *Shallow seismic trapping structure in the San Jacinto fault zone near Anza, California*, Geophys. J. Int. *162*, 867–881, doi:10.1111/j.1365-246X.2005.02684.x.

LINDVALL, S.C., ROCKWELL, T.K., DAWSON, T.E., HELMS, J.G., and BOWMAN, K.W. (2002), *Evidence for two surface ruptures in the past 500 years on the San Andreas fault at Frazier Mountain, California*, Bull. Seismol. Soc. Am. *92*, 7, 2689–2703.

LOCKNER, D., NAKA, H., TANAKA, H., ITO, H., and IKEDA, R. (1999), *Permeability and strength of core samples from the Nojima fault of the 1995 Kobe Earthquake*, In Proc. of Internat. *Wrkshp on the Nojima fault Core and Borehole Data Analysis*, Tsukuba, Japan, Nov. 22–23, USGS Open File Report 00–129 (H. Ito, K. Fujimoto, H. TANAKA, and D. LOCKNER, eds.) pp. 147–152.

LUTTER, W.J., FUIS, G.S., RYBERG, T., OKAYA, D.A., CLAYTON, R.W., DAVIS, P.M., PRODEHL, C., MURPHY, J.M., LANGENHEIM, V.E., BENTHIEN, M.L., GODFREY, N.J., CHRISTENSEN, N.I., THYGESEN, K., THURBER, C.H., SIMILA, G., and KELLER, G.R. (2004), *Upper crustal structure from the Santa Monica Mountains to the Sierra Nevada, Southern California: Tomographic results from the Los Angeles Regional Seismic Experiment, Phase II (LARSE)*, Bull. Seismol. Soc. Am. *94*, 619–632.

MATTHEWS, V. (1976), *Correlation of the Pinnacles and Neenach volcanic formations and their bearing on the San Andreas fault problem*, Am. Assoc. Petro. Geol. Bull. *60*, 2128–2141.

MCGUIRE, J. and BEN-ZION, Y. (2005), *High-resolution imaging of the Bear Valley section of the San Andreas fault at seismogenic depths with fault-zone head waves and relocated seismicity*, Geophys. J. Int. *163*, 152–164, doi: 10.1111/j.1365-246X.2005.02703.x.

MCGUIRE, J.J., ZHAO, L., and JORDAN, T.H. (2002), *Predominance of unilateral rupture for a global catalog of large earthquakes*, Bull. Seismol. Soc. Am. *92*, 3309–3317.

MAGISTRALE, H. and SANDERS, C. (1995), *P wave image of the Peninsular Ranges batholith, southern California*, Geophys. Res. Lett. *22*, 2549–2552.

MERIFIELD, P.M., ROCKWELL, T.K., and LOUGHMAN, C.C. (1991), *A slip rate based on trenching studies, San Jacinto fault zone near Anza, California*. In *Engineering Geology and Geotechnical Engineering 27* (James McCalpin, ed.), pp. 28-1–28-21.

MIZOGUCHI, K., HIROSE, T., and SHIMAMOTO, T. (2000), *Internal and permeability structures of Nojima fault zone: Data correlation from surface and core samples*, journal data.

MIZUNO, K., HATTORI, H., SANGAWA, A., and TAKAHASHI, Y. (1990), *Geology of the Akashi district, quadrangle-series* (in Japanese with English abstract), scale 1:50,000, 90 pp., Geol. Surv. Jpn., Tsukuba, Japan.

MORTON, D.M. and MATTI, J.C. (1987), *The Cucamonga fault zone, geologic setting and Quaternary history*, USGS Professional Paper, 1339

NOBLE, L.F. (1954), *Geology of the Valyermo Quadrangle and vicinity*, Map GQ-50, USGS.

OAKESHOFTT (1971)

OHTANI, T., FIJIMOTO, K., ITO, H., TANAKA, H., TOMIDA, N., and HIGUCHI, T. (2000) *Fault rocks and past to recent fluid characteristics from the borehole survey of the Nojima rupture in the 1995 Kobe earthquake, southwest Japan*, J. Geophys. Res. *105*, B7, 16,161–16,171.

PENG, Z., BEN-ZION, Y., MICHAEL, A. J., and ZHU, L. (2003), *Quantitative analysis of seismic trapped waves in the rupture zone of the 1992 Landers, California earthquake: Evidence for a shallow trapping structure*, Geophys. J. Int. *155*, 1021–1041.

PENG, Z. and BEN-ZION, Y. (2004), *Systematic analysis of crustal anisotropy along the Karadere-Düzce branch of the north Anatolian fault*, Geophys. J. Int. *159*, 253–274, doi: 10.1111/j.1365-246X.2004.02379.x.

POLIAKOV, A.N., DMOWSKA, B.R., and RICE, J.R. (2002), *Dynamic shear rupture interactions with fault bends and off-axis secondary faulting*, J. Geophys. Res. *107*.

POWELL, R.E. and WELDON, R.J. (1992), *Evolution of the San Andreas Fault*. Annu. Rev. Earth Planet. Sci., *20*, 431–468.

PRIEST, S.D., *Discontinuity Analysis for Rock Engineering* (Chapman and Hall, New York 1993) 473 pp.

RANJITH, K. and RICE, J.R. (2001), *Slip dynamics at an interface between dissimilar materials.*, J. Mech. Phys. Solids. *49*(2), 341–361.

RECHES, Z. and DEWERS, T.A. (2004), *Gouge formation by dynamic pulverization during earthquakes.* Submitted to EPSL.

RECHES, Z. and LOCKNER, D.A. (1994), *The nucleation and growth of faults in brittle rocks*, J. Geophys. Res. *99*, 18,159–18,173.

RICE, J.R., SAMMIS, C.G., and PARSONS, R. (2005), *Off-fault secondary failure induced by a dynamic slip pulse*, Seismol. Soc. Am. Bull. *95*, 109–134.

ROCKWELL, T.K., LOUGHMAN, C., and MERIFIELD, P. (1990), *Late Quaternary rate of slip along the San Jacinto fault zone, Southern California*, J. Geophys. Res. *95*, 6, 8593–8605.

ROCKWELL, T., SEITZ, G., RAGONA, D., DAWSON, T., FANEROS, G., VERDUGO, D., and ALTANGEREL, O. (2005), *Investigation of segment controls on the rupture history of the southern San Jacinto fault*, Seismol. Res. Lett. *76*, 254.

RUBIN, A. (2002), *Aftershocks of microearthquakes as probes of the mechanics of rupture*, J. Geophys. Res. *107*, 10.1029/2001JB0000496.

RUBIN, A. and GILLARD, D. (2000), *Aftershock asymmetry/rupture directivity along central San Andreas fault microearthquakes*, J. Geophys. Res. *105*, 19,095–19,109.

SANDERS, C.O. and KANAMORI, H. (1984), *A seismotectonic analysis of the Anza seismic gap, San Jacinto fault zone, Southern California*, J. Geophys. Res. *89*, 5873–5890.

SCHOLZ, C.H. *The Mechanics of Earthquakes and Faulting* (New York, Cambridge University Press, 2002) 2nd ed. 471 pp.

SCHULZ, S.E. and EVANS, J.P. (1998), *Spatial variability in microscopic deformation and composition of the Punchbowl fault, Southern California: Implications for mechanisms, fluid-rock interaction, and fault morphology*. Tectonophysics *295*, 223–244.

SCHULZ, S.E. and EVANS, J.P. (2000), *Mesoscopic structure of the Punchbowl fault, Southern California and the geologic and geophysical structure of active strike-slip faults*, J. Struct. Geol. *22*, 913–930.

SCHWARTZ, D.P. and WELDON, R.J. (1986), *Late Holocene slip rate on the Mojave segment of the San Andreas fault zone, Littlerock, CA; Preliminary results*, EOS Trans. Am. Geophys. Union *67*, Abstract.

SCOTT, J.S., MASTERS, T.G., and VERNON, F.L. (1994), *3-D velocity structure of the San Jacinto fault zone near Anza, California – I. P waves*, Geophys. J. Int. *119*, 611–626.

SENGOR, A.M.C., TUYSUZ, O., IMREN, C., SAKLNC, M., EYIDOGAN, H., GORUR, N., LE PICHON, X., and RANGIN, C. (2005), *The North Anatolian fault: A new look*, Annu. Rev. Earth Planet. Sci. *33*, 37–112.

SHARP, R. (1967), *The San Jacinto fault in southern California*, Geol. Soc. Am. Bull. *78*, 705–730.

SHAPIRO, N.M., CAMPILLO, M., STEHLY, L., and RITZWOLLER, M.H. (2005), *High-resolution surface-wave tomography from ambient seismic noise*, Science *307*, 1615–1618.

SHI, Z. and BEN-ZION, Y. (2006), *Dynamic rupture on a bimaterial interface governed by slip-weakening friction*, Geophys. J. Int., 165, doi: 10.1111/j.1365-246x.2006.02853.x.

SIBSON, R.H. (1977), *Fault rocks and fault mechanisms*, J. Geol. Soc. London *133*, 191–213.

SIBSON, R.H. (1989), *Earthquake faulting as a structural process*, J. Struct. Geol. *11*, 1–14.

SIBSON, R.H. (2003), *Thickness of the seismic slip zone*, Seismol. Soc. Am. Bull. *93*, 1169–1178.

SIEH, K.H. (1978a), *Slip along the San Andreas fault associated with the great 1857 earthquake*, Bull. Seismol. Soc. Am. *68*, 5, 1421–1447.

SIEH, K.H. (1978b), *Central California foreshocks of the great 1857 earthquake*, Bull. Seismol. Soc. Am. *68*, 6, 1731–1749.

STECK, A. and HUNZIKER, J. (1994), *The Tertiary structural and thermal evolution of the Central Alps; Compressional and extensional structures in an orogenic belt*, Tectonophysics *238*, 1–4, 229–254.

STEIN, R.S., BARKA, A.A., and DIETERICH, J.H. (1997), *Progressive failure on the North Anatolian fault since 1939 by earthquake stress triggering*, Geophys. J. Int. *128*, 594–604.

TANAKA, H., FUJIMOTO, K., OHTANI, T., and ITO, H. (2001), *Structural and chemical characteristics of shear zones in the freshly activated Nojima Fault, Awaji Island, southwest Japan*, J. Geophys. Res. *106*, B5, 8789–8810.

VAN DER PLUIJM, B.A., HALL, C.M., VROLIJK, P.J., PEVEAR, D.R., and COVEY, M.C. (2001), *The dating of shallow faults in the Earth's crust*, Nature *412*, 6843, 172–175.

VERMILYE, J.M. and SCHULZ, C.H. (1998), *The Process zone: A microstructural view of fault growth*, J. Geophys. Res. *103*, 6, 12,223–12,237.

VERMILYE, J.M. and SCHULZ, C.H. (1999), *Fault propagation and segmentation: insight from the microstructural examination of a small fault*, J. Struc. Geol. *21*, 1623–1636.

WEERTMAN, J. (1980), *Unstable slippage across a fault that separates elastic media of different elastic constants*, J. Geophys. Res. *85*, 1455–1461.

WELDON, R.J. and FUMAL, T.E. (2005) *Slip rate of the San Andreas fault near Littlerock, California*, 2005 SCEC Annual Meeting, Proc. and Abstracts, XV.

WELDON, R.J., MEISLING, K.E., and ALEXANDER, J. (1993), *A speculative history of the San Andreas fault in the central Transverse ranges, California*. In (Powell, R.E., WELDON, R.J. and J.C. MATTI, Eds), *The San Andreas Fault System: Displacement, Palinspastic Reconstruction, and Geologic Evolution*. Geol. Soc. Am. Memoir *178*, 161–198.

WELDON, R.J., II, FUMAL, T.E., POWERS, T.J., PEZZOPANE, S.K., SCHARER, K.M., HAMILTON, J.C. (2002), *Structure and Earthquake Offsets on the San Andreas Fault at the Wrightwood, California, Paleoseismicity Site*. Bull. Seis. Soc. Am., *92*, 2704–2725.

WILSON, B., DEWERS, T., RECHES, Z., and BRUNE, J. (2005), *Particle size and energetics of gouge from earthquake rupture zones*, Nature *434*, 749–752.

WILSON, J.E., CHESTER, J.S., and CHESTER, F.M. (2003), *Microfracture analysis of fault growth and wear processes, Punchbowl Fault, San Andreas system, California*, J. Struc. Geol. *25*, 11, 1855–1873.

WOODBURNE, M.O. (1975), *Cenozoic stratigraphy of the transverse ranges and adjacent areas, Southern California, GSA*, Special Paper.

YAMASHITA, T. (2000), *Generation of microcracks by dynamic shear rupture and its effects on rupture growth and elastic wave radiation*, Geophys. J. Int. *143*, 395–406.

(Received: 16 March 2005; accepted: 9 June 2005)
Published Online First: February 8, 2006

Pure appl. geophys. 163 (2006) 351–372
0033–4553/06/030351–22
DOI 10.1007/s00024-005-0019-5

Pure and Applied Geophysics

The Applicability of Modern Methods of Earthquake Location

Paul G. Richards,[1,2] Felix Waldhauser,[1] David Schaff,[1] and Won-Young Kim[1]

Abstract—We compare traditional methods of seismic event location, based on phase pick data and analysis of events one-at-a-time, with a modern method based on cross-correlation measurements and joint analysis of numerous events. In application to four different regions representing different types of seismicity and monitored with networks of different station density, we present preliminary results indicating what fraction of seismic events may be amenable to analysis with modern methods. The latter can supply locations ten to a hundred times more precise than traditional methods. Since good locations of seismic sources are needed as the starting point for so many user communities, and potentially can be provided due to current improvements in easily-accessible computational capability, we advocate wide-scale application of modern methods in the routine production of bulletins of seismicity. This effort requires access to waveform archives from well-calibrated stations that have long operated at the same location.

Key words: Earthquake location, waveform cross correlation, seismicity studies, California earthquakes, Charlevoix earthquakes, China earthquakes, New Madrid earthquakes.

Introduction

Seismic events are usually still located one-at-a-time by measuring the arrival times of different seismic signals (phase picks) and then interpreting these observations in terms of the travel times predicted for a standard depth-dependent Earth model. In this traditional approach, the differences between observed and calculated arrival times (based on a trial origin time and location) are reduced by a process of iteration (for each event separately) to a value deemed acceptable.

Many different studies of specific regions and particular data sets have demonstrated that by use of whole waveforms and locating groups of events all together, location estimates can be very significantly improved over the results obtained by the traditional approach. In this paper we loosely refer to analysis of waveforms, and joint location of many events, as "modern methods"—in contrast to "traditional methods" based on phase picks and location of events one-at-a-time. We

[1]Lamont-Doherty Earth Observatory, 61 Route 9W, Palisades, New York, 10964, USA.
[2]Department of Earth and Environmental Sciences, Columbia University, New York, USA.
E-mail: richards@LDEO.columbia.edu

examine practical aspects of the earthquake location problem for four different regions, and assess the merits of a modern method in which waveform cross correlation is applied to large data sets. We find the fraction of events in each region that are amenable to this type of relocation, and comment on the degree of location improvement.

In regions for which an extensive archive of waveforms can be obtained from stations with a long history of operation at the same site with the same instruments, modern methods of event location (i.e., those that combine waveform cross correlation with a multi-event location algorithm) may now be considered seriously for application to the routine publication of seismicity bulletins.

Sections that follow give background on traditional and modern methods, describe in general our method of analysis for the different regions studied, and present results for the different regions. We conclude with discussion and comment on traditional versus modern methods of event location.

This is not a review paper that compares different methods of cross correlation and event relocation to come up with a verdict about which combination works best in a particular area. The practical experience of network operators will typically be an important guide in determining that combination for different areas. Nor is this an interpretational paper, in which results from each of the four regions are discussed in terms of their implications on the tectonics or earthquake physics. Rather, it is our purpose to present results on relocation using a modern method using waveform cross correlation to the extent permitted by available data and relocating multiple events at the same time; and to comment upon the degree of location improvement, in so far as possible using a common framework of analysis but applied to a broad variety of tectonic regions at different scales, and using seismographic monitoring networks of varying station density.

Background

Seismic data derived from earthquakes and explosions are used in scientific studies of Earth's internal structure, tectonics, and the physics of earthquake processes; in engineering studies of earthquake hazard and efforts in mitigation and emergency management; and in monitoring of nuclear explosions, either as a military activity to evaluate the weapons-development program of a potential adversary, or as an arms-control initiative such as monitoring compliance with the Comprehensive Nuclear-Test-Ban Treaty.

In practice the great majority of those who work with seismic data do not use seismograms directly. Instead they typically use data products derived from seismograms. The most important of these products are bulletins of seismicity, which consist of catalogs of earthquake and explosion locations, measures of event size (such as magnitudes, and scalar and tensor moments), and associated data such as the arrival times of seismic waves at particular stations.

Since the late 1970s there have been enormous improvements in the quality and quantity of seismograms, associated with the deployment of broadband feedback sensors and techniques of digital recording that can faithfully capture signals with high dynamic range across wide bands of frequency; and there is ongoing revolutionary improvement in access to seismogram data, handicapped only by political barriers as reliable satellite communications and the Internet spread even to remote locations and become less expensive. It may therefore seem surprising that routinely published global and regional bulletins of seismicity have not yet seen a corresponding radical improvement that would greatly benefit any of the principal user communities, even though demonstrably successful modern methods of event location and source characterization have been developed and applied in numerous special studies. As discussed further, below, the main reason modern methods have yet to be widely applied is that it is necessary to build up major archives of well-calibrated and easily accessible waveforms from fixed stations operated over many years.

When teleseismic arrival times are used for event location, the resulting location estimates using traditional procedures are typically in error by several km for events detected at hundreds of stations, and by a few tens of km for events detected at tens of stations. Such errors in traditional event locations in standard bulletins are not always appreciated by users, but can be shown for example by relocation studies such as that of ENGDAHL *et al.* (1998), and by comparison of standard location estimates with the hypocenters of events whose ground truth location is known from non-seismic methods (BONDÁR *et al.*, 2004). It can also be demonstrated by study (SCHAFF and RICHARDS, 2004a) of waveform doublets that must be within about 1 km of each other, but that standard bulletins report as having been tens of km apart when the events are located in the usual way (i.e., one-at-a-time from phase picks).

When regional arrival times are used, there is practical experience in the western U.S. to indicate that location uncertainties (as given in conventional bulletins) are at the level of one or two km in areas with high station density (station separation on the order of 10–20 km); and are at the level of about five km in areas with fewer stations (station spacing, approximately 50 km) (personal communication, HAUKSSON, 2005). In broad areas for which events are located with regional signals on the basis of a very sparse set of stations, mislocations can routinely reach several tens of km (RICHARDS *et al.*, 2003).

There are two independent reasons why events can be significantly mislocated, even in situations where station coverage is not a problem. First, there is the difficulty of picking arrival times accurately. When signal-to-noise ratios are good, the error in traditional methods of measuring the arrival time of seismic waves is usually less than one second (and can be less than 0.1 second in favorable conditions). But signals

(*Pn*, for example) are often weak and/or emergent, and later arrivals have to be picked in time windows that include the coda of earlier arrivals. Thus regional *S*-wave phases may be picked with errors that in some cases reach up to a few tens of seconds. Second and often more important, there is the problem of errors in the travel-time model used to interpret measured arrival times. In many cases it is model error, not pick error, which dominates the overall error in absolute location. In such cases the overall goal of improving locations based on seismic arrival times can be achieved only by reducing the effect of model errors. Model errors can also lead to poor-quality relative locations of events that in fact lie near to each other, if the events are located one-at-a-time using data derived from different station sets.

At depths greater than about 200 km, the Earth's laterally-averaged velocity structure is known quite accurately (i.e., to within 1% at most depths, the biggest difference from actual velocities being in regions of subducting tectonic plates). The main difficulty is at shallower depths, within the crust and uppermost mantle, where the actual velocity of seismic waves may differ in unknown ways, perhaps by as much as 10%, from the velocity that is often assumed (such as the velocity given by *ak135* or some other standard Earth model). Thus the problem of substantial differences between actual and standard travel-time models is most significant for regional waves. If a seismic event is about 500 km from a station that detects a regional arrival, and if the arrival is misinterpreted with a velocity that is wrong by 5%, then the event will be estimated from that observation alone as having originated at a distance that is incorrect by about 25 km. Non-horizontal interfaces within the Earth, and phase misidentification (for example between *Pg* and *Pn*) compound these problems.

Experience with the traditional location method originated in the era of analog recording, with sensors of limited dynamic range and bandwidth. It is therefore remarkable that even after decades of experience with digital seismic data beginning in the 1970s, event location is usually still based upon analysis of one-event-at-a-time, and traditional picking of the arrival times for specific seismic phases. In practice this means that only a very small fraction of the available information contained in a modern seismogram is being used to locate events.

In this paper we use a recently-developed highly-efficient waveform cross-correlation algorithm for locally-observed phases (SCHAFF *et al.*, 2004) to measure accurate arrival-time differences for neighboring earthquakes observed at common stations. The waveform cross-correlation method has been applied on the local scale with considerable success. It uses much more of the information contained in seismograms, and can significantly reduce the error in relative arrival times derived from phase picks. We analyze cross-correlation measurements on local and regional signals, combined with ordinary phase picks from earthquake bulletins, using the double-difference technique (WALDHAUSER and ELLSWORTH, 2000; WALDHAUSER, 2001) to obtain precise relative locations for many earthquakes all at the same time.

Several other algorithms have been developed to reduce the effect of errors in the travel time model by locating many events at the same time. These include joint hypocenter determination (JHD) (DOUGLAS, 1967; and FROHLICH, 1979, among others), hypocentroidal decomposition (HDC) (JORDAN and SVERDRUP, 1981), or the use of source specific station terms (SSST) (RICHARDS-DINGER and SHEARER, 2000). All of these methods attempt to correct for correlated effects, due to three-dimensional Earth structure, in the arrival times generated by a cluster of events. The corrections are typically accomplished via use of residual-based station corrections, either in the form of a 1D (static) correction (JHD, HDC), or by using source specific station terms (SSST). The former reduces model errors due to unusual structure beneath a station, while the latter reduces model errors due to structure somewhere between the region of seismicity for which a correction factor is determined, and a particular station. With the double-difference method (DD), a network of events is built in which each event is linked to its nearest neighbors through travel-time differences observed at common stations, which are directly inverted for event separation. The method thus differs from earlier methods of joint hypocentral determination in that no station corrections are necessary, because unmodeled velocity structure is directly removed for each event pair individually. Thus double-difference results are best if events are densely distributed (a few km event separations when using local stations separated by several tens of km). For studies based on regional signals (distances typically in the range from a few hundred km up to about a thousand km), events can be more sparse although event pairs need to have approximately the same take-off direction for corresponding wave types recorded at any common station. The web of DD-relocated events can be expected to give precise relative locations for neighboring events, particularly when pick error is avoided by use of cross correlation. It is not our purpose here to compare these different algorithms, which we would expect to give similar results for clusters of events. Waveform cross-correlation methods together with multi-event location algorithms have been applied by many previous investigators (for example POUPINET *et al.*, 1984; ISRAELSON, 1990; HARRIS, 1991; GOT *et al.*, 1994; WALDHAUSER *et al.*, 1999 and 2004a,b; RUBIN *et al.*, 1999; WITHERS *et al.*, 1999; THURBER *et al.*, 2001; SCHAFF *et al.*, 2002; ROWE *et al.*, 2002; XIE *et al.*, 1997, and XIE, 2001; SHEARER, 1997, HAUKSSON and SHEARER, 2005, and SHEARER *et al.*, 2005). Most of these studies focused on local earthquakes, though a few used regional and teleseismic waveforms to locate small clusters of events. These papers have convincingly demonstrated major improvements in the precision of relative (and in some cases absolute) event locations. In the present paper we are interested in evaluating a modern method (i.e., waveform cross correlation to the extent permitted by available data, combined with multiple event location) on different scales, including some examples at quite large scales; and in commenting generally on the advantages for the modern approach that accrue at larger scales.

Method of Analysis for the Different Regions

For each of the four regions of seismicity to which we have applied modern methods of event location, we have taken the three steps of (a) data collection, (b) waveform analysis, and (c) event relocation using the double difference (DD) algorithm.

The four regions of seismicity for which we have obtained relocations and on which we comment in this paper are:

(1) A broad-area study of the seismicity of China and surrounding regions from 1985 to 2000, described in more detail by SCHAFF and RICHARDS (2004a). This study built upon a relatively small study of foreshocks, main shock, and aftershocks, over a period of several months for the Xiuyan region of Liaoning Province, China, described in more detail by SCHAFF and RICHARDS (2004b). The small study used regional seismograms in the distance range 600 to 1500 km archived by the IRIS Consortium's Data Management Center (IRIS DMC), and showed that complex *Lg* waveforms correlated well for a significant fraction of events for time windows of several hundred seconds duration. The larger study of China was based on 115 stations at distances up to 20° from about 14,000 events, with measurements made on about 130,000 seismograms. Cross correlation was based on window lengths from 60 to 400 s beginning prior to the first *P*-arrival, ending 40 s after *Lg*, and signals passed in the band from 0.5 to 5 Hz. 1.2 million cross correlations were made, for signals associated with event pairs that (according to the Annual Bulletin of Chinese Earthquakes) were not more than 150 km apart.

(2) A study of the New Madrid region of the central United States, for two different eras of station deployment. Data consisted of phase picks as well as waveforms (used for cross-correlation measurement), associated with seismicity bulletins of events located by traditional methods applied by local network operators. These were the Center for Earthquake Research Information of the University of Memphis, which deployed a PANDA network (42 stations, 918 events, 17,598 phases) from October 1989 to August 1992, and a different network (85 stations, 614 events, 16,461 phase picks) from January 2000 to October 2003. Window lengths of 1 s were used for cross-correlation measurements.

(3) A study of the Charlevoix region of Eastern Canada from January 1988 to December 2003 involving 2797 events recorded at 46 stations operated by the Geological Survey of Canada with 33,423 phase picks. Window lengths of 1 s were used for cross-correlation measurements. In practice, almost all of the highly cross-correlated seismograms derived from only 7 stations, closest to the Charlevoix region.

(4) A study of the region monitored by the Northern California Seismic Network (NCSN). This was by far the largest study, with about 225,000 events recorded by 900 stations from 1984 to 2003. Window lengths of only 1 and 2 seconds were used, and the NCSN seismicity bulletin (prepared using traditional methods) was used to select these time windows for all events whose inter-event distance was up to 5 km. Further details are given by SCHAFF and WALDHAUSER (2005).

The associated waveforms for sets of previously defined events were obtained for study (1) from the IRIS DMC; and for study (4) from the Northern California Earthquake Data Center (including phase picks in this case). Though large amounts of data were requested, these two acquisitions were straightforward in that both these data centers had archives previously prepared with the expectation that outside scientists would be using them. But for studies (2) and (3) the basic data sets of associated waveforms and phase picks entailed considerable effort to assemble (with the willing help of network operators). It was necessary to account for different formats, different station sets, and different practices of bulletin preparation over the periods of time during which the seismicity in these regions had been documented.

For some purposes, such as scientific study of earthquake interactions in a fault zone or seismic sources associated with magma conduits in a volcano, relative locations can be sufficient. But of course for seismic monitoring of nuclear explosions, or identification of earthquakes with particular fault structures, there is the need to move beyond relative locations and to estimate absolute hypocentral coordinates. In this paper we focus on improvements in precision, i.e., on relative locations, noting that absolute locations may be achievable at a later step in several different ways—for example by use of ground truth information for a subset of events, and/or by use of a good 3-D travel-time model.

Preliminary Results for Four Different Regions

(1) China and Surrounding Regions of East Asia

This project began with a small preliminary study of 28 earthquakes drawn from a sequence of 90 events during 1999–2000 in the Xiuyan region of Liaoning Province, and moved on to an analysis of about 14,000 events for all of China.

The small study surprisingly showed that in some cases the complex, highly scattered *Lg*-wave is similar at far-regional distances for clusters of events (SCHAFF and RICHARDS, 2004b). Figure 1 shows the high degree of similarity for more than 300 s of waveform, even at frequencies up to 5 Hz. Cross correlations provided highly accurate differential travel-time measurements. Their error estimated from the internal consistency is about 7 ms. These travel-time differences were then inverted by the double-difference technique to obtain epicenter estimates that have location precision on the order of 150 meters, as shown in Figure 2. The locations were computed with waveform data acquired by four or five regional stations 500 to 1000 km away. The relative epicenter estimates were not substantially affected by the paucity of stations or by large azimuthal gaps. For example, using four stations with an azimuthal gap of 133°, or using only three stations with an azimuthal gap of 220°, resulted in locations that differed by less than 150 m.

Regional event locations must often be based on a small number of phases and stations due to weak signal-to-noise ratios and sparse station coverage. This is

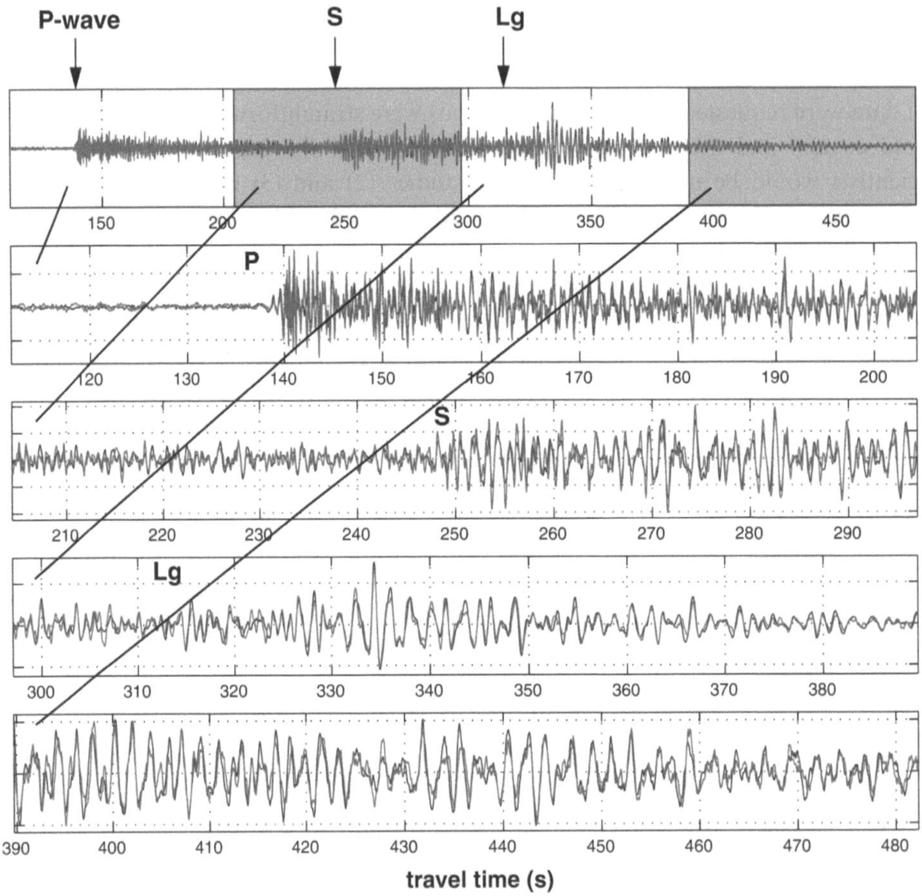

Figure 1
A pair of similar events in China filtered from 0.5 to 5 Hz and superimposed in gray and black. Vertical axes are normalized to unit amplitude. Lower subpanels are enlargements of the white and gray segments. The predicted *P* wave arrives at 143 s, the *S* wave at 256 s, and the *Lg* wave at 315 s. It is apparent that these waveforms are very similar. The quality of this waveform doublet is typical of the events described in more detail by SCHAFF and RICHARDS (2004ab), who argue that such events must be not more than about 1 km apart.

especially true for monitoring work that seeks to locate smaller magnitude seismic events with a handful of regional stations. Two primary advantages of using *Lg* for detection and location are that it is commonly the largest amplitude regional wave (enabling detection of smaller events); and it propagates more slowly than *P* waves or *Sn* (resulting in smaller uncertainty in distance, for a given uncertainty in travel time). This preliminary study demanded a high standard for identification of similar events (cross correlation ≥ 0.8 for a several hundred sec window of signal passed in the band from 0.5 to 5 Hz).

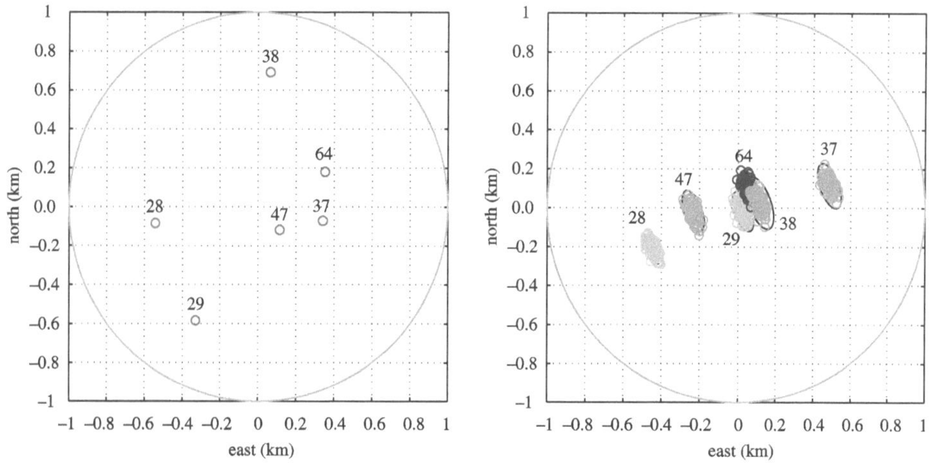

Figure 2
Comparison of double-difference relative locations for a subset of events in the sequence for a local/regional network (left) using only *P*-wave phase picks recorded at several hundred stations and for a sparse regional network (archived at IRIS) using *Lg* cross-correlation measurements (right). Event numbers are for identification. The RMS travel-time residuals are about 1 sec for the *P* waves and 0.02 sec for *Lg*. 95% confidence formal error ellipses and bootstrap errors (shaded small circles) are in good agreement (right). The epicenter in each case is taken as the centroid of the cluster. For further details see SCHAFF and RICHARDS (2004b).

In the larger study we found that about 9% of seismic events in and near China, from 1985 to 2000, were repeating events not more than about 1 kilometer from each other (SCHAFF and RICHARDS, 2004a). This conclusion was based on the stringent criterion for waveform similarity described in Figure 3 (see caption). We cross-correlated seismograms from about 14,000 earthquakes and explosions and measured relative arrival times to within about 0.01 seconds, enabling lateral location precision of about 100–300 meters. Recognition and measurement of repeating signals in archived data and the resulting improved locations enabled us to quantify the inaccuracy of current procedures for picking onset times and locating events. The fraction of cross-correlated events, though quite low (9%) in Figure 3 for the entire time period from 1985 to 2000, rose to a larger value (14%) in the later years, as shown in Figure 4. Table 1 shows too that the fraction increased significantly if the stringent criterion ($CC \geq 0.8$) is relaxed. Thus, relaxation to $CC \geq 0.6$ resulted in 23% of the events in the ABCE cross correlating successfully for the whole time period.

(2) New Madrid, Central United States

This region of intraplate seismicity in the Central United States experiences about 200–250 locatable earthquakes each year. It has been monitored by various networks since the 1970s, and currently we have preliminary relocation results for one three-year period and one four-year period.

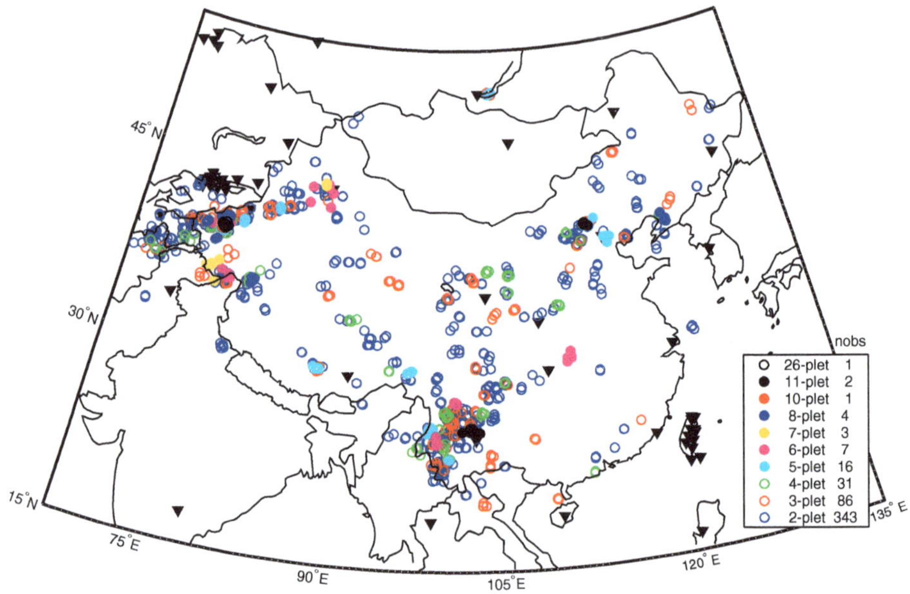

Figure 3
1301 events (9% of the Annual Bulletin of Chinese Earthquakes – ABCE), whose seismograms satisfy the criterion of cross-correlation coefficients greater than or equal to 0.8 with seismograms from at least one other earthquake, for long windows from 5 seconds before the *P* wave to 40 sec after the *Lg* wave on waveforms that are filtered from 0.5 to 5 Hz (as shown in Fig. 1). There are 494 multiplets here, and the inset shows the number of multiplets containing between 2 and 26 events. Recording stations archived by the IRIS Consortium are denoted with solid black triangles. Events are plotted at their ABCE absolute locations. For further details see SCHAFF and RICHARDS (2004a).

Thus, Figure 5 shows relocation of 783 events monitored by a 42-station network from 1989 to 1992. Although 707 of the events (90%) had five or more *P*-wave cross correlations (CC ≥ 0.7), these were not necessarily to the same event. It is possible that these events are relocated with high precision, but this is not assured. We found that a subset of 616 events (65%) had five or more *P*-wave cross correlations (CC ≥ 0.7) to a neighboring event and were thus relocated with high precision provided the azimuthal distribution of stations was adequate; 695 events (85%) cross-correlated at four or more stations; and 735 (90%) at three or more stations. Figure 6 shows relocation of 594 events that occurred in the period January 2000 to October 2003 and were monitored by an 82-station network still operating in 2005. In this case, 499 events (84%) had five or more *P*-wave cross correlations (CC ≥ 0.7), but these were not necessarily to the same event. 371 events (63%) had five or more *P*-wave cross correlations (CC ≥ 0.7) to a neighboring event and were thus relocated with highest precision in our study. Events that do not cross correlate are still located more precisely than in the traditional bulletin, because of the use of the double-difference algorithm applied to phase-pick data for the whole set of events.

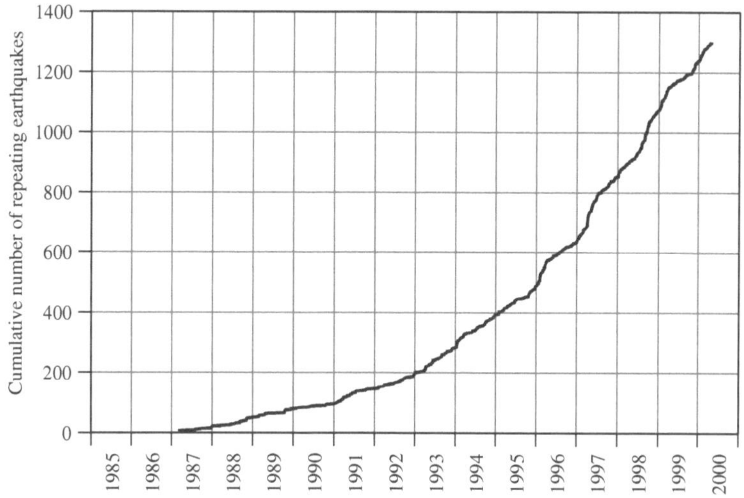

Figure 4
Cumulative number of events vs. time for the 1301 repeating events in China shown in Figure 3 for the period from 1985 to early 2000. Network coverage is more sparse for earlier years, resulting in underestimation of the actual percentage of repeating events over the entire period. If only the 908 repeating events later than 1994 are considered (out of the total of about 6400 events reported in the Annual Bulletin of Chinese Earthquakes), then about 14% of the events satisfy the stringent criterion of cross correlation not less than 0.8 for a window from the *P* arrival until 40 s into the *Lg* signal.

The general distribution of seismicity is very similar between Figures 5 and 6. Figure 7 shows the locations obtained by traditional methods for the same period and station set used in Figure 6. It is apparent especially in the cross sections and in the map view of events in the southwest subregion, that the relocated events in Figure 6 more clearly identify lineations than do the traditionally-obtained locations of Figure 7.

Table 1

The percentage of events in the Annual Bulletin of Chinese Earthquakes (ABCE) is shown for different values of the cross-correlation value. The effect of relaxing the stringent criterion ($CC \geq 0.8$) is substantial. (The analysis was done for 14,000 ABCE events, 130,000 seismograms, 1.2 million CC.)

cross-correlation value (CC)	% of events in ABCE
0.9	3
0.8	9
0.7	16
0.6	23
0.5	30
0.4	38
0.3	52
0.2	69
0.1	76

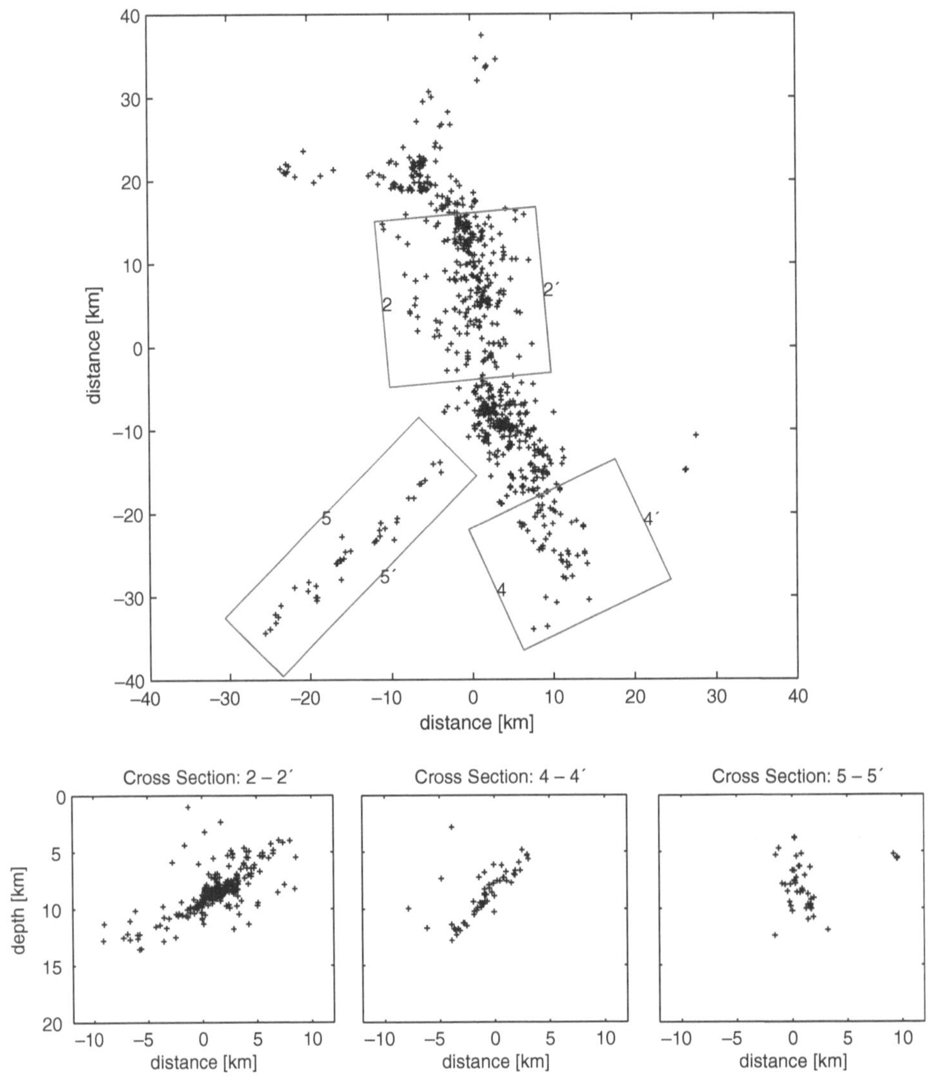

Figure 5
Map view and three cross sections for 783 relocated seismic events in the New Madrid region of the central United States from October 1989 to August 1992, using 42 stations of a PANDA network (see CHIU *et al.*, 1992). This application of the double-difference algorithm was based upon 58,807 phase-pick pairs and 80,697 cross correlations (CC ≥ 0.7) derived from *P* waves, plus 66,581 phase-pick pairs and 81,750 cross correlations (CC ≥ 0.7) derived from *S* waves. Cross sections show clear evidence for a westward-dipping fault plane, the dip increasing from about 30° (section 2–2′) to 45° further south (section 4–4′).

(3) Charlevoix, Eastern Canada

This too is a region of intraplate seismicity. The preliminary relocations shown in Figure 8, as well as the original bulletin locations for this region, indicate that

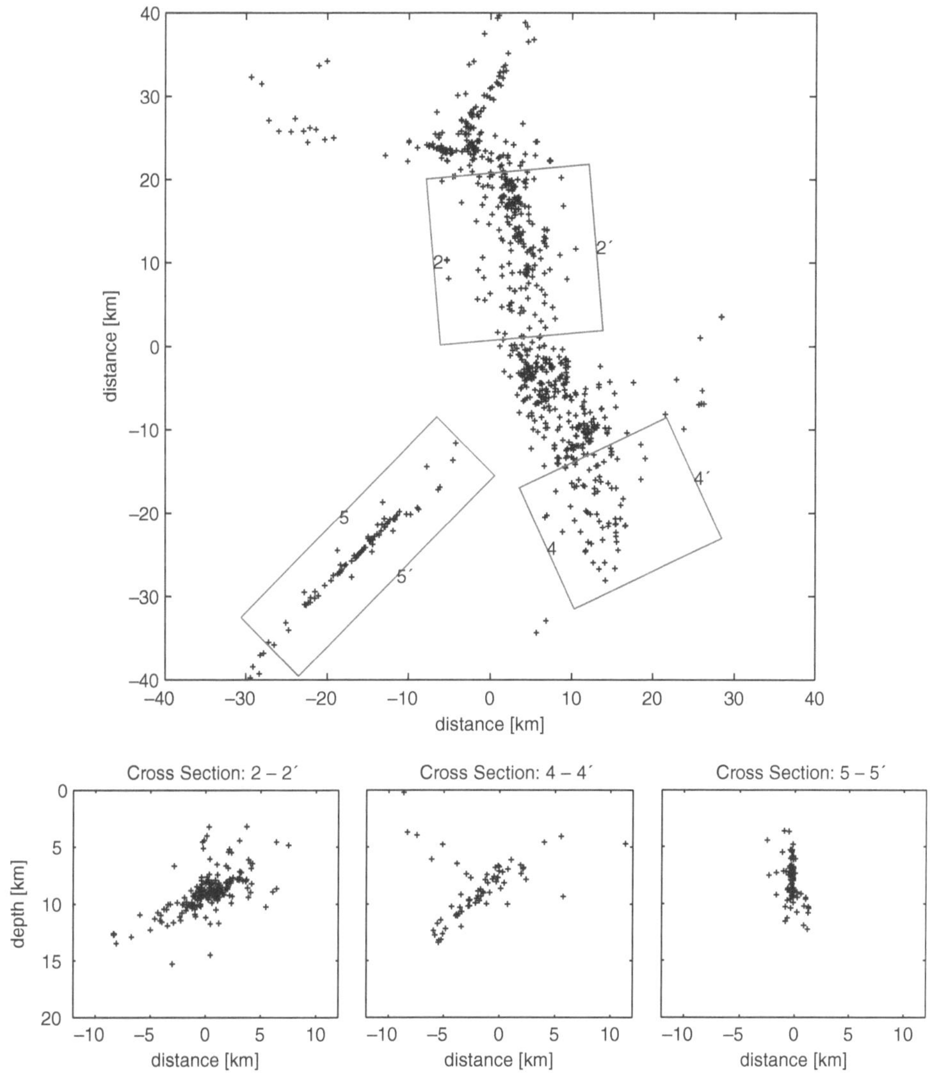

Figure 6

Map view and three cross sections for 594 relocated seismic events in the New Madrid region from January 2000 to October 2003, using 85 stations operated by the University of Memphis. This application of the double-difference algorithm was based upon 202,249 phase-pick pairs and 49,660 cross correlations (CC ≥ 0.7) derived from *P* waves, plus 192,277 phase-pick pairs and 93,009 cross correlations (CC ≥ 0.7) derived from *S* waves.

active faulting is simpler and more clearly defined in the northeastern part of the Charlevoix seismic zone, compared to more complex features in the southwestern part. The dominant faults are two parallel southwest-northeast running structures that dip to the southeast. Of these, the more eastern fault has a dip of about 50°

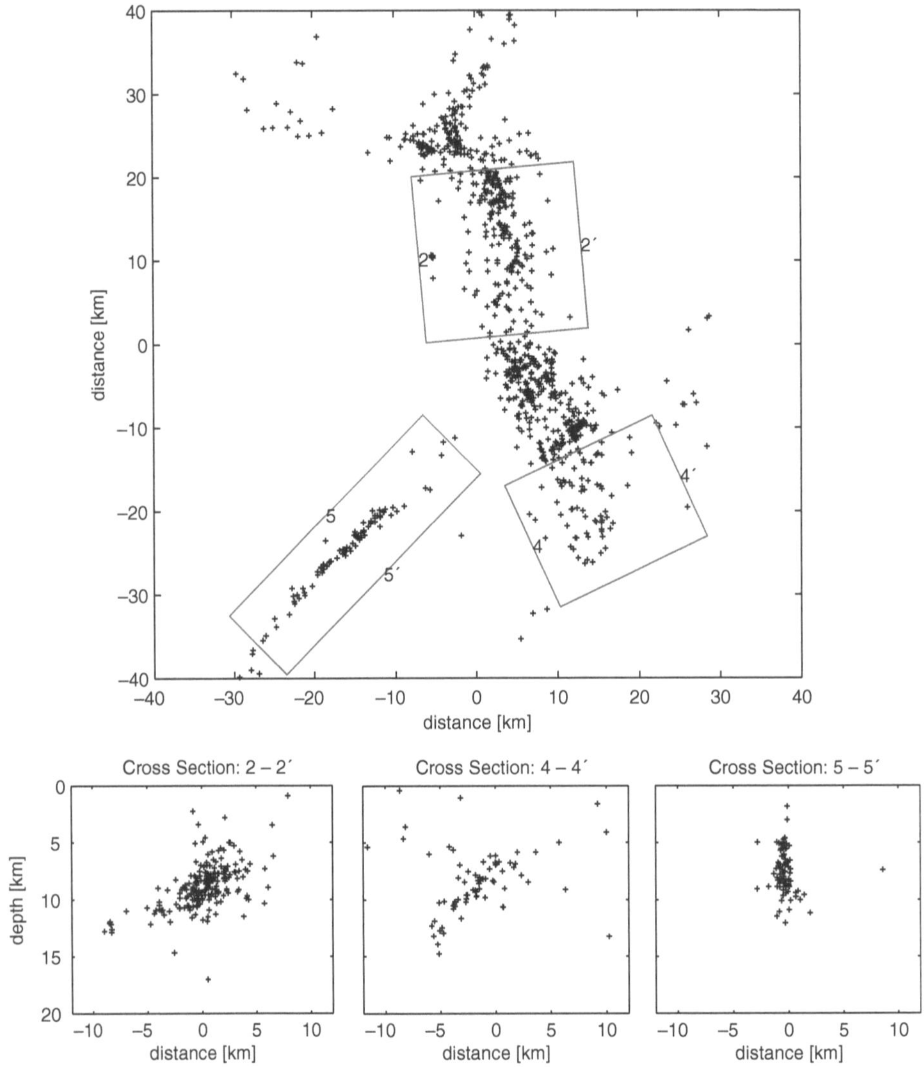

Figure 7
The traditionally-obtained event locations (catalog locations), in map view and for three cross sections, for the station set and time period of Figure 6.

(cross section 2–2′) and that to the northwest has a dip of about 75° (cross section 3–3′). In the northeast the seismicity starts at about 7 or 8 km depth; in the southwest it starts at about 4 km depth. Seismicity extends down to about 30–35 km depth. Larger events occur in the northeastern sub region (circled events have magnitude ≥ 4).

For the Charlevoix seismicity shown as 2272 relocated events in Figure 8, only 242 of them (10%) cross-correlated (with CC ≥ 0.7) at 5 or more stations; 622 (25%)

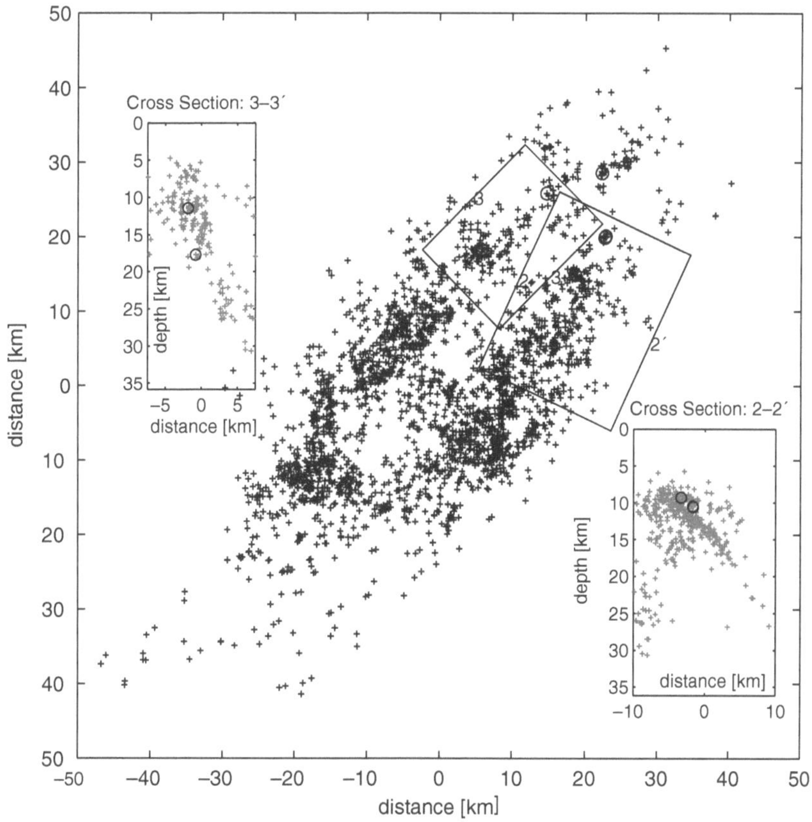

Figure 8
Relocation of 2272 events in the Charlevoix region of Eastern Canada from January 1988 to December 2003, using 46 stations operated by the Geological Survey of Canada. This application of the double-difference algorithm was based upon 495,347 phase-pick pairs and 127,600 cross correlations (CC ≥ 0.7) derived from *P* waves, plus 567,823 phase-pick pairs and 153,510 cross correlations (CC ≥ 0.7) derived from *S* waves. Circled events have magnitude ≥ 4.

at 4 or more stations; and 1439 (57%) at 3 or more stations. Nevertheless, the double-difference algorithm applied to a data set that was largely comprised of phase-pick differences rather than cross-correlation measurements still resulted in improved locations. Figure 9 compares a cross section of the seismicity located by the traditional method and by modern methods (phase picks plus cross correlation, and double difference). The latter cross section more clearly shows a lineation that presumably indicates faulting.

(4) Northern California

SCHAFF and WALDHAUSER (2005) describe results from an application of cross-correlation methods to process the complete digital seismogram database for northern

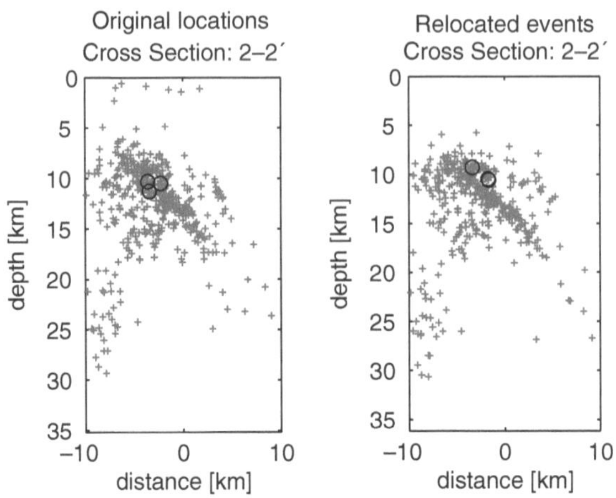

Figure 9
For the Charlevoix cross section 2–2′ of Figure 8, here are shown the locations obtained by traditional methods (on the left) and the relocations using phase picks plus a relatively small number of relative arrival times measured via cross correlation, and relocating events all together. The relocated events more clearly delineate a linear structure, dipping about 45° to the southeast.

California to measure accurate differential times for correlated earthquakes observed at common stations. The waveform database includes about 15 million seismograms from about 210,000 local earthquakes between 1984 and 2003. A total of 26 billion cross-correlation measurements were performed on a 32-node (64 processor) Linux cluster. All event pairs with separation distances of 5 km or less were processed at all stations that recorded the pair. A total of about 1.7 billion *P*-wave differential times had cross-correlation coefficients (CC) of 0.6 or larger. The *P*-wave differential times are often on the order of a factor of ten to a hundred times more accurate than those obtained from routinely picked phase onsets. 1.2 billion *S*-wave differential times were measured with CC > 0.6, a phase not routinely picked at the Northern California Seismic Network because of the generally weak onset of *S* phases, often obscured by *P*-wave coda. These results show a surprisingly high degree of waveform similarity for most of the Northern California catalog, which is very encouraging for improving earthquake locations. Overall statistics are that for each of about 200,000 events (95% of the total), waveforms have CC values that are greater than 0.7 for at least four stations with one or more other events. 90% of the events meet this criterion at eight or more stations, and 82% of the events in the catalog cross correlate at twelve or more stations. To illustrate the spatial distribution of correlated events, Figure 10 shows the percentage of events, within bins of 5 km $\times$ 5 km, that have CC > 0.7 for *P*-wave trains with at least one other event at four or more stations. Even tectonically complicated zones exhibit favorable statistics, such as Long Valley Caldera and Geysers Geothermal Field, where mechanisms are quite variable. Apparently, as long as the earthquake density is high enough there is a high probability that at least one other

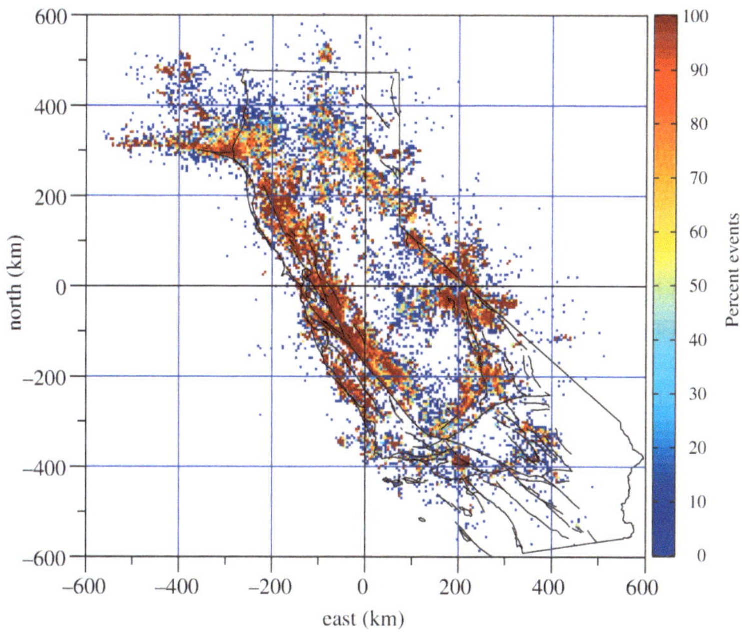

Figure 10
There are more than 200,000 events in the Northern California catalog since 1984 for which waveforms cross correlate (at CC > 0.7) with at least one other event at four or more stations. Here is shown the percentage of events in 5 km by 5 km bins that meet this criterion.

event occurs nearby with a similar focal mechanism, enabling very precise relative locations. To first order, the areas of highly correlated events agree with areas of dense seismicity, suggesting that the closeness of events is the main factor for producing similar waveforms. In a preliminary relocation of 80,000 of these Northern California events, WALDHAUSER *et al.* (2004b) obtained results exhibiting detailed features, such as streaks of seismicity, very similar to those obtained in previous studies of event subsets such as the Calaveras fault seismicity.

Discussion and Conclusions

We have assembled and analyzed waveform data sets and phase-pick data sets to assess the capability of modern methods of seismic event location, as compared to traditional methods, for studies of seismicity in China and North America. After making waveform cross-correlation (WCC) measurements within each station archive and combining them with phase-pick differences, the double-difference (DD) relocation algorithm has been applied for each of the study areas. The resulting location estimates are significantly improved over traditional estimates. In some cases, the improvement is very significant and is associated with a high percentage of events that cross correlate. But even without such CC measurements, there is improvement.

Many other authors have carried out related studies. Thus, RUBIN *et al.* (1999) applied modern methods to relocate about 75% of the events on a creeping section of the San Andreas fault, California. PENG and BEN-ZION (2005) have reported their analysis of about 18,000 aftershocks in the source regions (approximately 80 km by 40 km) of the 1999 Izmit and Duzce earthquakes in North Anatolia, Turkey. The events had been located quite accurately using traditional methods, and though these authors did not relocate the events they cross correlated signals at common stations of a ten-station PASSCAL deployment for those events estimated to be not more than 10 km apart. They found that about 60% of the events formed 2000 similar-event clusters (442 of which had five or more events).

For the four regions we examined in this paper, China was studied with far-regional signals and the other three regions were studied with local stations. Yet we found significant differences in the percentage of improved relocations even among the latter three. Specifically we found that about 85% of New Madrid events could be relocated with modern methods (using waveform cross correlation), and 95% of Northern California events; but only about 25% of Charlevoix events. It is interesting to find that New Madrid and Northern California have approximately the same station density (about one station per 100 km^2) and distance between stations (about 12 km) whereas Charlevoix is a factor of two smaller in density and a factor of two greater in station distance. Also, we found that primarily seven of the Charlevoix stations provided almost all the useful cross-correlation measurements. The other stations of the network in Eastern Canada are at greater distance, and typically can contribute only phase picks plus just a few correlations to the relocation of Charlevoix events. It is therefore easy to see why a criterion of "four or more" or "five or more" stations is hard to achieve for the events in this part of Eastern Canada: first, because greater epicentral distances are involved to reach four stations, and second because more than half the stations had to record the event with good signal-to-noise ratios and high similarity. For four out of the many more stations in the New Madrid region, good cross correlation is easier to achieve.

We conclude that the fraction of seismicity amenable to good relocations based on modern methods, ranges from about 10% across some broad regions to as much as 95% in other broad regions (major parts of Northern California). In some cases the percentage is influenced significantly by station density (and clearly this is the case for our study of China using a sparse network). We note that for purposes of monitoring a particular region—whether it be large such as North America or small such as a particular part of Liaoning Province, China—it is often true that archives of digital seismograms for events in that region are accumulating rapidly and will become more and more capable of supplying events that cross correlate. Modern methods are most suited to seismically active areas, since these allow cross-correlatable events to accumulate in a shorter time.

The choice of CC threshold has an influence on overall results, in that there is obviously a tradeoff between using as many relative arrival times measured from cross correlation as possible, but not using measurements of dubious quality. The choice of threshold is best made at a level that improves upon the accuracy of phase picks, and this resulted in thresholds of about 0.7 in our relocation studies using local stations. The DD algorithm eliminates outlier measurements, as long as they are not too numerous.

We note a conceptual inconsistency in that travel-time differences are determined by cross correlating with signals having a finite frequency band, and then measurements are passed to a location algorithm that assumes infinite-frequency ray theory. At event separations comparable to the shortest wavelength and comparable to the spatial scale of medium inhomogeneities, it may be more appropriate to use a location algorithm that explicitly recognizes the finite frequency of the underlying signals. A related point is that we have not explored the effects of material interfaces within the source region, and to the extent that such interfaces can substantially change the take-off directions of signals used for event location (as discussed for example in the case of fault-zone head waves by BEN-ZION and MALIN, 1991), there will be a distorting effect on relative locations, especially for events observed at a limited range of azimuths.

Although aspects of this work can now be regarded as quite routine, it continues to be of importance to acquire practical experience—for example, to provide feedback on questions such as: What station density is required to achieve enough CC measurements to locate the typical event; and what is the effect of archive longevity (how rapidly does the percentage of cross-correlatable events rise, as the data accumulate)? With such practical experience, it may then be appropriate to consider making modern methods of event location a part of standard operational practice, in support of publication of bulletins of seismicity which would then be enabled to attain considerable higher quality in seismically active areas where waveforms of neighboring events cross correlate.

The key to such an approach will be a commitment to long-term station operation, and building up waveform archives that can easily be used for large-scale searches. In practice, it is those network operators and their clients — who have already supported the effort to consistently collect, apply quality control, and build easily-usable archives for seismic waveforms — who are likely to benefit the most from the application of modern methods of event location.

Concerning computing power, we note that for a network consisting of tens of stations and thousands of events, a modern workstation is adequate. But for processing data in the amounts archived for example by the Northern California Earthquake Data Center (approximately a thousand stations, and a quarter of a million events), hardware with the necessary capability has only recently become widely available. The application of modern methods to major (terabyte scale) waveform archives, now being built for different networks and data centers, may require some further applied research at the scale of the full data set. A test bed that enables access to online archives could be the best facility to acquire such practical experience, prior to operational application of modern methods of seismic event location (waveform cross-correlation and multi-event location algorithms) on a large scale.

Once an archive has been established, and the signals of detected events have been associated and relocated by modern methods, new events can be accurately relocated relative to previously located events. After a period of time, say on the order of a year, all the events associated with the original waveform archive plus a year's worth of additional waveforms can be relocated all together to form a new archive, suitable as the reference set against which to locate the next year's events, and so on.

The context of this work is quite broad, since good locations are commonly the starting point for quantitative seismological studies including tomography and improved knowledge of Earth structure, seismic hazard analysis, earthquake physics and the interaction between neighboring events in a sequence, and the monitoring of explosions and earthquakes.

Finally we note that waveform cross correlation can be used not only to measure time differences and relative amplitudes very accurately, but to improve detection of small events, still allowing time differences to be measured even in cases where no phase picks can be made with confidence. See for example, GIBBONS and RINGDAL (2005). If waveform cross correlation can drop detection thresholds by about one magnitude unit, then approximately ten times more events may be detected, which in turn can lead (via multi-event location algorithms) to significant improvement in the location of all the events, including those of larger magnitude.

Acknowledgements

We greatly appreciate the assistance of network operators associated with the University of Memphis, the Geological Survey of Canada, the Northern California Seismic Network, and the Northern California Earthquake Data Center (NCEDC). In particular we thank David McCormack and Jim Lyons at the Geological Survey of Canada, Ottawa, who helped assemble phase and waveform data from earthquakes in the Charlevoix region; Jer-Ming Chiu of CERI, University of Memphis, and Jiakang Xie at Lamont-Doherty who helped us to acquire phase-pick and waveform data from the PANDA deployment at New Madrid. We also thank Mitch Withers at CERI, University of Memphis for New Madrid seismic network data and Doug Neuhauser for assistance with the NCEDC archive; and Yehuda Ben-Zion, Stephanie Ross and Malcolm Sambridge for constructive comments on the originally submitted manuscript. This work was supported by the Air Force Research Laboratory under contract F19628-03-C-0129. This is Lamont-Doherty Earth Observatory contribution number 6867.

References

BEN-ZION, Y., and MALIN, P. (1991), *San Andreas fault zone head waves near Parkfield, California*, Science *251*, 1592–1594.

BONDÁR, I., MYERS, S.C., ENGDAHL, E.R., and BERGMAN, E.A. (2004), *Epicenter accuracy based on seismic network criteria*, Geophys. J. Int. *156*, 483–496.

CHIU, J.M., JOHNSTON, A.J., and YANG, Y.T. (1992), *Imaging the active faults of the central New Madrid seismic zone using PANDA array data*, Seismol. Res. Lett. *63*, 375–393.

DOUGLAS, A. (1967), *Joint epicentre determination*, Nature *215*, 47–48.

ENGDAHL, E.R., VAN DER HILST, R., and BULAND, R. (1998), *Global teleseismic earthquake relocation with improved travel times and procedures for depth determination*, Bull. Seismol. Soc. Am. *88*, 722–743.

FROHLICH, C. (1979), *An efficient method for joint hypocenter determination for large groups of earthquakes*, Computers and Geosciences *5*, 3–4, 387–389.

GIBBONS, S.J., and RINGDAL, F. (2005), *The detection of rockbursts at the Barentsburg coal mine, Spitsbergen, using waveform correlation on SPITS array data*, NORSAR Semiannual Technical Summary, 1 July – 31 December, 2004, pp. 35–48.

GOT, J.L., FRÉCHET, J., and KLEIN, F.W. (1994), *Deep fault geometry inferred from multiple relative relocation beneath the south flank of Kilauea*, J. Geophys. Res. *99*, 375–393.

HARRIS, D.B. (1991), *A waveform correlation method for identifying quarry explosions*, Bull. Seismol. Soc. Am. *81*, 2395–2418.

HAUKSSON, E., and SHEARER, P. (2005), *Southern California hypocenter relocation with waveform cross correlation: Part 1. Results using the double-difference method*, Bull. Seismol. Soc. Am. *95*, 896–903.

ISRAELSON, H. (1990), *Correlation of waveforms from closely spaced regional events*, Bull. Seismol. Soc. Am. *80*, 2177–2193.

JORDAN, T.H., and SVERDRUP, K.A. (1981), *Teleseismic location techniques and their application to earthquake clusters in the South-Central Pacific*, Bull. Seismol. Soc. Am. *71*, 1105–1130.

PENG, Z., and BEN-ZION, Y. (2005), *Spatiotemporal variations of crustal anisotropy from similar events in aftershocks of the 1999 M7.4 Izmit and M7.1 Duzce, Turkey, earthquake sequences*, Geophys. J. Int. *160*, 1027–1043.

POUPINET, G., ELLSWORTH, W.L., and FRÉCHET, J. (1984), *Monitoring velocity variations in the crust using earthquake doublets: An application to the Calaveras fault, California*, J. Geophys. Res. *89*, 5719–5731.

RICHARDS, P.G., ARMBRUSTER, J., BURLACU, V., CORMIER, V.F., FISK, M.D., KHALTURIN, V.I., KIM, W.-Y., MOROZOV, I.B., MOROZOVA, E.A., SAIKIA, C.K., SCHAFF, D., STROUJKOVA, A., and WALDHAUSER, F. (2003), *Seismic Location Calibration for 30 International Monitoring System Stations in Eastern Asia: Final Results*, Proceedings, DOD/DOE Seismic Research Review, Tucson.

RICHARDS-DINGER, K.B., and SHEARER, P.M. (2000), *Earthquake locations in Southern California obtained using source-specific station terms*, J. Geophys. Res. *105*, 10,939–10,960.

ROWE, C.A., ASTER, R.C., PHILLIPS, W.S., JONES, R.H., BORCHERS, B., and FEHLER, M.C. (2002), *Using automated, high-precision repicking to improve delineation of microseismic structures at the Soultz Geothermal Reservoir*, Pure Appl. Geophys. *159*, 563–596.

RUBIN, A.M., GILLARD, D., and GOT, J.L. (1999), *Streaks of microearthquakes along creeping faults*, Nature *400*, 635–641.

SCHAFF, D.P., BOKELMANN, G.H.R., BEROZA, G.C., WALDHAUSER, F., and ELLSWORTH, W.L. (2002), *High-resolution image of Calaveras fault seismicity*, J. Geophys. Res. *107*, 2186, doi: 1029/2001JB000633.

SCHAFF, D.P., and RICHARDS, P.G. (2004a), *Repeating seismic events in China*, Science *303*, 1176–1178.

SCHAFF, D.P., and RICHARDS, P.G. (2004b), *Lg-wave cross correlation and double-difference location: Application to the 1999 Xiuyan, China, sequence*, Bull. Seismol. Soc. Am. *94*, 867–879.

SCHAFF, D.P., BOKELMANN, G.H.R., BEROZA, G.C., and ELLSWORTH, W.L. (2004), *Optimizing correlation techniques for improved earthquake location*, Bull. Seismol. Soc. Am. *94*, 705–721.

SCHAFF, D.P., and WALDHAUSER, F. (2005), *Waveform cross correlation based differential travel-time measurements at the Northern California Seismic Network*, Bull. Seismol. Soc. Am., *95*, 2446–2461.

SHEARER, P. (1997), *Improving local earthquake locations using the L1 norm and waveform cross correlation: Application to the Whittier Narrows, California, aftershock sequence*, J. Geophys. Res. *102*, 8269–8283.

SHEARER, P., HAUKSSON, E., and LIN, G. (2005), *Southern California hypocenter relocation with waveform cross correlation: Part 2. Results using source-specific station terms and cluster analysis*, Bull. Seismol. Soc. Am. *95*, 904–915.

THURBER, C., TRABANT, C., HASLINGER, F., and HARTOG, R. (2001), *Nuclear explosion locations at the Balapan, Kazakhstan, nuclear test site: The effects of high-precision arrival times and three-dimensional structure*, Phys. Earth Plan. Int. *123*, 283–301.

WALDHAUSER, F., and ELLSWORTH, W.L. (2000), *A double-difference earthquake location algorithm: Method and application to the Northern Hayward fault, California*, Bull. Seismol. Soc. Am. *90*, 1353–1368.

WALDHAUSER, F. (2001), *HypoDD: A program to compute double-difference hypocenter locations*, U.S.G.S. Open-file Report, 01–113, Menlo Park, California.

WALDHAUSER, F., ELLSWORTH, W.L., and COLE, A. (1999), *Slip-parallel seismic lineations along the Northern Hayward fault, California*, Geophys. Res. Lett. *26*, 3525–3528.

WALDHAUSER, F., SCHAFF, D.P., RICHARDS, P.G., and KIM, W.-Y. (2004a), *Lop Nor revisited: Underground nuclear explosion locations, 1976–1996, from double-difference analysis of regional and teleseismic data*, Bull. Seismol. Soc. Am. *94*, 1879–1889.

WALDHAUSER, F., SCHAFF, D.P., KIM, W.-Y., and RICHARDS, P.G. (2004b), *Improved characterization of seismicity and fault structure by wide area event relocation*, EOS, Trans. Am. Geophys. U., 85, Fall Meet. Suppl., Abstract S53C–07.

WITHERS, M., ASTER, R., and YOUNG, C. (1999), *An automated local and regional seismic event detection and location system using waveform correlation*, Bull. Seismol. Soc. Am. *86*, 657–669.

XIE, J., LIU, Z., CONG, L., HERRMANN, R.B., and CHIU, J.M. (1997), *Rupture properties of clustered microearthquakes near intersecting intraplate faults of the New Madrid Seismic Zone: Implications on fault weakening*, J. Geophys. Res. *102*, 8,187–8,202.

XIE, J. (2001), *Rupture characteristics of clustered microearthquakes and variations in fault properties in the New Madrid Seismic Zone*, J. Geophys. Res. *106*, 26,495–26,509.

(Received May 26, 2005; accepted September 27, 2005)

Pure appl. geophys. 163 (2006) 373–403
0033–4553/06/030373–31
DOI 10.1007/s00024-005-0021-y

Pure and Applied Geophysics

Development and Applications of Double-difference Seismic Tomography

HAIJIANG ZHANG,[1] and CLIFFORD THURBER[1]

Abstract—Double-difference (DD) tomography is a generalization of DD location; it simultaneously solves for the three-dimensional velocity structure and seismic event locations. DD tomography uses a combination of absolute and more accurate differential arrival times and hierarchically determines the velocity structure from larger scale to smaller scale. This method is able to produce more accurate event locations and velocity structure near the source region than standard tomography, which uses only absolute arrival times. We conduct a stability and uncertainty analysis of DD tomography based on a synthetic data set. Currently three versions of the DD tomography algorithms exist: *tomoDD*, *tomoFDD* and *tomoADD*. *TomoDD* assumes a flat earth model and uses a pseudo-bending ray-tracing algorithm to find rays between events and stations while *tomoFDD* uses a finite-difference travel-time algorithm and the curvature of the Earth is considered. Both codes are based on a regularly distributed inversion grid, with the former for a local scale and the latter for a regional scale. In contrast, *tomoADD* adapts the inversion mesh to match with the data distribution based on tetrahedral and Voronoi diagrams. We discuss examples of applying DD tomography to characterize fault zone structure, image high-resolution structure of subduction zones, and determine the velocity structure of volcanoes.

Key words: Double-difference, tomography, fault zone, subduction zone, volcano.

Introduction

We have recently developed a double-difference (DD) seismic tomography method that makes use of both absolute and more accurate relative arrival times (ZHANG and THURBER, 2003). The differential times can be calculated from cross-correlation (CC) techniques for similar waveforms and by directly subtracting catalog arrival times for pairs of events at common stations (WALDHAUSER and ELLSWORTH, 2000; ZHANG and THURBER, 2003). DD tomography is a generalization of DD location (WALDHAUSER and ELLSWORTH, 2000); it simultaneously solves for the three-dimensional (3-D) velocity structure and seismic event locations. DD tomography uses an evolving weighting scheme for the absolute and differential arrival times in order to determine the velocity structure from larger scale to smaller

[1]Department of Geology and Geophysics, University of Wisconsin-Madison, 1215 W. Dayton St., Madison, WI, 53706, USA. E-mail: hjzhang@geology.wisc.edu

scale. As shown by synthetic tests (ZHANG and THURBER, 2003, 2005; ZHANG *et al.*, 2004), this method yields more accurate event locations and velocity structure near the source region than standard tomography, which uses only absolute arrival times. It has the unique ability to sharpen the velocity image near the source region because of the combination of the higher accuracy of the differential time data and the concentration of the corresponding model derivatives in the source region. The latter results from the cancellation of model derivative terms where the ray paths overlap away from the source region.

In many situations, CC data are not available due to the lack of waveform data. There have been many applications using both DD location and tomography algorithms demonstrating that using just the catalog pick differences leads to significant improvement of event locations, as well as for the velocity structure for DD tomography. As discussed in ZHANG and THURBER (2003) and ZHANG *et al.* (2005), picking errors include two parts: random and systematic. Although random errors may increase through the differencing process, the systematic errors will be reduced or cancelled. The differencing process also causes the ray-path derivatives to concentrate near the source region, rather than covering the entire model region. This in turn reduces or removes the effect of the velocity model ambiguity outside the source region on the source region model. As a result, the source region structure will be better illuminated.

The DD tomography code *tomoDD* is built upon the double-difference location code *hypoDD* written by WALDHAUSER (2001). In the original *tomoDD* algorithm, we used an approximate pseudo-bending (ART-PB) ray-tracing algorithm (UM and THURBER, 1987) to find the rays and calculate the travel times between events and stations. The model is represented by velocity values specified on a regular set of 3-D nodes and the velocity values are interpolated by using the linear B-spline interpolation method. The hypocentral partial derivatives are calculated from the direction of the ray and the local velocity at the source (LEE and STEWART, 1981). The ray path is divided into a set of segments and the model partial derivatives (calculated in terms of fractional slowness perturbation, so that the derivatives are related to path length) are evaluated by apportioning the derivative to its eight surrounding nodes according to their interpolation weights on the segment midpoint (THURBER, 1983).

TomoDD assumes a flat earth model and is appropriate for local scale problems (10's to 100's of kilometers). At the regional scale (100's to 1000's of kilometers), however, sphericity of the earth should be taken into account. Major velocity discontinuities such as Conrad, Moho, and subducting slab boundary should also be considered. The ART-PB approach assumes a continuous velocity model and cannot deal properly with velocity discontinuities. KOKETSU and SEKINE (1998) have developed a spherical-earth 3-D ray-tracing method (including discontinuities) using pseudo-bending, extended from a method developed in Cartesian coordinates (ZHAO *et al.*, 1992). ZHAO *et al.* (1992) and KOKETSU and

SEKINE (1998) explicitly took into account velocity discontinuities by applying Snell's law to find the intersecting point between a ray path and a discontinuity. These methods have advantages over the original ART-PB approach if the positions of velocity discontinuities are known *a priori* from other independent studies. However, these velocity discontinuities are seldom known *a priori* or are not well constrained in many subduction zones. For this reason, we developed a regional DD seismic tomography method (*tomoFDD*) that deals effectively with discontinuous velocity structures without knowing them *a priori*. *TomoFDD* uses a finite-difference method for determining travel times and ray paths, and treats the spherical Earth by embedding it (in part or in whole) within a Cartesian "box." For the local and regional-scale problems, the ray-tracing accuracy for both pseudo-bending and finite-difference algorithms is carefully controlled to match with the higher accuracy of the differential times. For example, for pseudo-bending ray-tracing algorithm we do not stop the iteration finding rays until the travel-time differences between two close iterations are smaller than 1 ms.

The current versions of *tomoDD* and *tomoFDD* use a regular inversion grid. For typical seismic tomography studies, however, the ray distribution is highly uneven due to nonuniform station geometry, uneven distribution of seismic sources, missing data, and ray bending. Some nodes or cells may have few or even no rays sampling them, while others may have very dense rays sampling them. The regular grid spacing restriction makes it difficult to adapt the model to the uneven distribution of ray paths. The mismatch between the ray distribution and the grid chosen for the tomographic inversion results in instability of the inversion. To make the inversion more stable, some regularization method is required, such as damping or smoothing, which will inevitably bias the results. Ideally the inversion grid (or mesh) should be distributed adaptively to match with the resolving power of the data. The inversion problem will then be better conditioned and smaller damping and/or reduced smoothing constraints would thus be required. Our recently developed adaptive-mesh DD seismic tomography method (*tomoADD*), based on tetrahedral and Voronoi diagrams, can automatically match the inversion mesh to the data distribution (ZHANG and THURBER, 2005).

The following material first reviews the DD tomography method and conducts a stability and uncertainty analysis based on a simple 2-D synthetic model. Then we present the regional-scale and adaptive-mesh DD tomography methods, and discuss some of their applications for a variety of regions.

The Double-difference Tomography Method

The body wave arrival time T from an earthquake i to a seismic station k is expressed using ray theory as a path integral

$$T_k^i = \tau^i + \int_i^k u\, ds, \tag{1}$$

where τ^i is the origin time of event i, u is the slowness field and ds is an element of path length. For local earthquake tomography, the source coordinates (x_1, x_2, x_3), origin times, ray paths, and the slowness field are the unknowns. The relationship between the arrival time and the event location is nonlinear, thus a truncated Taylor series expansion is generally used to linearize eq. (1). If we also discretize the velocity model using a 3-D grid or mesh, we can write a linear equation relating the misfit between the observed and predicted arrival times r_k^i to the desired perturbations to the hypocenter and velocity structure parameters

$$r_k^i = \sum_{l=1}^{3} \frac{\partial T_k^i}{\partial x_l^i} \Delta x_l^i + \Delta \tau^i + \sum_{m=1}^{M_{ik}} \sum_{n=1}^{N} w_{mn}\, \Delta u_n\, \Delta s_m, \tag{2}$$

where M_{ik} indicates the number of segments of the ray path from event i to station k and w_{mn} are interpolating weights of nth mesh node on the mid-point of the mth segment with length Δs_m. Subtracting a similar equation for event j observed at station k from equation (2), we have

$$\begin{aligned} r_k^i - r_k^j = {} & \sum_{l=1}^{3} \frac{\partial T_k^i}{\partial x_l^i} \Delta x_l^i + \Delta \tau^i + \sum_{m=1}^{M_{ik}} \sum_{n=1}^{N} w_{mn}\, \Delta u_n\, \Delta s_m - \sum_{l=1}^{3} \frac{\partial T_k^j}{\partial x_l^j} \Delta x_l^j - \Delta \tau^j \\ & - \sum_{m=1}^{M_{jk}} \sum_{n=1}^{N} w_{mn}\, \Delta u_n\, \Delta s_m, \end{aligned} \tag{3}$$

where $r_k^i - r_k^j$ is the so-called double-difference (WALDHAUSER and ELLSWORTH, 2000). This term is the difference between observed and calculated differential arrival times for the two events, and can also be written as

$$r_k^i - r_k^j = (T_k^i - T_k^j)^{\text{obs}} - (T_k^i - T_k^j)^{\text{cal}}. \tag{4}$$

The observed differential arrival times $(T_k^i - T_k^j)^{\text{obs}}$ can be calculated from both waveform cross-correlation techniques for similar waveforms and absolute catalog arrival times.

Note that the ray paths from two nearby events will substantially overlap, meaning that the model derivative terms in eq. (3) will essentially cancel outside the source region. For this reason, we include the absolute arrival times in the inversion to resolve the velocity structure outside the source region. To combine absolute and differential systems together, we apply a hierarchical weighting scheme during the inversion to apply the appropriate relative weighting between them at different stages of inversion. We start the inversion by applying higher weighting to the catalog data (both differential and absolute catalog data) to establish the large-scale result (1 for absolute data, 0.1 for differential catalog data, and 0.01 for cross-correlation data) in a manner similar to *hypoDD* (WALDHAUSER, 2001). Then for later iterations the catalog differential data are weighted more to refine the event locations and the velocity structure near the source regions (1 for differential catalog data, 0.1 for absolute catalog data, and 0.01 for cross-correlation data). If waveform cross-correlation data are available, they will be weighted more heavily than the catalog differential data in the final iterations to further refine the event locations and the velocity structure near the source region (1 for cross-correlation data, 0.01 for differential catalog data, and 0.001 for absolute catalog data). We downweight the differential catalog data by a factor of 100 in the final stage of the inversion, because the cross-correlation data are one to two orders of magnitude more precise than the manual picks (WALDHAUSER and ELLSWORTH, 2000). The relationship between arrival time residuals and slowness model parameters is more linear than the event locations (THURBER, 1992), as can be seen from eq. (1) where the integral in the equation is linear in slowness but nonlinear in the hypocentral coordinates. This indicates that the convergence of velocity structure is faster than event locations (THURBER, 1992). Consequently, the simultaneous inversion of event locations and velocity structure alternates with the inversion of only event locations in practice.

WOLFE (2002) carried out a strict and complete analysis of the use of difference operators to relocate earthquakes. Next we conduct a similar analysis on using the absolute data (i.e., standard location and tomography), the differential data (i.e., DD location and tomography), and both of them to determine event locations and/or velocity structure. Most of the notation used in this section follows WOLFE (2002).

Consider a set of $p = 1, \ldots, P$ earthquakes, with N_p arrival times for each earthquake. For each individual earthquake, eq. (2) can be represented in matrix form as follows,

$$\mathbf{A_p}\Delta\mathbf{X_p} + \mathbf{C_p}\Delta\mathbf{M} = \Delta\mathbf{T_p}, \tag{5}$$

where $\mathbf{A_p}(N_p \times 4)$ is the partial derivative matrix corresponding to the hypocenter and origin time, $\Delta\mathbf{X_p}(4 \times 1)$ is the perturbation vector for earthquake location and origin time, $\mathbf{C_p}(N_p \times L)$ is the model derivative (path length times node weight) matrix corresponding to the slowness model, $\Delta\mathbf{M}(L \times 1)$ is the vector of slowness perturbations, and $\Delta\mathrm{T_p}(N_p \times 1)$ is the vector of arrival time residuals. A station term may be also included to take into account the velocity heterogeneity near the stations, which may not be well resolved during the inversion, as follows,

$$\mathbf{A_p}\Delta\mathbf{x_p} + \mathbf{C_p}\Delta\mathbf{M} + \mathbf{s_p} = \Delta\mathbf{T_p}, \tag{6}$$

where $\mathbf{s_p}(N_p \times 1)$ is the station correction vector (note that it is defined as the path anomaly in WOLFE (2002)).

Define $N_T(= \sum_{p=1}^{P} N_p)$ as the total number of arrival time data and M_T as the total number of unknown hypocenter and origin parameters $(= 4 \times P)$. We can combine all the equations for P earthquakes into one linear system, as follows,

$$\mathbf{A\Delta X} + \mathbf{C\Delta M} + \mathbf{S} = \mathbf{\Delta T}\,, \tag{7}$$

where

$$\mathbf{A} = \begin{bmatrix} \mathbf{A_1} & \mathbf{0} & \bullet & \bullet & \mathbf{0} \\ \mathbf{0} & \mathbf{A_2} & \bullet & \bullet & \mathbf{0} \\ \bullet & \bullet & \bullet & \bullet & \bullet \\ \bullet & \bullet & \bullet & \bullet & \bullet \\ \mathbf{0} & \mathbf{0} & \bullet & \bullet & \mathbf{A_p} \end{bmatrix}; N_T \times M_T \tag{8}$$

$$\mathbf{\Delta T} = \begin{bmatrix} \mathbf{\Delta T_1} \\ \mathbf{\Delta T_2} \\ \bullet \\ \mathbf{\Delta T_P} \end{bmatrix}; N_T \times 1 \tag{9}$$

$$\mathbf{C} = \begin{bmatrix} \mathbf{C_1} & \mathbf{0} & \bullet & \bullet & \mathbf{0} \\ \mathbf{0} & \mathbf{C_2} & \bullet & \bullet & \mathbf{0} \\ \bullet & \bullet & \bullet & \bullet & \bullet \\ \bullet & \bullet & \bullet & \bullet & \bullet \\ \mathbf{0} & \mathbf{0} & \bullet & \bullet & \mathbf{C_P} \end{bmatrix}; N_T \times L \tag{10}$$

$$\mathbf{\Delta M} = \begin{bmatrix} \Delta \mathrm{m_1} \\ \Delta \mathrm{m_2} \\ \bullet \\ \Delta \mathrm{m_L} \end{bmatrix}; L \times 1 \tag{11}$$

$$\mathbf{S} = \begin{bmatrix} \mathbf{s_1} \\ \mathbf{s_2} \\ \bullet \\ \mathbf{s_P} \end{bmatrix}; N_T \times 1. \tag{12}$$

For station corrections, we assume that each is a constant for all the arrivals observed at a given station. As a result, let $\mathbf{s_0}$ stand for the station corrections for the K_T stations, which can be represented as

$$\mathbf{S} = \mathbf{Bs_0}, \tag{13}$$

where

$$\mathbf{B} = \begin{bmatrix} \mathbf{B_1} \\ \mathbf{B_2} \\ \bullet \\ \mathbf{B_P} \end{bmatrix}; N_T \times K_T, \tag{14}$$

and each $\mathbf{B_p}$ is a $N_p \times K_T$ matrix such that

$$[B_p]_{ij} = \begin{cases} 1 \text{ when } [\Delta T_p]_i \text{is from station j} \\ 0 \text{ otherwise} \end{cases} . \tag{15}$$

The DD tomography eq. (3) is equivalent to the application of a difference operator $\mathbf{Q_{DD}}$ to equation (7), where

$$\mathbf{Q_{DD}} = \begin{bmatrix} 1 & -1 & \bullet & \bullet & \bullet & 0 \\ 1 & \bullet & \bullet & -1 & \bullet & \bullet \\ \bullet & \bullet & \bullet & \bullet & \bullet & \bullet \\ \bullet & \bullet & \bullet & \bullet & \bullet & \bullet \\ 0 & \bullet & 1 & \bullet & \bullet & -1 \end{bmatrix}. \tag{16}$$

Let w_k equal the number of earthquakes observed at station k, then the matrix $\mathbf{Q_{DD}}$ has the dimension of $\sum_{k=1}^{K_T} \frac{w_k(w_k-1)}{2} \times N_T$, corresponding to the possible combinations of differences of earthquake arrival times at a station k, $\Delta T_k^i - \Delta T_k^j$ (for $i<j$, to eliminate repetition).

It can be seen that the difference operator $\mathbf{Q_{DD}}$ will annihilate the station corrections as a result of $\mathbf{Q_{DD}Bs_0} = \mathbf{0}$. This is an advantage when comparing DD tomography with standard tomography. The matrix form for the differential time data of DD tomography is

$$\mathbf{Q_{DD}A\Delta X} + \mathbf{Q_{DD}C\Delta M} = \mathbf{Q_{DD}\Delta T}. \tag{17}$$

The difference operator $\mathbf{Q_{DD}}$ has the rank of $\sum_{k=1}^{K_T} (w_k - 1)$, because the operation removes one degree of freedom from the arrival times observed at one station (WOLFE, 2002). This rank is usually considerably larger than the total number of unknown parameters, including event hypocenter and origin time parameters and model (slowness) parameters. If both matrices $\mathbf{Q_{DD}A}$ and $\mathbf{Q_{DD}C}$ have full column rank, the DD tomography system will have the ability to determine absolute event locations (MENKE and SCHAFF, 2004) and slowness (or velocity) structure uniquely. Although the hypocenter partial derivatives for closely spaced events are very similar, they will be quite different for events spaced far apart. Thus, in theory, if we include all the possible combinations of pairs of arrival times observed at common stations, the matrix $\mathbf{Q_{DD}A}$ should have full column rank. $\mathbf{Q_{DD}C}$ should also have full column rank if we remove slowness model parameters not sampled by ray paths from the inversion.

For the DD location algorithm in which the path anomaly biases between events are not taken into account explicitly, only closely spaced events should be chosen in order to reduce the path anomaly bias. Distance weighting is applied to reduce the effect of event pairs that are far apart (WALDHAUSER and ELLSWORTH, 2000). However, for closely spaced events, the partial derivatives will be very similar or equal, thus the DD location algorithm is only capable of resolving relative locations between events (WOLFE, 2002). For the DD tomography system, since the path anomaly biases between event pairs are taken into account explicitly, we do not have to limit the inversion only to the closely spaced events. Distance weighting is an option in the DD tomography algorithm and is mainly designed to downweight differential times resulting from far apart events based on the fact that the similarity of waveforms is less for far apart events, as are the common errors associated with picks.

Equation (17) can be further transformed into a more compact form, as follows,

$$\mathbf{Q_{DD}EY} = \mathbf{Q_{DD}}\Delta\mathbf{T}, \tag{18}$$

where $\mathbf{E} = [\mathbf{A} \quad \mathbf{C}]$ has the dimension of $N_T \times (M_T + L)$, and $\mathrm{Y} = \begin{bmatrix} \Delta\mathrm{X} \\ \Delta\mathrm{M} \end{bmatrix}$ has the dimension of $(M_T + L) \times 1$. The unknown parameters including hypocenters, origin times, and slowness model parameters can be obtained, as follows,

$$\mathbf{Y} = (\mathbf{Q_{DD}E})^{-1}\mathbf{Q_{DD}}\Delta\mathbf{T}. \tag{19}$$

For origin times, since the associated partial derivatives are the same (equal to 1), only relative origin times can be recovered from this system (WOLFE, 2002).

If the differential arrival times are used to determine event locations only (WALDHAUSER and ELLSWORTH, 2000), and the hypocenter partial derivatives are evaluated at different locations, the absolute event locations can potentially be resolved, as follows (WOLFE, 2002),

$$\Delta\mathbf{X} = (\mathbf{Q_{DD}A})^{-1}\mathbf{Q_{DD}}\Delta\mathbf{T}. \tag{20}$$

The partial derivative operator used in DD location is $\mathbf{Q_{DD}A}$. Note that a row of the matrix $(\mathbf{Q_{DD}A})^{-1}\mathbf{Q_{DD}}$ can be used to indicate the dependence of the corresponding model parameter (location or origin time in this case) on all the travel times.

The perturbations to event locations can also be obtained from absolute arrival times by assuming the velocity structure is known and presumably correct,

$$\Delta\mathbf{X} = \mathbf{A}^{-1}\Delta\mathbf{T}. \tag{21}$$

Similarly, the system to determine event locations and velocity structure from absolute arrival times (standard tomography) is represented as

$$\mathbf{Y} = \mathbf{E}^{-1}\Delta\mathbf{T}. \tag{22}$$

For the combined system that uses both absolute and differential data, the formula to determine event locations is

$$\Delta\mathbf{X} = \begin{pmatrix} \mathbf{Q_{DD}A} \\ w\mathbf{A} \end{pmatrix}^{-1} \begin{pmatrix} \mathbf{Q_{DD}} \\ w\mathbf{I} \end{pmatrix} \Delta\mathbf{T}, \tag{23}$$

where w is the relative weighting between absolute and relative arrival times, and I is the identity matrix. The complete formula to solve the velocity structure simultaneously with event locations by using both absolute and differential data is

$$\mathbf{Y} = \begin{pmatrix} \mathbf{Q_{DD}E} \\ w\mathbf{E} \end{pmatrix}^{-1} \begin{pmatrix} \mathbf{Q_{DD}} \\ w\mathbf{I} \end{pmatrix} \Delta\mathbf{T}. \tag{24}$$

For simplicity, we construct a 2-D synthetic model (Fig. 1) to show the properties of these different systems of equations. There are 6 clusters of earthquakes (circles) and each cluster has 4 closely spaced events with inter-event distances of less than 200 m. In the X and Y directions, the distances between clusters are 10 and 15 km, respectively. There are 36 stations around these events, with all the events observed at each station. As a result, there are 894 travel times in total for all the event and station pairs. The velocity is assumed to be constant at 5 km/s, thus the rays between events and stations are all straight lines. The region is divided uniformly into 81 cells (or blocks). The model derivatives relative to slowness are just the path lengths crossing the cells.

Since this is relatively a small data set, we can use the singular value decomposition method to calculate the singular values, model resolution and

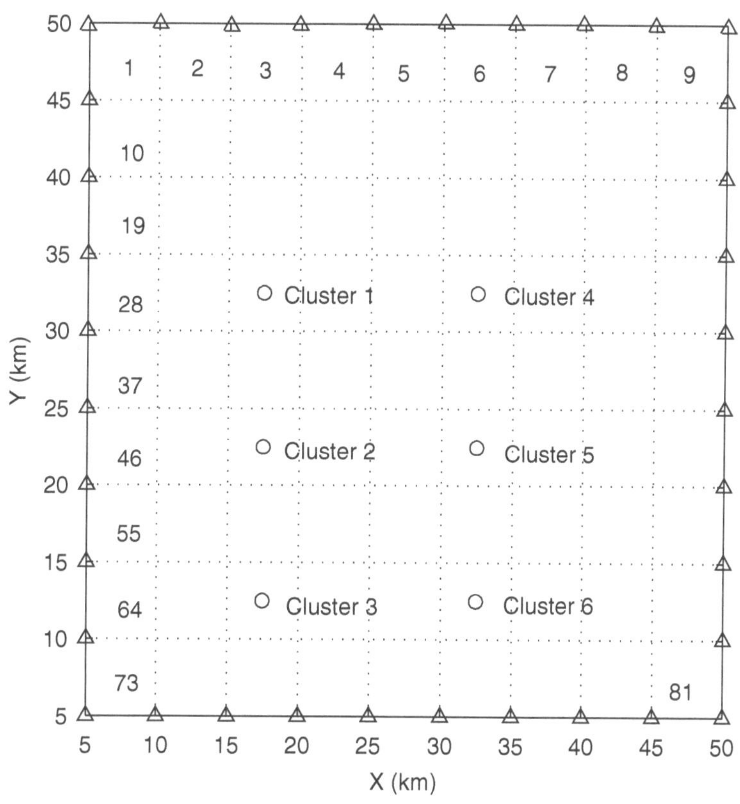

Figure 1
Map of stations (triangles) and earthquake clusters (circles) for the synthetic example. Each cluster has 4 closely spaced events, with interevent distances among them less than 200 m. The velocity model is assumed to be a constant 5 km/s. The model region is divided into 81 cells uniformly, with numbers in the cells indicating the model parameter indices.

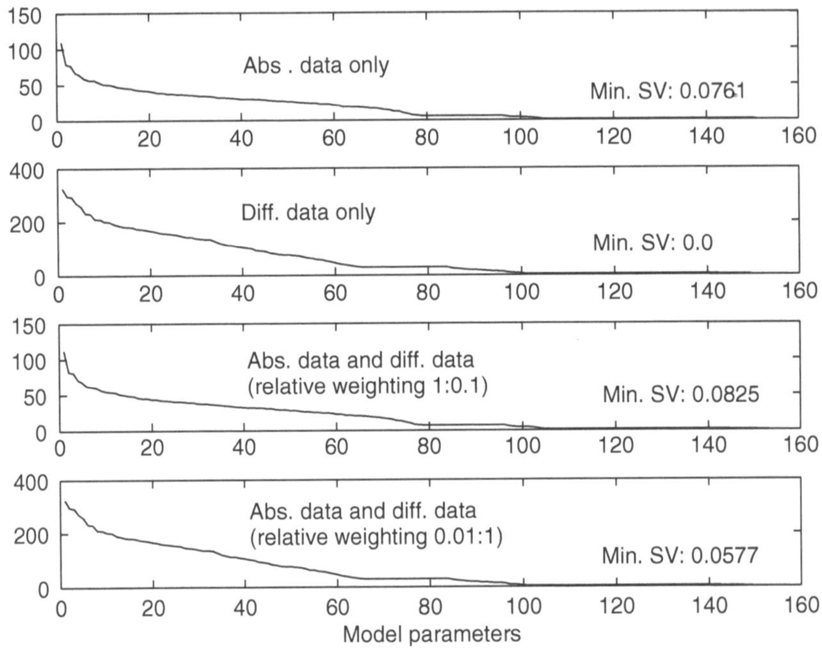

Figure 2
The singular value distribution of the partial derivative matrix from using different data types to determine event locations and velocity structure simultaneously.

uncertainties for the different systems (MENKE, 1989; ASTER *et al.*, 2005). Figure 2 shows the singular values for the operators $\mathbf{E}$, $\mathbf{Q_{DD}E}$, $(\mathbf{E}, 0.1\mathbf{Q_{DD}E})^{\mathbf{T}}$, and $(0.01\mathbf{E}, \mathbf{Q_{DD}E})^{\mathbf{T}}$ when using the absolute data only, the differential data only, and both of them (with relative weighting 1:0.1 and 0.01:1), respectively, to jointly determine event locations and slowness structure. The system using the differential data to jointly determine event locations and slowness structure has one zero singular value (Fig. 2). The zero singular value is due to the fact that the partial derivatives relative to the origin time for all events are equal and the differential operator has no ability to recover the absolute origin times, only relative ones. Including the absolute data with only a small weight is helpful to stabilize the system, by increasing the minimum singular value to 0.0577 (Fig. 2). This indicates that including the absolute data into the differential system gives it the ability to resolve the absolute origin times, as would be expected. Also notice that the largest singular value from the system having higher weighting on the differential data is about 2 times greater than that from the system having higher weighting on the absolute data (Fig. 2). To resolve these systems by using the damped least-squares algorithm, different damping values should be chosen to make their condition numbers similar. The damping value that makes the condition number of the system close to ~60 to 80 is consistent with the

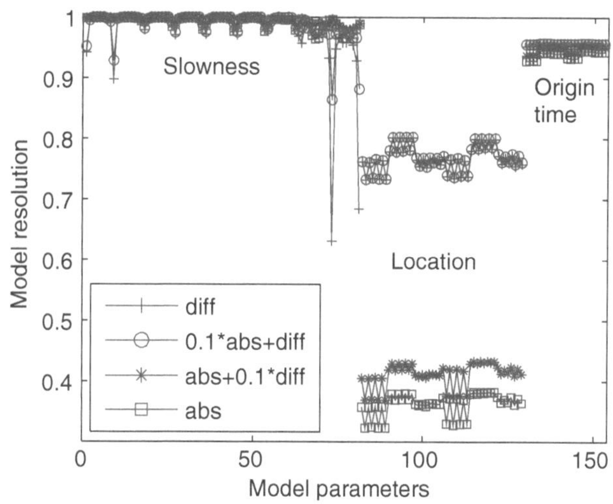

Figure 3
Model resolution for location and slowness model parameters when using different data types to determine event locations and velocity structure simultaneously by a damped least squares solution. The damping values are 1.5 for the systems of abs and abs + 0.1diff, and 4.4 for the systems of diff and diff + 0.01abs. For the model parameters, the slowness model cells are arranged at first in the sequence as shown in Figure 1, followed by the location and origin time parameters.

one chosen based on the trade-off curve between data variance and model norm. For this reason, we choose damping values of 1.5 and 4.4 for the systems having higher weighting on the absolute data and higher weighting on the differential data, which would make all the systems have condition numbers ~70.

Figure 3 shows model resolution for origin time, location and slowness model parameters. The systems with higher weighting on the differential data seem to better resolve event locations by having higher resolution values (~0.7 versus ~0.3). For slowness model parameters around the edge of the model, the systems having higher weighting on the absolute data have higher model resolution. We also note that the model resolution is generally higher for the slowness parameters than location parameters. By carefully checking the singular values of the Fréchet derivative matrices for slowness and locations parameters, we found that the largest singular value for the slowness derivative matrix is considerably larger than that for the location derivative matrix. For example, the largest singular value for the slowness matrix in the case of using just differential times is 325.3, compared to 29.5 for the location matrix. Since we use the same damping value of 4.4 for slowness and location parameters, the latter ones are overdamped. This is the reason why the model resolution for slowness parameters is greater than location parameters. This overdamping for location parameters is justified since the relationship between arrival time residuals and hypocenter perturbations is nonlinear. This is helpful for preventing large data residuals (or noise) from projecting into location parameters. We also note that events are generally relocated one more time based on the final velocity model.

We assume that the errors associated with both absolute and differential data are an uncorrelated Gaussian-distributed random process. The standard deviations for the absolute and differential data are 80 and 20 ms, respectively. This simulates the scenario that the differential data are more accurate than the absolute data. These values would be representative of absolute catalog picks and differential catalog picks; using the even more precise CC differential times would obviously make the error contrast greater. Figure 4 shows the model uncertainties (s/km for slowness parameters and km for locations) for different systems. The model uncertainties estimated from the system having higher weighting on the absolute data are substantially larger, especially for event locations and origin times. For all the systems, the model uncertainties associated with origin times are much smaller than those for hypocenters. This is caused by the different dimensionality between the origin time and hypocenter location (KLEIN, 1978). For example, a change of several km in hypocenter location is equivalent to a one second change in origin time. We also notice that the slowness model cells closer to two model region edges (X = 5 and 50 km) tend to have larger model uncertainties. This phenomenon results from the relatively fewer rays crossing through those cells. The characteristics of model uncertainties having higher weighting on the differential data (diff and

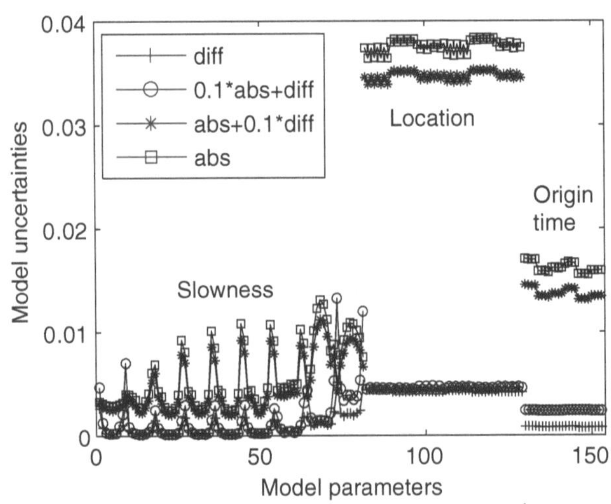

Figure 4
Model uncertainties for location and slowness model parameters when using different data types to determine event locations and velocity structure simultaneously by a damped least squares solution. The damping values for different systems are shown in Figure 3. The model parameters are arranged in the same way as Figure 3.

diff + 0.01*abs) are quite similar, with the system (diff + 0.01*abs) using the combination of data types having slightly larger uncertainties for some slowness model cells.

These results demonstrate the advantages of using both absolute and differential data. The model resolution for location model parameters is higher in the systems having higher weighting on the differential data. The system using both the absolute and differential data with the relative weighting 0.01 to 1 between them has higher model resolution and lower model uncertainties for location parameters than the system using the absolute data only. For some model-edge slowness model cells, the systems having higher weighting (abs and abs + 0.1*diff) on the absolute data have higher model resolution and lower model uncertainties.

Regional-scale Double-difference Tomography (tomoFDD)

The regional-scale DD tomography code *tomoFDD* uses the finite-difference travel-time algorithms developed by PODVIN and LECOMTE (1991) and HOLE and ZELT (1995) (modified from VIDALE 1990) to accurately calculate travel times in the presence of severe velocity changes and discontinuities. The algorithms solve a finite-difference approximation to the Eikonal equation on a regularly gridded velocity model through a systematic application of Huygens' principle. Both algorithms repeatedly compute the travel times at the grid points that are adjacent to those with known (previously computed) travel times. This procedure can explicitly take into account different propagation modes including transmitted and diffracted body waves and head waves, in addition to direct waves.

For the finite-difference travel-time methods, the station can be regarded as a source and travel times can be calculated from this station to all the grid nodes. According to the principle of reciprocity, the travel times from the event to the station are equal to those from the station to the event. After the travel-time field for each station is calculated, the travel time from an event to this station can be interpolated by the linear B-spline interpolation method through 8 neighboring grid nodes near this event. This method is more efficient than other ray-tracing methods that trace rays from a single event to a large number of stations in the case that there are far more events than stations.

The errors that arise from finite-difference ray-tracing methods mainly depend on the grid interval of the computation grid. The smaller the grid interval is, the more accurate the travel-time field will be. However, the computational efficiency is proportional to the size of the computation grid. Thus there exists a tradeoff between the computational efficiency and accuracy. In the finite-difference travel-time algorithms, it is assumed that the wavefronts are planar, which is not valid in the region near the source where the wavefronts have significant curvature. Normally the rays between the source and the grid nodes in the source region are assumed to be straight lines, and the travel-times are calculated based on these straight lines. When the grid interval is too large, the errors associated with this simplification will be considerable. PODVIN and LECOMTE (1991) used a recursive scheme to reduce this wavefront distortion to some extent. In comparison, the finite-difference scheme

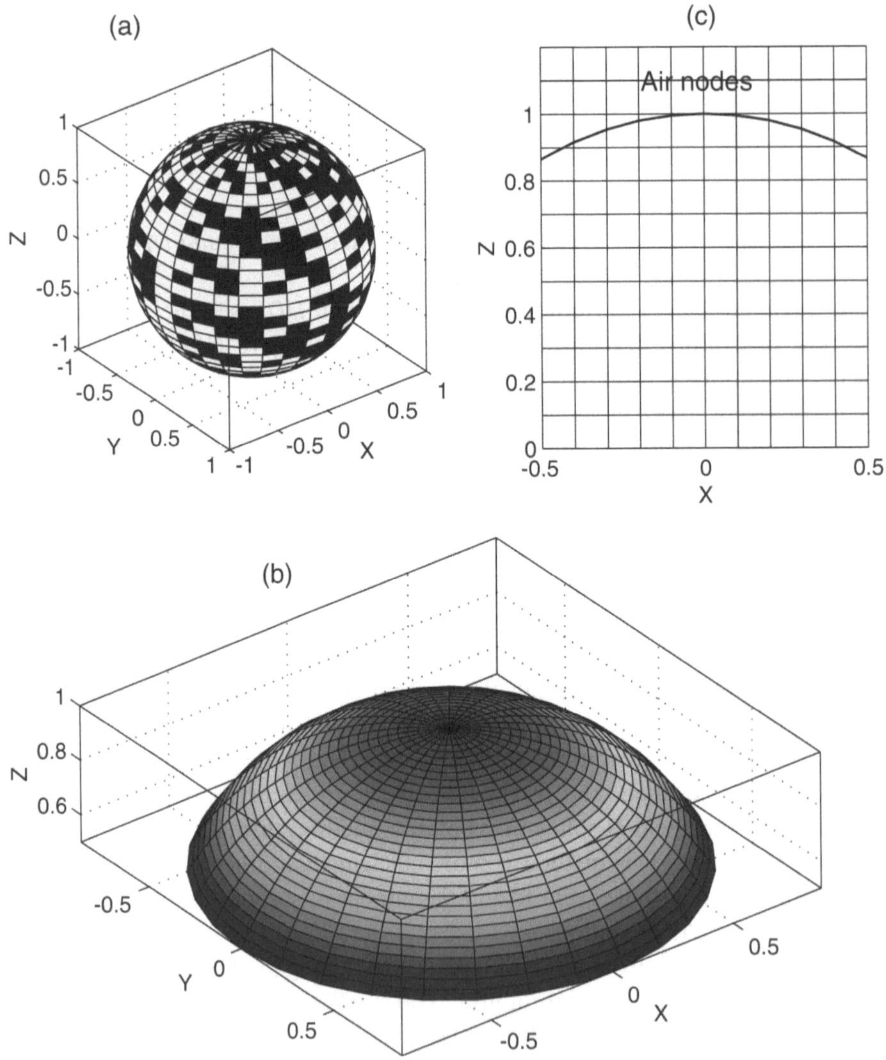

Figure 5
Grid setup for regional scale double-difference tomography. (a) The whole earth is inserted into a cubic box. (b) A rectangular box covers the region of interest. (c) The representation of (b) in the X-Z plane. The surface of the Earth is shown as a thick line. A regular grid is used for the finite-difference travel-time algorithm.

proposed by HOLE and ZELT (1995) does not have such a scheme to deal with the wavefront distortion near the source region.

Recently, ROECKER (pers. comm., 2003) proposed to calculate the travel-time field in the source region with a much finer grid, and then combine this source region travel-time field with that outside the source region. In this way, the effect of wavefront curvature in the source region will be greatly reduced and the accuracy in the travel-time field will also be improved without sacrificing too much computational time and memory space. This strategy will be incorporated in future versions of *tomoFDD*.

To adapt DD tomography to the regional scale, the curvature of the Earth must be taken into account. Following FLANAGAN *et al.* (2000), we solve this problem by parameterizing a spherical surface inside a Cartesian volume of grid nodes. Figure 5a shows the case of putting the Earth into a cube; this model can be used for global seismic tomography. For regional seismic tomography, we are only interested in a part of the Earth, so we construct a rectangular box covering the part of interest (Figs. 5b and c). The coordinate center is placed at the surface of the Earth, the positive X and Y axes point in the directions of North and West, respectively, and the positive Z axis points downward. The grid nodes above the Earth's surface (air nodes) are given the velocity for P waves traveling in air (Fig. 5c). As a result, all the rays travel inside the Earth.

There are two different grids in the algorithm: one is the inversion grid that can be nonuniform but regular and the other is the computational grid that must be uniform. The uniform grid is used to calculate the travel-time field using the finite-difference algorithms and the velocity values at its nodes are interpolated from the non-uniform inversion grid through linear B-spline interpolation. If an inversion grid node is 2 or more km above the Earth's surface (allowing for topography), then it is treated as an air node and its value is fixed during the inversion. First we treat each station as a source and calculate travel times to all velocity nodes in the volume. The travel time from a station to each earthquake is interpolated from its 8 neighboring nodes through linear B-spline interpolation. The ray path from the earthquake to the station is found iteratively with increments opposite to the travel-time gradient. After these set-ups, the ray paths and the related partial derivatives are calculated by the finite-difference method. The code has the option to choose the finite-difference method either by PODVIN and LECOMTE (1991) or by Hole and ZELT (1995). In our experience, the former method treats the source region problem better than the latter method, but it calculates the travel-time field more slowly. Our future work is to incorporate Roecker's multi-grid method (pers. Comm., 2003) into the *tomoFDD* algorithm to better deal with the source region problem.

Adaptive-mesh DD Tomography (tomoADD)

The algorithms *tomoDD* and *tomoFDD* have already proven to be valuable tools for tomography applications (see next section), but both of them are based on a regular inversion grid. We recently developed an adaptive-mesh DD tomography method based on tetrahedral and Voronoi diagrams to automatically match the inversion mesh to the data distribution (ZHANG and THURBER, 2005). Both tetrahedral and Voronoi diagrams have been shown to be powerful tools to represent spatial relationships in three dimensions and they allow a more flexible representation of the model. For example, model volumes of widely varying sizes with complex distributions are easily implemented, and it is more convenient to build parameterizations containing particular interfaces, on which the nodes can be distributed.

Linear and natural-neighbor interpolation methods can be derived for tetrahedral and Voronoi diagrams, respectively. Linear interpolation uses 4 tetrahedron nodes to interpolate the velocity/slowness values at any point and it has the advantage of calculating the interpolating basis functions easily and quickly. However, the linear interpolation function is not continuously differentiable, which is a desired property for some applications. In comparison, the natural-neighbor interpolation method interpolates the value at any point from its n natural neighbors. The natural-neighbor interpolation function guarantees continuity in first and second derivatives except at the nodes (SAMBRIDGE *et al.*, 1995). Our algorithm allows the use of either approach.

For the linear interpolation method, we use a direct method to calculate the partial derivatives of travel times with respect to the slowness model parameters in the same way as in the regular-grid DD tomography. That is, a ray is divided into many small segments, with the mid-point of each segment located in a specific tetrahedron. Each segment length can be attributed to the relevant irregular mesh nodes in proportion to interpolating weights at the mid-point. In the case of natural neighbor interpolation, however, it is extremely slow to calculate the interpolating weights and thus the partial derivative directly, due to the large number of nodes that contribute to the interpolated value at any given point. Instead, we use an indirect way to calculate the partial derivatives of travel times with respect to the model slowness parameters, by projecting them from the regular computational grid to the irregular inversion mesh. The finer the regular grid is, the more accurate the partial derivatives are.

The inversion is started from a regular inversion grid, equivalent to that of *tomoDD*. We then construct a tetrahedral or Voronoi diagram using the Quick Hull algorithm (BARBER *et al.*, 1996) around the starting regular inversion grid that is randomly perturbed by a very small amount so that the nodes are not located on the same plane. A regular computational grid that remains fixed during the inversion is constructed to be the same as the starting regular inversion grid or finer. Ray paths between events and stations are traced using the ART-PB ray-tracing algorithm (UM and THURBER, 1987) and saved for later use in defining the adaptive mesh. In the process, we calculate the derivative weight sum (DWS) values (THURBER and EBERHART-PHILLIPS, 1999) on the inversion mesh nodes using the saved rays. Threshold DWS values are set to add or remove nodes. Currently we add nodes by inserting one additional node into the middle of a tetrahedron when the sum of DWS values on its 4 nodes satisfies a predefined threshold value (ZHANG and THURBER, 2005). The threshold checking is repeated to determine the inversion mesh to be used for the current iteration of simultaneous inversion (ZHANG and THURBER, 2005). Once the inversion mesh is determined to be data adaptive, a new tetrahedral or Voronoi diagram is constructed and the partial derivatives of the travel times with respect to the new set of inversion mesh nodes are calculated using the saved rays for the construction of the seismic tomography equations. After each simultaneous inversion, the velocity values on the irregular inversion mesh nodes and the regular computational grid nodes are updated. For subsequent simultaneous inversions, the inversion mesh is again updated following the same procedure to better match with the ray distribution, which will change as the velocity model changes and hypocenters move.

The adaptive mesh method considers the different ray distribution between the P and S waves. Due to the relative simplification of picking P arrivals compared to S, normally there will be more data and hence more inversion mesh nodes for P waves than S waves. At each simultaneous inversion, the same procedure as discussed above is used to construct different inversion meshes for P and S waves.

The above procedure can be applied directly for absolute data with the inversion mesh built according to the absolute ray paths. In the case of using a combination of absolute and differential data, however, the model derivatives from the differential data are included in the mesh refinement process, so the inversion mesh will be distributed more finely near the source region where these derivatives are largest. In this case, we keep those irregular inversion mesh nodes outside the source region and fix their velocity values if their DWS values are smaller than a threshold value (ZHANG and THURBER, 2005). In this way, we preserve the velocity structure outside the source region that is determined by the absolute data information while refining the velocity structure near the source region.

Once the inversion mesh is set up, we can calculate the partial derivatives of travel times with respect to the event locations and slowness model parameters. In a similar way to the case of the regular inversion grid, we can then construct the linear system of equations to solve for the event locations and slowness perturbations at the irregular mesh nodes.

Review of DD Tomography Applications

DD tomography has been applied to a suite of problems including characterizing fault zone structure, imaging high-resolution structure of subduction zones, and determining the velocity structure of volcanoes. In these studies, we have also carried out a variety of tests of the DD tomography method using synthetic data sets. We discuss some examples of each type of application below, including references to associated synthetic tests.

1. Characterizing Fault Zone Structure

Our first application of DD tomography (*tomoDD*) was to study the Hayward fault in California (ZHANG and THURBER, 2003). This study used 17,955 differential arrival times calculated from waveform CC, 767,127 catalog differential times constructed directly from the absolute catalog arrival times, and 20,257 absolute catalog times for P waves (ZHANG and THURBER, 2003). In comparison to standard tomography using absolute arrival times, the DD tomography model shows a sharper velocity contrast across the fault with faster velocity to the west and slower velocity to the east of the fault. DD tomography also produces a sharp picture of event locations similar to DD location (WALDHAUSER and ELLSWORTH, 2001) but with different absolute locations. We also noticed some differences in relative locations between the two DD methods and attribute the differences to the velocity heterogeneity not considered in DD location. Without considering velocity complexity in DD location, partial derivatives of arrival times with respect to event locations may be biased and thus relative event locations may also be biased (MICHELINI and LOMAX, 2004). A synthetic test based on an idealized model of the velocity structure of the San Andreas fault in central California was also presented to show DD tomography is able to better characterize the velocity structure near the source region and the absolute event locations were more accurately located (ZHANG and THURBER, 2003). THURBER *et al.* (2004) applied *tomoDD* to study the San Andreas fault zone near Parkfield using catalog picks. Compared to the simul2000 algorithm, *tomoDD* resulted in more tightly clustered earthquake locations. Though the model results from both algorithms were comparable in most areas, the *tomoDD* model showed more detailed features near the fault zone (THURBER *et al.*, 2004).

TomoDD has been applied to the study of a number of fault zones in Japan. OKADA *et al.* (2004) used *tomoDD* to image the detailed structure of focal areas of the 1995 M7.3 southern Hyogo (Kobe) earthquake, 2000 M7.3 western Tottori earthquake and 2003 M6.4 northern Miyagi earthquake. They found that large slip areas (asperities) corresponded to the high velocity zones for all three earthquakes. An independent study by ENESCU and MORI (2004) also found high velocity anomalies around the 2000 M7.3 western Tottori mainshock area using *tomoDD* and waveform cross correlation. The application of *tomoDD* to the recent 2004 M6.6 mid-Niigata prefecture earthquake showed lower velocity for the hanging wall (to the west of the focal area), higher velocity for the footwall (to the east of the focal area), and aftershocks are distributed along a zone where seismic velocity changes abruptly. The results suggest the earthquake sequence occurred along faults that were normal in the Miocene and then recently reactivated as reverse faults under compression (OKADA *et al.*, 2005). This study used manually picked arrival times and associated differential times obtained from a temporary dense network composed of 54 stations around the focal area. KATO *et al.* (2005) independently applied *tomoDD* to the arrival times from 716 aftershocks collected from 14 temporary seismic stations immediately deployed in and around the source region of the 2004 mid-Niigata prefecture earthquake. Their results also showed the aftershocks were distributed along a clear boundary between the low and high velocity structures corresponding

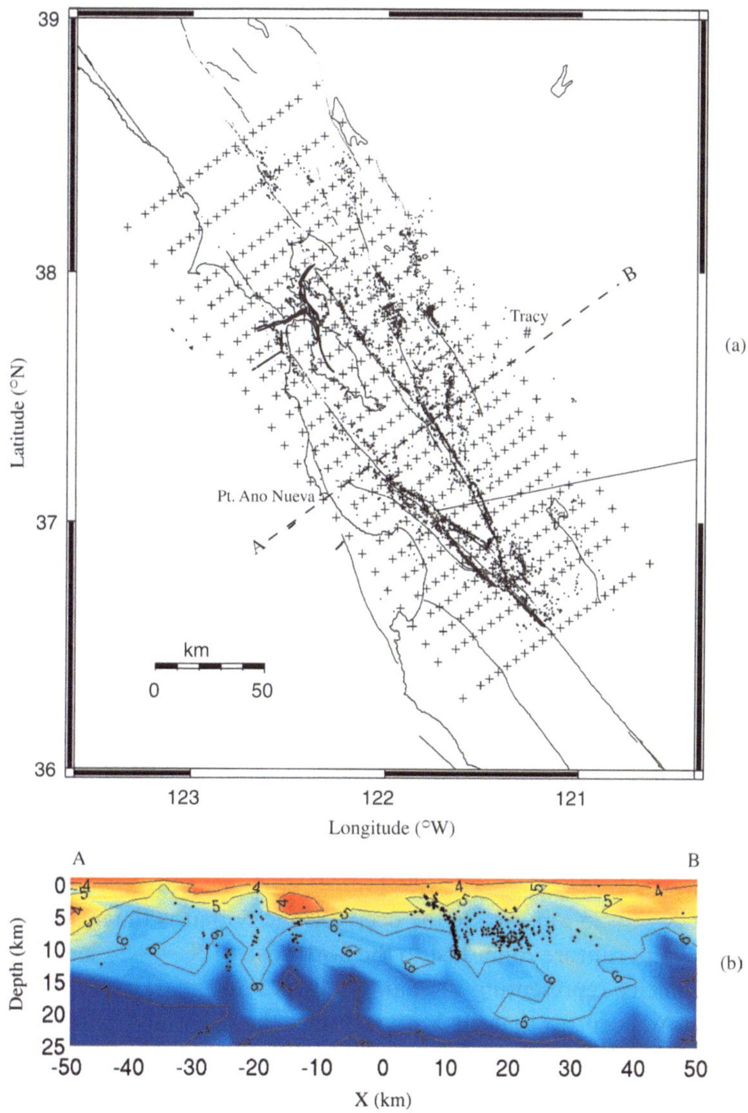

Figure 6
(a) Event and inversion grid distribution map of Northern California. Events are indicated by dots and inversion grid nodes are indicated by crosses. Thin lines indicate fault traces. The coordinate center is located at latitude 31.5° N and longitude 122° W. (b) NE-SW cross section of P-wave velocity model from Pt. Ano Nueva in the southwest to near the town of Tracy in the northeast.

to the hanging wall and footwall, respectively. TAKEDA *et al.* (2004) studied the Atotsugawa fault in central Japan using *tomoDD*. They found a low-velocity zone up to 10 km depth in the cross section along the fault and the earthquakes distributed along the upper boundary of this low-velocity zone.

We also applied the adaptive-mesh version of DD tomography (*tomoADD*) to the Parkfield, California data set (ZHANG and THURBER, 2005) using both linear and natural neighbor interpolation methods. Both methods resulted in more clustered earthquakes than standard tomography, and the events are concentrated where the velocity contrast is sharpest. The velocity models using both interpolation methods are characterized by a clear velocity contrast across the fault and they are comparable in most areas, with the velocity model using natural-neighbor interpolation being slightly smoother. We also found a velocity reversal at ~4 km depth to the southwest of the fault when using only the absolute arrival times, but this anomaly disappears when including the differential data in the inversion. Synthetic tests indicate that this velocity reversal could be caused by the noisier absolute data. Thus including the differential data in the inversion may be helpful for removing some velocity features that could be artifacts resulting from the noisier absolute data.

Our current work involves the development of a regional crustal model for Northern California (Fig. 6a). The model will be used for improving earthquake locations and for generating synthetic waveforms and estimating ground motions for the 1906 San Francisco earthquake. Approximately 6000 earthquakes with arrival data extracted from the Northern California Earthquake Data Center (NCEDC) and about 5,500 shots or airgun blasts are used in this study. 466,368 absolute catalog times and 1,629,973 differential times were used in the inversion. The inversion grid was rotated so that the +X and +Y axes are approximately SAF-normal and SAF-parallel, respectively. The grid intervals were 5 km in the X direction and 10 km in the Y direction. In the Z direction, grid nodes were positioned at −1, 0, 1, 3, 6, 9, 12, 15, 20, 25 and 30 km. For this study, we used the *tomoFDD* algorithm considering the relatively large size of the study region. An example cross section through a preliminary model with 5 × 10 km horizontal gridding is shown in Figure 6b. The section runs from Pt. Ano Nueva in the southwest to near the town of Tracy in the northeast. At X = −25 km, we see the deep seismicity that is the northwestern edge of the Loma Prieta rupture zone. Just to the northeast, at X = −20 km, is the diffuse, shallower seismicity representing the San Andreas fault, marked by a distinct low-velocity zone. Adjacent to that, at X = −15 km, the deep, low-velocity, Cupertino Basin is evident, with an active fault situated beneath it. On the northeast side of the section, the northeast-dipping Calaveras Fault seismicity is evident at X = +10 km, associated with the edge of a dipping high-velocity block. There is earthquake activity on both sides of the Calaveras here — the shallower activity to the southwest appears to mark the beginning of the Hayward fault, whereas the deeper activity to the northeast (X = +15 to +25 km) is on an obliquely striking fault connecting the Calaveras and Greenville faults.

These applications of DD tomography to a variety of fault zones showed that DD tomography is able to better relocate events and resolve the finer details of fault zone structure. The relationship between the event distribution and velocity contrast across the fault for some fault zones is more clearly shown. More accurate event locations and finer velocity structure will be more helpful for understanding the factors controlling the occurrence of earthquakes in the fault zone. We note that by itself, however, DD tomography is not a solution for characterizing sharp material interfaces. One can try to model secondary arrivals that provide evidence for an interface (reflections or refractions) but that is a separate issue from what differential times can provide (BEN-ZION *et al.*, 1992).

2. Imaging High-resolution Structure of Subduction Zones

Subduction zones are one of the most important components of the Earth's plate tectonic system. Our first application of the regional-scale version of DD tomography (*tomoFDD*) was to image high-resolution subducting-slab structure beneath northern Honshu, Japan using catalog picks (ZHANG *et al.*, 2004). Previous seismic tomography studies for the same region (ZHAO *et al.*, 1992; NAKAJIMA *et al.*, 2001)

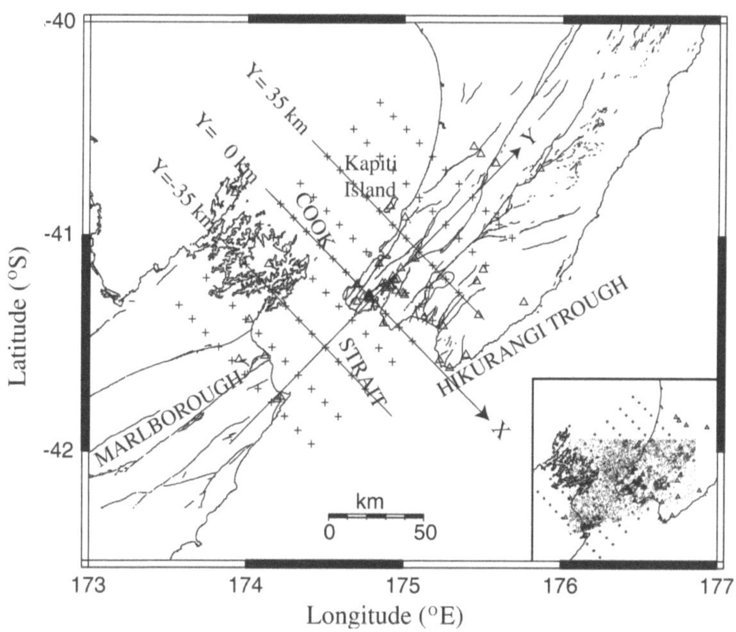

Figure 7
Event and station map for our New Zealand study. Events are indicated by dots in the inset map. Stations are indicated by triangles and inversion grid nodes are indicated by crosses. Positive Y points northeast, positive X points southeast, and positive Z points downward.

could not resolve the detailed internal structure of the down-going slab using standard tomography; their models showed the slab as a relatively homogeneous high velocity anomaly. Our model, however, showed for the first time that the interior of the slab actually has a highly heterogeneous velocity structure. The upper plane of earthquakes lies in a region with relatively low Vp and Vs values and average to high Vp/Vs ratios (1.72–1.85) that may correspond to the transformations of metabasalt and metagabbro to blueschist. The lower plane of earthquakes exists in a region of relatively low Vp and high Vs values and low Vp/Vs ratios that could be due to dehydration reactions of serpentine and other hydrous minerals. The region between the two planes of earthquakes has relatively high Vp/Vs ratios that indicate partial hydration of the normal dry mantle. Synthetic tests with and without key slab features found in this study using the same data distribution as the real data showed the model was well resolved. SHELLY *et al.* (2004) found similar results when they applied *tomoDD* and waveform cross correlation to the subduction zone near Ibaraki, Japan that is located to the south of ZHANG *et al.* (2004)'s study region.

We further applied *tomoFDD* to study the transition zone from oblique subduction to active continent-continent collision in the Wellington region, New Zealand, where a double seismic zone is also present (Fig. 7). We used 5471 earthquakes spanning the transition region between the North and South Islands that were recorded on 56 stations during the period 1990–2001. For the chosen events and stations, there are 58,082 catalog P data, 299,948 P differential catalog data, and 427,771 WCC P differential times that were verified using a bispectrum analysis method (DU *et al.*, 2004). We adopt a coordinate center of latitude −41.3°, longitude 174.8°, and depth 0 km (with respect to sea level). The coordinate system is rotated in the horizontal plane by 45° clockwise to be consistent with the strike of the subducting slab. The inversion grid intervals are 10 km in the X direction and 15 to 20 km in the Y direction (Fig. 7). The node spacing in the vertical direction is 4.5 km. In addition to a low-velocity anomaly present in the upper part of the slab corresponding to the subducting oceanic crust and sediments, our model also shows varying features deeper inside the slab (Fig. 8). For the cross sections beneath the North Island, the P-wave velocity is relatively high between the two planes, and it is relatively low in the lower part of the slab, partly in association with the lower plane of seismicity (Fig. 8). Similar to the subducting slab beneath northern Honshu, Japan (ZHANG *et al.*, 2004), the low Vp region associated with the lower plane of seismicity might be caused by dehydration reactions of serpentine. However, the features are quite different between the cross sections beneath the North versus the South Island (Fig. 8). In the South Island cross sections, the velocity is relatively low and there is almost no seismicity between the two planes of seismicity (Fig. 8). The lower plane of seismicity is associated with a relatively high velocity region (Fig. 8). One possible cause of such features is that the water released from the dehydration reactions of serpentine escaped from the lower plane and entered into the region between the two planes. As a result, the minerals in the lower plane may be mainly composed of

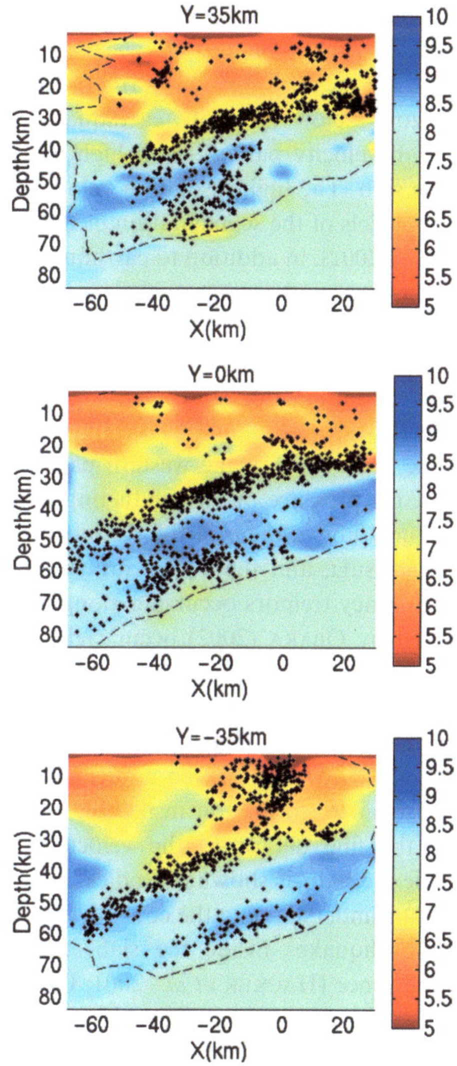

Figure 8
NW-SE cross sections of the New Zealand Vp structure at Y = −35, 0, and 35 km. Structure above the dashed line is meaningful, where the DWS for each node exceeds 15.

forsterite and enstatite, which have relatively high Vp (JI and WANG, 1999). The water released into the region between the two planes decreases Vp. The different features may be related to the transition from subduction to a strike-slip plate boundary from northeast to southwest, and stagnation of the subducting slab in the southwest.

We also found velocity variations in and around the subducting slab for the Kodiak, Kenai, and McKinley blocks for the Alaska subduction zone using *tomoFDD* (Zhang *et al.*, 2004). For the Kodiak block, a low velocity zone is present within the slab right below the high velocity zone where most of the Wadati-Benioff Zone (WBZ) earthquakes are located. In comparison, for the Kenai block, the velocity is higher in the zone below the WBZ earthquakes. The McKinley block shows a more complex slab velocity pattern, with high and low velocity zones at different depths within the slab. The high-resolution velocity structure may provide additional constraints for models of the segmentation of the Alaska subduction zone (Ratchkovski and Hansen, 2002), in addition to earthquake locations, volcanic arc geometry and composition, etc., and may provide a possible explanation for the segmentation.

Another application of *tomoFDD* is to image crust, mantle, and slab structure around the Shikoku district, Japan, and to study its relationship to the occurrence of nonvolcanic deep tremors (Nakajima *et al.*, 2004). High-velocity zones are associated with the deep seismic zone beneath the western part of Shikoku and the Kii Channel that may indicate the presence of the subducting Philippine Sea plate. Low velocity zones lie immediately above them in the western part of Shikoku and the central part of the Kii Peninsula, and the Vp/Vs ratios are greater than 1.8. The nonvolcanic deep, low-frequency tremors occurring along the strike of the subducting Philippine Sea detected by Obara (2002) occur right in and around the low-velocity and high Vp/Vs zones, lending support to the idea that fluids are involved in the tremor. This study, along with those discussed above, shows strong evidence for dehydration reactions in the subducting slab.

Thanks to the ability of DD tomography to resolve the finer structure near the source region and the seismicity distribution inside the subducting slab, it is now possible to characterize the fine details of the velocity structure and accurate earthquake locations inside the slab, as shown in studies discussed above. This will in turn be extremely helpful for understanding the constitution of the slab, the cause of the intermediate depth earthquakes inside the slab, the fluid distribution and recycling, and tremor occurrence (Hacker *et al.*, 2001; Obara, 2002).

3. Determining the Velocity Structure of Volcanoes

Brown *et al.* (2004) applied *tomoADD* to study Mt. Spurr Volcano, Alaska, located approximately 120 km west of Anchorage in the Cook Inlet, using a data set from 1991 to 2004. They found a low velocity body beneath the crater from $\sim$1 km above sea level to $\sim$10 km below sea level. This low velocity body is directly associated with the steeply dipping 1992 eruption seismicity that may reflect the magmatic conduit at depth and the presence of a shallow hydrothermal system southeast of Crater Peak. Monteiller *et al.* (2004) imaged a detailed structure of the Kilauea Volcano magmatic system using DD tomography based on a Bayesian approach.

Another example is the imaging of the Mt. Etna plumbing system using *tomoDD*. We chose 278 earthquakes that were recorded by 41 stations of Instituto Nazionale di

Geofisica e Vulcanologia - Sezione di Catania (INGV-CT) seismic network (Fig. 9a) during the July 12–18 eruption period, similar to the data set used by PATANÉ *et al.* (2002). A hypothesis for this eruption is that magma was continuously injected from a depth of 6 to 10 km into the shallow magma reservoir and accumulated in the upper part of Mt. Etna's plumbing system during 1994 to 2001 (PATANÉ *et al.*, 2003). This study used ~5200 catalog P and S times and ~35,000 catalog P and S differential times, among which S arrival times are about 1/6 the number of P arrival

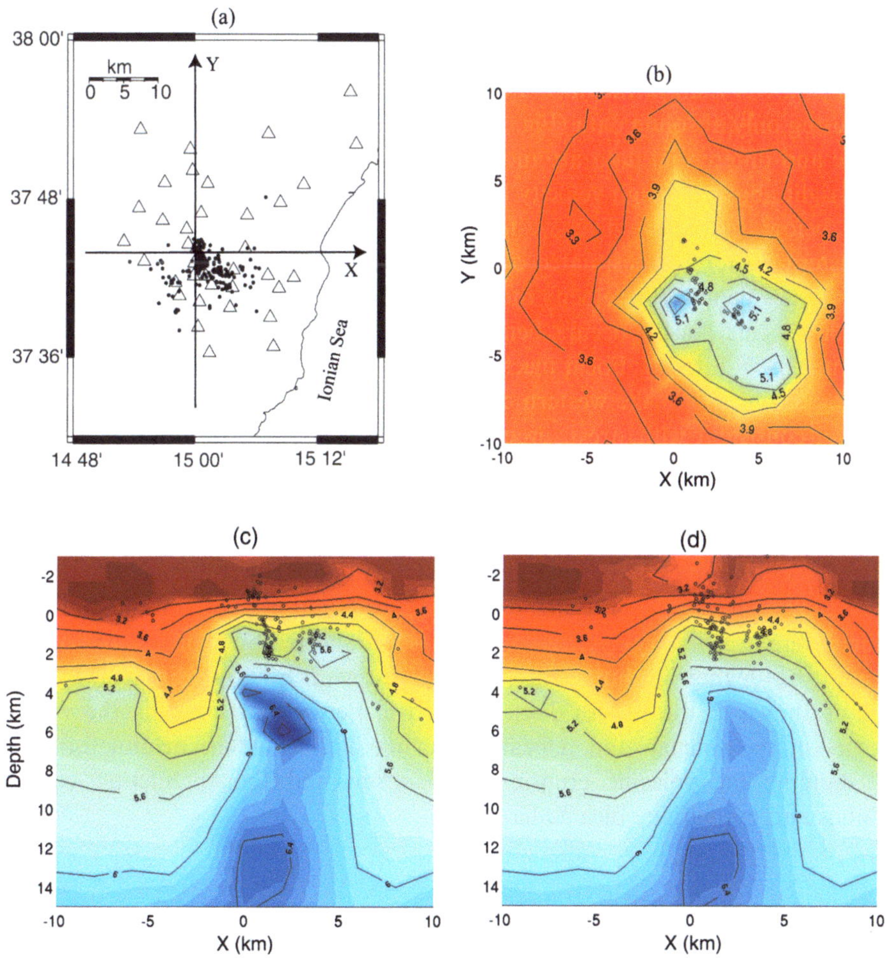

Figure 9
(a) Event and station map for Mt. Etna. The summit crater is located at latitude 37.75°N and longitude 15.0°E. (b) Horizontal section of P-wave velocity model at depth of 1 km with earthquakes within 0.5 km plotted as dots. (c) W-E cross section of P-wave velocity model at Y = −2 km determined using both absolute and differential data. (d) W-E cross section of P-wave velocity model at Y = −2 km using only the absolute data. For (c) and (d), earthquakes within 1 km of the cross section are plotted as dots.

times. The inversion grid intervals in the horizontal and vertical directions are 2 km and 1 km, respectively. In the horizontal section at depth of 1 km (Fig. 9b), a high P-wave velocity zone of 4.8 – 5.1 km/s exists beneath the southeastern part of the summit area, similar to the results of PATANÉ *et al.* (2002). This high-Vp body extends to a depth up to 15 km as shown in the W-E cross-section of Y = −2 km in the velocity model (Fig. 9c), and is interpreted as a mainly solidified intrusive body (PATANÉ *et al.*, 2003).

For comparison, Figure 9d shows the same W-E cross section of Y = −2 km from the velocity model calculated using only absolute data. The velocity model is more sharply defined in the inversion using both absolute and differential times. In particular, the high velocity body shows a greater velocity contrast and more detailed features (Fig. 9c) than in the absolute-data-only model (Fig. 9d). Compared to the results using only absolute data (Fig. 9d), the event locations determined using both absolute and differential data are more clustered. Two clusters of earthquakes are present right above the high velocity body from depths of ~1 km above sea level to ~3 km below sea level. The western cluster is more concentrated than the southeastern cluster and is situated on the edge of a knob of high velocity (Fig. 9c), which is not evident in the velocity model using only absolute data (Fig. 9d; PATANÉ *et al.* 2002). This cluster occurs beneath a surface fractures field and shows an approximately N-S trend. Focal mechanism analysis showed that the P axes and T axes for earthquakes in the western cluster are mostly oriented approximately N-S and W-E, consistent with stress directions induced by a dike intrusion (MUSUMECI *et al.*, 2004). The southeastern cluster shows an approximately NNW-SSE alignment and might be related to an aborted intrusion or active NNW-SSE faults (MUSUMECI *et al.*, 2004).

For most volcanoes in the world, it is generally difficult to image the detailed structure of their plumbing systems, due to small numbers of stations and poor station coverage geometry. DD tomography is less affected by poor station coverage because it has higher resolution near the source region by using differential arrival times, as shown in the studies discussed above. Better characterizing the velocity structure of the volcano and relocating earthquakes would be helpful for better understanding the volcanic eruption mechanisms and thus predicting volcanic eruptions and mitigating volcanic hazards.

Summary and Future Work

We conducted a resolution and uncertainty analysis based on a 2-D synthetic model for different location and tomography systems of equations, using absolute data (i.e., standard location and tomography), the differential data (i.e., DD location and tomography), and both of them to determine event locations and/or velocity structure. The study shows that the system using the differential data has lower model uncertainties for both location and most slowness model parameters, but has lower model resolution for some boundary slowness model parameters. Including the absolute data with a relatively small weighting into the differential system is useful to stabilize the system without sacrificing the low model uncertainties and high model resolution that the differential system can provide.

Methods and applications of three versions of the DD tomography algorithm are presented: *tomoDD*, *tomoFDD* and *tomoADD*. *TomoDD* is a local-scale version that assumes a flat Earth model and uses an ART-PB ray-tracing algorithm. The regional-scale version *tomoFDD* considers the curvature of the Earth and uses a finite-difference travel-time algorithm that can better deal with sharp velocity contrasts and discontinuities. Both *tomoDD* and *tomoFDD* are based on a regularly distributed inversion grid. In contrast, *tomoADD* adapts the inversion mesh to match with the data distribution based on tetrahedral and Voronoi diagrams, such that the ray sampling densities on inversion mesh nodes are more uniform than the regular grid case. This approach provides an automated way to take full advantage of the ability to resolve fine-scale structure in the source region using the differential times. We then discuss some examples of applying DD tomography to characterize fault zone structure, image high-resolution structure of subduction zones, and determine the velocity structure of volcanoes.

There are some issues to be considered in our future work, as follows:

1. Currently we have finished the implementation of the adaptive-mesh scheme on the flat-model version of DD tomography (ZHANG and THURBER, 2005). We will implement an adaptive-mesh scheme for the regional-scale version of DD tomography.

2. All three versions of the DD tomography algorithm, *tomoDD*, *tomoFDD* and *tomoADD*, directly solve for P- and S-wave velocities using absolute and differential arrival times. Knowing 3-D Vp/Vs variations is valuable in order to have a more complete characterization of the mechanical properties and geological identity of crust and upper mantle materials (THURBER, 1993). Vp/Vs variations could be determined directly from the Vp and Vs models if they have essentially identical quality. It has been observed that in cases where S-wave arrival data are less numerous and of poorer quality than P-wave data, however, that Vs would be not as well resolved as Vp, making the interpretation of Vp/Vs variations difficult (EBERHART-PHILLIPS, 1990; WAGNER *et al.*, 2005). For this reason, it is generally not appropriate to obtain Vp/Vs ratios by dividing Vp by Vs directly. Alternatively, Vp/Vs variations can be determined by the inversion of S-P time differences (WALCK, 1988; THURBER, 1993). We plan to add this functionality to the DD tomography algorithm to directly invert Vp/Vs ratios using both absolute and differential S-P time differences.

3. The current DD tomography algorithms cannot directly estimate model resolution and uncertainty due to the substantial amount of data used for the system. In ZHANG and THURBER (2003), the model resolution and uncertainty were estimated

by means of the solution technique in the *simul2000* algorithm (THURBER and EBERHART-PHILLIPS, 1999), by inputting the velocity model inverted from DD tomography and using only the absolute times. This technique provides a realistic but conservative estimate of the DD tomography solution quality. Another way of assessing the tomography solution quality is to use a synthetic test or restoration test employing the same data distribution and inversion scheme as the real data (ZHAO *et al.*, 1992; ZHANG *et al.*, 2004). Some sensitivity tests such as a checkerboard test and statistical analysis methods such as "jackknifing" and "bootstrapping" can also be used to estimate the model resolution and uncertainty, respectively (NOLET *et al.*, 1999). However, the sensitivity tests suffer the shortcomings of measuring the sensitivity only with respect to fixed cell or grid patterns (NOLET *et al.*, 1999). The "jackknifing" and "bootstrapping" methods are computationally very expensive and of questionable use for tomographic systems (NOLET *et al.*, 1999). We are now investigating the applicability of several resolution and uncertainty estimation methods for massive tomographic systems proposed by ZHANG and MCMECHAN (1995), MINKOFF (1996), NOLET *et al.* (1999), and YAO *et al.* (1999).

4. MIYAZAWA and KATO (2004) raise questions about the differences between interpolation using velocity versus slowness in grid-based tomography algorithms. We will investigate the effect of the choice of velocity versus slowness on DD tomography both theoretically and empirically.

Acknowledgements

We are extremely grateful to our collaborators for sharing their results with us. We thank INGV-CT for allowing us to use their Mt. Etna 2001 eruption data. We especially thank Felix Waldhauser and Steve Roecker for their assistance and helpful comments in the early stage of *tomoDD* algorithm development. Megan Flanagan helped us to realize the *tomoFDD* algorithm using part of her code. Certain figures were constructed using GMT (WESSEL and SMITH, 1991). This material is based upon work supported by National Science Foundation grants EAR-9814192 (Continental Dynamics), EAR-0125164 (Geophysics), EAR-0346105 (EarthScope), and EAR-0409291 (Geophysics).

REFERENCES

ASTER, R.C., BORCHERS B., and THURBER C.H., *Parameter Estimation and Inverse Problems*, (Elsevier Academic Press, Burlington, MA, 2005) 301 pp.

BARBER, C.B., DOBKIN, D.P., and HUHDANPAA, H.T. (1996), *The Quickhull algorithm for convex hulls*, Trans. Mat. Software *22*, 469–483.

BEN-ZION, Y., KATZ, S., and LEARY, P. (1992), *Joint inversion of fault zone head waves and direct P arrivals for crustal structure near major faults*, J. Geophys. Res. *97*, 1943–1951.

BROWN, J., PREJEAN, S., ZHANG, H., POWER, J., and THURBER, C. (2004), *Double difference earthquake relocation and tomography at Mount Spurr Volcano, Alaska, 1991 to 2004,* EOS Trans. AGU *85*(47), Fall Meet. Suppl., Abstract S51A–0140.

DU, W., THURBER, C. H., and EBERHART-PHILLIPS, D. (2004), *Earthquake relocation using cross-correlation time delay estimates verified with the bispectrum method,* Bull. Seismol. Soc. Am. *94*, 856–866.

EBERHART-PHILLIPS, D. (1990), *Three-dimensional P and S velocity structure in the Coalinga region, California,* J. Geophys. Res. *95*, 15343–15363.

ENESCU, B., and MORI, J. (2004), *Relocations and 3-D velocity structure for aftershocks of the 2000 W. Tottori (Japan) earthquake and 2001 Gujarat (India) earthquake, using waveform cross correlations,* EOS Trans. AGU *85*(47), Fall Meet. Suppl., Abstract S54A–05.

FLANAGAN, M.P., MYERS, S.C., SCHULTZ, C.A., PASYANOS, M.E., and BHATTACHARYYA, J. (2000), *Three-dimensional a priori model constraints and uncertainties for improving seismic location,* Proceedings of 22nd Seismic Research Review: Technologies for Monitoring the CTBT, 2–15.

HACKER, B.R., ABERS, G.A., and PEACOCK, S.M. (2001), *Subduction factory 1, Theoretical mineraology, densities, seismic wave speeds, and H2O contents*, J. Geophys. Res. *108*, B12030, doi:2001JB001129.

HOLE, J.A., and ZELT, B.C. (1995), *3-D finite-difference reflection travel times,* Geophys. J. Int. *121*, 427–434.

JI, S., and WANG, Z. (1999), *Elastic properties of forsterite-enstatite composites up to 3.0 Gpa,* Geodynamics *28*, 147–174.

KATO, A., KURASHIMO, E., HIRATA, N., SAKAI, S., IWASAKI, T., and KANAZAWA, T. (2005), *Imaging the source region of the 2004 mid-Niigata prefecture earthquake and the evolution of a seismogenic thrust-related fold,* Geophys. Res. Lett. *32*, L07307, doi:10.1029/2005GL022366.

KLEIN, F.W. (1978), *Hypocenter location program HYPOINVERSE,* U.S. Geol. Surv. Open File Rep. *78–694*, 114 pp.

KOKETSU, K., and SEKINE, S. (1998), *Pseudo-bending method for three-dimensional seismic ray tracing in a spherical earth with discontinuities,* Geophys. J. Int. *132*, 339–346.

LEE, W.H.K., and STEWART, S.W. (1981), *Principles and applications of microearthquake networks,* Adv. Geophys. Suppl. *2*, 293 pp.

MONTEILLER, V., GOT, J., VIRIEUX, J., and OKUBO, P.G. (2004), *Imaging Kilauea Volcano magmatic system using an efficient algorithm for double-difference tomography and cross-spectral time delays,* EOS Trans AGU *85*(47), Fall Meet. Suppl., Abstract S11D–03.

MENKE, W., (1989), *Geophysical Data Analysis: Discrete Inverse Theory,* revised edition, vol. 45 of International Geophysics Series (Academic Press, San Diego) 289 pp.

MENKE, W., and SCHAFF, D. (2004), *Absolute earthquake locations with differential data,* Bull. Seismol. Soc. Am. *94*, 2254–2264.

MICHELINI, A., and LOMAX, A. (2004), *The effect of velocity structure errors on double-difference earthquake location,* Geophys. Res. Lett. *31*, L09602, doi:10.1029/2004GL019682.

MINKOFF, S.E. (1996), *A computationally feasible approximate resolution matrix for seismic inverse problems,* Geophys. J. Int. *126*, 345–359.

MIYAZAWA, M., and KATO, M. (2004), *On interpolation functions in travel-time tomography,* Geophys. J. Int. *158*, 169–178.

MUSUMECI, C., COCINA, O., GORI, P.D., and PATANÉ, D. (2004), *Seismological evidence of stress induced by dike injection during the 2001 Mt. Etna eruption,* Geophys. Res. Lett. *31*, L07617, doi:10.1029/2003GL019367.

NAKAJIMA, J., HASEGAWA, A., ZHANG, H., and THURBER, C.H. (2004), *Imaging crust, mantle and slab structure around the Shikoku district, Japan, by double-difference tomography and its relationship to the occurrence of nonvolcanic deep tremors,* EOS Trans. AGU *85*(47), Fall Meet. Suppl., Abstract S51B–0161.

NAKAJIMA, J., MATSUZAWA, T., HASEGAWA, A., and ZHAO, D. (2001), *Three-dimensional structure of Vp, Vs, and Vp/Vs beneath northeastern Japan: Implications for arc magmatism and fluids,* J. Geophys. Res. *106*, 21,834–21,857.

NOLET, G., MONTELLI, R., and VIRIEUX, J. (1999), *Explicit, approximate expressions for the resolution and a posteriori covariance of massive tomographic systems,* Geophys. J. Int. *138*, 36–44.

OBARA, K. (2002), *Nonvolcanic deep tremor associated with subduction in southwest Japan*, Science *296*, 1679–1681.

OKADA, T., HASEGAWA, A., SUGANOMATA, J., ZHAO, D., ZHANG, H., and THURBER, C. (2004), *Imaging the fault plane and asperities of the 1995 southern Hyogo (Kobe) earthquake (M7.3) by double-difference tomography*, Eos Trans. AGU *85*(47), Fall Meet. Suppl., Abstract S53C–01.

OKADA, T., UMINO, N., MATSUZAWA, T., NAKAJIMA, J., UCHIDA, N., NAKAYAMA, T., HIRAHARA, S., SATO, T., HORI, S., KONO, T., YABE, Y., ARIYOSHI, K., GAMAGE, S., SHIMIZU, J., SUGANOMATA, J., KITA, S., YUI, S., ARAO, M., HONDO, S., MIZUKAMI, T., TSUSHIMA, H., YAGINUMA, T., HASEGAWA, A., ASANO, Y., ZHANG, H., and THURBER, C. (2005), *Aftershock distribution and seismic velocity structure in and around the focal area of the 2004 mid-Niigata prefecture earthquake obtained by applying double-difference tomography to dense temporary seismic network data*, Earth Planets Space *57*, 435–440.

PATANÉ, D., CHIARABBA, C., COCINA, O., and GORI, P.D. (2002), *Tomographic images and 3D earthquake locations of the seismic swarm preceding the 2001 Mt. Etna eruption: Evidence for a dyke intrusion*, Geophys. Res. Lett. *29*, doi:10.1029/2001GL014391.

PATANÉ, D., GORI, P.D., CHIARABBA, C., and BONACCORSO, A. (2003), *Magma ascent and the repressurization of Mount Etna's volcanic system*, Science *29*, 2061–2063.

PODVIN, P., and LECOMTE, I. (1991), *Finite difference computation of travel times in very contrasted velocity models: A massively parallel approach and its associated tools*, Geophys. J. Int. *105*, 271–284.

RATCHKOVSKI, N.A., and HANSEN, R.A. (2002), *New evidence for segmentation of the Alaska subduction zone*, Bull. Seismol. Soc. Am. *92*, 1754–1765.

SAMBRIDGE, M., BRAUN, J., and MCQUEEN, H. (1995), *Geophysical parameterization and interpolation of irregular data using natural neighbors*, Geophys. J. Int. *122*, 837–857.

SHELLY, D.R., BEROZA, G.C., ZHANG, H., THURBER, C.H., and IDE, S. (2004), *High-resolution subduction zone seismicity and velocity structure in Ibaraki, Japan*, EOS Trans. AGU *85*(47), Fall Meet. Suppl., Abstract S43D–04.

TAKEDA, T., KUWAHARA, Y., MIZUNO, T., IMANISHI, K., OKADA, T., ITO, K., WADA, H., and HARYU, Y. (2004), *Crustal structure and micro-seismic activity along the Atotsugawa Fault system, central Japan*, Eos Trans. AGU *85*(47), Fall Meet. Suppl., Abstract S13D–1103.

THURBER, C.H., *Local earthquake tomography: Velocities and Vp/Vs theory*, in *Seismic Tomography: Theory and Practise* (eds. H.M. Iyer and K. Hirahara) pp. 563–583 (Chapman and Hall, New York 1993)

THURBER, C.H. (1992), *Hypocenter velocity structure coupling in local earthquake tomography*, Phys. Earth. Planet. Int. *75*, 55–62.

THURBER, C.H. (1983), *Earthquake locations and three-dimensional crustal structure in the Coyote Lake area, central California*, J. Geophys. Res. *88*, 8226–8236.

THURBER, C.H., and EBERHART-PHILLIPS, D. (1999), *Local earthquake tomography with flexible gridding*, Computers and Geosciences *25*, 809–818.

THURBER, C., ROECKER, S., ZHANG, H., Baher, S., and ELLSWORTH, W. (2004), *Fine-scale structure of the San Andreas fault and location of the SAFOD target earthquakes*, Geophys. Res. Lett. *31*, doi:10.1029/2003GL019398.

UM, J. and THURBER, C.H. (1987), *A fast algorithm for two-point seismic ray tracing*, Bull. Seism. Soc. Am. *77*, 972–986.

VIDALE, J.E. (1990), *Finite-difference calculation of travel-times in three dimensions*, Geophysics *55*, 521–526.

WAGNER, L.S., BECK, S., and ZANDT, G. (2005), *Upper mantle structure in the south central Chilean subduction zone (30° to 36°S)*, J. Geophys. Res. *110*, B01308, doi:10.1029/2004JB003238.

WALCK, M.C. (1988), *Three-dimensional Vp/Vs variations for the Coso region, California*, J. Geophys. Res. *93*, 2047–2051.

WALDHAUSER, F. (2001), *hypoDD: A computer program to compute double-difference hypocenter locations*, U.S. Geol. Surv. Open File Rep. *01–113*, 25 pp.

WALDHAUSER, F., and ELLSWORTH, W.L. (2000), *A double-difference earthquake location algorithm: Method and application to the Northern Hayward fault, California*, Bull. Seismol. Soc. Am. *90*, 1353–1368.

WESSEL, P., and SMITH, W.H.F. (1991), *Free software helps map and display data*, EOS Trans. AGU *72*, 441 and 445–446.

WOLFE, C. J. (2002), *On the mathematics of using difference operators to relocate earthquakes,* Bull. Seismol. Soc. Am. *92*, 2879–2892.

YAO, Z.S., ROBERTS, R.G., and TRYGGVASON, A. (1999), *Calculating resolution and covariance matrices for seismic tomography with the LSQR method,* Geophys. J. Int. *138*, 886–894.

ZHANG, J., and MCMECHAN, G.A. (1995), *Estimation of resolution and covariance of large matrix inversions,* Geophys. J. Int. *121*, 409–426.

ZHANG, H., and THURBER, C. (2005), *Adaptive-mesh seismic tomography based on tetrahedral and Voronoi diagrams: Application to Parkfield, California,* J. Geophys. Res. *110*, B04303, doi:10.1029/2004JB003186.

ZHANG, H., RATCHKOVSKI, N., THURBER, C., and HANSEN, R. (2004), *High-resolution seismic velocity structure of the Alaska subduction zone revealed by double-difference tomography*, EOS Trans. AGU *85*(47), Fall Meet. Suppl., Abstract S51B–0168.

ZHANG, H., THURBER, C.H., SHELLY, D., IDE, S., BEROZA, G.C., and HASEGAWA, A. (2005), *High-resolution subducting-slab structure beneath northern Honshu, Japan, revealed by double-difference tomography, Geology 32*, 361–364, doi: 10.1130/G20261.

ZHANG, H., and THURBER, C.H. (2003), *Double-difference tomography: The method and its application to the Hayward fault, California,* Bull. Seismol. Soc. Am. *93*, 1875–1889.

Zhao, D., Hasegawa, A., and Kanamori, H. (1992), *Tomographic imaging of P- and S-wave velocity structure beneath northeastern Japan,* J. Geophys. Res. *97*, 19, 909–19,928.

(Received April 1, 2005; accepted September 15, 2005)
Published Online First: February 8, 2006

Pure appl. geophys. 163 (2006) 405–430
0033–4553/06/030405–26
DOI 10.1007/s00024-005-0022-x

Pure and Applied Geophysics

Estimating Crustal Heterogeneity from Double-difference Tomography

JEAN -LUC GOT,[1] VADIM MONTEILLER,[1] JEAN VIRIEUX,[2] and PAUL OKUBO[3]

Abstract—Seismic velocity parameters in limited, but heterogeneous volumes can be inferred using a double-difference tomographic algorithm, but to obtain meaningful results accuracy must be maintained at every step of the computation. MONTEILLER *et al.* (2005) have devised a double-difference tomographic algorithm that takes full advantage of the accuracy of cross-spectral time-delays of large correlated event sets. This algorithm performs an accurate computation of theoretical travel-time delays in heterogeneous media and applies a suitable inversion scheme based on optimization theory. When applied to Kilauea Volcano, in Hawaii, the double-difference tomography approach shows significant and coherent changes to the velocity model in the well-resolved volumes beneath the Kilauea caldera and the upper east rift. In this paper, we first compare the results obtained using MONTEILLER *et al.*'s algorithm with those obtained using the classic travel-time tomographic approach. Then, we evaluated the effect of using data series of different accuracies, such as handpicked arrival-time differences ("picking differences"), on the results produced by double-difference tomographic algorithms. We show that picking differences have a non-Gaussian probability density function (pdf). Using a hyperbolic secant pdf instead of a Gaussian pdf allows improvement of the double-difference tomographic result when using picking difference data. We completed our study by investigating the use of spatially discontinuous time-delay data.

Key words: Double-difference tomography, earthquake multiplets, Kilauea volcano

1. *Introduction*

Following the pioneering work of AKI and LEE (1976), earthquake travel-time tomography has been widely used to infer the spatial distribution of seismic velocities at various scales. Although this technique has been successfully applied to the modeling of large-scale patterns of crustal heterogeneity, its utility has often been severely limited by inadequate earthquake and seismic station distributions, inaccurate travel-time data, approximate travel-time computations in heterogeneous media, inadequate inversion algorithms and computer performance.

[1]Laboratoire de Géophysique Interne et Tectonophysique, Université de Savoie, 73376, le Bourget-du-Lac, France.
[2]Géosciences Azur, Sophia-Antipolis, 06560, Valbonne, France.
[3]USGS - Hawaiian Volcano Observatory, Hawaii National Park, 97560, Hawaii, U.S.A.

The description of local-scale heterogeneities in crustal structure using seismic tomography requires adequate analysis tools and high-quality data, i.e., accurate time measurements and favorable geometry. Only limited volumes of crust can be studied using natural earthquake data from permanent seismic networks; however, a great advantage of these permanent networks is they allow the accurate determination of time-delays through the application of cross-spectral analysis or waveform cross correlation (time delays hereafter referred to as "cross-spectral time-delays") to repeating earthquakes over long periods. Travel-time delay residuals are second-order quantities; they are often small and they can only be interpreted if they are accurately estimated. Double-difference relocation methods (e.g., JORDAN and SVERDRUP, 1981; POUPINET *et al.*, 1984; ITO, 1985; GOT *et al.*, 1994; SLUNGA *et al.*, 1995; SHEARER, 1997; RUBIN *et al.*, 1999; WALDHAUSER and ELLSWORTH, 2000. Also see WOLFE, 2002 for a discussion of double-difference algorithms) have shown that some knowledge of crustal structures may be gained from the accurate relocation of earthquakes. It was therefore tempting to take advantage of the accuracy provided by cross correlation in a specific, double-difference tomographic algorithm. A simple, double-difference tomography algorithm was developed by ZHANG and THURBER (2003), based on work carried out by WALDHAUSER and ELLSWORTH (2000). However, there is an apparent contradiction in using a cross correlation method with tomographical analysis, as cross correlation methods use clusters of events that occurred in a small area, whereas tomographical analysis requires widely and homogeneously distributed series of earthquakes. To resolve this contradiction, regularization has to be carried out extremely carefully. Therefore, to be effective, a double-difference tomographic algorithm based on cross-spectral time-delays requires: (1) accurately computed theoretical time-delays; (2) data with a carefully chosen spatial distribution; and (3) optimization of the inversion process.

MONTEILLER *et al.* (2005) recently set up an accurate and stable algorithm for double-difference tomography that provides (1) a unique and coherent framework for both the tomographic and double-difference relocation processes and (2) a criterion for choosing the final tomographic model, based on the introduction of an optimal amount of *a priori* information.

Cross-spectral time-delays are high-quality data from which we can expect high-quality results; however, it can be difficult to obtain large data series. It is much easier to compute travel-time differences from handpicked arrival-times (hereafter referred to as "picking-differences"), although the accuracy of the results is relatively low. In this paper, we investigate whether meaningful results can be obtained by applying double-difference tomography methods to picking-differences and whether a L_2-norm tomographic algorithm can provide accurate estimates of seismic velocities when using such data. We also determined the effect on the results of introducing various quantities of high-quality data and the extent to which the distribution of these data influences the result. Finally, we examined the relations between correlation length (between velocity samples), inter-event distance and wavelength content in the velocity model. This work has enabled us to present new solutions and new results that may help to further our understanding and use of double-difference algorithms.

2. *Method*

For the purpose of clarity, the following section summarizes the most important features of the recent developments presented in MONTEILLER *et al.* (2005).

2.1 Solving the Double-difference Equation System

WALDHAUSER and ELLSWORTH (2000) followed a linearized approach to double-difference location in which the travel-time difference residual ("double difference") r_{ij}^k for the events i and j recorded at station k is related to perturbations $\delta\mathbf{m}_i$ and $\delta\mathbf{m}_j$ of the model parameter vectorm by a linear equation (hereafter referred to as the "double-difference equation"):

$$\frac{\partial t_i^k}{\partial \mathbf{m}}\delta\mathbf{m}_i - \frac{\partial t_j^k}{\partial \mathbf{m}}\delta\mathbf{m}_j = r_{ij}^k, \tag{1}$$

where t_i^k and t_j^k are the theoretical travel times computed between station k and events i and j respectively.

However, double-difference equations are non-linear, as $\frac{\partial t_i^k}{\partial \mathbf{m}} = \mathbf{s}_i^k = \mathbf{s}_i^k(\mathbf{m})$, where $\mathbf{s}_i^k$ is the slowness vector computed at the hypocenter of event i for station k. The double-difference equation (1) may also be poorly conditioned (Fig. 1). Figure 1 shows that, under favorable conditions (i.e., sufficiently large ρ/Δ), slowness vectors may be sufficiently different and conditioning sufficient to allow the solution of the double-difference equations. From a seismic tomography point of view, it is well known that poor conditioning of equation (1) arises from the poor sampling of the medium by seismic rays. Consequently, there are several reasons why double-difference tomographic problems may be nonlinear, overdetermined and poorly conditioned.

Solutions of such nonlinear problems have been discussed in TARANTOLA and VALETTE (1982) and TARANTOLA (1987). The double-difference tomographic problem may be expressed by the equation:

$$g(\mathbf{m}) = \mathbf{d} \tag{2}$$

where g is a functional representing the direct problem, and $\mathbf{d}$ is the time-delay data.

In order to take into account the uncertainty both in the data and in the model parameters, and to attempt to find the minimal norm model, TARANTOLA and VALETTE (1982) proposed minimizing a cost function derived from the weighted sum of the data misfit and a penalty function (the *a posteriori* model variance):

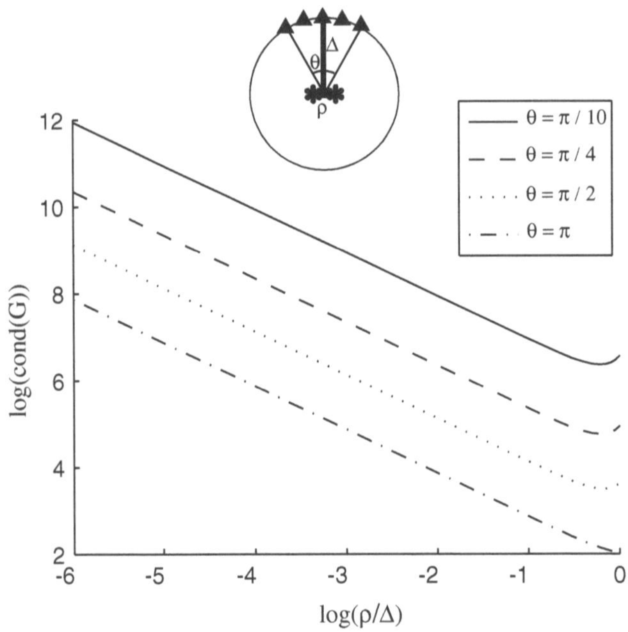

Figure 1
Conditioning of the partial derivative matrix **G** as a function of the aperture ρ/Δ of an earthquake doublet, where ρ is the inter-event distance and Δ the hypocentral distance. Conditioning is given for various circular seismic station network geometries, which are characterized by their aperture θ, in the interval $\rho/\Delta < \theta$. In the upper diagram, seismic stations are shown as triangles and earthquakes are shown as stars.

$$(g(\mathbf{m}) - \mathbf{d})^{\mathrm{T}}\mathbf{C}_{\mathbf{d}}^{-1}(g(\mathbf{m}) - \mathbf{d}) + (\mathbf{m} - \mathbf{m}_0)^{\mathrm{T}}\mathbf{C}_{\mathbf{m}}^{-1}(\mathbf{m} - \mathbf{m}_0) \tag{3}$$

where $\mathbf{C_d}$ is the data covariance matrix, $\mathbf{C_m}$ is the *a priori* model covariance matrix and $\mathbf{m}_0$ is the *a priori* model vector. MONTEILLER *et al.* (2005) showed that minimizing equation (3) using a Gauss-Newton scheme is equivalent to iteratively solving the system:

$$\begin{pmatrix} \mathbf{C}_{\mathbf{d}}^{-1/2}\mathbf{G}_k \\ \mathbf{C}_{\mathbf{m}}^{-1/2} \end{pmatrix} \delta\mathbf{m}_k = \begin{pmatrix} \mathbf{C}_{\mathbf{d}}^{-1/2}\mathbf{d} - \mathbf{g}(\mathbf{m_k})) \\ \mathbf{C}_{\mathbf{m}}^{-1/2}(\mathbf{m}_0 - \mathbf{m}_k) \end{pmatrix}, \tag{4}$$

where $\mathbf{G}_k$ is the matrix containing the partial derivatives of the travel-time differences (differences of travel-time derivatives), and $\mathbf{m}_k$ is the model vector at iteration k.

By introducing *a priori* information, the double-difference equation system can be solved for all forms of initial conditioning, i.e., for all geometric conditions. The amount of *a priori* information needed to solve the double-difference equation system depends on how the data constrain the model. Hence, it is necessary to follow an optimization process to determine the relative importance of *a priori* information and actual data. As a result, the introduction of a fixed geometric center constraint or the simultaneous resolution of absolute and double-difference equations (see respectively WALDHAUSER and ELLSWORTH, 2000; ZHANG and THURBER, 2003) is unnecessary: absolute data information is present in the *a priori* information.

2.2 *Estimating Accurate Model Parameters from Double-difference Tomography*

Travel-time computation

Travel times are computed using the finite-difference solutions to the Eikonal equation of PODVIN and LECOMTE (1991), which provides a robust estimation in the case of heterogeneous structures. *A posteriori* ray-tracing was performed by computing the travel-time field gradient. Recomputed travel times achieve the $O(10^{-4})$ accuracy needed to use cross-spectral time-delays in double-difference equations at low computational expense (MONTEILLER *et al.*, 2005).

Inverse problem:

1) A priori model covariance matrix

In order to solve the double-difference equations (equation (4)), $\mathbf{C}_{\mathrm{m}}^{-1/2}$ has to be computed. MONTEILLER *et al.* (2005) have shown that

$$\mathbf{C}_{\mathrm{m}}^{-1/2} = \mathbf{U}\mathbf{S}^{-1/2}\mathbf{U}^{\mathrm{T}}, \tag{5}$$

where $\mathbf{C}_{\mathrm{m}}$ is a symmetric positive definite matrix, $\mathbf{U}$ is an orthogonal matrix, $\mathbf{U}^{\mathrm{T}} = \mathbf{U}^{-1}$, and $\mathbf{S}$ is a diagonal matrix containing the eigenvalues of $\mathbf{C}_{\mathrm{m}}$.

$\mathbf{C}_{\mathrm{m}}$ is diagonal for hypocentral parameters. Due to the large number of velocity model parameters and the limited memory of computers, $\mathbf{C}_{\mathrm{m}}^{-1/2}$ cannot be computed directly. To overcome computing limitations, MONTEILLER (2005) searched for and found a direct approximation of $\mathbf{C}_{\mathrm{m}}^{-1/2}$ when the correlation between samples of the velocity model, which are located respectively at positions ρ and ρ', is expressed by the L_1-norm correlation function:

$$\mathbf{C}_v(\boldsymbol{\rho}, \boldsymbol{\rho}') = \sigma_v^2 e^{\frac{-|\rho-\rho'|}{\lambda}}, \tag{6}$$

where σ_v^2 is the variance of the velocity for the node located at ρ, and λ is a correlation length.

At small scales, $\mathbf{C}_{\mathrm{v}}^{-1/2}$ exhibits a band-diagonal structure and the value of its coefficients is independent of the number of model parameters. This feature allows the construction of $\mathbf{C}_{\mathrm{v}}^{-1/2}$ at large scales at low computational expense. In practical terms, $\mathbf{C}_{\mathrm{v}}^{-1/2}$ is computed and stored for various values of λ and σ_v. The rapid solution of equation (4) allows the exploration of numerous values of λ and σ_v and the choice of an optimal model.

2) Inversion algorithm
Inversion of equation (4) is performed by using the conjugate gradient algorithm LSQR (PAIGE and SAUNDERS, 1982). LSQR provides an iterative least-squares solution to the linear ill-conditioned problem $\mathbf{Ax} = \mathbf{b}$ for sparse matrices. $\mathbf{C}_m^{-1/2}$ is a sparse matrix and the formulation presented in equation (4) is suitable for use in the LSQR algorithm.

Checking the algorithm

We performed a double-difference tomography using the synthetic travel-time differences computed in the propagation medium shown in Figure 2a. This medium exhibits velocity perturbations with various wavelengths. Station and source coverage is homogeneous. All possible event pairs were used in the computation, with minimum and maximum inter-event distances of 2.5 and 15 km, respectively. The distance between velocity nodes was 200 m. Tomographic inversion was first performed for various *a priori* velocity standard deviations and an optimal standard deviation was found. In a second step, inversion was performed for 13 different correlation lengths (Figs. 2c-d). This process was used to determine an optimal correlation length. The result (Fig. 2e) is shown for the optimal value of the correlation length ($\lambda = 1$ km). It should be noted that: (1) using "large" correlation lengths (larger than the optimum) does not change the cost function (equation (3)) and model variance far from their optimal values; (2) cost function and model variance reach their minimum for the same value of the correlation length, taken as the optimal value, and a turning point appears in the cost function $= f$(model variance) curve; and (3) wavelengths are recovered in the velocity model even though they are smaller than the optimal correlation length. Property (2) shows that the minimal-norm model optimizes the cost function. These properties may be dependent on the estimator chosen for representing the correlation between velocity parameters. It seems that the L_1-norm correlation function exhibits favorable characteristics, which may not be shared by all possible correlation functions. It decreases rapidly with distance and therefore preserves short wavelengths, nonetheless it spreads a minimum amount of *a priori* information over considerable distances. Only a small amount of *a priori* information is required to stabilize the computation in cases where the data are insufficient to constrain the model. Loosely speaking, correlating velocity parameters might be better understood in terms of "stacking" the information and the noise carried by different thick rays, rather than in terms of "smoothing" the model. This quality is highly desirable in (double-difference) tomography, especially when modelling heterogeneous media.

2.3 Investigating the Resolving Power of Different Algorithms and Data Sets

We first used our algorithm to compute a reference solution using a reference time-delay data set. Subsequently, this reference solution was compared with the results produced by four different data sets:

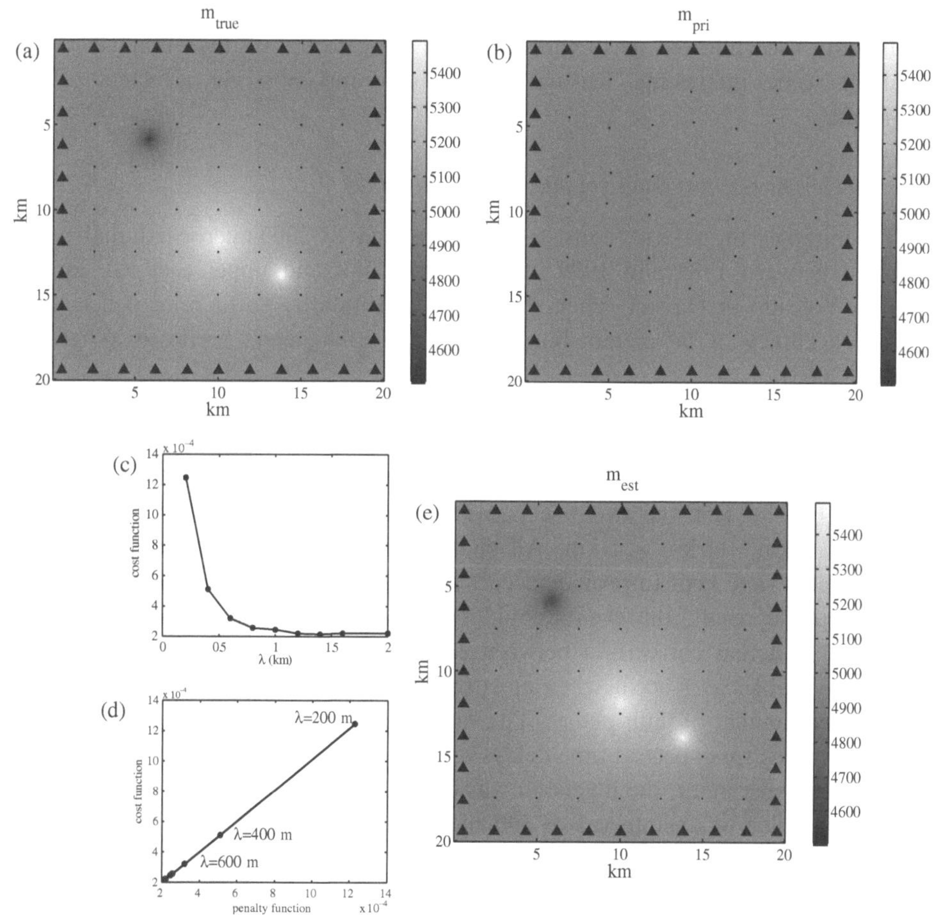

Figure 2
Result of the double-difference tomographic inversion using travel times computed in a synthetic propagation medium, composed of three Gaussian heterogeneities of various characteristic sizes and velocities. The sampling rate of the velocity model is 200 m. (a) Synthetic propagation medium; (b) initial model; (c) cost function as a function of the correlation length λ between velocity parameters; (d) cost function as a function of the *a posteriori* model variance, for various correlation lengths; (e) resulting model, as estimated by double-difference tomographic inversion using the optimal correlation length.

(1) Travel-time tomography (minimal-norm model),
(2) Double-difference tomography with time delays (non-minimal norm model),
(3) Double-difference tomography with picking-differences (minimal-norm model), with and without considering the probability density function of the picking-differences,
(4) Double-difference tomography with time delays of numerous subclusters, with and without correlation between the subclusters.

In each case the initial model is the one shown in Figure 5. The quality and quantity of the data used allowed us to thoroughly investigate the results obtained and to highlight some interesting features, limitations and possible artefacts of the algorithm.

Reference event set, inversion volume and initial model

The initial earthquake set consisted of approximately 2000 microearthquakes that occurred between 1988 and 1999 in the south flank and upper east rift zone of Kilauea Volcano in Hawaii. They show enough similarity to be relocated with an average accuracy of 50 meters horizontally and 100 meters vertically (Got and Okubo, 2003).

The reference event set used by Monteiller *et al.* (2005) consisted of 614 well-recorded microearthquakes with similar waveforms to their neighbors that occurred in four extended clusters of the upper east rift zone and the south flank of Kilauea Volcano (Fig. 3). The four event sets covered volumes of between 7 km × 10 km × 4 km and 4 km × 1 km × 1 km. All clusters were declustered, although enough similar events were kept to provide accurate cross-spectral time-delays and to avoid blunders. As a result, the events could be widely correlated across the data set (Fig. 4). The actual correlation between earthquakes may depend on the tectonic setting; favorable cases occur when most of the deformation occurs along one, or a very limited number of fault planes. The various numerical experiments described in this paper used cross-spectral time-delays or arrival-time pickings from this reference set. In each experiment, the inversion was carried out in a 12 km × 16 km × 12 km volume, sampled at an interval of 500 meters and comprising 20,625 nodes. Our initial model only covered a small area of the regional 3D tomographic model computed by Monteiller *et al.* (2005) (Fig. 5).

Reference cross-spectral time-delay tomography results

Time-delays were computed using a cross-spectral analysis performed on 2.56 s windows of the 100-sample per second digitized signal. 614 events were used, providing 953,265 time-delays from 87,798 event pairs showing long-range interactions (Fig. 4). A coherency threshold of 80% was used to avoid blunders in this data set. Double-difference relocation in our initial model (Monteiller *et al.*, 2005) was performed as a preliminary step. In our initial model, the resulting locations were used as *a priori* information in the double-difference tomographic inversion. From the optimization procedure, the *a priori* model standard deviation was chosen to be $\sigma_h = 10{,}000$ m for the hypocenters, and $\sigma_v = 500$ m/s for the velocity parameters. A parametric study of the cost function = f (model variance) gave an optimal value of $\lambda = 1$ km (Fig. 6). The RMS decreased from 35 to 13 milliseconds. A checkerboard test (Fig. 7) was used to show where the model can be correctly reconstructed from the reference data set.

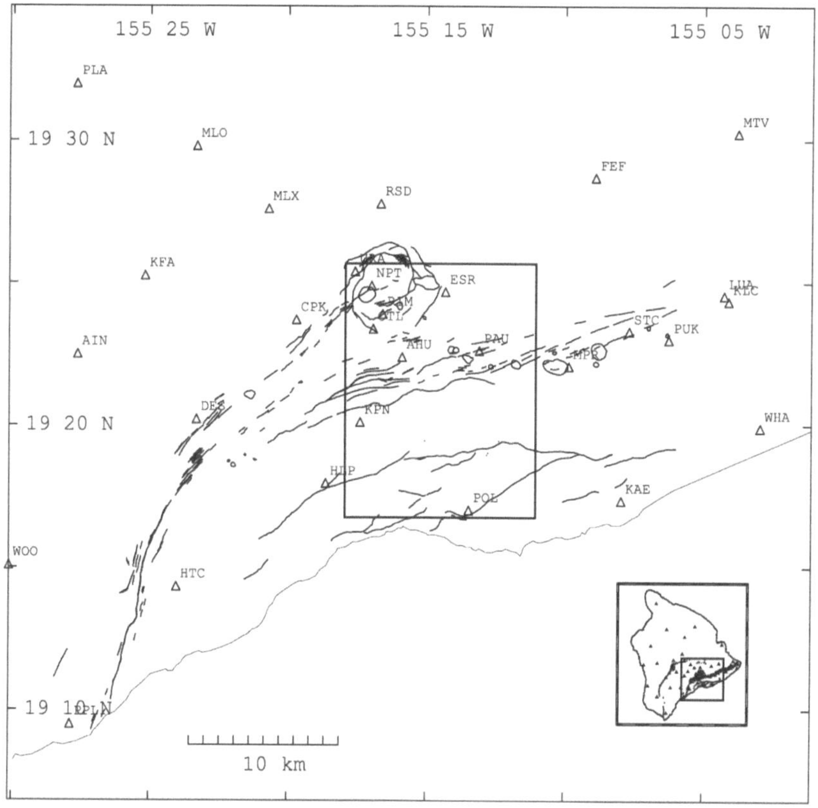

Figure 3

Map of Kilauea Volcano, showing the main topographic features, the catalog epicenters of the earthquakes used for the double-difference tomography (black dots) and the location of the HVO short-period seismic stations (open triangles).

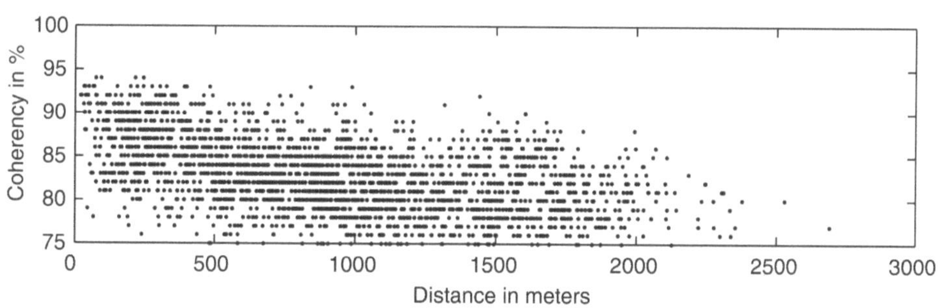

Figure 4

Correlations (represented by the averaged coherency) between events used in the double-difference tomography, as a function of distance. Coherency is frequency-averaged between 5 and 15 Hz for each station, and spatially-averaged using a set of stations for which the coherency is greater than 75%. Therefore, the station set changes with the event pair: The coherency of an event pair having a large interevent distance and large magnitude is best at large hypocentral distances.

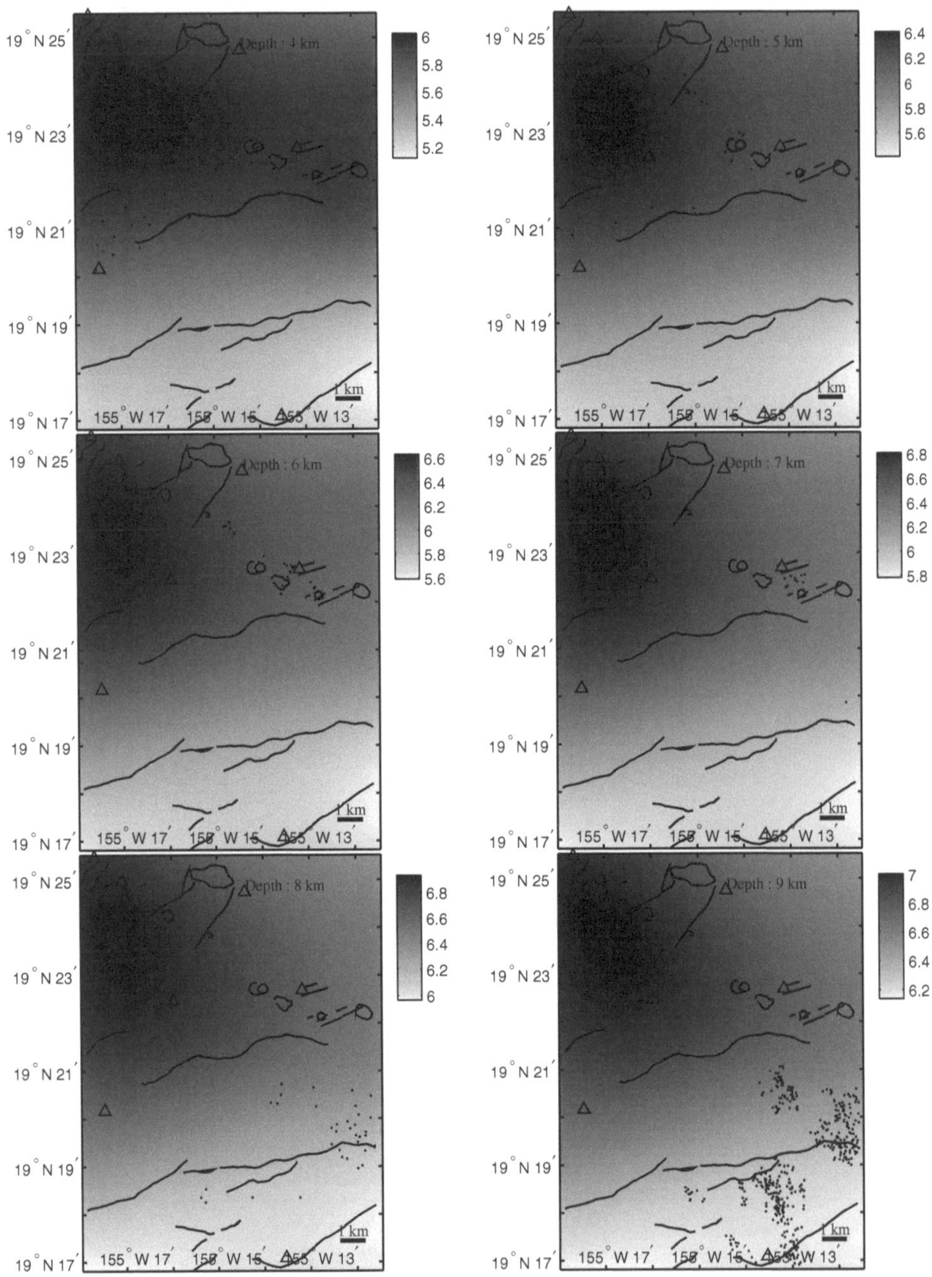

Figure 5

P-wave velocity map after travel-time tomographic inversion at a regional scale: Detail of the regional tomography, in the volume investigated using double-difference tomography, for various depths (from Monteiller *et al.*, 2005; $\lambda = 5$ km, $\sigma = 1$ km/s). This model will be used as the initial model in further local tomographic inversions. The grey scale indicates P-wave velocity in km/s.

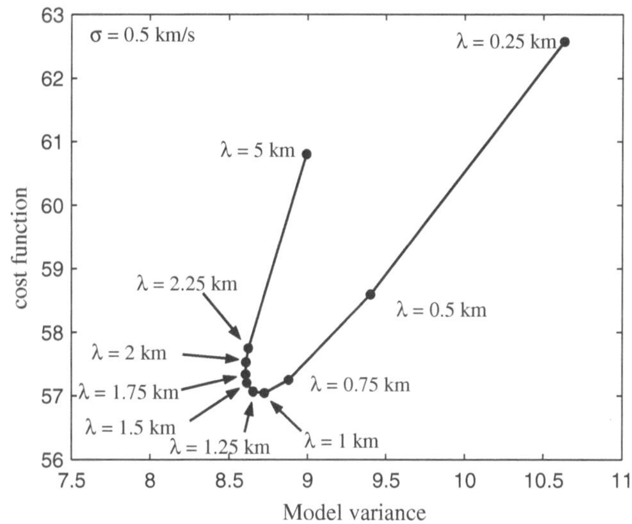

Figure 6
Cost function shown as a function of the *a posteriori* model (velocity parameters) variance, for various correlation lengths λ and $\sigma = 0.5$ km/s (double-difference tomography).

The double-difference tomographic inversion significantly modified the velocity image of the volume (Fig. 8b), introducing high wave number variations in the initially smooth model (Fig. 5) in places where the model was well resolved (Fig. 7). It can be first compared with the classic travel-time tomography results obtained from the same event set (Fig. 8a). The change introduced by the double-difference tomographic process was stable and the image exhibited few random fluctuations. The model obtained by this method was a minimal-norm model. It showed the least possible random fluctuations and fit the data. Its norm could be considered minimal in that each parameter showed the least possible variation (while fitting the data) and because it had the lowest possible number of degrees of freedom (the number of independent parameters) while still fitting the data. This guarantees that the result comprises the least possible components of the kernel. The high wave-number variations that are introduced in the minimal-norm model are more likely to be real than those found using a non-minimal norm solution and inaccurate data.

Comparison with the solution provided by a classic double-difference tomographic algorithm

In order to find an objective way of producing images of the earth's heterogeneity, we analyzed a single data set using different methods and compared the results. We computed solutions for the double-difference equations using the classic tomographic algorithm, solving the equations

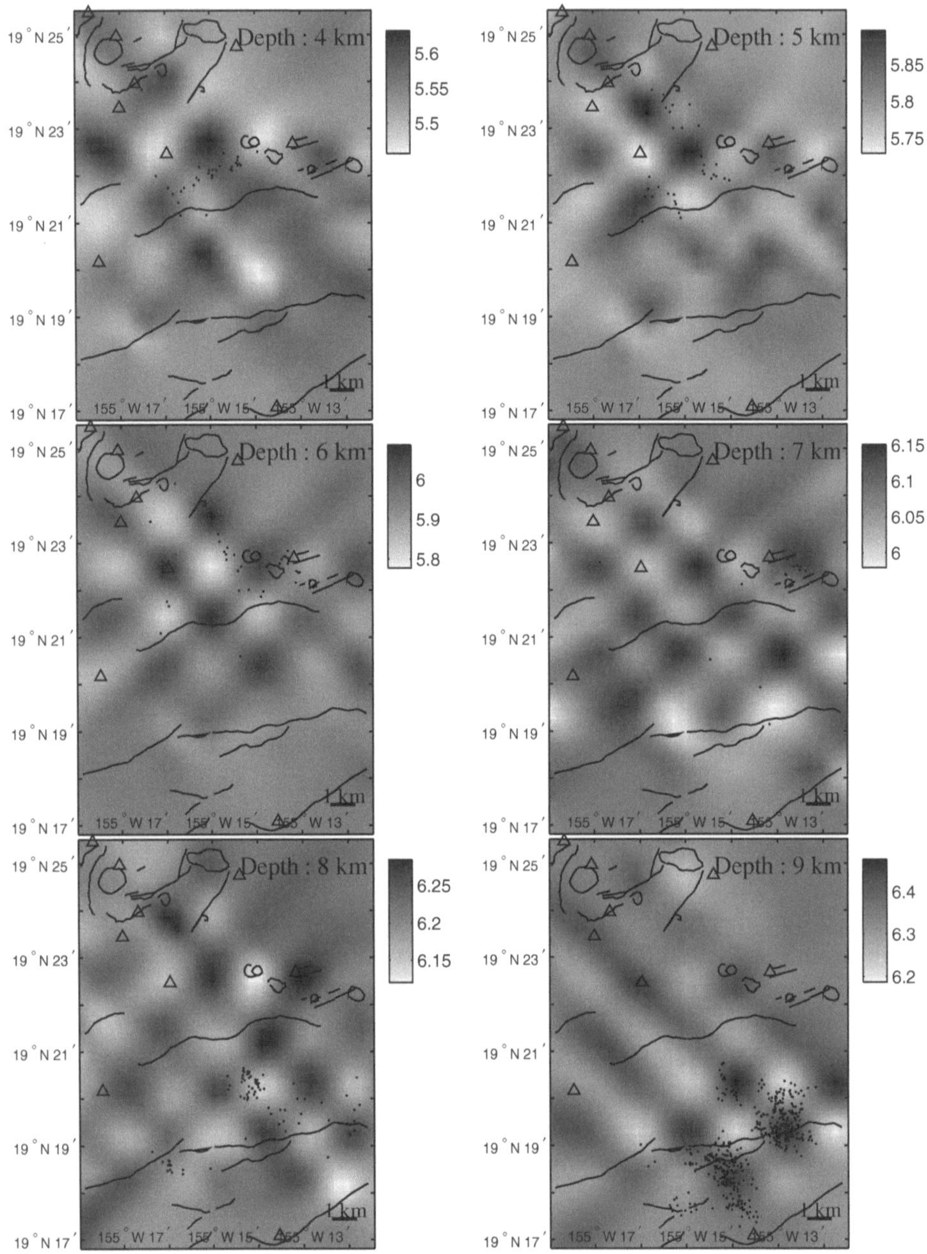

Figure 7
Checkerboard test (double-difference tomography) for the southeastern part of Kilauea caldera at various depths. Velocity anomalies are 5% amplitude and 4 km wavelength 2-D sinusoidal functions (each positive or negative circular pattern has a 1 km radius). Synthetic earthquake data match the real distribution of hypocenters. The grey scale indicates P-wave velocity in km/s.

$$\begin{pmatrix} \mathbf{C}_{\mathbf{d}}^{-1/2}\mathbf{GP} \\ \mathbf{C}_{nn}^{-1/2} \\ \lambda\mathbf{I} \end{pmatrix} \delta_k = \begin{pmatrix} \mathbf{C}_{\mathbf{d}}^{-1/2}(\mathbf{d} - g(\mathbf{m}_k)) \\ 0 \\ 0 \end{pmatrix}, \tag{7}$$

$$\delta \mathrm{m}_k = P\delta_k, \tag{8}$$

where $\mathbf{P}$ is a diagonal preconditioning matrix, $\mathbf{C}_{nn}^{-1/2}$ is the matrix corresponding to nearest neighbor smoothing, λ is a damping coefficient, and δ_k is the perturbation of the model in non-physical units (see e.g., NOLET, 1993, equation (9.7)). The lowest two matrices of the left-hand side of equation (7) take the place of the term $\mathbf{C}_{\mathrm{m}}^{-1/2}$ in equation (4). In this paragraph, we use the word "classic" to indicate that the algorithm does not use the probabilistic Tarantola-Valette approach, though it is used to inverse double-difference equations.

Practically, both the direct and inverse problems are computed using the same solvers (i.e., Podvin-Lecomte FD travel-time computation and LSQR). This is the same algorithm as the one used by LATORRE *et al.* (2004), and GAUTIER *et al.* (2006, this issue), except that the double-difference tomographic equations (and only them) were solved instead of the travel-time tomographic equations. As is common practice in classic tomography, four preconditioning coefficients, damping and a nearest-neighbor smoothing constraint were used to stabilize the system. The values of the preconditioning coefficients were found by trial and error using sensitivity tests (e.g., reconstructing a checkerboard). However it is not proven that preconditioning coefficients leading to correct sensitivity tests will necessarily provide correct velocity models. The data were cross-spectral time-delays computed from the reference event set. The use of nearest-neighbor smoothing does not allow optimization of the correlation length, or the choice of a minimal-norm solution; however, the damping coefficient can be chosen using an optimization process. It is difficult to choose a solution, as this one is heavily depending on the choice of the preconditioning coefficients. The selected result (Fig. 9c) shows a positive velocity anomaly in the vicinity of the east rift zone, revealing differences with the minimal-norm solution (Fig. 8b). A more thorough investigation of these differences revealed that the actual smoothing applied by the nearest-neighbor operator is not homogeneous throughout the model. Computation of the $\mathbf{C}_{\mathbf{nn}}$ matrix corresponding to the nearest-neighbor operator $\mathbf{C}_{nn}^{-1/2}$ (see equation (7)) showed that the coefficients of $\mathbf{C}_{\mathbf{nn}}$ are not homogeneously distributed: taking a simple and homogeneous form for $\mathbf{C}_{nn}^{-1/2}$ led to a heterogeneous distribution of the $\mathbf{C}_{\mathbf{nn}}$ coefficients. This heterogeneity in the pseudo-covariance matrix $\mathbf{C}_{\mathbf{nn}}$ produces an artificial heterogeneity in the tomographic image.

Minimal-norm double-difference tomography using picking-differences

Cross-spectral time-delays provide high-quality data for double-difference tomography, however they are more difficult to obtain than picking-differences. Although picking-differences are far less accurate than cross-spectral time-delays, it

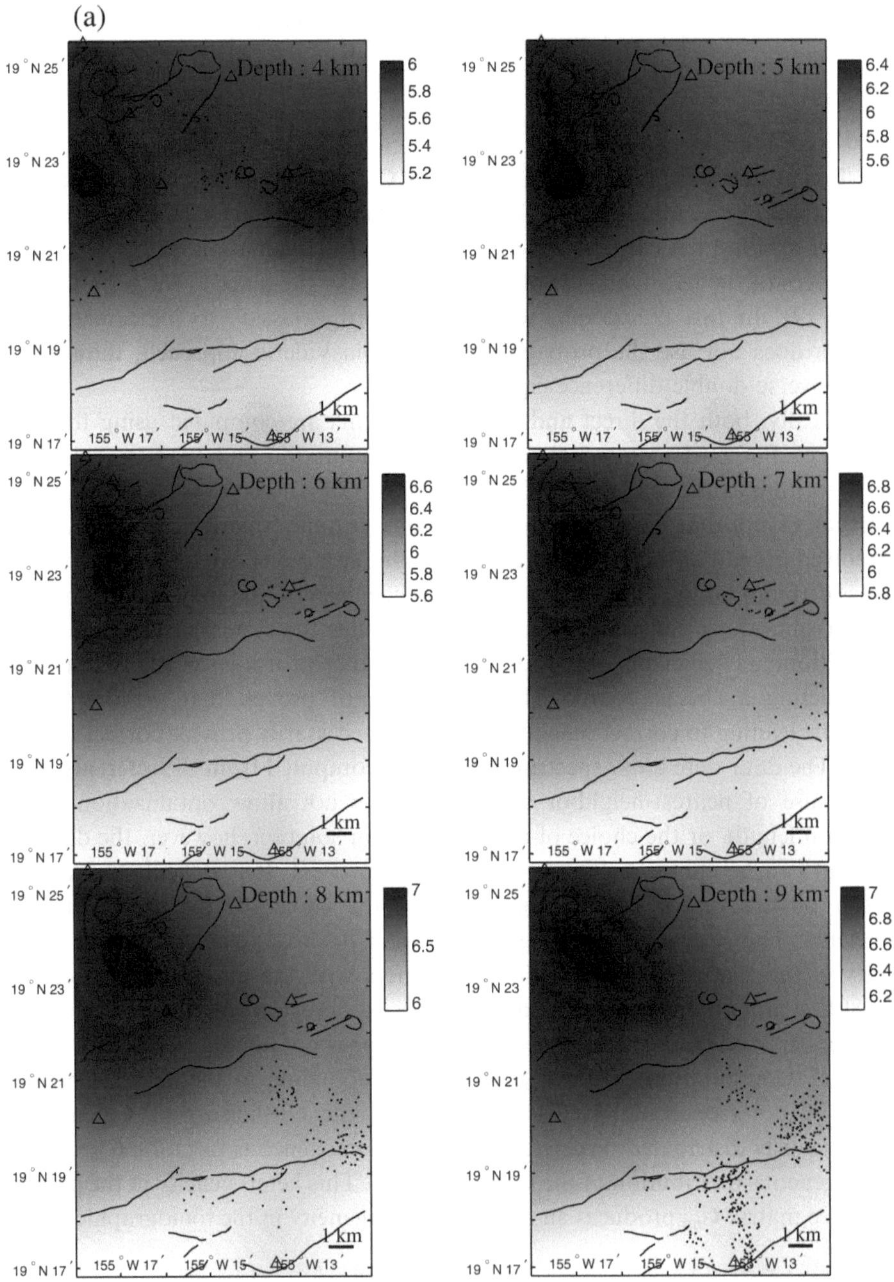

Figure 8

Maps of the P-wave velocity around Kilauea east rift zone at various depths (a) after local travel-time tomography (using 21,954 arrival times from the 614 events recorded by an average of 35 stations used in the double-difference tomography); (b) after double-difference tomography (using 953,265 cross-spectral time-delays). λ = 1 km, σ = 0.5 km/s in both cases. Geographical labels are in the upper left subplot. KC: Kilauea caldera; UERZ: upper east rift zone; KSF: Kilauea south flank. The grey scale indicates P-wave velocity in km/s.

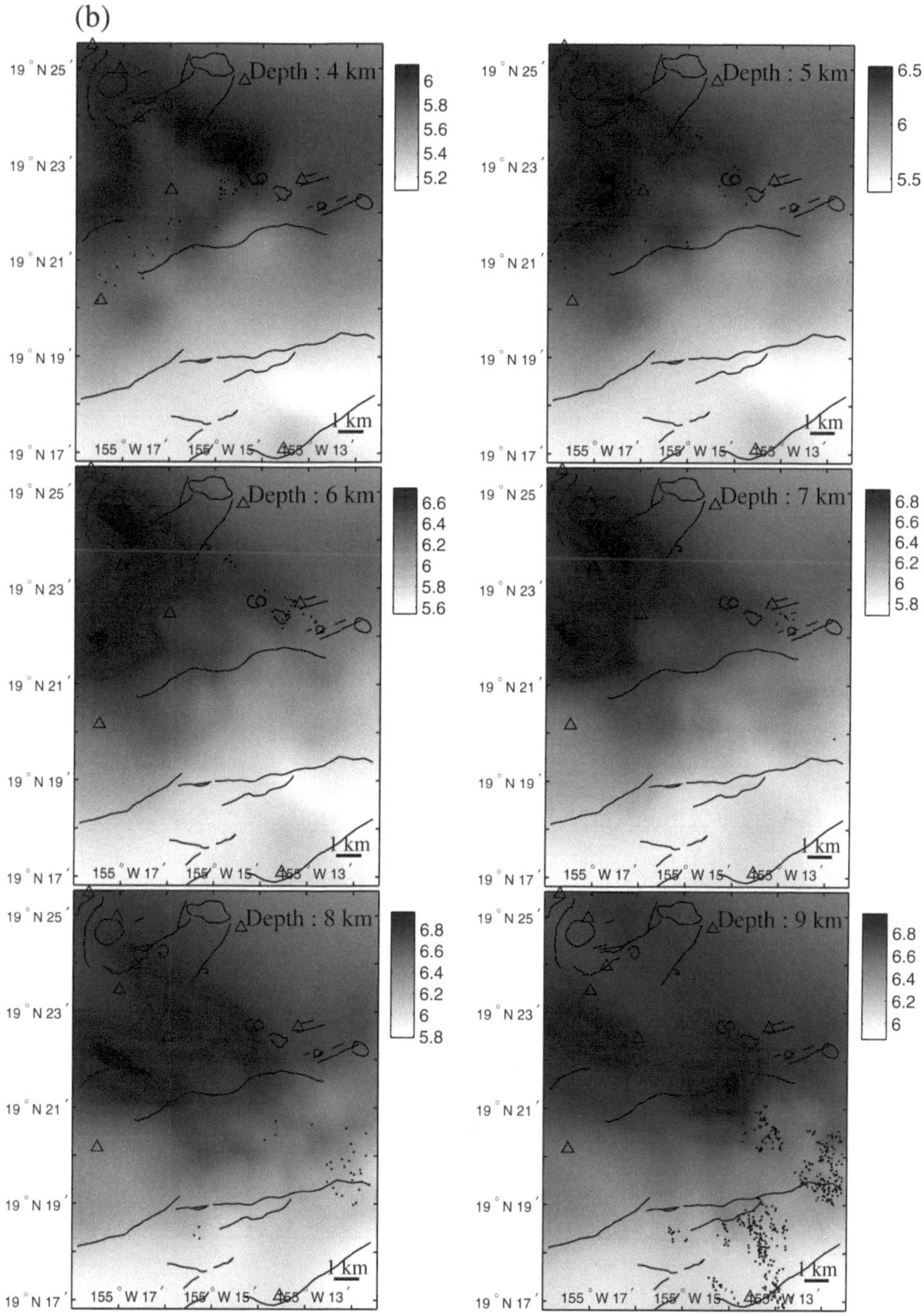

Figure 8
Continued

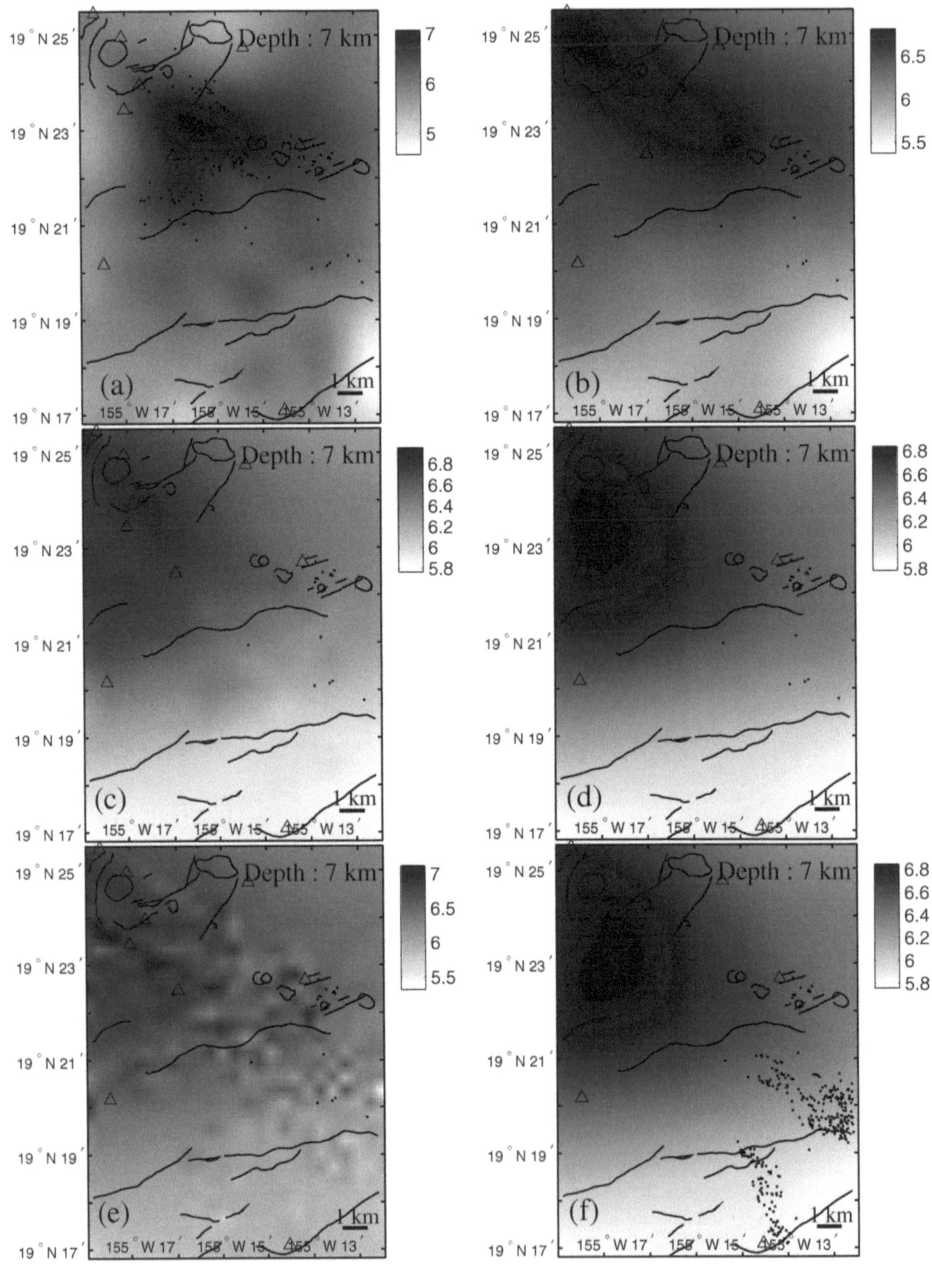

is tempting to use them with double-difference tomographic methods. In order to investigate the picking-difference error distribution, we took a subset of 476 events, providing 1,218,350 high-quality (coherency > 90%) cross-spectral time-delays, and

◄

Figure 9
Maps of the P-wave velocity around Kilauea east rift zone at 7 km depth (see text for more details), after double-difference tomography: (a) from picking-differences, minimal-norm solution (λ = 2 km, σ_v = 1 km/s), without taking into account the picking-difference error distribution; (b) from picking-differences, minimal-norm solution (λ = 2km, σ_v = 1 km/s), taking into account the picking-difference error distribution; (c) from cross-spectral time-delays, using the classic tomographic algorithm (equation (7)); (d) from cross-spectral time-delays of subclusters, minimal-norm solution (λ = 2 km, σ_v = 1 km/s), with a small amount of handpicked data (2,430 picking-differences added to 90,470 cross-spectral time-delays) to provide inter-cluster correlation; (e) from cross-spectral time-delays of uncorrelated subclusters, using λ = 0 km and σ_v = 1 km/s. RMS is 6 ms; (f) from cross-spectral time-delays of uncorrelated subclusters, using λ = 2 km and σ_v = 1 km/s; RMS is 10 ms. The grey scale indicates P-wave velocity in km/s.

plotted the distribution of the difference between picking-differences and cross-spectral time-delays (Fig. 10). This distribution may be considered to be very close to the error distribution of picking-differences, as the error in cross-spectral time-delays is far lower than the error in picking-differences. It showed that a significant proportion of the travel-time differences may be more or less correctly estimated by picking-differences. However, picking-differences have a long-tailed L_1-type distribution with strong blunders, rather than a Gaussian distribution (Fig. 9). Errors may

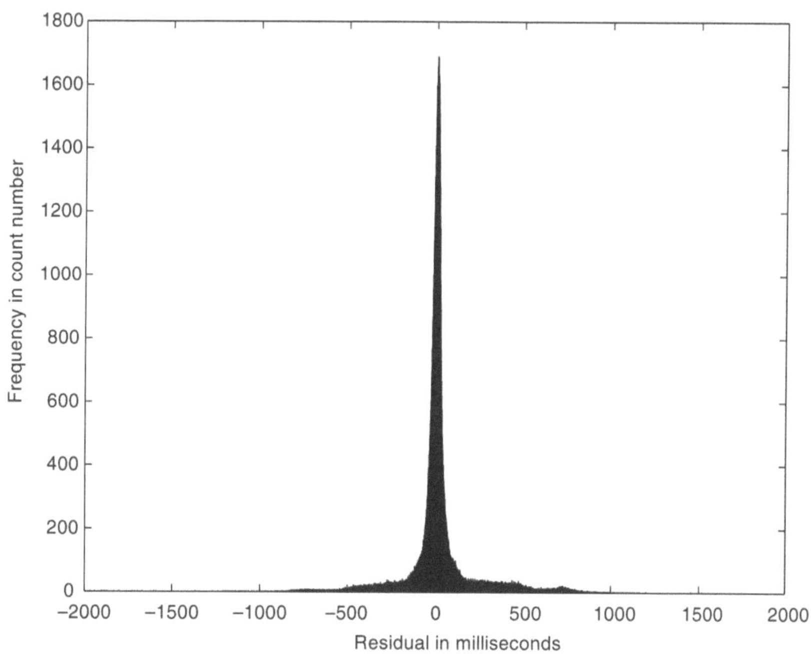

Figure 10
Distribution of the difference between cross-spectral time-delays (coherency > 90%) and picking-differences (1,218,350 data from 476 well-recorded microearthquakes on the south flank of Kilauea volcano).

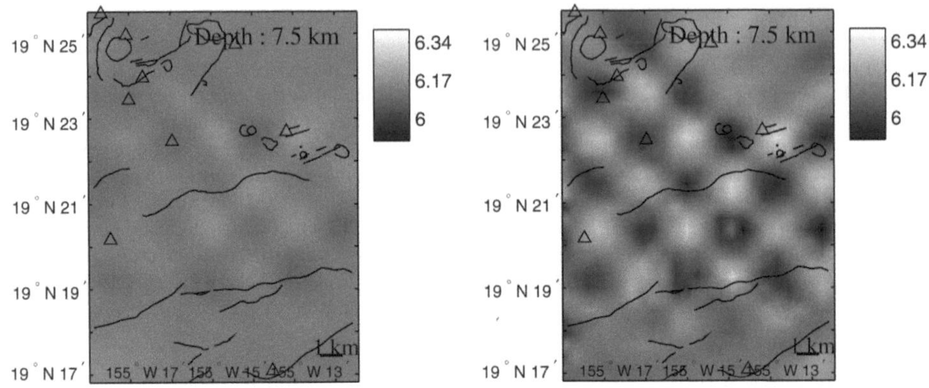

Figure 11
Checkerboard test (double-difference tomography) for the southeastern part of Kilauea caldera at 7 km depth. Velocity anomalies are 5% amplitude and 4 km wavelength 2D sinusoidal functions (each positive or negative circular pattern has a 1 km radius). Synthetic earthquake data match the real distribution of hypocenters. Noise (rectangular pdf, see text) is added to the synthetic travel-time differences. The grey scale indicates P-wave velocity in km/s. (a) Result obtained when using a Gaussian model of noise. (b) Result obtained when using a hyperbolic secant model of noise.

be larger than the theoretical travel-time differences. This distribution is not favorable to L_2-norm inversion algorithms. An *a posteriori* MOSTELLER and TUKEY (1977) weighting scheme (which uses statistics on residuals) is not appropriate for processing such data, as the non-weighted first iteration provides a biased solution (and therefore biased residuals). To investigate the use of such picking-difference data with our L_2-norm algorithm, we used picking-differences obtained from the reference event set. The standard deviation of the picking-difference data was greater than 70 ms; hence, differences of less than 70 ms could not be differentiated from the noise. Therefore, we used event pairs with more than 10 picking-differences larger than 100 ms in order to identify data with signal-to-noise ratios that were greater than 1.

No weighting was used in this application. The results (Fig. 9a) show that strong fluctuations perturbated the initial model, although the computation was made with a 2 km correlation length and a 1 km/s *a priori* standard deviation. The RMS was approximately 70 ms. To investigate the effect of noise and to determine whether the results remain interpretable, we carried out a checkerboard test using non-Gaussian noise added to the synthetic travel times. The noise, which had a rectangular probability density function of between −0.1 s and 0.1 s, was added to 10% of the synthetic data. The resulting checkerboard was poorly reconstructed (Fig. 11a), demonstrating that even the use of an optimization approach will not guarantee an accurate L_2 estimation of the model parameters for data with non-Gaussian noise. In the following section, we show that the estimation in the case of non-Gaussian data may be greatly improved using the Tarantola-Valette probabilistic approach.

Introducing the actual picking-difference error distribution into the minimal-norm double-difference tomographic solution.

The probabilistic point of view used in the Tarantola-Valette approach allows non-Gaussian distributions to be introduced into the inverse problem solution by using the auxiliary datum

$$x' = \mathrm{Erf}^{-1}\left(2\int_{x_0}^{x} f(u)du\right), \tag{9}$$

where f(x) is the probability density of the datum x, of mean x_0, and Erf(x) is the error function.

As a result, x' is a centered Gaussian datum with variance ½.

Given the form of the picking-difference error distribution, it is tempting to choose an L_1 probability density function for f(x). However, experience has shown that numerical instabilities appear when small standard deviations are used with such a distribution. Alternatively, the hyperbolic secant probability density function

$$f(x) = \frac{1}{\pi\sigma}\sec\mathrm{h}\left(\frac{x - x_0}{\sigma}\right) \tag{10}$$

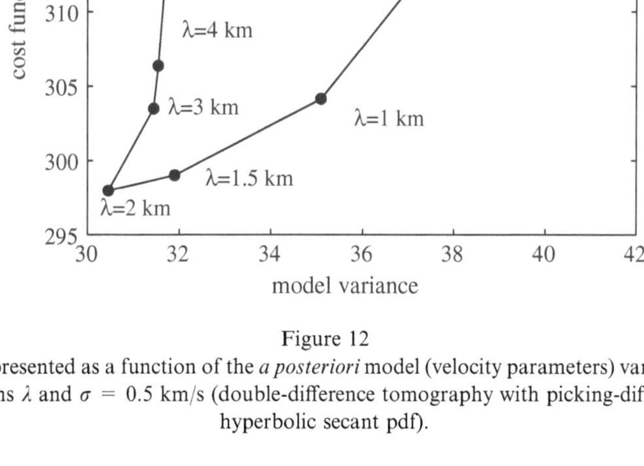

Figure 12
Cost function represented as a function of the *a posteriori* model (velocity parameters) variance, for various correlation lengths λ and $\sigma = 0.5$ km/s (double-difference tomography with picking-differences using the hyperbolic secant pdf).

(see e.g. CRASE *et al.*, 1990) behaves like the L_1 distribution for large values of the residual, and like a Gaussian distribution for small values. It is stable when used with small standard deviations σ and suitable for representing the long-tailed picking-difference error distribution. In this case, the auxiliary datum is (VALETTE and LESAGE, 2005; B. Valette, pers. comm.):

$$x' = \mathrm{Erf}^{-1}\left(\frac{2}{\pi}\arctan\left(\sinh\left(\frac{x - x_0}{\sigma}\right)\right)\right) \tag{11}$$

x' exhibits a Gaussian distribution and can therefore be used in a L_2-norm inversion algorithm to provide non-biased estimates of the model parameters. The inversion of auxiliary data x' using a Gauss-Newton scheme allows us to solve a system similar to equation (4):

$$\begin{pmatrix} D\mathbf{G}_k \\ \mathbf{C}_{\mathbf{m}}^{-1/2} \end{pmatrix} \delta\mathbf{m}_k = \begin{pmatrix} x' \\ \mathbf{C}_{\mathbf{m}}^{-1/2}(\mathbf{m}_0 - \mathbf{m}_k) \end{pmatrix} \tag{12}$$

where $\mathbf{D}$ is the matrix containing the partial derivatives:

$$\frac{\partial x'}{\partial x} = \frac{e^{x'^2}}{\sigma\sqrt{\pi}\cosh\left(\frac{x-x_0}{\sigma}\right)}. \tag{13}$$

This approach was used with the picking-difference data obtained from the reference event set. We found an optimal correlation length $\lambda = 2$ km (Fig. 12). Interestingly, the strong fluctuations observed in Figure 9a do not appear in the results of this computation (Fig. 9b). A stable coherent pattern, revealing a sharply defined, high-velocity body along the east rift zone of Kilauea, can be seen. This pattern may be compared to the results obtained with: (i) travel-time data (Fig. 8a); (ii) cross-spectral time-delays (Fig. 8b); and (iii) picking-differences (L_2-norm solution, Fig. 9a) at the same depth. A checkerboard test computed from synthetic travel times, 10% of which are perturbated with the rectangular pdf noise used later, is best reconstructed (Fig. 11b) using the auxiliary datum (equation (11)). Notice that using the transformation (11) with cross-spectral time-delays is equivalent to using a coherency threshold and a Mosteller and Tukey weighting scheme. This equivalence is probably related to the low number of cross-spectral time-delay blunders computed with a high coherency.

Effect of the heterogeneous distribution of the event correlation

In a number of cases, events could not be homogeneously correlated in similar pairs throughout the event set, and data sets appeared as a heterogeneous collection of cross-spectral time-delays and picking-differences. To investigate the effect of such heterogeneity in the event correlation on the double-difference tomographic result, we first studied an event set that had been broken into several small, uncorrelated subclusters.

Uncorrelated subclusters

To check the effect of using uncorrelated small clusters, we artificially broke our 614 event sets into 16 subclusters, mimicking the massive rupture of "weak critical links" in our initial event correlation set. Therefore, there was no longer any inter-cluster correlation, although intra-cluster correlation was sufficiently high to allow the computation of cross-spectral time-delays. Computation of cross-spectral time-delays was, therefore, limited to short-wavelength correlation between events. The absolute locations of small subcluster events and the velocity parameters were less constrained by cross-spectral time-delays and more dependent on the *a priori* information. In this case, tomographic images strongly depend on the *a priori* correlation length used. If the correlation length is too short, or null, the final tomographic model (Fig. 9e) will exhibit short-wavelength and strong velocity fluctuations, mimicking the heterogeneity of the seismic velocities. A checkerboard test indicated that these data could not be used to reconstruct the velocity model. When an optimization approach was followed, inversion using the optimal correlation length led to a final model that was very close to the initial model (Fig. 9f). Data provided by the numerous uncorrelated subclusters were coherent but too poor and insufficient to constrain a perturbation of the tomographic model. The stability of the model can be maintained by using *a priori* information. Hence, uncorrelated small subclusters appear to provide inefficient data for double-difference tomography at the scale of our model.

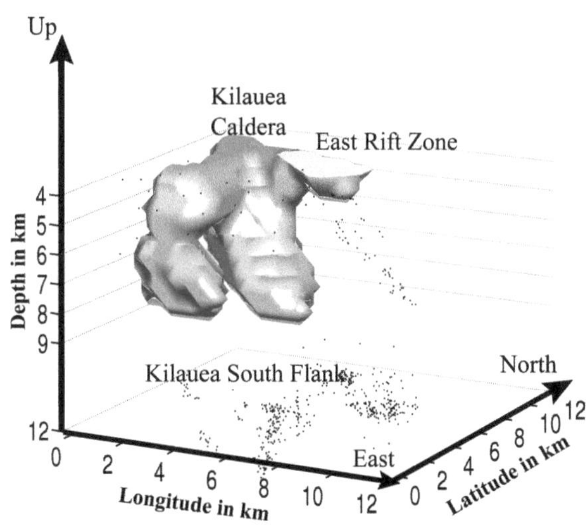

Figure 13
3D representation of the high-velocity anomaly inferred from double-difference tomography in the southern caldera – upper east rift zone between 4 and 9 km depth. Earthquakes are shown by black dots.

Correlating subclusters with picking-differences

In this experiment, we introduced inter-cluster correlation by adding picking-differences computed for pairs of events belonging to different multiplets in the 16 subcluster data set. A sech error distribution was assumed for the whole set of data, and the minimal-norm inversion scheme was used. When introducing a small amount of handpicked data (2,430 picking-differences added to 90,470 cross-spectral time-delays), the inferred model (Fig. 9d) remained very close to the initial model. When larger numbers (up to 65,610) of picking-differences were introduced, the computed model converged rapidly to the result obtained with the handpicked data only, both using the sech error distribution and the minimal-norm inversion (Fig. 9b).

3. Discussion

The tomographic images produced by travel-time and by double-difference tomographic inversion of the same events in the same volume (Fig. 8) show a number of significant differences. The double-difference tomographic image shows coherent structures and it does not display random high-frequency velocity changes (see MONTEILLER *et al.*, 2005 for a detailed discussion of the structures). This is especially true at the depths where the southern caldera – east rift zone region is well reconstructed (Fig. 7). Most notably, double-difference tomography has identified a fast (dense) magmatic body between 4 and 9 km below the east rift zone. This body, which forms a complex, vertically elongated cylinder (Fig. 13), may be part of the Kilauea Volcano magma feeding system.

In this study we focused on the use of double-difference tomographic algorithms to estimate crustal heterogeneity. For a given set of earthquakes, we compared the results produced using a variety of methods and sets of time data.

The results provided by cross-spectral time-delay (double-difference) tomography and local travel-time tomography (Fig. 8) were compatible, given their different levels of accuracy. The travel-time tomography results, which were computed according to the minimal-norm model from the reference set of events, appear to give an "averaged" image of the double-difference tomographic results.

Using a classic non-minimal norm tomographic algorithm with a nearest-neighbor smoothing constraint and the same data set provides a wide class of models, with the result being heavily dependent on the choice of the values for the four preconditioning coefficients. In real cases, where ray distribution is non-optimal and noise is non-Gaussian, choosing a solution is a difficult task.

In order to obtain accurate model parameters, it is necessary to combine a stable inversion algorithm and accurate data. Thus, it is important to know how accurate the data needs to be to achieve a given level of accuracy for the model parameters. When cross-spectral time-delays are replaced by picking-differences in a double-

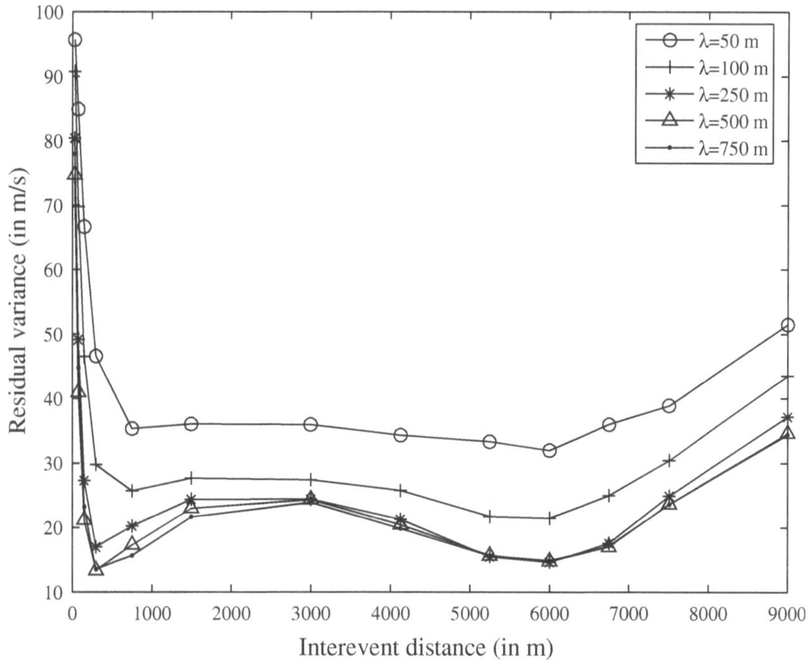

Figure 14
Model residual variance as a function of the inter-event distance, for various correlation lengths λ between velocity parameters. Residual variance of the estimated model is computed by reference to the propagation medium. The propagation medium consists of a 2-D Gaussian heterogeneity embedded in a homogeneous 5 km/s vertical slice, comprising 20 stations along a 20 km line at the surface and 800 earthquakes along a 12 km line at 10 km depth. Maximum amplitude of the Gaussian heterogeneity is 500 m/s and characteristic size is 800 m. Each double-difference tomography was performed by selecting event pairs of a given inter-event distance, for a given correlation length.

difference tomographic algorithm with a simple MOSTELLER and TUKEY (1977) weighting scheme (Fig. 9a), the resulting velocity model displays strong spatial fluctuations. These fluctuations can be attributed to the large data standard deviation (Fig. 10) and the incorrectly modeled picking-difference error distribution. A checkerboard using non-Gaussian noise added to the synthetic travel-times is poorly reconstructed (Fig. 11a). In addition, the L_2-norm optimization approach itself tends to fail: conditioning remains good, due to the favorable geometric distribution of event pairs, proving that data geometrically constrain the model, however the data do not constitute a coherent set that can be interpreted using a single physical model.

A model without strong fluctuations can be produced by using a probabilistic approach that takes into account the picking-difference error distribution by applying the hyperbolic secant probability density function inside a minimal-norm inversion. Sensitivity tests with additional noise show that this approach can correctly reconstruct the model over a significant area (Fig. 11b). Therefore, it may still be possible to keep an acceptable confidence level when using picking-differences in double-difference tomography, although the accuracy is lower than when cross-spectral time-delays are used.

Using cross-spectral time-delay data from numerous small, uncorrelated multiplets (well intra-correlated clusters), instead of a homogeneous and unique set of well-correlated event pairs does not induce significant changes to the initial velocity model and it removes the long-wavelength correlation between events. The addition of a sufficiently large quantity of picking-differences to these data leads to convergence with the model obtained using picking-differences alone. The longer-wavelength correlation borne by picking-differences (intercluster correlation) outweighs the short-wavelength correlation borne by cross-spectral time-delays (intra-cluster correlation). This suggests that the longer-wavelength correlation bears most of the information needed to reconstruct the velocity model. To investigate this problem further, we performed a series of double difference tomographies for various inter-event correlation lengths, using a synthetic propagation medium. This allowed us to show the residual variance between the estimated model perturbation and the true (synthetic) perturbation as a function of the inter-event correlation length. Residual variance is high when resolution is poor, and may be taken to be a very rough estimate of the inverse of the resolution (which may be represented as a transfer function between the true propagation medium and the model). Figure 14 shows that the model residual variance strongly varies with the inter-event correlation length. The residual variance is lower (resolution is better) when the inter-event correlation length is of the order of the characteristic size of the heterogeneity to be reconstructed. Large-scale heterogeneities cannot be reconstructed from a set of short-distance inter-event correlations: double-differences only resolve differences between parameters; they do not give the absolute values of these parameters. Alternatively, large-distance inter-event correlations under-sample small-scale heterogeneities and cannot be used to reconstruct them. In natural situations, earthquake swarms have characteristic sizes and their inter-event distances often show a unimodal distribution. In such cases, the use of all possible event pairs leads to the filtering of the velocity field around the wavelength corresponding to the characteristic size of the earthquake swarm. Therefore, each earthquake set has its own transfer function, but this transfer function may be naturally dominated by intermediate wavelengths. Thus, an ideal data set should be selected in order to have the whitest possible spectrum of inter-event correlation lengths. It should be noted that correlating velocity parameters (increasing the correlation length λ, Fig. 14): (1) increases the resolution (decreases the residual variance); (2) tends to widen the resolution spectrum; and (3) preserves the short wavelength content of the velocity model (residual variance decreases and resolution increases with λ for a given inter-event distance). The search for a minimal-norm model concentrates the model perturbation on a minimum number of parameters.

Conclusion

In this paper, the double-difference tomographic algorithm developed by MONTEILLER *et al.* (2005) was investigated and used with various data sets. This algorithm allows us to find the minimal-norm model fitting the double-difference data, whereas classic algorithms produce a large range of possible models, between which it is very difficult to decide without *a priori* knowledge. Combined with (accurate) cross-spectral time-delay data, the double-difference tomographic algorithm has been used to reconstruct the velocity model in the southern caldera-upper east rift zone of Kilauea Volcano, Hawaii. In this paper, we examined the effects of using less accurate data and of the distribution of the interevent correlation. Picking-differences are noisy data for which the relative error may reach 100%. Using these data in a L_2-norm algorithm without taking into account the error distribution does not produce a reliable velocity model. The model is best reconstructed when adequate picking-difference error statistics are used. We show that searching for a minimal-norm model and taking into account the error distribution in a double-difference tomographic algorithm allows us to improve velocity models. Data of varying degrees of accuracy, for example, cross-spectral time-delays and picking-differences, may be combined in an attempt to improve spatial coverage or the accuracy of the data set. However, the results are strongly dependent on the proportion of each type of data, and on the distribution of the inter-event correlation.

REFERENCES

AKI, K., and LEE, W. H. K. (1976), *Determination of three-dimensional velocity anomalies under a seismic array using first P arrival times from local earthquakes 1. A homogeneous initial model*, J. Geophys. Res. *81*, 4381–4399.

CRASE, E., PICA, A., NOBLE, M., MCDONALD, J., and TARANTOLA, A. (1990), *Robust elastic nonlinear waveform inversion: Application to real data*, Geophysics *55*, 527–538.

GOT, J.-L., FRÉCHET, J., and KLEIN, F.W. (1994), *Deep fault plane geometry inferred from multiplet relative relocation beneath the south flank of Kilauea*, J. Geophys. Res. *99*, 15375–15386.

GOT, J.-L., and OKUBO, P.G. (2003), *New insights into Kilauea's volcano dynamics brought by large-scale relative relocation of microearthquakes*, J. Geophys. Res. *108*, 2337, doi:10.1029/2002JB002060.

ITO, A. (1985), *High resolution relative hypocenters of similar earthquakes by cross-spectral analysis method*, J. Phys. Earth. *33*, 279–294.

JORDAN, T.H., and SVERDRUP, K.A. (1981), *Teleseismic location techniques and their application to earthquake clusters in the south-central Pacific*, Bull. Seismol. Soc. Am. *71*, 1105–1130.

LATORRE, D., VIRIEUX, J., MONFRET, T., MONTEILLER, V., VANORIO, T., GOT, J.-L., and LYON-CAEN, H. (2004), *A new seismic tomography of Aigion area (Gulf of Corinth – Greece) from a 1991 data set*, accepted to Geophys. J. Int.

MICHELINI, A., and LOMAX, A. (2004), *The effect of velocity structure errors on double-difference earthquake location*, Geophys. Res. Lett. *31*, doi:10.1029/2004GL019682.

MONTEILLER, V. *Tomographie des décalages temporels*, Thèse de l'Université de Savoie, Chambéry (2005).

MONTEILLER, V., GOT, J.-L., VIRIEUX, J., and OKUBO, P.G. (2005), *An efficient algorithm for double-difference tomography and location in heterogeneous media, with an application to the Kilauea volcano*, J. Geophys. Res. *110*, B12306, doi: 10.1029/7004JB003466.

MOSTELLER, F., and TUKEY, J.W., *Data Analysis and Regression* (Addison-Wesley, Reading, Mass., 1977).

NOLET, G., *Solving large linearized tomographic problems*. In *Seismic Tomography: Theory and Practice* (eds. Iyer and Hirahara), (Chapman and Hall, London, 1993).

PAIGE, C.C., and SAUNDERS, A. (1982), *LSQR: sparse linear equations and least squares problems*, ACM Trans. Math. Softw. *8*, 195–209.

PODVIN, P., and Lecomte, I. (1991), *Finite difference computation of travel times in very contrasted velocity models: A massively parallel approach and its associated tools*, Geophys. J. Int. *105*, 271–295.

POUPINET, G., ELLSWORTH, W.L., and FRÉCHET, J. (1984), *Monitoring velocity variations in the crust using earthquake doublets: An application to the Calaveras fault, California*, J. Geophys. Res. *89*, 5719–5731.

SLUNGA, R., RÖGNVALDSSON, S., and BÖDVARSSON, R. (1995), *Absolute and relative locations of similar events with application to microearthquakes in southern Iceland*, Geophys. J. Int. *123*, 409–419.

RUBIN, A.M., GILLARD, D., and GOT, J.-L. (1999), *Streaks of microearthquakes along creeping faults*, Nature *400*, 635–641.

SHEARER, P.M. (1997), *Improving local earthquake locations using the L_1 norm and waveform cross-correlation: application to the Whittier Narrows, California, aftershock sequence*, J. Geophys. Res. *102*, 8269–8283.

TARANTOLA, A., *Inverse Problem Theory* 613 pp (Elsevier, New York, 1987).

TARANTOLA, A., and VALETTE, B. (1982), *Generalized nonlinear inverse problems solved using the least-squares criterion*, Rev. Geophys. Space Phys. *20*, 219–232.

VALETTE, B., and LESAGE, P. (2005), *Inferring mean earth mechanical models from normal modes, mass and inertia: 1. Theoretical developments*, submitted to Geophys. J. Int.

WALDHAUSER, F., and ELLSWORTH, W.L. (2000), *A double difference earthquake location algorithm: Method and application to the northern Hayward fault, CA*. Bull. Seismol. Soc. Am. *90*, 1353–1368.

WOLFE, C.J. (2002), *On the mathematics of using difference operators to relocate earthquakes*, Bull. Seismol. Soc. Am. *92*, 2879–2892.

ZHANG, H., and THURBER, C.H. (2003), *Double difference tomography: The method and its application to the Hayward fault, California*, Bull. Seismol. Soc. Am. *93*, 1875–1889.

(Received April 4, 2005; accepted October 15, 2005)

Pure appl. geophys. 163 (2006) 431–453
0033–4553/06/030431–23
DOI 10.1007/s00024-005-0033-7

Pure and Applied Geophysics

A New Passive Tomography of the Aigion Area (Gulf of Corinth, Greece) from the 2002 Data Set

S. Gautier,[2] D. Latorre,[3] J. Virieux,[1] A. Deschamps,[1] C. Skarpelos,[4] A. Sotiriou,[4] A. Serpetsidaki,[4] and A. Tselentis[4]

Abstract—We present the results of a tomographic study performed in the framework of the 3F-Corinth project. The aim of this work is to better understand the rifting process by imaging the crustal structure of the western Gulf of Corinth. Forty-nine stations were deployed for a period of six months, allowing us to monitor the microseismicity. Delayed P and S first-arrival times have been simultaneously inverted for both hypocenter locations and 3-D velocity distributions. We use an improved linearized tomography method based on an accurate finite-difference travel-time computation to invert the data set. The obtained Vp and Vs models confirm the presence of a two-layer vertical structure characterized by a sharp velocity gradient lying at 5–7 km depth, which may be interpreted as a lithological contrast. The shallower part of the crust (down to 5 km depth) is controlled by the N-S extension and lacks seismicity. The deeper part (7–13 km depth) matches the seismogenic zone and is characterized by faster and more heterogeneous anomalies. In this zone, the background seismicity reveals a low-angle active surface dipping about 20° toward the north and striking WNW-ESE. The position of this active structure is consistent with both high Vp/Vs and low Vp.Vs anomalies identified at 8–12 km depth and suggesting a highly fracturated and fluid-saturated zone. Both the geometry of the active structure beneath the gulf and the presence of fluids at 8–12 km depth are in accordance with a low-angle detachment model for the western part of the Gulf of Corinth.

Key words: Gulf of Corinth, rifting, low-angle detachment, seismic tomography, Vp/Vs, Vp.Vs.

1. Introduction

The Gulf of Corinth (Greece) is a continental rift, which has long been recognized as one of the most active regions in Europe. The historical seismicity is very high: five events of magnitude larger than 5.8 have been observed in this area within the last 40 years. The last event was the Ms = 6.2, 15 June 1995, Aigion earthquake

S. Gautier and D. Latorre formerly at Géosciences Azur

[1]Géosciences Azur, Université de Nice - Sophia Antipolis/CNRS, 06560, Valbonne, France.
[2]Laboratoire de Tectonophysique, Université de Montpellier 2/CNRS, Place Eugène Bataillon, CC049, 34095, Montpellier cedex 5, France.
[3]Instituto Nazionale di Geofisica e Vulcanologia, CNT, Roma, Italy.
[4]Patras Seismological Laboratory, University of Patras, Greece.

(BERNARD *et al.*, 1997), located in the western part of the gulf where we focus our study.

Despite a large amount of recent tectonic, geodetic and seismic observations, there is considerable debate concerning the rifting process. The presence of a low-angle detachment zone (RIGO *et al.*, 1996; SOREL, 2000) and the rise of the brittle-ductile transition up to 10 km depth (HATZFELD *et al.*, 2000) are the two main tectonic models proposed to explain both the distribution of the seismicity at depth and the focal mechanisms. The understanding of the rifting process first requires a better-constrained structural model of the mid-crust for the Corinth region.

In 1991 a two-month-long passive seismological experiment took place in the western part of the gulf, around the city of Aigion (Fig. 1). The collected data set was used by RIGO *et al.* (1996) for seismotectonic interpretation. This data set has been used as well to perform tomographic studies. First results were obtained by LE MEUR *et al.* (1997) and new velocity models were constructed by LATORRE *et al.* (2004b) using an improved tomographic technique. LATORRE *et al.* (2004b) found a two-layer vertical structure and the presence of a high Vp/Vs anomaly at depth, suggesting a possible role of fluids in the rifting process. Despite important results, the quality of the 1991 data is limited because of many vertical single component (1C) stations, imposing a major limitation in constraining the mid-crust structure and the seismicity distribution.

A new passive seismological experiment (3F-Corinth European Project) was performed from November 2001 until June 2002 around the Aigion area. The new data set has been collected with a homogeneous and efficient instrumental network (Fig. 1). The aim of this seismological experiment is to provide 3-D and high-resolved images of the mid-crust in order to shed light on our understanding of the rifting process. In this paper, we present the results of a tomographic study obtained from the large 2002 data, using the technique presented by LATORRE *et al.* (2004b), VANORIO *et al.* (2005), and MONTEILLIER *et al.* (in press).

2. *Geologic Setting*

The Gulf of Corinth is located in the Aegean area, in the western part of the Mediterranean region. The Aegean is bounded at the south and west by the Hellenic arc. The African plate subducts beneath the Aegean along the Hellenic arc with a rate of $\sim$0.5 cm.yr^{-1} (MCCLUSKY *et al.*, 2000). At the east of the Aegean, the Anatolian block is laterally extruded westward along the North Anatolian fault.

Extension in the Aegean region started in the Miocene times (LE PICHON and ANGELIER, 1979; JOLIVET *et al.*, 1994; ARMIJO *et al.*, 1996). This extension could be related to both a gravitational collapse of the Hellenides mountains (LE PICHON *et al.*, 1995; JOLIVET, 2001) and a lithospheric thinning in the Aegean backarc zone (DOUTSOS *et al.*, 1988). A series of periodic graben structures seems to accommodate

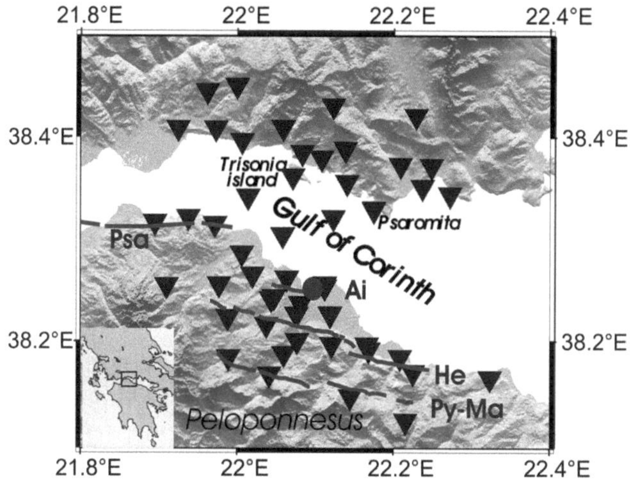

Figure 1
Map of the western Gulf of Corinth. The selected area is shown on the left bottom corner. The location of the town of Aigion is indicated with the dark dot. The complete network used during the 2001–2002 experiment is represented with triangles and major faults are drawn in dark grey (Psa: Psathopyrgos fault; Ai: Aigion fault; He: Heliki fault and Py-Ma: Pyrgaki-Mamoussia fault).

a part of this deformation (ARMIJO *et al.*, 1996; TIBERI *et al.*, 2001). In this paper, we focus on one of the most active of these structures: the Gulf of Corinth.

The Gulf of Corinth, which separates the Peloponnesus from continental Greece, is a 110 km-long, N120°E oriented asymmetrical graben, bounded by quaternary E-W normal faults (Fig. 1). The extension is N10° and GPS measurements show that its rate is more rapid in the west (1.6 cm.yr^{-1} at Aigion) than in the east (1.1 cm.yr^{-1} at Xilocastro, 50 km eastward from Aigion) (BRIOLE *et al.*, 2000; AVALLONE *et al.*, 2004). The most active faults crop out on the southern uplifted side of the Gulf with a dip of 50–60° toward the north. These active faults correspond to a well-organized *en echelon* system, with a characteristic length of 15–25 km (ARMIJO *et al.*, 1996). On the contrary, some smaller antithetic faults can be found on the northern subsided coast.

The geological structure around the gulf region results from the E-W Cenozoic Alpine compression which gave rise to the Hellenic mountains (JACOBSHAGEN *et al.*, 1978). The Hellenic mountains are comprised of a stack of Mesozoic nappes with a NNW-SSE trend. In our studied area, the Pindos series corresponds to the shallowest unit. It consists of deep-water carbonates with a minimum thickness of 3.5 km (AUBOUIN *et al.*, 1962). Magneto-telluric modelling suggests that the Pindos unit is highly fractured up to 3 km depth (PHAM *et al.*, 2000). This nappe overthrusts the Gavrovo-Tripolitza series composed of about 2.5-km-thick carbonate platform sediments (AUBOUIN *et al.*, 1962). Both the Pindos and the Gavrovo-Tripolitza series extend on the southern and the northern edges of the Gulf (DOUTSOS and POULIMENOS, 1992). Below these nappes, a metamorphic unit called the Phyllite series is likely be encountered (JACOBSHAGEN *et al.*, 1978; DORNSIEPEN *et al.*, 2001). The Phyllite series consists of an alternation of quartzites, phyllites and schists (DORNSIEPEN *et al.*, 2001). This unit crops out on a tectonic window located on the southern side of the studied area (XYPOLIAS and KOUKOUVELAS, 2001) while the extent of this nappe toward the north is not clearly known. Finally, field observations point out a 1-km-thick synrift sedimentary sequence in north Peloponnesus (DOUTSOS and POULIMENOS, 1992; FLOTTÉ and SOREL, 2001) whereas no evidence of synrift deposits was found in the northern edge of the Gulf (DOUTSOS and POULIMENOS, 1992). Moreover, east of our studied area, reflection seismic images show that the synrift sediment basin beneath the Gulf reaches a depth of 2.5 km (SACHPAZI *et al.*, 2003).

Gravity data suggest the presence of a NW-SE trending crustal thinning beneath the northern side of the Gulf of Corinth, probably related to the Miocene extension (TIBERI *et al.*, 2001). Some authors assume that the Gulf structure has been reactivated in Quaternary times ($\sim$1 Ma) by the southwestward propagation of the North Anatolian fault (ARMIJO *et al.*, 1996). This led to a westward propagation of the rift (CLARK *et al.*, 1997), which can be a possible cause of the high extension rate observed in this area (AVALLONE *et al.*, 2004). Both gravity and seismological data (TIBERI *et al.*, 2000, 2001) also indicate an offset between the crustal thinning and the location of the rift at the surface that becomes maximum in the western part of the Gulf. This suggests a modification of the rupture propagation between the Miocene and the Quaternary times. Changes in boundary conditions or pre-existing crustal heterogeneities may account for this offset (TIBERI *et al.*, 2001).

In the western part of the Gulf, the microseismicity is mainly distributed along a subhorizontal north-dipping trend, located at about 6–12 km depth (RIGO *et al.*, 1996; LATORRE *et al.*, 2004b; LYON-CAEN *et al.*, 2004). RIGO *et al.* (1996) interpret the seismogenic zone as a north-dipping detachment structure that could accommodate the deformation in this area. Fault plane solutions indicating the presence of an active low-angle dipping structure at depth are consistent with this tectonic model (RIETBROCK *et al.*, 1996; BERNARD *et al.*, 1997). Some authors also hypothesize a connection between this low-angle structure and the steeper north-dipping normal faults that crop out on the southern edge of the Gulf (RIETBROCK *et al.*, 1996; RIGO *et al.*, 1996; BERNARD *et al.*, 1997; SOREL, 2000; FLOTTÉ and SOREL, 2001). HATZFELD *et al.* (2000) propose an alternative model suggesting that the background microseismicity marks a brittle-ductile transition zone, without any peculiar structure. In their interpretation, microearthquake mechanisms with a dip steeper than 15° may be related to small faults located at the base of the brittle-crust.

3. The 2002 Passive Tomographic Experiment

In the framework of the 3F-Corinth project, a dense passive seismological experiment took place around the Aigion area, in the western part of the Gulf of

Corinth, from November 2001 until June 2002. From February until June 2002, the network was made up of 30 3C L22 2 Hz digital and 19 3C CMG40 60 s digital stations using GPS time (temporary stations and CRL permanent stations (LYON-CAEN *et al.*, 2004)) deployed on both sides of the Gulf. The complete network is represented in Figure 1. The distance between two stations is around 5 km. During March-April, three 3C ocean-bottom seismometers were deployed in the center of the Gulf for a one-month period, allowing an increase of the ray coverage in the study area. Compared with the 1991 experiment, the quality of the new collected data set was improved thanks to a geographic and instrumental homogeneity, improved timing, improved station locations, and 3C sensors.

A mean of 400 events per month was recorded during this dense passive experiment. Both P- and S-first arrival times were handpicked and an initial location was obtained using the 1-D model proposed by RIGO *et al.* (1996), assuming an *a priori* Vp/Vs ratio of 1.8. In order to obtain a more reliable and uniform tomographic data set, we have performed a selection using different criteria. First, we selected first arrival times that have high-quality readings by following the pick qualities given by the operator ($w_P \leq 2$, $w_S \leq 3$). Then, we removed events with a greatest angle without observation (GAP) higher than 180° and with a focal depth lower than 17 km, i.e., the maximum depth of the tomographic velocity model. Finally, only events with more than 8P and 4S picked phases and with a rms travel-time residual value lower than 0.5 s were kept for this study. The resulting reduced data set consists in 451 events with 9236 P- and 7523 S-first arrival times.

Figures 2 and 3 show the initial velocity distribution and hypocenters localized in the 1-D model of RIGO *et al.* (1996). In accordance with previous results, we note that the microseismicity lies between 4 km and 15 km depths and is particularly concentrated between 6 km and 12 km depths (RIGO *et al.*, 1996; LATORRE *et al.*, 2004b; LYON-CAEN *et al.*, 2004). Pictures also indicate a lack of seismicity above 4 km depth.

4. *The 3-D Delayed Travel-time Tomography*

4.1 Methodology

We use a 3-D delayed travel-time tomography method to invert simultaneously the velocity distribution and the hypocenter parameters (AKI and LEE, 1976; SPAKMAN and NOLET, 1988; BENZ *et al.*, 1996; SPENCER and GUBBINS, 1980; THURBER, 1992). The inversion method provides a smooth velocity model estimated on a 3-D, regularly spaced, rectangular grid. Sharp discontinuity imaging may be performed by converted phase analysis (LATORRE *et al.*, 2004a) allowing detection of small-scale discontinuities, as active fault planes or other structural interfaces (BEN-ZION and MALIN, 1991). A comprehensive description of the tomographic method and its linearized iterative scheme is given by LATORRE *et al.* (2004b).

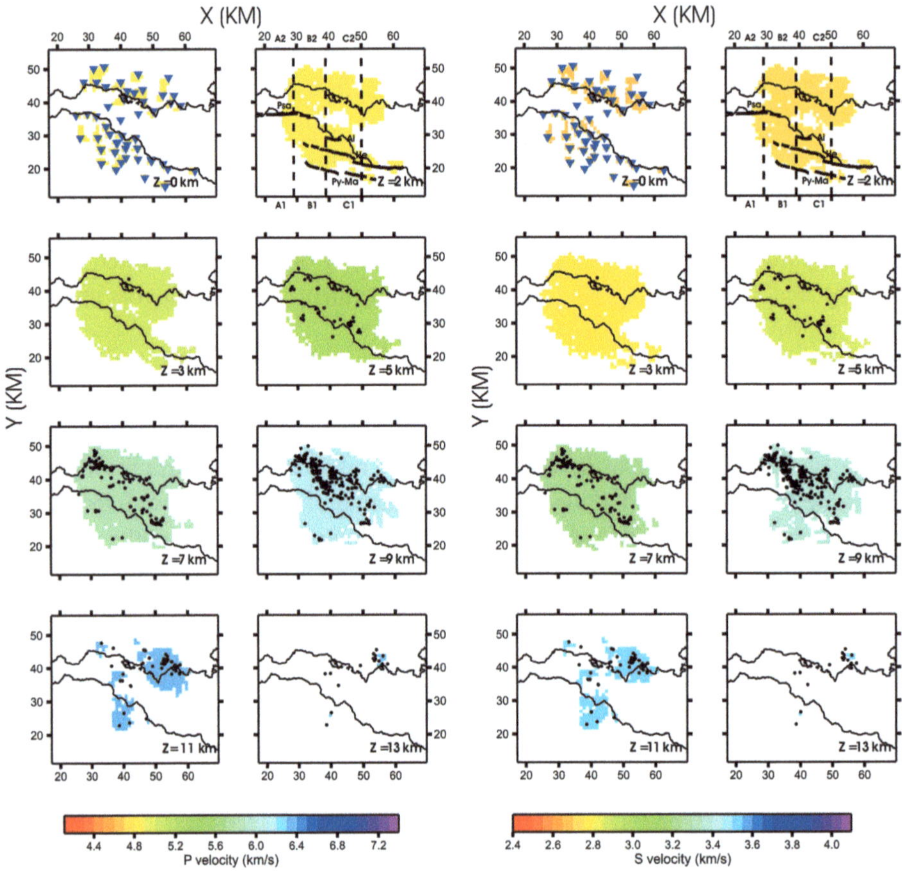

Figure 2
Hypocenters localized in the 1-D model (Vp on the left and Vs on the right) and used for the tomographic study are presented on map views at different depths ranging from 0 km to 13 km. Distances are represented in the X-Y coordinate system derived from the Universal Transverse Mercator (UTM) projection (zone 34) and a local origin point (longitude = 21.6°E and latitude = 38°N) is used for plotting presentation. Events located in a range of 1 km around each layer are displayed (dark dots). Stations are plotted on the first section (blue triangle) whereas the major faults are drawn on the second map. 1 km^3 cells with no ray coverage have been masked. The color scale indicates the absolute velocity from the lowest value in red to the highest value in blue.

At each iteration, we compute travel-times by solving the Eikonal equation with a finite-difference algorithm (PODVIN and LECOMTE, 1991). Rays are obtained using an a *posteriori* ray-tracing method that is based on time gradients. More precise travel-times and partial derivatives, both for slowness fields and for hypocenter parameters, are evaluated along the ray paths. As proposed by LE MEUR *et al.* (1997), a normalization/scaling of the derivative matrix is performed in order to control the quality of the recovered parameters. The aim of this operation is to

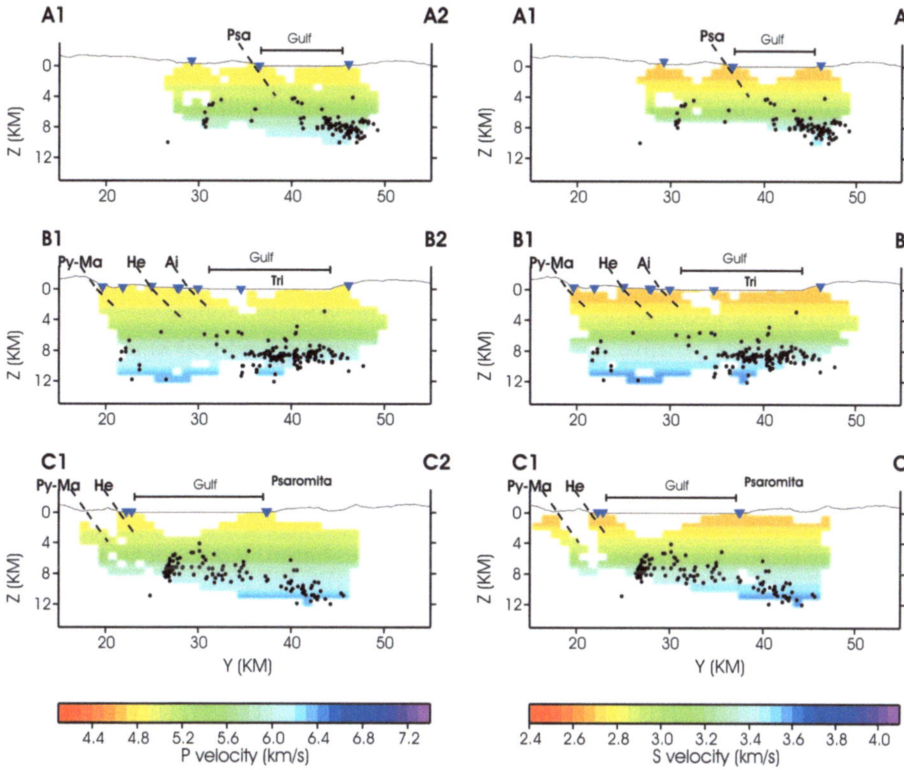

Figure 3
S-N sections of the initial microseismicity localized in the 1-D model (Vp on the left and Vs on the right). The location of each S-N section is indicated on Figure 2. Earthquakes (dark dots) that are located in a 3.5 km wide area around each section are plotted. Major normal faults are displayed on the pictures. We used the same mask as in Figure 2.

remove influences of parameter units and also to take into account the sensitivity of the data to each class of parameters. Because this procedure is configuration dependent, synthetic tests with the real source-receiver distribution were performed to estimate values of the weightings. Residual weighting of the data set is also used in our tomography to eliminate the outliers. Finally, the scaled and weighted linear system is solved by means of the *LSQR* method (PAIGE and SAUNDERS, 1982) and both the velocity models and the hypocenter parameters are updated. For inversion regularization, we use a damping factor determined through synthetic tests. For *LSQR*, 70 internal iterations allow a good accuracy of recovered parameters. The total number of iterations for the global tomographic procedure with new ray tracing is fixed to 25 iterations. An *a posteriori* analysis of both misfit and model perturbation functions shows that the convergence is usually reached after 10 iterations (Fig. 7).

4.2 Parameterization for the 2002 Experiment

The tomographic inversion is applied over a 56 km × 56 km area, which includes all the stations. The velocity model is defined up to 17 km depth in order to include all the ray paths. In order to compare the 2002 tomographic results with those of 1991, we use the same grid as the optimal one defined for the 1991 tomographic study (LATORRE *et al.*, 2004b). The distance between grid points is 7 km in the two horizontal X and Y directions and 2 km in the vertical Z direction.

We fix the different weightings and the damping parameter through synthetic tests using the real event-station geometry. Synthetic velocity models were obtained by adding a central positive anomaly (400 km.s^{-1} for Vp and 200 km.s^{-1} for Vs) within a homogeneous distribution (4000 km.s^{-1} for Vp and 2000 km.s^{-1} for Vs). Theoretical travel-times were computed in these velocity models in order to generate the synthetic data set. We added a noise term to the synthetic arrivals from a uniform distribution between −0.05 s and 0.05 s. This simulates systematic errors associated with the arrival times as the picking ones. In order to obtain new hypocenter locations, events were relocated in the initial model defined as the homogeneous velocity distribution. Tomographic inversions were performed for different sets of weightings and damping parameter. For the estimation of the optimal parameterization, we analyze the quality of both the reconstructed velocity models and the retrieved hypocenter parameters. A low final rms value is an additional criterion. Analyzing the test results, we estimated that the optimal set of weightings for the 2002 tomographic study is 1 for P waves, 2 for S waves, 5 for both the location and the origin time of earthquakes and finally 0.5 for the damping parameter. These results are consistent with the set of parameters used by LATORRE *et al.* (2004b) for the 1991 data set corresponding to a damping value 0.5 and weights equal to 1 for P waves, 1.25 for S waves and 5 for the hypocenter parameters. The strong difference of the weighting factor for S waves illustrates the more homogeneous and dense S-coverage (very high number of S-pickings) of the 2002 data set with respect to that of 1991.

4.3 The Initial Velocity Model

We start with a 1-D velocity model: the P-velocity model proposed by RIGO *et al.* (1996). The corresponding initial S-velocity model is directly obtained from the P-velocity distribution assuming a Vp/Vs ratio of 1.8.

Because the initial model has a strong influence in linearized iterative inversions (KISSLING *et al.*, 1994), we analyze the sensitivity of the 2002 tomographic results with respect to the starting model. No specific estimation of the model uncertainty was quantified in this study because the medium is insufficiently sampled to give relevant results (MOSEGAARD and TARANTOLA, 1995). The procedure consists in testing many possible starting 1-D models to assess the robustness of the tomographic results. We defined an upper limit

(+1000 km.s^{-1}) and a lower limit (–1000 km.s^{-1}) around the initial P-velocity model, after which, inversions are performed for 400 1-D models randomly obtained in this given domain, using the real station-event configuration and the optimal parameterization.

Results are presented in Figure 4. The dense heterogeneous distribution of initial rms values (from 0.16 s to 0.75 s) confirms a good sampling of the 1-D starting model space. The 400 final weighted rms values are almost identical, suggesting that there are no important differences in the recovered models. We may deduce that the tomographic results are rather independent of the initial reference in the range of 1-D tested models. Because the model proposed by RIGO *et al.* (1996) gave both the lowest initial and the lowest final weighted rms values, it seems to be the most suitable model among the 400 tested to provide reliable tomographic images.

4.4 Resolution Estimation

We perform checkerboard and spike anomaly tests in order to assess resolution of the tomographic models. Spike tests may be used for the resolution matrix reconstruction for both velocity and earthquake parameters, whereas checkerboard tests provide a global and quick insight of the local resolution length by examining the degree of smearing of the velocity patterns. Checkerboard tests are a practical way of defining the minimum anomaly size that is expected to be resolved in tomographic inversions. This is an *a posteriori* procedure because the final tomographic model is required for performing the checkerboard tests.

First, an input model is constructed by adding a velocity perturbation to the final tomographic velocity models (400 km.s^{-1} for Vp and 200 km.s^{-1} for Vs). We verify that the velocity perturbation is strong compared to the numerical noise level and also small enough to avoid noticeable disturbances in the ray coverage. Synthetic travel-times are computed in this new input model using the 2002 source-receiver geometry. In addition, a noise term is added to the synthetic data set from a uniform distribution between –0.05 s and 0.05 s. This simulates errors in the arrival times such as for example picking errors. The resulting synthetic data set is then inverted using the same procedure and parameterization that was used for the real data. Finally, the recovered velocity is compared to the input model in order to estimate the model resolution for some parameters like the amplitude, the location, the size and the shape of the reconstructed anomalies, as well as earthquake parameters.

We performed several checkerboard tests for input anomalies of different sizes and locations. The smaller synthetic pattern anomaly corresponds to $7 \times 7 \times 2$ km^3, i.e., the spacing of the inversion grid. Figure 5 displays the results related to this smaller pattern in order to illustrate that the data set is able to reconstruct the checkerboard anomaly. The velocity perturbation beneath the seismic network is well resolved for both Vp and Vs between the depths of 1 km and 9 km. The

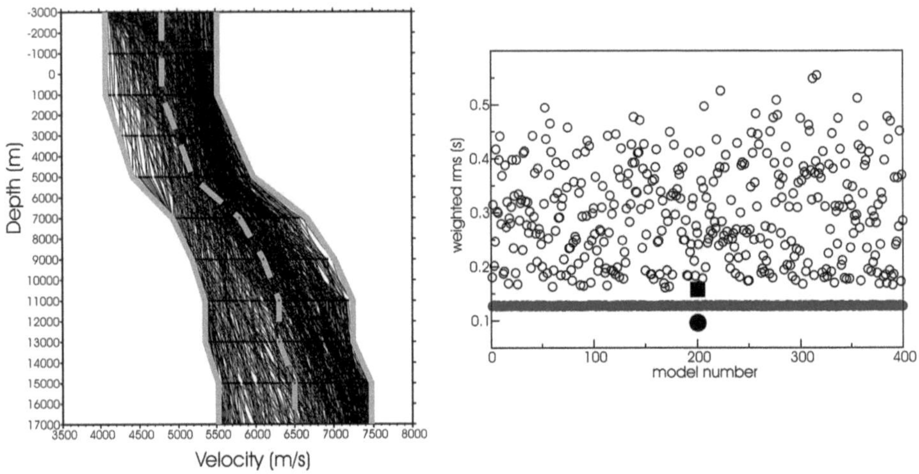

Figure 4
Sensitivity test performed for the initial 1-D model. The panel on the left shows the 400 random 1-D initial models. The dashed line indicates our selected 1-D-model (RIGO *et al.*, 1996) and the grey lines delineate the sampling space. The panel on the right presents the initial weighted rms corresponding to the 400-tested 1-D models (black circles) and the final weighted rms values (grey circles). The rms values for our selected 1-D model are indicated with a square for the initial one and with a circle for the final one.

images show a poor resolution below 11 km depth and at the surface (the −1 km layer), which may be a consequence of the weaker ray coverage. The smearing detected at 11 km depth is probably also due to poor ray sampling. One has to be aware of this maximum depth resolution for geodynamic interpretations.

To illustrate the resolution of our tomographic results, spike tests corresponding to different areas of the inversion grid are performed. Ideally, the resolution matrix is estimated by performing a spike test on each node of the inversion grid (LATORRE *et al.*, 2004b). In our case, the resolution matrix was not entirely computed but the analysis was focused on specific nodes related to the main anomalies observed in the tomographic model. This approach allows us to assess the resolution of different areas for the interpretation of the tomographic images. Taking into account the previous results, spikes have been distributed at different depth ranging from 3 km to 11 km depths. Figure 6 presents results for positive spikes but similar results have also been obtained for negative ones. Spikes located at 7 km depth are well recovered in shape and in position. At 9 km and 11 km depths, the location of the heterogeneities is correctly resolved, although the amplitude is ~50% lower in average. We found that the poor-resolved spike located at 3 km depth is smeared up. This demonstrates the capability of the data set to resolve velocity anomalies located beneath the seismic network between 3 km to 11 km depth.

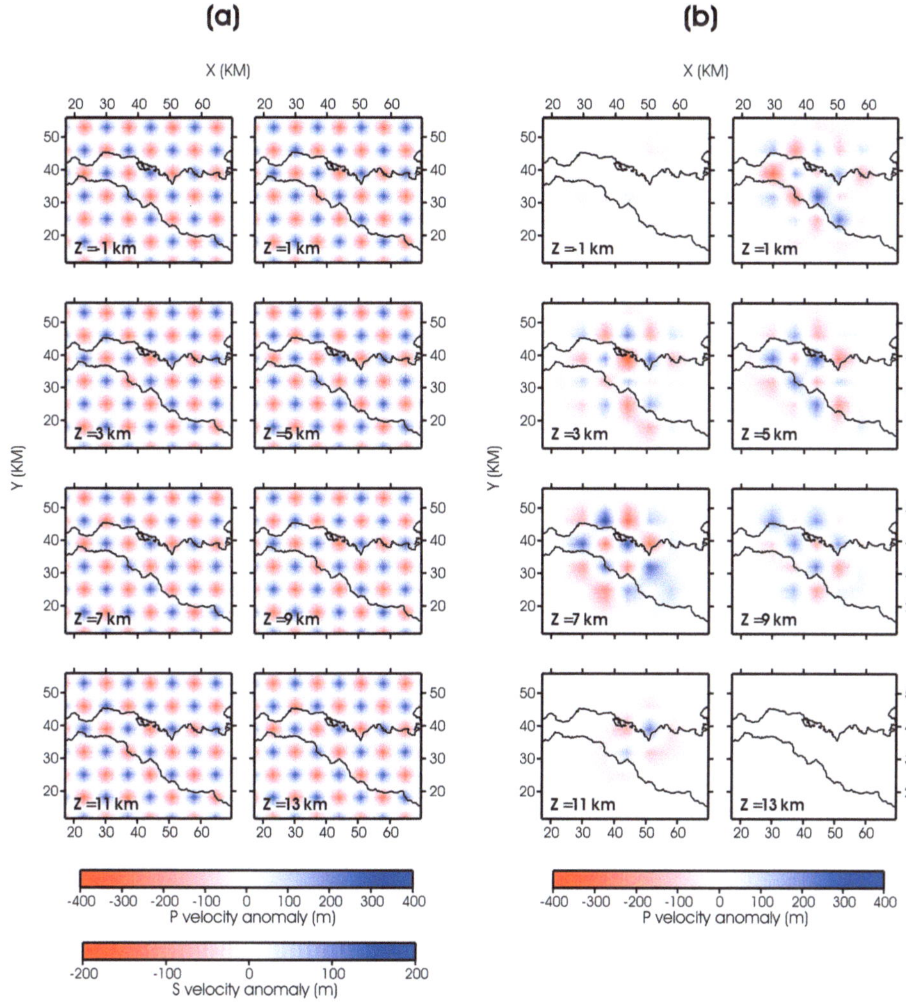

Figure 5
Checkerboard tests for both P- and S-velocity models. (a) Velocity perturbation for P waves (± 400 m/s) and S-waves (± 200 m/s) added to the 3-D tomographic models. The P- and S-reconstructed patterns are shown respectively in panels (b) and (c).

5. Results of the Tomography

We present the tomographic solution obtained after ten iterations. The convergence history (Fig. 7) points out that few modifications of the tomographic model occur after this iteration. The weighted root-mean-squares residual decreases from 162 ms to 106 ms, i.e., a reduction of about 34% of the weighted rms. Both the velocity models (Vp and Vs) and the earthquake locations are shown in Figure 8. In

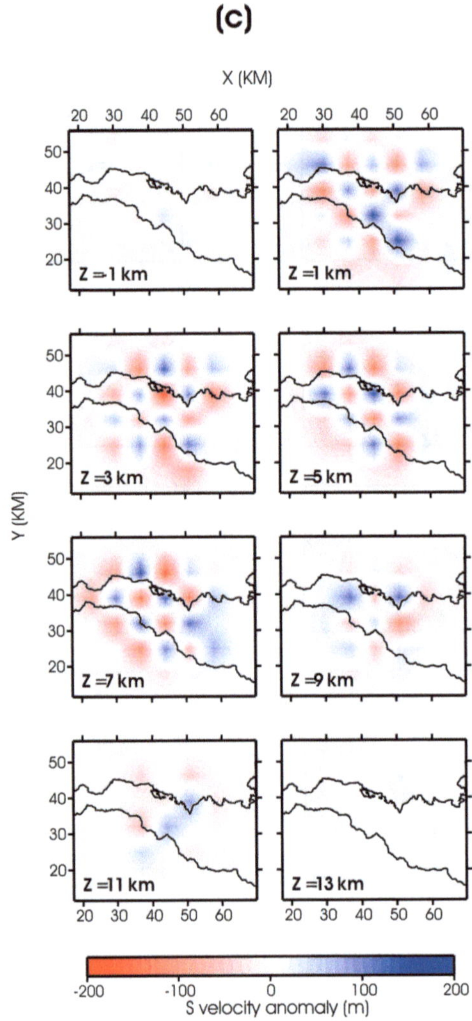

Figure 5
(Contd.)

addition to the map views, the results are also presented along three S-N vertical sections (Fig. 9), corresponding to different areas of the Gulf region, from the west to the east. Their locations are indicated by dotted lines on the map views.

5.1 The P- and S-velocity Models

The velocity models show a two-layer vertical structure occurring at a regional scale. A comparison between initial and final mean velocity profiles (Fig. 10) confirms an important increase of the velocity gradient at 5–7 km depth (from

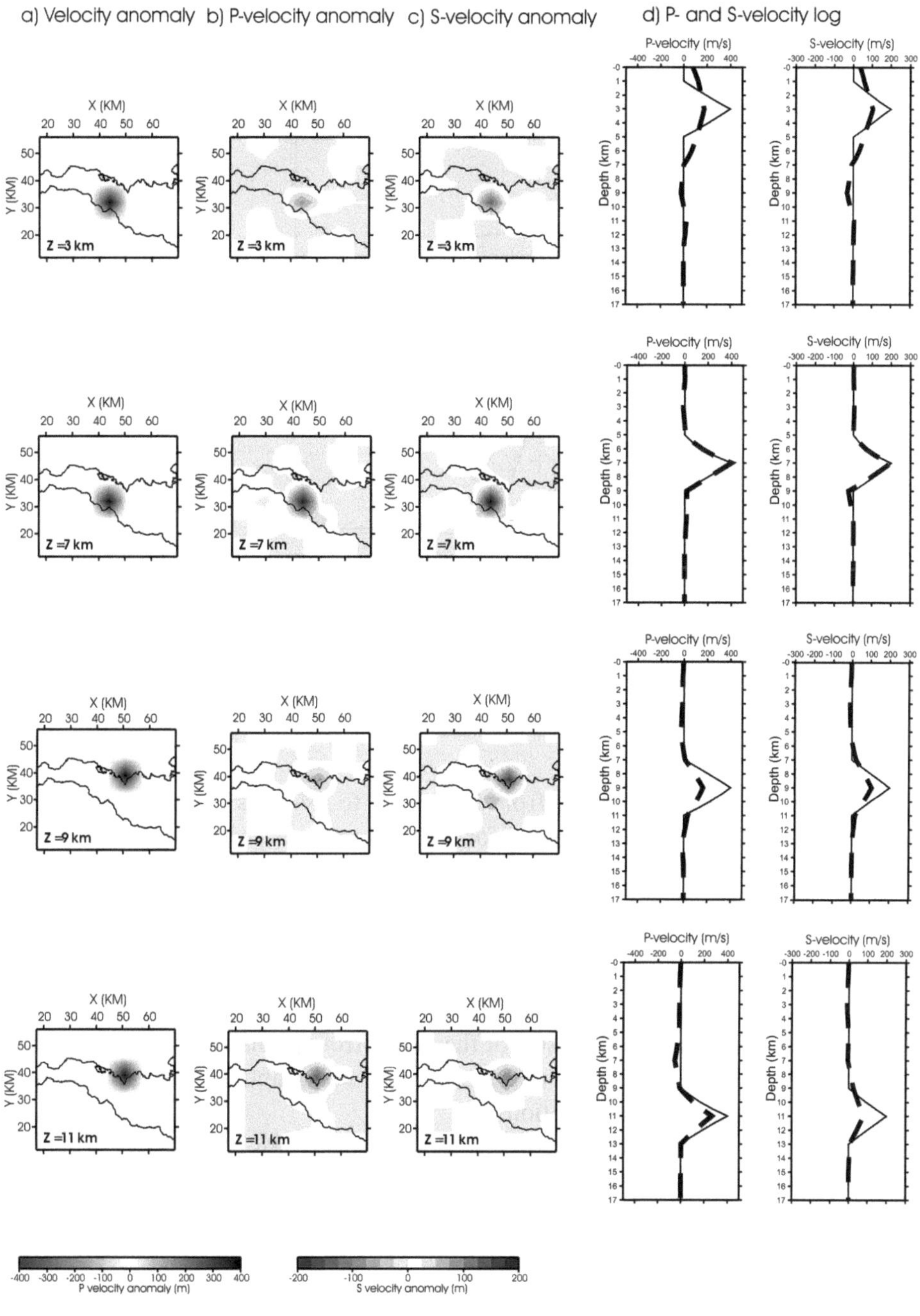

Figure 6
Spike tests for both P- and S- velocity models. Spikes are positioned at different depths ranging from 3 km (top) to 11 km (bottom). The different panels show: (a) Velocity perturbation for P-waves (± 400 m/s) and S-waves (± 200 m/s) added to the 3-D tomographic models, (b) and (c) the P- and S-reconstructed patterns, and (d) the P- and S-velocity logs corresponding to the spike positions.

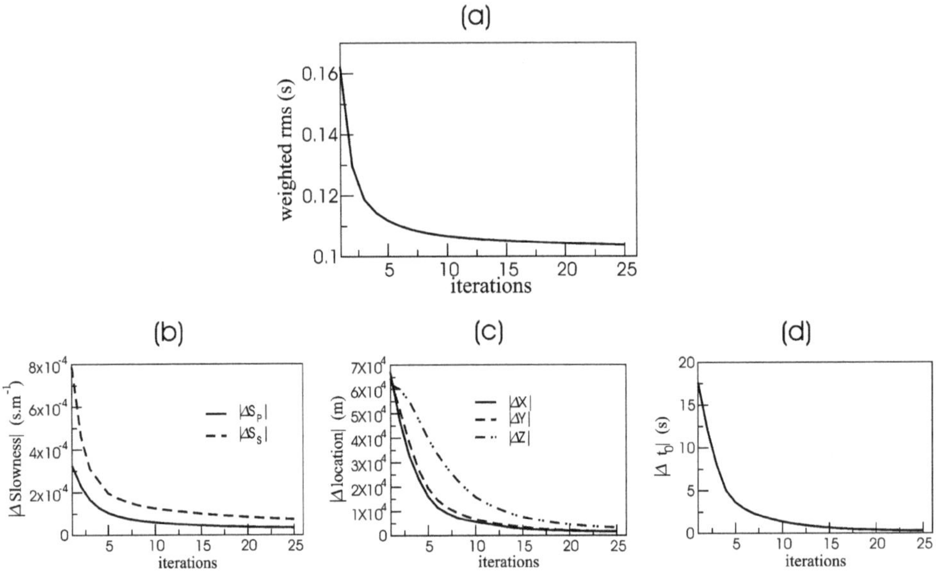

Figure 7
Representation of the convergence history of the tomographic inversion: (a) The misfit function, (b) the total model perturbation norm by step, (c) and (d) the total hypocenter parameters perturbation norm, respectively the event locations and the origin time. We assume that the convergence is reached when the perturbation of the different class of parameters is less than ~20%.

5.1 km.s^{-1} to 6 km.s^{-1} for Vp and from 2.9 km.s^{-1} to 3.4 km.s^{-1} for Vs). This sharp increase seems to be more accentuated for Vp. S-N sections point out a slight lateral variation of the location of this vertical transition.

Between depths of 0 km and 5 km, the reconstructed P and S models reflect similar velocity patterns. In particular, images point out a low-velocity anomaly (4.3 km.s^{-1} for the P waves and km.s^{-1} for the S waves) which correlates well with the graben axis orientation. This zone is bounded by the Heliki fault and the Pyrgaki-Mamoussia fault on the southern side of the Gulf. Around this anomaly, the velocity distribution is more homogeneous (5.2 km.s^{-1} for P waves and 3 km.s^{-1} for S waves) with less pronounced lateral variations.

Between 7 km and 13 km depth, the velocity fields are completely distinct in shape, in values and in position from the shallower one. The velocity distribution manifests lateral variations that are uncorrelated with the surface structures. We also note that the P- and S-velocity models display different patterns. For example, at 7 km depth, images show that high velocity anomalies are located on the northern side of the Gulf for the P fields whereas high velocity anomalies are mainly localized on the southern side for the S fields. However, the three S-N sections reveal a low P- and S- velocity anomaly located at 8–12 km depth beneath the northern side of the Gulf, near the Psaromita peninsula.

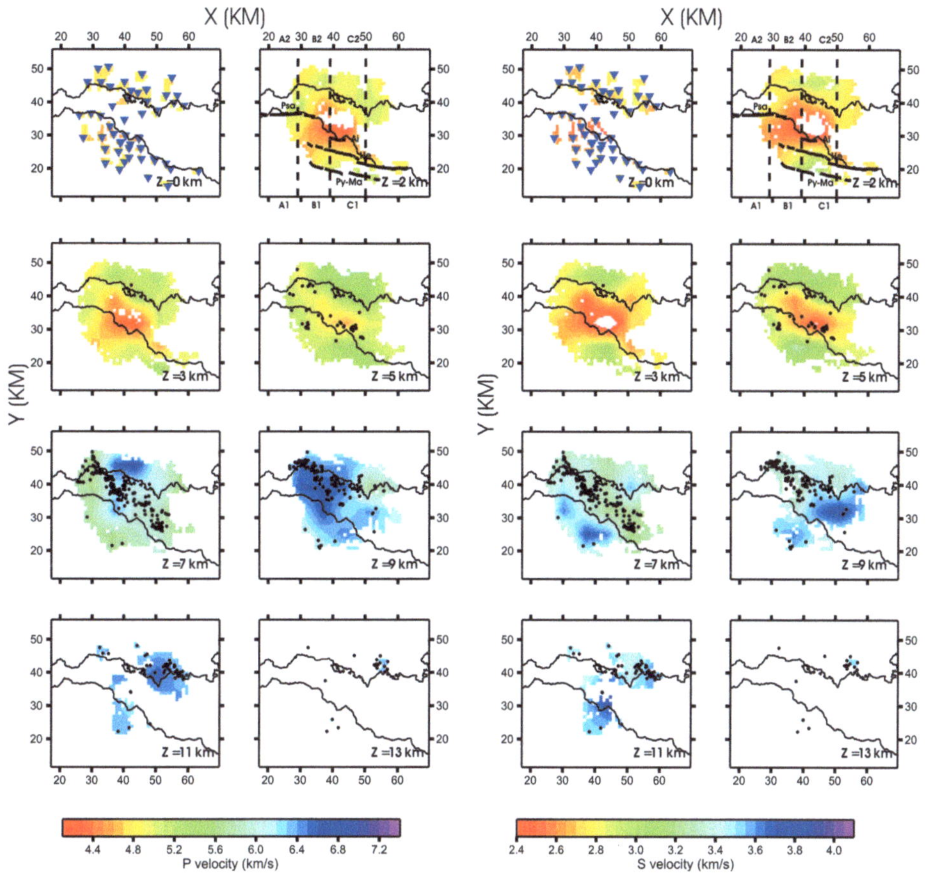

Figure 8
Presentation of the final tomographic results on map views at different depths ranging from 0 km to 13 km with the same caption as in Figure 2. Vp is presented on the left side and Vs on the right side.

5.2 Event Locations

The final location of the events is indicated on both map views (Fig. 8) and S-N sections (Fig. 9). In comparison with the initial locations, the tomographic results provide a narrow picture of the microseismicity. Event collapse and certain trends in the microseismicity distribution can be identified.

The first result is a lack of events in the shallower zone, between 0 km and 5 km depths. Both map views and S-N sections indicate that the microseismicity is concentrated in the deeper part of the model, between 6 km and 12 km depth. On the map view, at 7 km depth, we observe that earthquakes are concentrated in the center of the Gulf and aligned along a 120° NW-SE direction similar to the orientation of the Gulf. Event distribution moves from S-W to N-E with almost

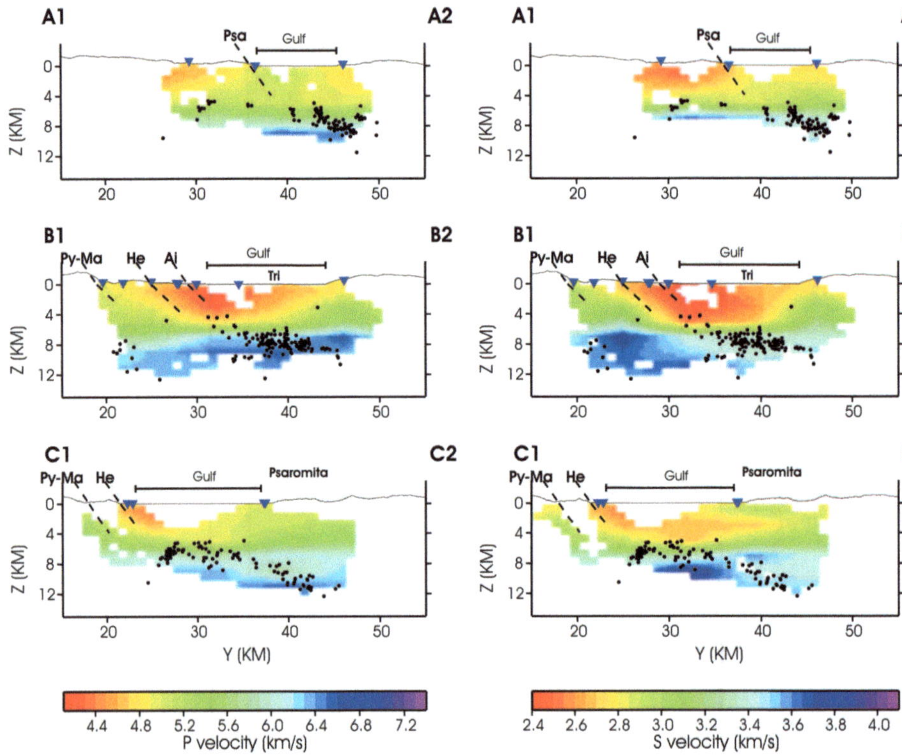

Figure 9

S-N sections of the tomographic results, respectively the P-velocity distribution on the left and the S-velocity model on the right. The location of each S-N section is indicated on Figure 8. Major normal faults are displayed on the pictures. We used the same mask as for the map views.

the same strike when the depth increases. The S-N sections also suggest that events trend along a low-angle dipping surface. This trend is clearly highlighted by projecting earthquakes on a vertical plane orthogonal to the 120° NW-SE direction given by the map view. Figure 11 points out that the microseismicity is aligned along a well-defined 20° north-dipping plane. It is a major result because previous studies did not precisely image this low-angle structure in three dimensions. Contrary to 1991, the seismic activity during the 2002 experiment was very quiet, allowing us to monitor the background microseismicity of the area. Finally, the C1-C2 section (Fig. 9) reveals that the microseismicity is localized in the low Vp and Vs anomalies observed at 8–12 km depth beneath the Psaromita area. Spike tests indicate that theses anomalies do not correspond to an artefact. The position is well resolved but the amplitude can be underestimated.

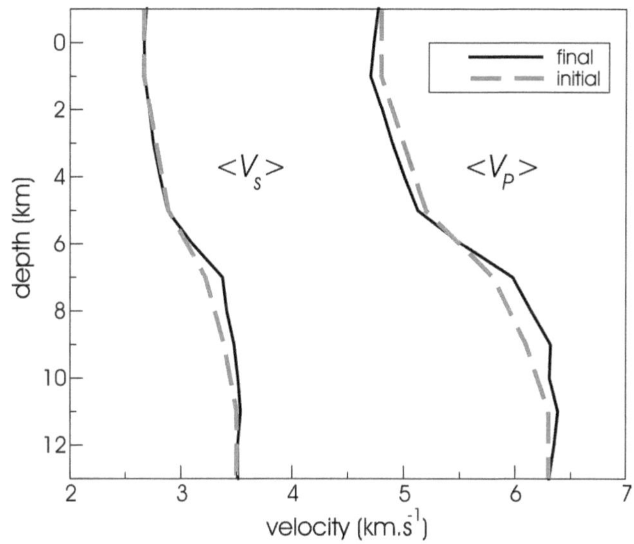

Figure 10
Mean velocity profiles, Vs on the left and for Vp on the right, estimated for both the initial model and the tomographic results, taking into account the mask of uncovered areas shown on Figure 8.

5.3 Three-dimensional Deduced Vp.Vs and Vp/Vs Models

Images of Vp.Vs and Vp/Vs parameters have been deduced from the tomographic models Vp and Vs. As previously done by LATORRE *et al.* (2004b), we have checked that these parameters are also at the minimum rms misfit by performing an independent inversion using Vp/Vs and Vp.Vs as original parameters. Results are presented on the three S-N sections shown in Figure 12. These two new models bring qualitative information about the distribution of rock properties at depth. For a given lithology, these two parameters are sensitive to porosity and fluid saturation (SANDERS *et al.*, 1995; LEES and WU, 2000; LATORRE *et al.*, 2004b; VANORIO *et al.*, 2005). In particular, low Vp.Vs values may be related to a high fractured zone whereas high Vp/Vs values can be explained by fluid saturation. Taking into account the intense seismicity and a possible role of fluids suggested by some authors (RIETBROCK *et al.*, 1996; BERNARD *et al.*, 1997; LATORRE *et al.*, 2004b), these two parameters might provide new constraints for the geodynamic interpretation in the Corinth region.

Vp.Vs images emphasize the two-layer vertical structure with a high Vp.Vs gradient lying at 5–7 km depth, increasing from ~15 $(km.s^{-1})^2$ to ~19 $(km.s^{-1})^2$. The shallower part is characterized by a low Vp.Vs anomaly (~11 $(km.s^{-1})^2$) probably related to the structure of the graben. Strong lateral variations of Vp.Vs are observed in the deeper part. In particular, the S-N sections point out a low Vp.Vs anomalous zone located northward between 8 km and 12 km depth. This anomaly is clearly

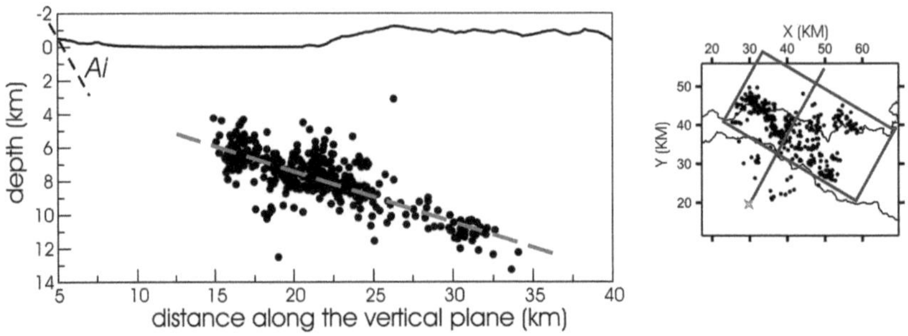

Figure 11
Projection of events located in a given area along a vertical plane orthogonal to the N-120° E orientation of the half-graben. The dashed grey line shows the result of a linear regression. The orientation of the vertical plane and the given area is indicated on the right on a map view. We chose this grey box in order to remove events located beneath the Peloponnesus, which are probably not related to the rifting process. The cross represents the origin of the distance computed along the vertical plane.

observed beneath the Psaromita area (Fig. 12, section C1-C2), where the microseismicity is concentrated in the low anomalous area. The Vp/Vs pictures show a high anomaly beneath the northern edge of the Gulf at 8–12 km depth (Fig. 12, sections A1–12 and B1–B2), while another positive anomaly is observed at 3–4 km depth (Fig. 9, section C1–C2). We note that events are localized inside the high Vp/Vs and low Vp.Vs zone at around 8–12 km depth.

6. *Discussion and Conclusion*

The 2002 data set has provided 3-D and high-resolved images of the mid-crust for the Aigion area. The tomographic models point out a two-layer vertical structure of the mid-crust, existing at a regional scale. The presence of this structure is emphasized by a high velocity gradient lying at 5–7 km depth. These results are in agreement with previous tomographic images obtained from the 1991 data set (LATORRE *et al.*, 2004b). As proposed by LATORRE *et al.* (2004b), this abrupt transition may describe a lithological contact between two different tectonic units of the Hellenic orogenic belt: the Gavrovo-Tripolitza carbonate unit in the upper part and the Phyllite series in the deeper part. By following this interpretation, our tomographic models show that the Gavrovo-Tripolitza nappes may extend down to 5–7 km depth, in accordance with the thickness estimation of the upper series (AUBOUIN *et al.*, 1962). Taking into account the structural position of the Phyllite series beneath the Gavrovo-Tripolitza unit in the Hellenides nappes system (e.g., JACOBSHAGEN *et al.*, 1978) and its presence in a tectonic window, south of the studied area (XYPOLIAS and KOUKOUVELAS, 2001), many authors assume the presence of the Phyllite unit beneath the Gulf of Corinth (PHAM *et al.*, 2000; LATORRE *et al.*, 2004b). In the deeper part of the tomographic models, velocity values are consistent with

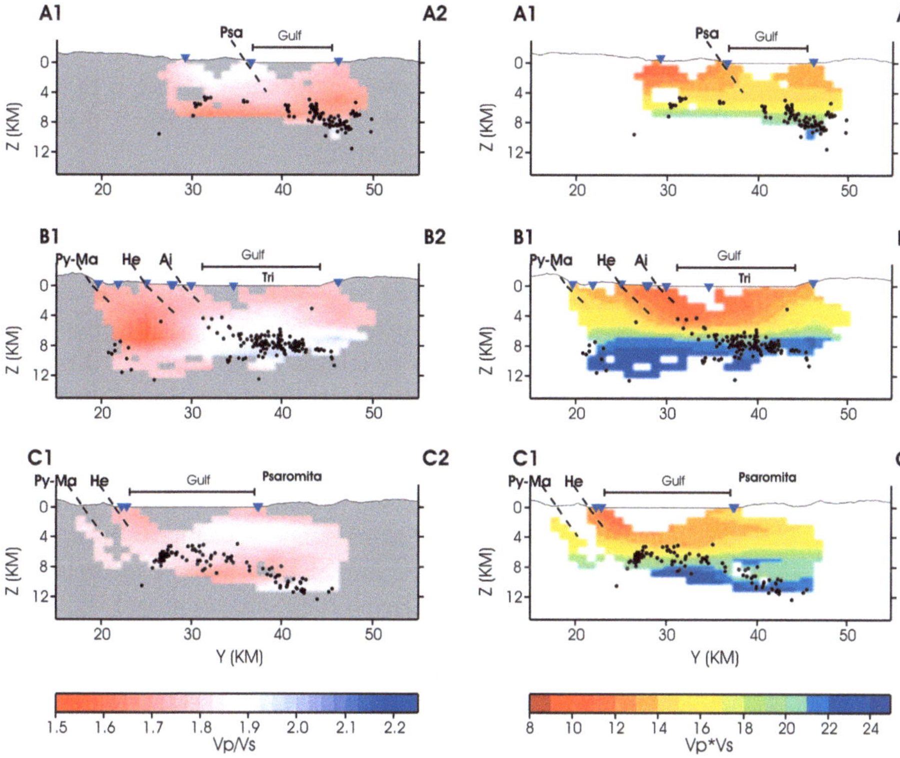

Figure 12

S-N sections of the deduced parameters Vp/Vs ratio (left) and Vp.Vs (right). See Figure 8 for location of the three sections. As in previous figures, main faults are indicated by dashed lines, stations are represented by blue triangles and events by dark dots.

velocity estimations of Phyllite rocks obtained from seismic refraction experiments and core measurements (Christensen and Mooney, 1995; Christensen, 1996). This supports the assumption of the presence of the Phyllite unit at depth greater than 7 km.

In the shallow crust, the velocity distribution seems to be controlled by the still active extension regime. The low velocity anomaly located beneath the Gulf is found parallel to the graben structure. On the southern side, the low-velocity zone is spatially correlated with the major faults, in particular the Heliki fault. Because this result is consistent with the 1991 tomographic images (Latorre *et al.*, 2004b), we can consider this low-velocity anomaly as a well-constrained structure. The trend, the position and the velocity values (Vp, Vs and Vs.Vs) of this anomaly might be explained both by the presence of thick synrift deposits and the existence of highly fracturated carbonates in the first 3 km. This assumption is supported by geological observations (Doutsos and Poulimenos, 1992; Flotté and Sorel, 2001), reflection seismic images (SACHPAZI *et al.*, 2003) and magneto-telluric modelling (PHAM *et al.*, 2000). Moreover, the 2002 tomographic results presented in this paper clearly confirm the lack of microseismicity in the shallower part, in accordance with long-term seismological observations (LYON-CAEN *et al.*, 2004). This is probably due to a low level of stress in this layer, except near major faults.

The background microseismicity marks a WNW-ESE surface dipping 20°N and lying at 5–13 km depth beneath the Gulf. The event distribution moves from the S-W to the N-E along this surface when the depth increases. Low-dip angle (≤ 20°) normal structures with a W-E strike were already observed by previous seismological studies (RIGO *et al.*, 1996; RIETBROCK *et al.*, 1996), but no SW-NE-trending distribution was detected. Differences in direction and in dip can be related to differences in the data set and in the velocity models. In order to determine the events location, these authors used a 1-D-velocity model (RIGO *et al.*, 1996), which does not take into account the lateral variations of the medium. On the contrary, the high quality of the 2002 data set and the 3-D tomography provide a well-constrained image of the low-angle structure, which might be interpreted as a low-angle detachment zone. The extension of the shallower anomaly down to 4–5 km depth is the sole information suggesting that the outcropped faults, here the Heliki fault, are connected with this seismogenic zone (RIGO *et al.*, 1996).

The identified low-angle dipping structure is parallel to the graben axis, indicating a contribution in the deformation of the Gulf region. This result would favor a geodynamic interpretation with two major tectonic phases (TIBERI *et al.*, 2000, 2001): 1) The Miocene extension involving crustal thinning and 2) a reactivation of the rift 1 Ma ago, in a more westward direction. The observed WNW-ESE low-dip surface is in good agreement with this scenario, which postulates that a low-angle detachment could accommodate the offset between the maximum crustal thinning and the graben structure in the western part of the Gulf. Moreover, the SW-NE distribution of the events along the low-dip surface might be related to the westward propagation of the rift.

The tomographic images also point out that the event distribution is consistent with the position of the low P- and S-velocity anomalies observed in the deeper layer (Fig. 9, section C1-C2). This indicates that the low-angle structure does not match the transition zone between the shallower and the deeper layer (HATZFELD *et al.*, 2000; LATORRE *et al.*, 2004b), but crosses this discontinuity. The low-velocity anomalies coupled with a decrease of the Vp.Vs parameter, may be associated with the presence of a highly fractured zone at 8–12 km depth. The presence of a highly fracturated zone can support the assumption of a circulation of deep fluids. This could explain the high Vp/Vs anomalies observed in the deeper layer on the tomographic images. Fluids may play an important role in the mechanism of the rupture because they may reduce the friction coefficient, allowing seismic slip along the low-angle detachment. This cause has been previously advocated by RIETBROCK *et al.* (1996) to explain the existence of the low-angle normal fault zone beneath the Gulf of Corinth.

Our results provide new insights on the geodynamic framework of this region. The 3-D velocity models will be used to locate events recorded by the permanent network CRLnet (LYON-CAEN *et al.*, 2004) in order to obtain a precise description of the active structures between Aigion and Trizonia Island. The next step will be a 4-D time-lapse imaging of the area. Deploying the same geometrical network of sensors will help us to design differential tomography if natural sources are found in proximity between these two experiments. Our tomographic models could then be used as the reference frame and one may hope to point out possible time-variations of specific parameters as was done for reservoir engineering. Consequently, comparative studies may bring further key elements for geodynamic modelling.

Acknowledgements

We are grateful to two anonymous reviewers for their critical reviews, which have improved the original manuscript. This work was performed through funding of the European 3F-Corinth Project (ENK6-CT-2000-00056), the French GDR Corinthe and the Egide Projects. Partial support has been provided by the CNRS with a post-doctoral fellowship.

REFERENCES

AKI, K. and LEE, W. (1976), *Determination of three-dimensional velocity anomalies under a seismic array using first P-arrival times from local earthquakes*, J. Geophys. Res. *81*, 4381–4399.

ARMIJO, R., MEYER, B., KING, G., RIGO, A., and PAPANASTASSIOU, D. (1996), *Quaternary evolution of the Corinth Rift and its implication for the late Cenozoic evolution of the Aegean*, Geophys. J. Int. *126*, 11–53.

AUBOUIN, J., BRUNN, J., CELET, P., DERCOURT, J., GODFRIAUX, I., and MERCIER, J. (1962), *Esquisse de la géologie de la Grèce*, Mém. Hors Sér. Soc. Géol. France *2*, 583–610.

AVALLONE, A., BRIOLE, P., AGATZA-BALODIMIOU, A., BILLIRIS, H., CHARADE, O., MITSAKAKI, C., NERCESSIAN, A., PAPAZISSI, K., PARADISSIS, D., and VEIS, G. (2004), *Analysis of eleven years of deformation measured by GPS in the Corinth Rift laboratory area*, C. R. Geoscience *336*, 301–311.

BENZ, H.M., CHOUET, S.A., DAWSON, P.B., LAHR, J.C., PAGE, R.A., and HOLE, J.A. (1996), *Three-dimensional P- and S- wave velocity structure of Redoubt Volcano, Alaska*, J. Geophys. Res. *101*, 8111–8128.

BERNARD, P., BRIOLE, P., MEYER, B., LYON-CAEN, H., GOMEZ, J.-M., TIBERI, C., BERGE, C., CATTIN, R., HATZFELD, D., LACHET, C., LEBRUN, B., DESCHAMPS, A., COURBOULEX, F., LARROQUE, C., RIGO, A., MASSONET, D., PAPADIMITRIOU, P., KASSARAS, J., DIAGOURTAS, D., MAKROPOULOS, K., VEIS, G., PAPAZISI, E., MITSAKAKI, C., KARAKOSTAS, V., PAPADIMITRIOU, E., PAPANASTASSIOU, D., CHOULIARAS, M., and STAVRAKAKIS, G. (1997), *A low angle normal fault earthquake: the Ms = 6.2, June, 1995 Aigion earthquake (Greece): Evidence for a low-angle normal faulting in the Corinth Rift*, J. Seismol. *1*, 131–150.

BEN-ZION, Y. and MALIN, P. (1991), *San Andreas fault zone head waves near Parkfield, California*, Science *251*, 1592–1594.

BRIOLE, P., RIGO, A., LYON-CAEN, H., RUEGG, J.-C., PAPAZISSI, K., MISTAKAKI, C., BALODIMOU, A., VEIS, G., HATZFELD, D., and DESCHAMPS, A. (2000), *Active deformation of the Gulf of Korinthos, Greece: Results from repeated GPS surveys between 1990 and 1995*, J. Geophys. Res. *105*, 25605–25625.

CHRISTENSEN, N. and MOONEY, W.D. (1995), *Seismic velocity structure and composition of the continental crust: A global view*, J. Geophys. Res. *100*, 9761–9788.

CHRISTENSEN, N. (1996), *Poisson's ratio and crustal seismology*, J. Geophys. Res. *101*, 3139–3156.

CLARK, P.J., DAVIES, R.R., ENGLAND, P.C., PARSONS, B.E., BILLIRIS, H., PARADISSIS, D., VEIS, G., DENYS, P.H., CROSS, P.A., ASHKENAZI, V., and BINGLEY, R. (1997), *Geodetic estimate of seismic hazard in the Gulf of Korinthos*, Geophys. Lett. *24*, 1303–1306.

DORNSIEPEN, U.F., MANUTSOGLU, E., and MERTMANN, D. (2001), *Permian-Triassic paleogeography of the external Hellenides*, Paleogeog. Paleoclimat. Paleoecol. *172*, 327–338.

DOUTSOS, T., KONTOPOULOS, N., and POULIMENOS, G. (1988), *The Corinth-Patras rift as the initial stage of the continental fragmentation behind an active island arc (Greece)*, Basin Research *1*, 177–190.

DOUTSOS, T. and POULIMENOS, G. (1992), *Geometry and kinematics of active faults and their seismotectonic significance in the western Corinth-Patras rift (Greece)*, J. Struct Geol. *14*, 689–699.

FLOTTÉ, N. and SOREL, D. (2001), *Structural cross sections through the Corinth-Patras detachment fault-system in Northern Peloponnesus (Aegean Arc, Greece)*, Bull. Geol. Soc. Greece *XXXIV*, 235–241.

HATZFELD, D., KARAKOSTAS, V., ZIAZIA, M., KASSARAS, I., PAPADIMITRIOU, E., MAKROPOULOS, K., VOULGARIS, N., and PAPAIOANNOU, C. (2000), *Microseismicity and faulting geometry in the Gulf of Corinth (Greece)*, Geophys. J. Int. *141*, 438–456.

JACOBSHAGEN, V., DURR, V., KOCKEL, F., KOPP, K.O., KOWALCZYK, G., BERCKHEMER, H., and BUTTNER, D. (1978), *Structure and geodynamic evolution of the Aegean Region*. In *Alpes Apennines Hellenides* (eds. Closs H., Roeder D., Schmidt K.) (Schweiz-Verlag, 1978) pp. 537–564.

JOLIVET, L., BRUN, J.P., GAUTIER, P., LALLEMEANT, S., and PATRIAT, M. (1994), *3D-kinematics extension in the Aegean region from the early Miocene to the present, insight from the ductile crust*, Bul. Soc. Geol. France *165*, 195–209.

JOLIVET, L. (2001), *A comparison of geodetic and finite strain pattern in the Agean, geodynamic implications*, Earth Planet Sci. Lett. *187*, 95–104.

KISSLING, E., ELLSWORTH, W.L., EBERHART-PHILLIPS, D., and KRADOLFER, U. (1994), *Initial reference models in local earthquake tomography*, J. Geophys. Res. *99*, 19635–19646.

LATORRE, D., VIRIEUX, J., MONFRET, T., and LYON-CAEN, H. (2004a), *Converted seismic wave investigation in the Gulf of Corinth from local earthquakes*, C.R Geosciences *336*, 259–267.

LATORRE, D., VIRIEUX, J., MONFRET, T., MONTEILLER, V., VANORIO, T., GOT, J.-L., and LYON-CAEN, H. (2004b), *A new seismic tomography of Aigion area Gulf of Corinth, Greece from a 1991 data set*, Geophys. J. Int. *159*, 1013–1031.

LEE, J.M. and WU, H. (2000), *Poisson's ratio and porosity at Coso geothermal area, California*, J. Volcanol. Geotherm. Res. *95*, 157–173.

Le MEUR, H., VIRIEUX, J., and PODVIN, P. (1997), *Seismic tomography of the Gulf of Corinth: A comparison of methods*, Ann. Geophys. *XL*, 1–25.

LE PICHON, X. and ANGELIER, J. (1979), *The Hellenic arc and trench system: A key to the neotectonic evolution of the eastern Mediterranean area*, Tectonophysics *60*, 1–42.

LE PICHON, X., CHAMOT-ROOKE, N., LALLEMANT, S. (1995), *Geodetic determination of the kinematics of central Greece with respect to Europe: Implications for eastern Mediterranean tectonics*, J. Geophys. Res. *100*, 12675–12690.

LYON-CAEN, H., PAPADIMITRIOU, P., DESCHAMPS, A., BERNARD, P., MAKROPOULOS, K., PACCHIANI, F., and PATAU, G. (2004), *First results of the CRLN seismic array in the western Corinth rift: Evidence for old fault reactivation*, C. R. Geoscience *336*, 343–351.

MCCLUSKY, S., BALASSANIAN, S., BARKA, A., DEMIR, C., ERGINTAV, S., GEORGIEV, I., GURKAN, O., HAMBURGER, M., HURST, K., KAHLE, H., KASTENS, K., KEKELIDZE, G., KING, R., KOTZEV, V., LENK, O., MAHMOUD, S., MISHIN, A., NADARIYA, M., OUZOUNIS, A., PARADISSIS, D., PETER, Y., PRILEPIN, M., REILINGER, R., SANLI, I., SEEGER, H., TEALEB, A., TOKSÖZ, M.N., and VEIS, G. (2000), *Global Positioning System constraints on plate kinematics and dynamics in the eastern Mediterranean and Caucasus*, J. Geophys. Res. *105*, 5695–5719.

MONTEILLER, V., GOT, J.L., VIRIEUX, J., and OKUBO, P., *An efficient algorithm for double-difference tomography and location in heterogeneous media, with an application to Kilauea volcano*, J. Geophys. Res., in press.

MOSEGAARD, K. and TARANTOLA, A. (1995), *Monte Carlo sampling of solutions to inverse problems*, J. Geophys. Res. *100*, 12431–12447.

PAIGE, C.C. and SAUNDERS, M.A. (1982), *LSQR: An algorithm for sparse linear equations and sparse least squares*, ACM Trans. Math. Software *8*, 43–71.

PHAM, V.N., BERNARD, P., BOYER, D., CHOULIARAS, G., MOUEL, J.L.L., and STAVRAKAKIS, G.N. (2000), *Electrical conductivity and crustal structure beneath the Central Hellenides around the Gulf of Corinth (Greece) and their relationship with the seismotectonics*, Geophys. J. Int. *142*, 948–969.

PODVIN, P. and LECOMTE, I. (1991), *Finite difference computation of travel-time in very contrasted velocity model: A massively parallel approach and its associated tools*, Geophys. J. Int. *105*, 271–284.

RIETBROCK, A., TIBERI, C., SCHERBAUM, F., and LYON-CAEN, H. (1996), *Seismic slip on a low angle normal fault in the Gulf of Corinth: Evidence from high-resolution cluster analysis of microearthquakes*, Geophys. Res. Lett. *23*, 1817–1820.

RIGO, A., LYON-CAEN, H., ARMIJO, R., DESCHAMPS, A., HATZFELD, D., MAKROPOULOS, K., PAPADIMITRIOU, P., and KASSARAS, I. (1996), *A microseismic study in the western part of the Gulf of Corinth (Greece): Implications for large-scale normal faulting mechanisms*, Geophys. J. Int. *126*, 663–688.

SACHPAZI, M., CLÉMENT, C., LAIGLE, M., HIRN, A., and ROUSSOS, N. (2003), *Rift structure, evolution, and earthquakes in the Gulf of Corinth, from reflection seismic images*, Earth Plant. Sci. Lett. *216*, 243–257.

SANDERS, C.O., PONKO, S.C., NIXON, L.D., and SCHARTZ, E.A. (1995), *Seismological evidence for magmatic and hydrothermal structure in Long Valley caldera from local earthquake attenuation and velocity tomography*, J. Geophys. Res. *100*, 8311–8326.

SOREL, D. (2000), *A Pliocene and still-active detachment fault and the origin of the Corinth-Patras rift, Greece*, Geology *28*, 83–86.

SPAKMAN, W. and NOLET, G., *Imaging algorithms, accuracy and resolution, In Mathematical Geophysics* (ed. Vlaar N.) (D. Reidel Publ. Co., Dordrecht 1988) pp. 155–187.

SPENCER, C.P. and GUBBINS, D. (1980), *Travel-time inversion for simultaneous earthquake location and velocity structure determination in laterally varying media*, Geophys. J. R. Astron. Soc. *63*, 95–116.

TIBERI, C., LYON-CAEN, H., HATZFELD, D., ACHAUER, U., KARGIANNI, E., KIRATZI, A., LOUVARI, E., PANAGIOTOPOULOS, D., KASSARAS, I., KAVIRIS, G., MAKROPOULOS, K., and PAPADIMITRIOU, P. (2000), *Crustal and upper mantle structure beneath the Corinth rift (Greece) from a teleseismic tomography study*, J. Geophys. Res. *105*, 28159–28171.

TIBERI, C., DIAMENT, M., LYON-CAEN, H., and KING, T. (2001), *Moho topography beneath the Corinth rift area (Greece) from inversion of gravity data*, Geophys. J. Int. *145*, 797–808.

THURBER, C.H. (1992), *Hypocenter-velocity structure coupling in local earthquake tomography*, Phys. Earth Planet Inter. *75*, 55–62.

VANORIO, T., VIRIEUX, J., CAPUANO, P., and RUSSO, G. (2005), *Three-dimensional seismic tomography from P-wave and S-wave microearthquake travel times and rock physics characterization of the Campi Flegrei Caldera*, J. Geophys. Res. *110*, B03201,10.1029/2004JB003102.

XYPOLIAS, P. and KOUKOUVELAS, I. (2001), *Kinematics vorticity and strain rate patterns associated with ductile extrusion in the Chelmos Shear Zone (External Hellenides, Greece)*, Tectonophysics *338*, 59–77.

(Received April 4, 2005; accepted September 23, 2005)
Published Online First: February 8, 2006

Pure appl. geophys. 163 (2006) 455–473
0033–4553/06/030455–19
DOI 10.1007/s00024-005-0026-6

Pure and Applied Geophysics

The Theory of Coda Wave Interferometry

Roel Snieder[1]

Abstract—Coda waves are sensitive to changes in the subsurface because the strong scattering that generates these waves causes them to repeatedly sample a limited region of space. Coda wave interferometry is a technique that exploits this sensitivity to estimate slight changes in the medium from a comparison of the coda waves before and after the perturbation. For spatially localized changes in the velocity, or for changes in the source location, the travel-time perturbation may be different for different scattering paths. The coda waves that arrive within a certain time window are therefore subject to a distribution of travel-time perturbations. Here I present the general theory of coda wave interferometry, and show how the time-shifted correlation coefficient can be used to estimate the mean and variance of the distribution of travel-time perturbations. I show how this general theory can be used to estimate changes in the wave velocity, in the location of scatterer positions, and in the source location.

Key words: Monitoring, scattering, coda waves.

1. Introduction

The seismic coda constitutes the tail of strongly scattered waves in a seismogram. Aki (1985) was one of the pioneers in using the seismic coda. He used the temporal decay of the seismic coda as a measure of the scattering in the Earth, and proposed to use changes of coda Q to monitor changes in the stress in the subsurface (Aki and Chouet, 1975; Jin and Aki, 1986). This approach considers the amplitude of the coda waves, but does not use the phase information in the coda.

Here I present the theory of coda wave interferometry, a technique to monitor time-lapse changes based on the phase and amplitude information of coda waves. In a strongly-scattering medium, the waves repeatedly sample the same region in space. Such a medium therefore works as a natural interferometer. Just as in a man-made interferometer (Lauterborn *et al.*, 1995) the multiply-scattered waves are extremely sensitive to minute changes in the medium. In coda wave interferometry we exploit this sensitivity to measure small changes in the medium.

[1]Center for Wave Phenomena and Department of Geophysics, Colorado School of Mines, Golden CO, 80401, U.S.A. E-mail: rsnieder@mines.edu

This idea is not new, the sensitivity of coda waves has been used to estimate velocity changes in fault zones (POUPINET *et al.*, 1984), in volcanoes (RATDOMOPURBO and POUPINET, 1995; MATSUMOTO *et al.*, 2001), and in ultrasound experiments (ROBERTS *et al.*, 1992; SNIEDER *et al.*, 2002; GRÊT *et al.*, 2006). Temporal changes in the coda waves within a couple of days have been observed in a volcano (GRÊT *et al.*, 2005). In the physics community a related technique called diffusing wave spectroscopy (WEITZ and PINE, 1993; COWAN *et al.*, 2002) has been used to monitor fluidized suspensions (COWAN *et al.*, 2000; PAGE *et al.*, 2000).

In this work I present a general theory of coda wave interferometry. Because of the generality of the derivation, it can be applied to a number of different applications of coda wave interferometry. The theory is based on path summation (section 2). This is a formulation of scattering that states that the total wave field is the superposition of the waves that propagate along all possible scattering paths. In section 3 I show how the changes in the coda waves can be characterized with the time-shifted correlation coefficient. This quantity is related to the distribution of the travel time perturbation when averaged over all scattering paths (section 4). In principle, the distribution of the travel-time perturbation can be obtained from the time-shifted correlation coefficient, but I show in section 5 how in practical application the mean and variance of the travel-time perturbation can be obtained from the time-shifted correlation coefficients. In section 6 I apply the theory to three examples; a change in the velocity, uncorrelated perturbations of the scatterer locations, and a change in the source position.

2. The Path Summation and the Change in the Waves

The theory of coda wave interferometry is based on the path summation where the wavefield at a given location is written as a sum of the waves that propagate along all possible paths (SNIEDER, 1999):

$$u(t) = \sum_P S_P(t). \tag{1}$$

This expression does not specify how the waves are being scattered, it only states that the wavefield is the superposition of the waves propagate along all possible scattering paths. These paths include the direct wave, the singly-scattered waves, and the multiply-scattered waves. The direct wave arrives first, and, depending on scatterer strength, the wavefield is dominated by multiply-scattered waves at later times. By choosing a time window at a sufficiently late time one can ensure that the wavefield is dominated by multiply-scattered waves. The scattering series (1) follows by iterating the Lippman-Schwinger equation to give the Neumann series (MERZBACHER, 1970). For point scatterers, or for scattering caused at boundaries, one can count the number of scattering paths, and the Neumann series reduces to the discrete sum (1) over all scattering paths. This means that the theory presented here presumes that the scattering is caused by point scatterers or by reflection at boundaries. In the Earth, this assumption may not be valid. Note, however, that in the theory presented here, we do not need to know the scattering paths, or the waveforms $S_P(t)$ associated with each path P.

I use a scalar notation, but expression (1) is also valid for elastic waves; in that case equation (1) can be used for each of the components of the wave motion. For elastic waves, conversions between P and S waves occur at the scatterers; in that case the path summation includes a sum P and S waves for each segment of the scattering path, so that $\sum_P$ includes all the possible wave conversions along each path as well.

A perturbation of the medium leads to a perturbation of the waves. I assume that the scattering properties of the scatterers in the medium do not change, but that either the propagation velocity, the position of the scatterers, or the source position is slightly perturbed. These perturbations cause a change in the phase of the wave, the geometrical spreading, and the scattering angle for every scattering event. Let us consider a wave that propagates over a distance l between scatterers. The wave propagation between the scatterers is in the frequency domain described by the Green's function associated with the medium between the scatterers. For example, for scalar waves in three dimensions this Green's function is given by $G(\omega, l) = -\exp(i\omega l/v)/4\pi l$, with v the wave velocity. The average distance l between the scatterers, as seen by the waves, is given by the scattering mean free path (LAGENDIJK and VAN TIGGELEN, 1996; VAN ROSSUM and NIEUWENHUIZEN, 1999).

In the frequency domain the propagation over this distance corresponds to a phase shift $\exp(ikl)$. As an example, let us first consider a change δl in the path length, associated with a change in the location of scatterers. This change in the location of scatterers leads to a perturbation in the phase, the geometrical spreading, and the scattering amplitude. This change corresponds to a phase change $\exp(ik\delta l) \approx 1 + ik\delta l$, so that a change in the path length gives the following change in the wavefield:

$$|\delta u^{phase}| \sim |k\delta l u|. \tag{2}$$

In three dimensions the geometrical spreading for each segment of propagation varies as $1/l$, hence the change in the wavefield due to the geometrical spreading associated with a change in path length is given by

$$|\delta u|^{spreading} \sim \left|\frac{\delta l}{l} u\right|. \tag{3}$$

A change δl in the scatterer position leads to a change in the scattering angle that is of the order $\delta\theta \sim \delta l/l$. When the scattering amplitude, or source radiation pattern, varies as $\cos m\theta$ or $\sin m\theta$, the change in the wavefield due to this change in scattering angle is of the order

$$|\delta u|^{angle} \sim |m\delta\theta| u = \left| \frac{m\delta l}{l} u \right|. \tag{4}$$

The change in the phase of the wavefield dominates when the right-hand side expression (2) is larger than equations (3) and (4). Since in general $|m| > 1$ this conditions implies that the change in the phase is dominant when

$$\frac{l}{\lambda} > \frac{|m|}{2\pi}, \tag{5}$$

where the wavelength is given by $\lambda = 2\pi/k$. This means that the change in the phase dominates the change of the wavefield when the scattering mean free path is considerably larger than a wavelength. If this condition is not satisfied the waves are localized (VAN TIGGELEN, 1999) and the theory of this paper does not hold.

For a change in the velocity, the wavenumber k is perturbed, this affects only the phase of the wave. This means that when the distance between the scatterers is changed or when the velocity changes, the change in the phase is the most important change. If the perturbation is nondispersive in the employed frequency band, this causes a change in the arrival time of the wave. I denote the change in the travel time of the wave that propagates along path P by τ_P. This means that the perturbed wavefield is given by

$$\tilde{u}(t) = \sum_P S_P(t - \tau_P). \tag{6}$$

This expression should be compared with equation (1) for the unperturbed wave. Expression (6) may be an oversimplification because it implies that the wave $S_P(t)$ that propagates along path P does not change shape. This is not true when the perturbation causes a dispersive change in these waves. This would be the case, for example, when the attenuation changes. In that situation, equation (6) may still be a good approximation for the perturbed wave provided that it has been band-pass filtered over a frequency band that is sufficiently small for the dispersion to be ignored.

Equation (6) is only valid when the excitation of the unperturbed and perturbed waves is identical (although I do consider the special case of identical sources that are displaced in space). Controlled sources that do not damage the source region, such as vibrators or airguns (e.g., WEGLER *et al.*, 1999), transducers in ultrasound experiments (e.g., ROBERTS *et al.*, 1992; SNIEDER *et al.*, 2002; GRÊT *et al.*, 2005), are, in general, repeatable. Natural earthquakes can also be highly repeatable (e.g., POUPINET *et al.*, 1984; RATDOMOPURBO and POUPINET, 1995; BAISCH and BOKELMAN, 2001; SCHAFF and BEROZA, 2004). GRÊT *et al.* (2005) show that the bubble burst at a lava lake can act as a repeatable seismic source, while SNIEDER and HAGERTY (2004) show that repeatable waveforms can be retrieved by deconvolving the pressure and displacement recorded during volcanic tremors.

The travel-time perturbation τ_P depends, in general, on the scattering path P. Consider the example of a hydrocarbon reservoir whose properties during production change in a certain part of the reservoir. For the paths P that traverse the region of change the travel time changes, whereas for paths P that do not cross this region the travel time is unchanged. As another example consider a change in the source location. Waves that leave the source in the same direction as the source displacement arrive earlier, while waves that leave the source in the opposite direction arrive later. The coda waves recorded in a given time window consist of the interference of all the waves $S_P(t)$ that arrive in that time window. Since the travel time τ_P depends, in general, on the scattering path P, the travel-time perturbation for the waves arriving in a given time window has a certain distribution. It is the goal of this paper to extract meaningful properties of this distribution based on the coda waves recorded before and after the perturbation. Specifically, I derive expressions for the mean and variance of this distribution.

3. *Measuring the Change in the Wavefield*

The unperturbed and perturbed waves can be compared using the time-shifted correlation coefficient defined as

$$R(t_s) = \frac{\int_{t-T}^{t+T} u(t')\tilde{u}(t'+t_s)dt'}{\sqrt{\int_{t-T}^{t+T} u^2(t')\,dt' \int_{t-T}^{t+T} \tilde{u}^2(t')dt'}}. \tag{7}$$

In this expression t_s is the time shift of the unperturbed and perturbed waves in the correlation. The correlation is computed using a finite time-window with center-time t and window length $2T$. Let us first analyze the numerator of this expression that is given by

$$N(t_s) = \int_{t-T}^{t+T} u(t')\tilde{u}(t'+t_s)dt'. \tag{8}$$

Inserting expressions (1) and (6) in this equation gives a double sum $\sum_{PP'}$over paths. Such a double sum can be divided into diagonal terms, for which $P = P'$, and cross-terms $P \neq P'$:

$$\sum_{PP'}(\cdots) = \sum_{P=P'}(\cdots) + \sum_{P\neq P'}(\cdots), \tag{9}$$

with equations (1) and (6) this gives:

$$N(t_s) = \sum_P \int_{t-T}^{t+T} S_P(t')S_P(t' + t_s - \tau_P)\, dt' + \sum_{P \neq P'} \int_{t-T}^{t+T} S_P(t')S_{P'}(t' + t_s - \tau_{P'})\, dt'. \tag{10}$$

If the cross terms are uncorrelated, their contribution integrates to zero. Ignoring the cross terms amounts to making the ladder approximation (VAN ROSSUM and NIEUWENHUIZEN, 1999) that only retains pairs of scattering paths that are identical ($P = P'$). DE ROSNY *et al.* (2004) show that this approximation underlies time reversal focusing. SNIEDER (2004) investigated the validity of the ladder approximation by comparing the magnitude of the cross terms $\sum_{P \neq P'}$ to the diagonal terms $\sum_{P=P'}$, and showed that when the DC-component of the signal vanishes that

$$\frac{|\text{cross terms}|}{|\text{diagonal terms}|} \sim \sqrt{\frac{T_{\text{corr}}}{2T}}, \tag{11}$$

where T_{corr} is the width of the autocorrelation of the signal in the time domain. This width is proportional to $1/\Delta f$, with Δf the bandwidth of the signal. This means that the ratio in expression (11) can also be written as

$$\frac{|\text{cross terms}|}{|\text{diagonal terms}|} \sim \sqrt{\frac{1}{\Delta f\, 2T}}. \tag{12}$$

The bandwidth times the window length is the number of degrees of freedom in the signal (LANDAU, 1967; BUCCI and FRANCESCHETTI, 1989), hence the ratio of the cross terms to the diagonal terms is for a single realization equal to $1/\sqrt{n}$, with n the number of degrees of freedom in the data. The important point of expression (12) is that the cross terms decrease when the window length is increased. Note that this argument does not hold for monochromatic data, because in that case $\Delta f = 0$ and the right-hand side of expression (12) cannot be reduced by increasing the window length. This means that in order to ignore the cross terms both the window length and the bandwidth must be sufficiently large.

In the following I assume that the cross terms can be ignored. In this approximation

$$N(t_s) = \sum_P \int_{t-T}^{t+T} S_P(t')S_P(t' + t_s - \tau_P)\, dt'. \tag{13}$$

This integral can be written as a sum of the cross correlations of the waves that have propagated along the individual paths that is defined as

$$C_P(t_s - \tau_P) \equiv \int_{t-T}^{t+T} S_P(t')S_P(t' + t_s - \tau_P)\, dt', \tag{14}$$

so that

$$N(t_s) = \sum_P C_P(t_s - \tau_P). \tag{15}$$

A similar treatment can be applied to the terms in the denominator of equation (7). This gives

$$\int_{t-T}^{t+T} u^2(t')\, dt' = \int_{t-T}^{t+T} \tilde{u}^2(t')\, dt' = \sum_P C_P(0). \tag{16}$$

In deriving this result the same approximations are used as in the derivation of expression (15), specifically the cross terms $\sum_{P \neq P'}$ are ignored. Inserting the expressions (15) and (16) into the definition (7) of the correlation coefficient gives

$$R(t_s) = \frac{\sum_P C_P(t_s - \tau_P)}{\sum_P C_P(0)}. \tag{17}$$

Before we analyze this correlation coefficient I introduce another approximation. In the frequency domain the definition (14) for the correlation of S_P corresponds to

$$C_P(\omega) = |S_P(\omega)|^2. \tag{18}$$

The waves that propagate along the different paths are excited by the same source, they experience the same attenuation, and propagate through the same scattering medium. Because of focusing, the overall amplitude of these waves may differ. For this reason I assume that the waves that propagate along the different trajectories have a power spectrum with the same shape, but that the overall amplitude of each of these waves may be different. Note that this does not imply that the waves $S_P(t)$ are the same, because the phase spectrum may be different. In fact, the phase spectrum will be different because these waves have different arrival times and may have different wave shapes. Since the autocorrelation is the Fourier transform of the power spectrum, the assumption that the shape of the the power spectrum of the waves is the same implies that up to a constant the autocorrelation also is the same, so that

$$C_P(t) = I_P C(t) \quad \text{with} \quad C(0) = 1. \tag{19}$$

In this expression $C(t)$ is the autocorrelation of the $S_P(t)$ normalized at its maximum at $t = 0$, while I_P measures the intensity of the wave that has propagated along path P. Using these results the time-shifted correlation coefficient is given by

$$R(t_s) = \frac{\sum_P I_P C(t_s - \tau_P)}{\sum_P I_P}. \tag{20}$$

The power spectrum of the scattered waves is, in general, different from the power spectrum of the source-time function because scattering and diffraction are usually frequency-dependent. The assumption that the shape of the power spectrum $|S_P(\omega)|^2$

of the waves that arrive a time window is identical, can only be true if the scattered waves that arrive in that window have experienced the same number of scattering events. For this reason the time window should not be too long. The factor I_P in expression (19) accounts for the fact that the different arrivals may have a different amplitude.

4. *The Probability Density Function of the Travel-time Perturbation*

The time-shifted cross-correlation coefficient can be determined from the recorded waves using expression (7). The goal is to infer properties of the distribution of the travel-time perturbation, as described in equation (6), from this measurement. One way to achieve this is to define the normalized energy of the arrivals in the employed time window with a travel-time shift between τ and $\tau + d\tau$ as $P(\tau)d\tau$. This definition implies that

$$P(\tau)d\tau = \frac{\sum_P \text{ such that } \tau<\tau_P<\tau + d\tau I_P}{\sum_{\text{all } P} I_P}. \tag{21}$$

The sum in the numerator is over the paths that have a travel-time change between τ and $\tau + d\tau$. Since the denominator contains a sum over all paths, and hence all relevant values of τ, the function $P(\tau)$ is normalized:

$$\int_{-\infty}^{\infty} P(\tau)\, d\tau = 1. \tag{22}$$

The definition (21) consists of a ratio of positive numbers, therefore $P(\tau)$ is positive and normalized. For this reason it has the same properties as a probability density function. Using statistical jargon, I use the following definition of an expectation value:

$$\langle f(\tau)\rangle \equiv \int_{-\infty}^{\infty} f(\tau)P(\tau)\, d\tau. \tag{23}$$

Using expressions (21) this can also be written as

$$\langle f(\tau)\rangle \equiv \frac{\sum_P f(\tau_P)I_P}{\sum_P I_P}. \tag{24}$$

This expectation value corresponds to an average weighted by the intensity of the coda waves.

With the definitions (21) and (24), equation (20) can be written as

$$R(t_s) = \langle C(t_s - \tau)\rangle = \int C(t_s - \tau)P(\tau)\, d\tau. \tag{25}$$

The first identity shows that coda wave interferometry leads to a weighted average of a function of the travel-time perturbation. The coda waves travel along all possible paths, and the contribution of the travel-time perturbation is averaged over all possible paths with a weight function given by the intensity of the waves for each path.

The second identity of expression (25) states that the time-shifted correlation coefficient is given by the convolution of $P(\tau)$ with $C(\tau)$. The time-shifted correlation coefficient $R(t_s)$ follows from the recorded waves. The function $C(t)$ follows from the power spectrum of the recorded waves, and is known as well. In principle $P(\tau)$ can then be obtained from expression (25) by deconvolution. This provides direct information on the distribution of the travel-time perturbations over the paths that have arrivals within the employed time window. Note that in this approach we do not need to assume that the travel-time perturbation is small. Because of the instabilities of the deconvolution process, I take a different approach and extract the first and second moments of the travel-time perturbation from expression (25).

5. *Extracting the Moments of the Travel-time Perturbation from the Correlation*

In this section I assume that the travel-time perturbation does not change considerably among all the different paths with arrivals within the employed time window. Specifically, I assume that we can use a second-order Taylor expansion of $C(t)$. The autocorrelation is an even function, with $C(0) = 1$, hence the second-order Taylor expansion is given by

$$C(t) = 1 + \frac{1}{2}\ddot{C}(t = 0)t^2, \tag{26}$$

where the dots denote the second time derivative. In the following I consider first the case $\tau_P = 0$, and expression (20) gives

$$C(t_s) = R(t_s), \tag{27}$$

where $R(t)$ is the correlation coefficient defined in expression (7) with the unperturbed state equal to the perturbed state: $u = \tilde{u}$. Using this in expression (27) and replacing t_s by t gives

$$C(t) = \frac{\int u(t')u(t' + t)\, dt'}{\int u^2(t')\, dt'}, \tag{28}$$

where the integration is over the time window under consideration. Note that it follows from this expression that $C(0) = 1$, as required in equation (19).

Differentiating expression (28) twice with respect to time gives

$$\ddot{C}(t) = \frac{\int u(t') \dfrac{d^2u(t'+t)}{dt^2}\, dt'}{\int u^2(t')\, dt'} = \frac{\int u(t') \dfrac{d^2u(t'+t)}{dt'^2}\, dt'}{\int u^2(t')\, dt'}, \tag{29}$$

where the second derivative in the last term is with respect to t'. Setting $t = 0$, and using an integration by parts gives under the assumption that u vanishes at the end of the integration interval:

$$\ddot{C}(0) = -\frac{\int \dot{u}^2(t')\, dt'}{\int u^2(t')\, dt'}. \tag{30}$$

In practice, one tapers the integrand in the correlation coefficient (7) in order to suppress truncation artifacts. This taper ensures that the integrand of expression (7) indeed vanishes at the endpoints of the interval. The right-hand side of expression (30) has the physical dimension frequency^2; for this reason I introduce the following definition of the mean-squared angular frequency:

$$\overline{\omega^2} \equiv \frac{\int \dot{u}^2(t')\, dt'}{\int u^2(t')\, dt'}. \tag{31}$$

Note that this quantity can directly be computed from the recorded data. Using this result in the expressions (26) and (30) gives the following second-order Taylor expansion:

$$C(t) = 1 - \frac{1}{2}\overline{\omega^2}t^2. \tag{32}$$

Inserting this result in expression (20) gives

$$R(t_s) = 1 - \frac{1}{2}\overline{\omega^2}\frac{\sum_P (t_s - \tau_P)^2 I_P}{\sum_P I_P}. \tag{33}$$

It follows by differentiation that the correlation coefficient attains its maximum when

$$0 = \frac{dR(t_s)}{dt_s} = -\overline{\omega^2}\frac{\sum_P (t_s - \tau_P) I_P}{\sum_P I_P}. \tag{34}$$

This maximum is reached for

$$t_s = t_{\max} = \frac{\sum_P \tau_P I_P}{\sum_P I_P}. \tag{35}$$

This equation states that the correlation coefficient attains its maximum for a time shift that is equal to the intensity-weighted travel-time perturbation. According to the notation of expression (24) this result can also be written as

$$t_{\max} = \langle \tau \rangle. \tag{36}$$

The value $R_{\max}$ of the correlation coefficient at its maximum follows by replacing t_s in expression (33) by the average $\langle\tau\rangle$, thus

$$R_{\max} = 1 - \frac{1}{2}\overline{\omega^2}\frac{\sum_P(\tau_P - \langle\tau\rangle)^2 I_P}{\sum_P I_P}. \tag{37}$$

Using expression (24) the ratio in the last term satisfies

$$\frac{\sum_P(\tau_P - \langle\tau\rangle)^2 I_P}{\sum_P I_P} = \left\langle(\tau - \langle\tau\rangle)^2\right\rangle = \sigma_\tau^2\,, \tag{38}$$

where σ_τ^2 is the variance of the travel-time perturbation. Combining this result with expression (37) gives

$$R_{\max} = 1 - \frac{1}{2}\overline{\omega^2}\sigma_\tau^2. \tag{39}$$

The time-shifted correlation coefficient can for every employed time window be computed from expression (7) given the data before and after the perturbation. According to expression (36) the time-shifted correlation coefficient attains its maximum for a shift time that is equal to the mean travel-time perturbation. Following equation (39) the value of the time-shifted correlation coefficient gives the variance of the travel-time perturbation. This means that if the unperturbed and perturbed waveforms are known, the mean and variance of the travel-time perturbation can be computed. Computing the two lowest moments of the distribution of the travel-time perturbation is a less ambitious goal than determining the distribution $P(\tau)$ of the travel-time perturbations from equation (25), but as we will see in the examples of the next section, the mean and variance of the travel-time perturbation are useful in a number of practical applications.

The data may be contaminated with noise. The noise has two effects on the time-shifted cross-correlation coefficient. First, the noise introduces fluctuations in the estimated coefficient. Without knowing the noise there is no way to eliminate these fluctuations other than using nonoverlapping time windows of the coda to obtain independent estimates of the cross correlation. Second, noise leads to a bias because noise lowers the value of the cross correlation. This bias can be estimated given the energy in the noise ($\langle n^2\rangle$) and the energy of the noise-contaminated data ($\langle u^2\rangle$). DOUMA and SNIEDER (2006) show that the bias in the correlation coefficient can be accounted for by using the corrected correlation coefficient that is related to the uncorrected coefficient by the following relation

$$R_{\text{corr}} = R/\sqrt{1 - \frac{\langle n^2\rangle}{\langle u^2\rangle}}\sqrt{1 - \frac{\langle\tilde{n}^2\rangle}{\langle\tilde{u}^2\rangle}}. \tag{40}$$

The noise levels $\langle n^2\rangle$ and $\langle\tilde{n}^2\rangle$ before and after the perturbation can be estimated from the data recorded before the first-arriving waves.

6. Examples

In this section I consider three different perturbations that are of relevance for practical applications; a velocity perturbation, random displacement of scatterers, and a perturbation in the source position.

6.1 A Velocity Perturbation

Suppose that in a medium the velocity $v(\mathbf{r})$ is perturbed with a perturbation $\delta v(\mathbf{r})$, and that the relative velocity perturbation $\delta v/v$ is the same at every location in space. The unperturbed travel time is given by

$$t_P = \int_P \frac{1}{v}\, ds, \tag{41}$$

where the integration is along path P. The perturbed travel time is to first order in the velocity perturbation given by

$$t_P + \tau_P = \int_P \frac{1}{v + \delta v}\, ds = \int_P \left(\frac{1}{v} - \frac{\delta v}{v^2}\right) ds. \tag{42}$$

With expression (41) this gives

$$\tau_P = -\int_P \frac{1}{v}\frac{\delta v}{v}\, ds = -\left(\frac{\delta v}{v}\right)\int_P \frac{1}{v}\, ds, \tag{43}$$

since I used in the last identity that the relative velocity perturbation is assumed to be constant. With expression (41) the travel-time perturbation can be written as

$$\tau_P = -\left(\frac{\delta v}{v}\right) t. \tag{44}$$

In this expression I have replaced t_P by the center time t, of the employed time window, because this is the average arrival time of the waves that arrive in that time window. Note that the velocity v is not necessarily constant. According to expression (44), the travel-time perturbation depends on the arrival time of the wave only, but is independent of the particular path. This means that in a small time window the mean travel-time perturbation is also given by

$$t_{\max} = \langle \tau \rangle = -\left(\frac{\delta v}{v}\right) t. \tag{45}$$

Since the travel-time perturbation is the same for all trajectories with arrivals in a short time window, the variance of the travel-time perturbation vanishes

$$\sigma_\tau^2 = 0. \tag{46}$$

As shown in expression (36), the mean travel-time perturbation is equal to the shift time that gives the maximum of the time-shifted correlation coefficient, this quantity can easily be retrieved from the data. According to expression (45) the relative velocity change then follows from

$$\left(\frac{\delta v}{v}\right) = -\frac{t_{\max}}{t}. \tag{47}$$

This has been used by SNIEDER *et al.* (2002) to measure the velocity change in a granite sample with temperature. In their laboratory experiment coda wave interferometry is sufficiently sensitive to detect a velocity change of about 0.1%. This change in the velocity can be inferred from different nonoverlapping time windows in the coda. This redundancy serves as a consistency check on the method, and can be used for error estimation. In these measurements the estimated error in the velocity change was about 0.02%. GRÊT *et al.* (2006) applied this technique also to measure the velocity change in rocks due to changes in the stress state in laboratory conditions. Changes in the velocity before and after the Loma Prieta earthquake have been observed by BAISCH and BOKELMANN (2001), and by SCHAFF and BEROZA (2004). POUPINET *et al.* (1984) and NISHIMURA *et al.* (2000) also provide examples of temporal changes in the seismic velocity that are detected with coda waves.

For elastic waves there are two wave velocities. It follows from the intensity-averaged travel-time perturbation (35) that for elastic waves the inferred velocity change is a weighted average of the change in the *P*-wave and *S*-wave velocities (SNIEDER, 2002):

$$\frac{\delta v}{v} = \frac{\beta^3}{2\alpha^3 + \beta^3}\frac{\delta\alpha}{\alpha} + \frac{2\alpha^3}{2\alpha^3 + \beta^3}\frac{\delta\beta}{\beta}. \tag{48}$$

where α and β are the velocities of *P* waves and *S* waves, respectively. Since $\beta < \alpha$ the perturbation in the shear velocity dominates the perturbation in the *P*-wave velocity. For example, for a a Poisson medium ($\alpha = \sqrt{3}\beta$):

$$\frac{\delta v}{v} = 0.09\frac{\delta\alpha}{\alpha} + 0.91\frac{\delta\beta}{\beta} . \tag{49}$$

This confirms that coda waves are primarily sensitive to the shear velocity (AKI and CHOUET, 1975).

The theory of this section is valid when the relative velocity change is independent of location. In realistic situations this is not necessarily the case. The theory can be extended to situations where the relative velocity change depends on position. In that case the mean travel-time change is linearly related to the velocity change:

$$\langle\tau\rangle = \int K(\mathbf{r}, t)\delta v(\mathbf{r})\, dV, \tag{50}$$

where the kernel $K(\mathbf{r}, t)$ depends on the intensity-weighted average over all scattering paths. This expression can be used as the basis of a standard linear inversion for the velocity change $\delta v(\mathbf{r})$ given the mean travel-time change observed for different source-receiver pairs and different time windows. The kernel $K(\mathbf{r}, t)$ has been derived both for single-scattered waves (PACHECO and SNIEDER, 2006), as well as for strongly scattered waves (PACHECO and SNIEDER, 2005).

6.2 Random Displacement of Scatterers

As a second example I consider a perturbation that consists of uncorrelated movement of the scatterers. This is of relevance for studying the motion of particles in colloidal suspensions (HECKMEIER and Maret, 1997) and of bubbles in a turbulent fluid (COWAN *et al.*, 2000; PAGE *et al.*, 2000). The theory presented here is equivalent to diffusing wave spectroscopy (WEITZ and PINE, 1993; COWAN *et al.*, 2002), although the derivation is different.

Let us consider scatterers that move independently in three dimensions and that have a root-mean-squares displacement δ between the two measurements of the waves that are used to study the motion of the scatterers. On average the path length for each scattered wave does not change, hence the mean perturbation of the travel time vanishes:

$$\langle \tau \rangle = 0. \tag{51}$$

Some scattering paths are longer, while other scattering paths are shorter, therefore the variance of the travel-time perturbation is nonzero.

I compute the variance of the travel time by using that for a wave that has scattered n times, the variance in the path length is given by SNIEDER and SCALES (1998):

$$\sigma_L^2 = 2n\left(1 - \overline{\cos\theta}\right)\delta^2. \tag{52}$$

In this expression $\overline{\cos\theta}$ is the average of the cosine of the scattering angle over all paths in the employed time window. The number of scatterers encountered is related to the travel time by $n = vt/l$, where v is the wave velocity and l the scattering mean free path. Using this in expression (52) gives

$$\sigma_L^2 = \frac{2vt\delta^2}{l_*}, \tag{53}$$

where l_* is the transport mean free path that is defined as $l_* = l/(1 - \overline{\cos\theta})$. This is the distance of propagation over which the scattered wave has lost all information about its direction of propagation (LAGENDIJK and VAN TIGGELEN, 1996; VAN ROSSUM and NIEUWENHUIZEN, 1999). For a constant velocity the variance in the travel time is related to the variance in the path length by $\sigma_\tau = \sigma_L/v$, so that

$$\sigma_\tau^2 = \frac{2t\delta^2}{vl_*}. \tag{54}$$

With expression (39) this means that the measured maximum of the cross correlation is related to the mean displacement of the scatterers by

$$R_{\max} = 1 - \frac{\overline{\omega^2}\delta^2 t}{vl_*}. \tag{55}$$

Given a measured value of the maximum of the cross correlation, one can infer the mean scatterer displacement δ from this expression if the wave velocity and the transport mean free path are known. SNIEDER *et al.* (2002) show in a numerical example that the mean scatterer displacement can correctly be retrieved from the coda waves. The scatterer displacement can be computed from several nonoverlapping time windows of the coda, and they show how this redundancy can be used to compute error bounds on the scatterer displacement inferred from the coda waves.

6.3 A Displaced Source Position

The relative distance between earthquakes can be found from the absolute location of the events, but in this approach errors in the employed velocity model may lead to large errors in the distance between the events (PAVLIS, 1992). An alternative approach is the double-difference method where the relative event location is computed from the differential travel time of the direct arrivals (SHEARER, 1997; ASTIZ and SHEARER, 2000; WALDHAUSER and ELLSWORTH, 2002). Coda wave interferometry can also be used to determine the relative distance between earthquakes, provided the source mechanisms of the events are identical. This provides additional information on the relative position between events that can be used in addition to the constraints obtained from the double-difference method.

Let us consider two nearby seismic sources with the same source mechanism. When the source position is perturbed over the distance **r**, the distance from the source to the first scatterer along every scatterer path is perturbed. The distance between the scatterers is unperturbed, so we only need to account for the change in the distance to the first scatterer along every scattering path. For this reason it does not matter if the wave travels from the first scatterer along every path directly to the receiver, or whether the wave visits many other scatterers first. The theory of this section therefore is applicable both for single scattering as well as for multiple scattering.

For a scattering path P with take-off direction $\hat{\mathbf{t}}_P$ at the source, the perturbation in the travel time is given by

$$\tau_P = -\frac{(\hat{\mathbf{t}}_P \cdot \mathbf{r})}{v}. \tag{56}$$

For a scattering path where the wave leaves the source as a P wave, v is the P-wave velocity at the source location, whereas for a path where the wave leaves the source as an S wave, v denotes the shear velocity. When the scatterers are distributed homogeneously, some paths are longer when the source location is perturbed, while others are shorter. The resulting mean travel-time perturbation vanishes (SNIEDER and VRIJLANDT, 2005). This can be shown by integrating the travel-time perturbation (56) over all take-off directions.

The variance of the travel-time perturbation, however, is nonzero, because some paths are longer when the source location is perturbed while others are shorter. The variance of the travel-time perturbation depends on the type of source (explosion, point force, double couple), as well as on the orientation of the perturbation of the source location relative to the source mechanism (SNIEDER and VRIJLANDT, 2005). An important application is the location of aftershocks. SNIEDER and VRIJLANDT (2005) show that when aftershocks with identical source mechanisms are confined to a plane, and their slip vector is aligned with this plane, that the variance of the travel time is given by

$$\sigma_\tau^2 = \frac{\left(\frac{6}{\alpha^8} + \frac{7}{\beta^8}\right)}{7\left(\frac{2}{\alpha^6} + \frac{3}{\beta^6}\right)} r^2. \tag{57}$$

In this expression α is the P-wave velocity, and β denotes the shear-wave velocity. This expression depends on both velocities, because the path summation includes waves that leave the source as a P-wave, as well as waves that leave the source as an S-wave. The P- and S-wave velocities in expression (57) are raised to high powers. Since $\beta < \alpha$, the terms with the shear velocity dominate. Note that expression (57) depends on the distance r between the events, but not on their relative orientation in the fault plane. The reason for this is that the averaging (38) in this application involves an integration over all take-off directions. In this directional averaging information on the direction of the event separation is lost.

Expression (57) can be used to estimate the source separation in the following way. For several time windows in the coda the time-shifted correlation coefficient can be computed from expression (7). Using expression (39), the maximum of this function can be equated to the variance of the travel-time perturbation given by equation (57). The resulting expression can then be solved for the event separation r. In the presence of significant noise, the correction factor of equation (40) can be used to eliminate the bias in the cross correlation due to noise.

SNIEDER and VRIJLANDT (2005) applied this technique to events on the Hayward fault, California, and showed that the event separation obtained from the coda waves agrees with the event separation of the same events determined by WALDHAUSER and ELLSWORTH (2002) with the double-difference method.

Note that the estimation of the event separation can be carried out with a single station. An error estimate of the event separation can be obtained by comparing the event separation inferred from different nonoverlapping time windows in the coda.

7. *Discussion*

The main result of the theory of this paper is that the time-shifted correlation coefficient between perturbed and unperturbed waves can be related to the mean and variance of the travel-time perturbation. According to expression (36), the mean travel-time perturbation follows from the shift time that gives the maximum of the time-shifted correlation coefficient, and equation (39) relates the maximum of the cross correlation coefficient to the variance of the travel-time perturbation. This means that the mean and variance of the travel-time perturbation can be obtained from a comparison of the coda waves before and after the perturbation. Additive random noise leads to a bias in the maximum of the correlation coefficient, this bias can be removed using the correction factor of expression (40).

The present theory can be applied to a constant change in the velocity, to uncorrelated perturbations in the locations of the scatterers, and to changes in the source position. For a change in the velocity (section 6.1), the mean travel-time perturbation is nonzero (equation (45)), but the variance of the travel-time perturbation is zero (expression (46)). For random perturbations in the scatterer location the mean travel-time perturbation vanishes (expression (51)), but according to equation (52) the variance of the travel time is nonzero and depends linearly on time. This contrasts the case of a perturbation in the source position where the mean travel-time perturbation also vanishes and where according to expression (57) the variance is nonzero and independent of time. In general, it may not be obvious how a medium is perturbed. As shown above, the three different perturbation leave a different imprint on the mean and the variance of the travel-time perturbation. Since these quantities can be estimated for several independent windows of the coda waves using the time-shifted correlation coefficient, it is possible to discriminate between these different perturbations using the recorded coda waves.

Acknowledgments

I appreciate the critical and constructive comments of Michel Campillo, Yehuda Ben-Zion, and an anonymous reviewer.

References

AKI, K. (1986), *Theory of earthquake prediction with special reference to monitoring of the quality factor of lithosphere by the coda method*, Earthquake Res. Bull. *3* 219–230.

AKI, K., and CHOUET, L.B. (1975), *Origin of coda waves: source, attenuation, and scattering effects*, J. Geophys. Res. *80*, 3322–3342.

ASTIZ, L., and SHEARER, P.M. (2000), *Earthquake locations in the inner Continental Borderland, offshore southern California*, Bull. Seism. Soc. Am. *90*, 425–449.

BAISCH, S., and BOKELMANN, G.H.R. (2001), *Seismic waveform attributes before and after the Loma Prieta earthquake: Scattering change near the earthquake and temporal recovery*, J. Geophys. Res. *106*, 16323–16337.

BUCCI, O.M., and FRANCESCHETTI, G. (1989), *On the degrees of freedom of scattered fields*, IEEE Trans. on Antennas and Propagation *37*, 918–926.

COWAN, M.L., JONES, I.P., PAGE, J.H., and WEITZ, D.A. (2002), *Diffusing acoustic wave spectroscopy*, Phys. Rev. E *65*, 066605–1/11.

COWAN, M.L., PAGE, J.H., and WEITZ, D.A. (2000), *Velocity fluctuations in fluidized suspensions probed by ultrasonic correlation spectroscopy*, Phys. Rev. Lett. *85*, 453–456.

DE ROSNY, J., TOURIN, A., DERODE, A., VAN TIGGELEN, B., and FINK, M. (2004), *Relation between time reversal focusing and coherent backscattering in multiple scattering media: A diagrammatic approach*, Phys. Rev. E *70*, 046601.

DOUMA, H., and SNIEDER, R. (2006), *Correcting for bias due to noise in seismic interferometry*, Geophys. J. Int., *164*, 99–108.

GRÊT, A., SNIEDER, R., ASTER, R.C., and KYLE, P.R. (2005), *Monitoring rapid temporal changes in a volcano with coda wave interferometry*, Geophys. Res. Lett. *32*, L06304, 10.1029/2004GL02114.

GRÊT, A., SNIEDER, R., and SCALES, J. (2006), *Time-lapse monitoring of rock properties with coda wave interferometry*, submitted to J. Geophys. Res. 2004JB003354

HECKMEIER, M., and MARET, G. (1997), *Dark speckle imaging of colloidal suspensions in multiple light scattering media*, Progr. Colloid. Polym. Sci. *104*, 12–16.

JIN, A., and AKI, K. (1986), *Temporal changes in the coda Q before the Tangshan earthquake of 1976 and the Haicheng earthquake of 1975*, J. Geophys. Res. *91*, 665–673.

LAGENDIJK, A., and VAN TIGGELEN, B.A. (1996), *Resonant multiple scattering of light*, Phys. Rep. *270*, 143–215.

LANDAU H.J. (1967), *Necessary density conditions for sampling and interpolation of certain entire functions*. Acta Mathematica, *117*, 37–52.

LAUTERBORN, W., KURZ, T., and WIESENFELDT, M. *Coherent Optics, Fundamentals and Applications* (Springer-Verlag, Berlin, 1995).

MATSUMOTO, S., OBRA, K., YOSHIMOTO, K., SAITO, T., ITO, A., and HASEGAWA, A. (2001), *Temporal changes in P-wave scatterer distribution associated with the M6.1 earthquake near Iwate volcano, northeastern Japan*, Geophys. J. Int. *145*, 48–58.

MERZBACHER, E., *Quantum Mechanics (2nd Ed.)* (Wiley, New York, 1970).

NISHIMURA, T., UCHIDA, N., SATO, H., OHTAKE, M., TANAKA, S., and HAMAGUCHI, H. (2000), *Temporal changes of the crustal structure associated with the M6.1 earthquake on September 3, 1998, and the volcanic activity of Mount Iwate, Japan*, Geophys. Res. Lett. *27*, 269–272.

PACHECO, C., and SNIEDER, R. (2006), *Localizing time-lapse changes with single scattered waves*, Submitted to Geophys. J. Int.

PACHECO, C., and SNIEDER, R. (2005), *Localizing time-lapse changes with multiply scattered waves*, J. Acoust. Soc. Am. *188*, 1300–1310.

PAGE, J.H., COWAN, M.L., and WEITZ, D.A. (2000), *Diffusing acoustic wave spectroscopy of fluidized suspensions*, Physica B *279*, 130–133.

PAVLIS, G. (1992), *Appraising relative eartquake location errors*, Bull. Seism. Soc. Am. *82*, 836–859.

POUPINET, G., ELLSWORTH, W.L., and FRÉCHET, J. (1984), *Monitoring velocity variations in the crust using earthquake doublets: An application to the Calaveras Fault, California*, J. Geophys. Res. *89*, 5719–5731.

RATDOMOPURBO, A., and POUPINET, G. (1995), *Monitoring a temporal change of seismic velocity in a volcano: Application to the 1992 eruption of Mt. Merapi (Indonesia)*, Geophys. Res. Lett. *22*, 775–778.

ROBERTS, P.M., SCOTT PHILLIPS, W., and FEHLER, *Development of the active doublet method for measuring small velocity and attenuation changes in solids*, J. Acoust. Soc. Am. *91*, 3291–3302.

SCHAFF, D.P., and BEROZA, G.C. (2004), *Coseismic and postseismic velocity changes measured by repeating earthquakes*, J. Geophys. Res. *109*, B10302, doi:10.1029/2004JB003011.

SHEARER, P.M. (1997), *Improving local earthquake locations using the L_1-norm and waveform cross-correlation: Application to the Whittier Narrows California aftershock sequence*, J. Geophys. Res. *102*, 8269–8283.

SNIEDER, R. *Imaging and averaging in complex media*. In (J.P. Fouque, ed.), *Diffuse Waves in Complex Media*, pages 405–454 (Kluwer, Dordrecht, 1999).

SNIEDER, R. (2002), *Coda wave interferometry and the equilibration of energy in elastic media*, Phys. Rev. E *66*, 046615.

SNIEDER, R. (2004), *Extracting the Green's function from the correlation of coda waves: A derivation based on stationary phase*, Phys. Rev. E *69*, 046610.

SNIEDER, R., GRÊT, A., DOUMA, H., and SCALES, J. (2002), *Coda wave interferometry for estimating nonlinear behavior in seismic velocity*, Science *295*, 2253–2255.

SNIEDER, R., and HAGERTY, M. (2004), *Monitoring change in volcanic interiors using coda wave interferometry: Application to Arenal Volcano, Costa Rica*, Geophys. Res. Lett. *31*, L09608, doi:10.1029/2004GL019670.

SNIEDER, R., and SCALES, J.A. (1998), *Time reversed imaging as a diagnostic of wave and particle chaos*, Phys. Rev. E *58*, 5668–5675.

SNIEDER, R., and VRIJLANDT, M. (2005), *Constraining relative source locations with coda wave interferometry: Theory and application to earthquake doublets in the Hayward Fault, California*, J. Geophys. Res. *110*, B04301, 10.1029/2004JB003317.

VAN ROSSUM, M.C.W., and NIEUWENHUIZEN, Th.M. (1999), *Multiple scattering of classical waves: Microscopy, mesoscopy and diffusion*, Rev. Mod. Phys. *71*, 313–371.

VAN TIGGELEN, B.A. (1999), *Localization of waves*. In (J.P. Fouque, ed.), *Diffuse Waves in Complex Media*, pages 1–60 (Kluwer, Dordrecht, 1999).

WALDHAUSER, F., and ELLSWORTH, W.L. (2002), *Fault structure and mechanics of the Hayward Fault, California, from double-difference earthquake locations*, J. Geophys. Res. 107:10.1029/2000JB000084.

WEGLER, U., LÜHR, B.G., and RATDOMOPURBO, A. (1999), *A repeatable seismic source for tomography at volcanoes*, Ann. di Geofisica *42*, 565–571.

WEITZ, D.A., and PINE, D.J. (1993), *Diffusing wave spectroscopy*. In (W. Brown, ed.), Dynamic light scattering, The method and some applications, pages 652–720 (Clarendon Press, Oxford, 1993).

(Received March 23, 2005; accepted September 19, 2005)
Published Online First: February 8, 2006

Pure appl. geophys. 163 (2006) 475–502
0033–4553/06/030475–28
DOI 10.1007/s00024-005-0032-8

Pure and Applied Geophysics

Phase and Correlation in 'Random' Seismic Fields and the Reconstruction of the Green Function

Michel Campillo[1]

Abstract—We first present a summary of recent results on coda interpretation. We emphasize the observation of the stabilization of P to S energy ratio indicating the modal equipartition of the wavefield. This property clearly shows that the coda waves are in the regime of multiple scattering. Numerical solutions of the elastic radiative transfer equation are used to illustrate the evolution of the wave-field towards P-to-S energy stabilization, and asymptotically to complete isotropy. The energy properties of the coda have been widely studied but the phase properties have often been neglected. The recently observed coherent backscattering enhancement, an expression of the so-called 'weak localization', demonstrates that interference effects still persist for multiple diffracted waves. Another manifestation of the persistence of the phase is the possibility to reconstruct the Green function between two stations by averaging the cross correlation of coda waves produced by distant earthquakes and recorded at those two stations. This reconstruction is directly related to the properties of reciprocity and time reversal of any wavefield. Using broadband seismic coda waves, we show that the dominant phases of the Green function in the band 2 s–10 s, namely fundamental mode Rayleigh and Love waves, are reconstructed. We analyze the time symmetry of the cross correlation and show how the level of symmetry evolves with the isotropization of the diffuse field with lapse time. Similarly we investigate the correlation in continuous ambient noise records. Whereas the randomness of the coda results from multiple scattering by randomly distributed scatterers, we assume that the seismic noise is random mostly because of the distribution of sources at the surface of the Earth. Surface waves can be extracted from long time series. The dispersion curves of Rayleigh waves are deduced from the correlations. On paths where measurements from earthquake data are also available, we show that they are in good agreement with those deduced from noise correlation. The measurement of velocities from correlation of noise along paths crossing different crustal structures opens the way for a 'passive imaging' of the Earth's structure.

Key words: Random fields, coda, seismic noise, imaging.

1. Introduction

Keiiti Aki played a fundamental role in teaching generations of seismologists the need to consider that the actual complexity of the Earth is translated into the complexity of the seismograms. The studies presented here have long-ranging roots

[1]Laboratoire de Géophysique Interne et Tectonophysique, Observatoire de Grenoble, UniversitéJoseph Fourier and CNRS, BP 53, 38041 Grenoble, France. E-mail: michel.campillo@obs.ujf-grenoble.fr

in Aki's works and remarks. His pioneering 1957 study of noise has been the foundation of numerous applications in earthquake engineering and has paved the way for the present developments. On the other hand, Aki also continuously encouraged the investigation of the properties of coda waves. This paper summarizes certain results on multiple scattering in seismology and related fields. The emphasis is given to the works done in Grenoble where Aki made several visits and had a profound influence.

2. *Coda*

AKI (1969, 1980) interpreted coda waves as scattered waves from the inhomogeneities within the Earth. A major advance was made by defining coda Q (AKI and CHOUET, 1975). The envelope decay of coda waves was found to be in the form

$$E(t) = t^{-n} \exp\left(\frac{-\omega t}{Q_c}\right), \tag{1}$$

where ω is the circular frequency, t is time, and Q_c describes the rate of decay of the seismogram envelopes. n is another phenomenological parameter that is found to be between 1 and 2. The exponential decay is therefore controlled by Q_c which is a regional parameter, mostly independent from source mechanism, depth and epicentral distance when a sufficient lapse time is considered. Q_c has been measured in many regions of the world (see e.g., HERRAIZ and ESPINOSA, 1987) and shown to be dependent of the tectonic regime (e.g., SINGH and HERRMANN, 1983). The physical meaning of Q_c in terms of medium properties is still debated.

It is important to notice that the model expressed in (1) does not intend to describe the elastic displacement field but a scalar integrated variable such as the kinetic energy, or the intensity. Two end-member models were proposed by AKI and CHOUET: the single scattering model, which is pertinent only for the early coda as we will see later, and the diffusion model, defined as a solution of a diffusion equation in a full space. An extensive description of the numerous observations and theoretical developments that followed Aki's pioneering work is given in the book by SATO and FEHLER (1998) (see also WU and AKI, 1988, 1990).

We show in the following that coda waves are relevant to multiple scattering, a regime which can be approximated by diffusion for long lapse times. At shorter lapse time the energy can be described by the radiative transfer equation (RTE), an integro-differential expression of the local balance of energy taking into account all orders of scattering (e.g., CHANDRASEKHAR, 1960). It assumes an incoherent summation of waves, i.e., it neglects the phase of the waves and the related effects of interference. WU (1985) introduced radiative transfer in seismology. MARGERIN (2005) gives a comprehensive review of the subject. The radiative transfer equation can be derived from an ensemble average of the wave equation (e.g., RYZHIK *et al.*,

1996). It can only be solved numerically except in very specific cases (WU, 1985; ZENG, 1991; SATO, 1993). Numerical simulations have been done with the Monte-Carlo method (GUSEV and ABUBAKIROV, 1990; HOSHIBA, 1991, 1995, 1997; SATO, 1995; MARGERIN *et al.*, 1998; YOSHIMOTO, 2000). It relies on the analogy between radiative transfer and Boltzmann's equation in the kinetic theory of gases. The propagation of energy is represented as the random walk of a very large number of particles. The elastic treatment requires allowing for the polarized nature of P and S waves and numerical solutions can be computed (MARGERIN *et al.*, 2001).

The main parameter that describes the propagation in the multiple scattering regime is the mean free path l, or its counterpart the mean free time (that is simply l divided by V the wave velocity). The scattering mean free path l is the central quantity to describe the propagation in a scattering medium. It is the typical length scale of attenuation by scattering of a wave propagating in a given direction. It is related to the more common scattering quality factor Q_{sc} as follows:

$$l = \frac{Q_{sc} V}{\omega}. \tag{2}$$

The way the initial direction of a wave is affected by scattering is related to the 'transport mean free path' l^*, which is equal to l only for isotropic scattering (see e.g., SATO and FEHLER, 1998).

The different regimes of propagation are illustrated in Figure 1 where the full numerical solution of the RTE is compared with single scattering and diffusion approximations. It shows that the diffusion approximation is accurate to give the decay at large lapse times. We will discuss this point later from a more detailed example. The dashed line gives the solution for the single scattering approximation. The validity of the single scattering approximation is limited to the very beginning of the coda, i.e., within one or two mean free times. In practice it is difficult to know exactly the value of the mean free time. The measurement of the mean free path is actually ambiguous in most cases since both scattering and dissipation may contribute to the decay of a propagating wave. To overcome this difficulty, HOSHIBA (1991, 1993) proposed to use the RTE to interpret energy measurements performed for different lapse times and different source-station distances.

Since the Earth's structure is characterized by strong velocity variations with depth, wave propagation is strongly affected by the stratification. Assuming that the Moho is separating the heterogeneous crust and the more homogeneous mantle, a leakage of the diffuse energy of the crust into the mantle can be postulated. The strength of this leakage is controlled by the velocity contrast at the base of the crust. A wave resulting from a diffraction in the crust and incident upon the Moho will be trapped if its incidence angle is supercritical, while it mostly leaks into the transparent mantle if it is subcritically incident. The process was theoretically described under the diffusion approximation by MARGERIN *et al.* (1998, 1999) who advanced an interpretation of coda Q in the form $Q_c = Q_0 f$ with

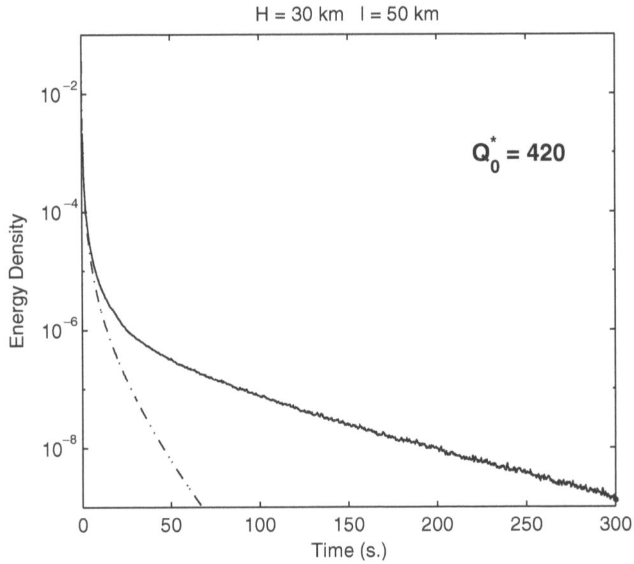

Figure 1
Coda decay obtained in a model consisting of a heterogeneous crust of thickness $H = 30$ km over a transparent mantle. The mean free path of the crust l is equal to 50 km. The solid line shows the numerical solution of the Radiative Transfer Equation obtained by Monte-Carlo modeling. The thin dashed line is the result of the single scattering approximation. The black circles indicate the behavior expected from a diffusion approximation (from CAMPILLO *et al.* 1999).

$$Q_0 = \frac{6\pi H^2}{Vl\xi^2}, \tag{3}$$

where H is the thickness of the crust, V the S-wave velocity, and ξ a parameter representing the effect of reflection at the Moho, dependent on the mean free path and the crustal thickness. Note that the frequency $f = 1$ Hz is assumed in the definition of Q_0. This formula is valid when the mean free path is smaller than the thickness of the crust. Otherwise one must rely on numerical or approximate RTE solution to infer Q_0. In this model Q_c can be regarded as the time of residence of diffuse waves in the crust. The same type of conceptual stratified model was used by LACOMBE *et al.* (2003) who modelled simultaneously the spatial decay of direct Lg waves and the envelopes of regional coda waves with only the mean free path and absorption length in the crust.

3. Diffusion and Equipartition

The validity of the Q_c model above is strongly limited by the assumption of the diffuse character of the wavefield that we have not yet demonstrated. It is difficult to assess the regime of propagation of the waves in a medium with absorption. With elastic waves we can rely on the concept of equipartition that leads eventually to a measurable marker of the diffusive regime, namely the ratio of P to S wave energies. The predominance of S waves in the coda was discussed in AKI (1992). He demonstrated that the scattering from P to S is much stronger than from S to P. Using an elegant reciprocity argument, he showed that the ratio of scattering coefficients is

$$\frac{g_{PS}}{g_{SP}} = \frac{\alpha^4}{\beta^4}, \tag{4}$$

where α and β are the velocity of P and S waves, respectively. We therefore expect a predominance of S waves in the coda. This was also observed by DAINTY and TOKSÖZ (1990) who have shown using array analysis of NORSAR data, that the coda is dominated by waves with apparent velocities less than 4 km/sec, that is, by S waves. More generally, in the diffusive regime that emerges after several scatterings, the wavefield is expected to consist of contributions of all possible modes of propagation. For example in a full space, modes of propagation are P and S plane waves.

In this case equipartion means that the wavefield consists of waves propagating in all directions and polarizations, with equal weight in average. A fixed weight implies that the relative contribution of P and S waves to the local total energy stabilizes to a constant ratio, whereas the total energy is continuously decaying due to the spatial expansion. A simple mode counting argument allows to compute the ratio of P to S energy at equipartition in a full space (WEAVER, 1982, 1990; RYZHIK *et al.*, 1996):

$$\frac{E_S}{E_P} = 2\frac{\alpha^3}{\beta^3}. \tag{5}$$

Assuming a ratio of velocities of $\sqrt{3}$, the energy ratio is 10.4. When there is preferential absorption of one of the modes (P or S), a stabilization of the energy ratio occurs in the multiple scattering regime (MARGERIN *et al.*, 2001). The effect of absorption is to shift the ratio in favor of the mode that is less absorbed. With realistic values of dissipation in rocks, this effect is not expected to strongly affect the observations. It is useful to further clarify the concept of equipartition and the definition of the underlying set of modes. The set of modes we consider is the one of a reference model on which the disorder is added. The disorder is assumed to be large enough to provoke the coupling between the modes but not to change drastically the structure of the spectrum and eigenfunctions. In other words, the modes we consider are not formally the ones of the actual Earth with all its complexity, but the ones of a fictitious model close enough to essentially share the same spectrum, i.e., essentially the same propagation properties. For a first-order computation of the energy ratio, it could be a simple stratified model. Furthermore, we deal here with 'local' propagating modes of the lithosphere, that is fundamental and higher modes of surface waves, and body waves leaking into the mantle. It is important to notice that equipartition is expected in the phase space so that it does not mean that the local distribution of energy is the same everywhere in the real space. Locally the ratio is governed by the eigenfunctions of the modes. For example, in the equipartition regime, the contribution of surface waves at the free surface will be larger than at depth due to the decay of fundamental mode eigenfunctions (see HENNINO *et al.*, 2001).

What is the significance of the stabilization of S-to-P energy ratio with respect to the validity of the diffusion approximation and the equipartition itself? Is there an unambiguous relation between stabilization of energy ratio and equipartition? To answer these questions, Figure 2 sets forth numerical solutions of the elastic radiative transfer equation computed with the Monte Carlo approach of MARGERIN *et al.* (2001) compared with solutions of the diffusion equation (PAUL *et al.*, 2005). They indicate that the stabilization occurs when the total energy is well approximated by a solution of the diffusion equation (Fig. 2, upper panels). On the contrary, when considering the angular dependence of the energy flow (Fig. 2, lower panel), it is clear that the stabilization occurs very early in the evolution of the wavefield toward isotropy. At a time when the stabilization is reached (60 s in Fig. 2), the anisotropy of the field is very strong with a ratio of more than four between energies propagating in the forward and backward directions. The diffusion solution itself includes a flow of energy from the source, and therefore an anisotropy of the field. Figure 2 also indicates that, when using the diffusion approximation, the anisotropy is underestimated with respect to the radiative transfer equation. We must conclude that the stabilization of S-to-P energy ratio is a good indication that the field is entering a regime in which the total energy is described by the diffusion equation and therefore will evolve towards equipartition and isotropy. Note that although the computations we just presented are performed in a full space, the argument of the remnant flow still will be valid in a stratified medium like the Earth.

It is possible to evaluate the S-to-P energy ratio from field records from a small aperture array. The energies are measured from the divergence and curl of the field. Divergence and curl can be evaluated from an array at the surface by using the extra information given by the free surface conditions. Such an experiment was performed in Mexico (CAMPILLO *et al.*, 1999; SHAPIRO *et al.*, 2000). The energy ratio was computed in moving windows along the seismograms. The P to S ratio stabilizes in the coda at a value of about 7.3, independently of source depth and epicentral distance (Fig. 3). This value is indeed far from the full space expectation (10.4). This theoretical ratio is obtained by neglecting reflections and the generation of Rayleigh waves while the measurement is made at the free surface itself. Considering a model that includes Rayleigh waves, HENNINO *et al.* (2001) showed that the energy ratio at the free surface of an elastic half space is 7.2, a value very close to the average observation. Note that at depth the ratio tends to the value obtained in a full space

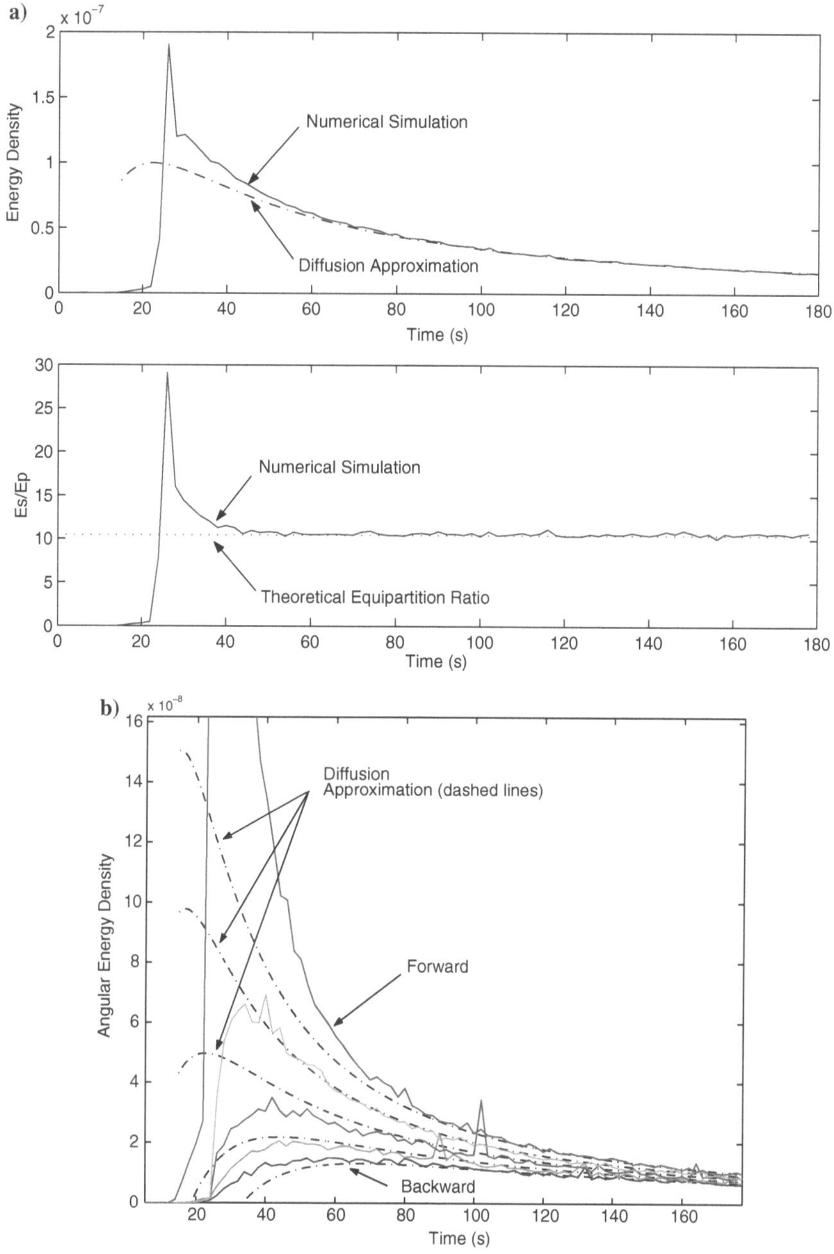

Figure 2

Comparison between numerical (Monte-Carlo) solutions of the radiative transfer equation, and analytical solutions of the diffusion equation. Energy density (top), P-to-S energy ratio (middle) and angular distribution of elastic energy flux (bottom). The dashed and solid lines show the results of the diffusion approximation and radiative transfer equation, respectively. The energy flux decreases monotonically from $\theta = 0$ (forward direction) to $\theta = \pi$ (backward direction), where θ denotes the angle between the propagation direction and the source-observer vector. The results for $\theta = \pi/4$, $\pi/2$ and $3\pi/4$ are also plotted.

(10.4). In field data the stabilization occurs in the early coda (Fig. 3), indicating that the late coda is clearly diffusive. This result is another indication that coda waves have the robust properties of diffusive waves and are particularly well adapted for measuring earthquake magnitudes (e.g., MAYEDA and WALTER, 1996) and site effects (e.g., PHILLIPS and AKI, 1986).

4. Interference Effects with Multiply-scattered Waves

One of the underlying assumptions with the use of RTE or diffusion equation is that the energy of the signal is made up of a summation of incoherent contributions. These theories account for the macroscopic behavior of averaged quantities such as the energy density. At the same time, the displacement is always a local solution of wave equations, and therefore the existence of phase effects, such as interferences, occurring at the scale of the wavelength must not be ruled out. In this sense, multiply-scattered seismic waves are relevant to mesoscopic physics. The effect of interferences within a diffuse field is illustrated by the so called coherent backscattering enhancement, also named weak localization. This effect appears only in the multiple scattering regime. It is a consequence of reciprocity that results in the constructive interference between long reciprocal paths in wave scattering. The probability to return to the source is increased by a factor of exactly 2 with respect to other paths. Eventually it results in a local energy density enhancement by the same factor 2 (e.g., AKKERMANS *et al.*, 1986).

In seismic experiments performed at the surface, weak localization appears as an enhancement of seismic energy in the vicinity of a source for long lapse times. MARGERIN *et al.* (2001) studied the scalar case in the configuration of a seismic experiment. Note that special care must be given to rules of reciprocity with polarized waves. The elastic case was treated by VAN TIGGELEN *et al.* (2001). A field experiment was performed in a volcanic environment using a sledge hammer as a source recorded along a profile (LAROSE *et al.*, 2004). The energy of the signal is computed in different time windows of 0.4 s duration. In each window, the energy is normalized by the maximum over the array, and then averaged over 12 different configurations. Figure 4 shows the average energy enhancement along the profile computed in time windows at different lapse times after the passing of the direct waves. As expected from the theory, the enhancement shows up progressively for the longer lapse times. This simple experiment indicates a nonintuitive mesoscopic effect that demonstrates that the phase effect cannot be neglected for seismic waves in the multiply-reflected regime. The characteristic time for the onset of the enhancement spot (approximately 0.7 s in the case of Fig. 4) is the scattering mean-free time, which is a measurement of the heterogeneity of the medium. This type of experiment is a way to measure the mean free time independently of dissipation.

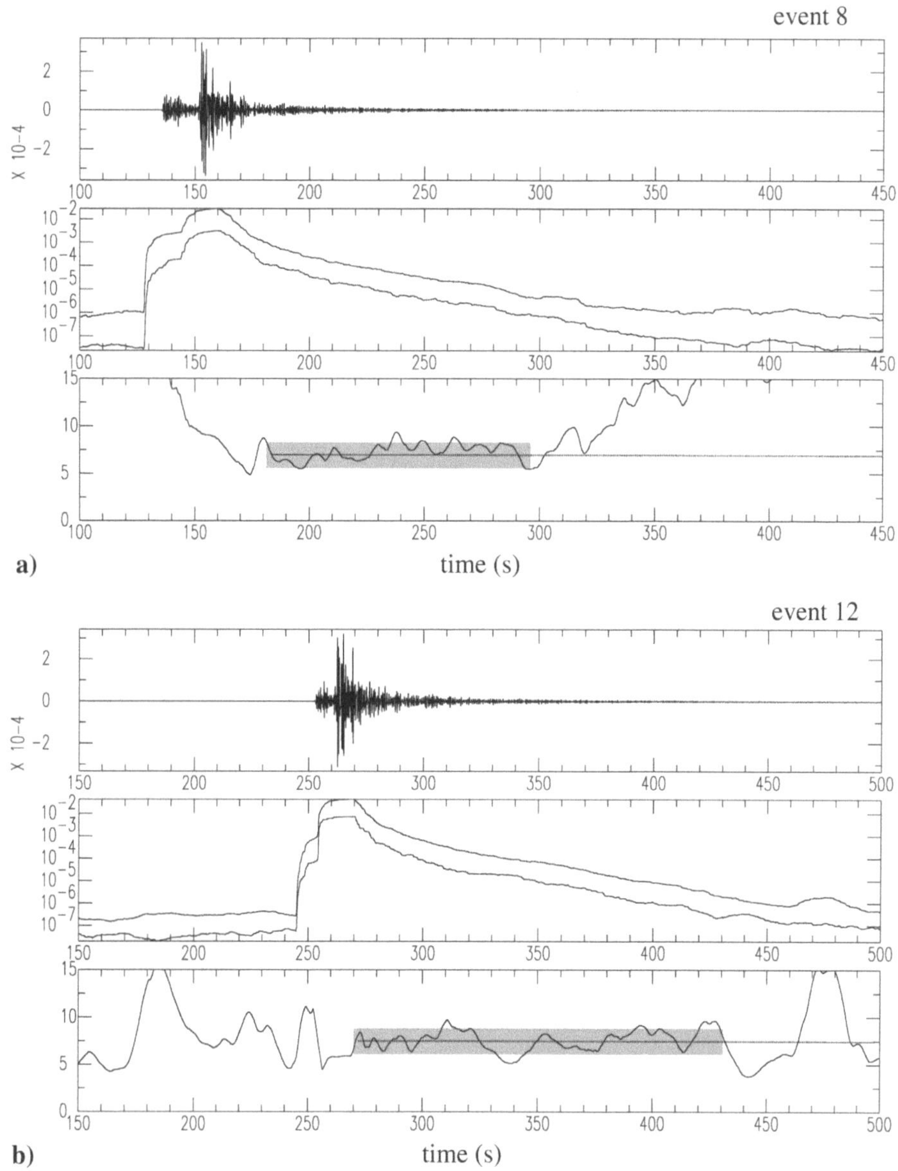

Figure 3
Examples of stabilization of the energy ratio for two earthquakes at different epicentral distances. In each case the panel shows the seismograms on the vertical component (top), the P and S energies computed in the frequency range 1 to 3 Hz (middle) and S-to-P energy ratio (bottom). Note that the energies are plotted on a logarithmic scale while the ratio is plotted on a linear scale (from CAMPILLO *et al.*, 1999).

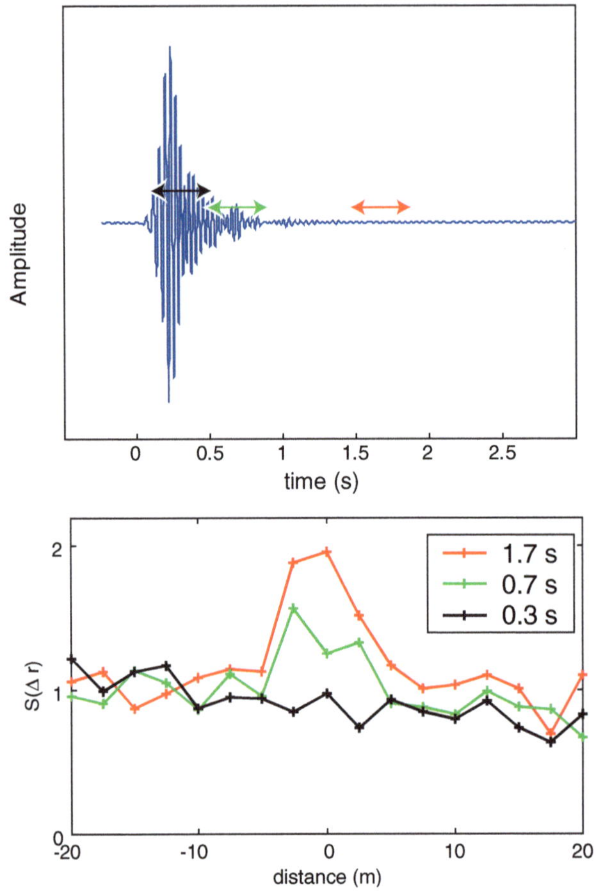

Figure 4
Emergence of weak localization. The upper panel shows an example of a signal produced by a hammer strike. The normalized average energy in the different time windows is plotted as a function of the distance from the source (lower panel). Note the absence of an enhancement spot for the early coda and the progressive onset of the weak localization.

5. Correlation and Diffuse Field

A consequence of interference effects such as the weak localization is that we can expect to extract deterministic wave propagation between two points where diffuse fields are recorded. Such an approach follows a long trend of attempts to reconstruct the Green's function from complex fields such as diffuse fields or noise, that were otherwise considered as random (e.g., AKI, 1957; TOKSÖZ, 1964; CLEARBOUT, 1968). The reconstruction of deterministic arrivals from the correlation of wavefields is not specific to diffuse waves. We will discuss this in the next section, however let us first consider the implications of multiple scattering. Formally, an implication of the presence of all modes is that the information regarding any possible path is represented in the coda records. Let us follow a conceptual approach to understand further the role of scattering and equipartition. Mathematically, a wavefield inside a finite elastic body can be expressed in modal form

$$u(\mathbf{x}, t) = \sum_n a_n \phi_n(\mathbf{x}) e^{i\omega_n t}, \tag{6}$$

where $\mathbf{x}$ is position, t is time, ϕ_n are eigenfunctions, ω_n are eigenfrequencies, and a_n are modal excitation functions that depend only on the source. As in the section on equipartition, the modes correspond to a reasonable reference model, close to the actual medium, and to which a disorder is added in the form of a perturbation of the elastic parameters for example. When disorder is added and after a sufficiently long time, the field becomes diffuse and the coefficients a_n become random functions of time. Equipartition means that the modal amplitudes are uncorrelated random variables

$$\langle a_n a_m^* \rangle = \delta_{nm} F(\omega_n), \tag{7}$$

where $F(\omega)$ is the spectral energy density of the source in the frequency band $[\omega - \delta\omega, \omega + \delta\omega]$. An implication of equation (7) is that all modes in a narrow frequency range are excited at the same energy level. Note that here we make the drastic hypothesis that, except for the total energy emitted, the signature of the source properties (mechanism, precise location, ..) has completely disappeared in the diffusion process. The brackets in equation (7) mean either an ensemble average over the disorder or a time/frequency average of a single realization. Another type of average can be performed by considering the fields produced by a set of sources inside the body. Since we can compute the correlation over limited time windows only in practice, source averaging is required. We will consider this kind of averaging in the following sections.

The average correlation between the fields at locations x and y simply becomes

$$\langle C(\mathbf{x}, \mathbf{y}, \tau) \rangle = \sum_n F(\omega_n) \phi_n(\mathbf{x}) \phi_n(\mathbf{y}) e^{-i\omega_n \tau}, \tag{8}$$

since the cross terms disappear on average due to (2). This expression, actually its time derivative, is close to the Green function between x and y defined for positive times as

$$G(\mathbf{x}, \mathbf{y}, t) = \Re\left(\sum_n \frac{\phi_n(\mathbf{x}) \phi_n(\mathbf{y})}{-i\omega_n} e^{-i\omega_n t} \right). \tag{9}$$

This reasoning must be regarded as a plausibility argument. LOBKIS and WEAVER (2001) gave a complete argument that does not rely on the equipartition hypothesis. In this case, the correlation depends also on the source whose influence was discarded when equipartition was assumed. The correlation is actually proportional to $G(\mathbf{x}, \mathbf{y}, t)$ convolved by the local response at the source $G_{ss} = G(\mathbf{s}, \mathbf{s}, t)$. The same formal result was already in DRAEGER and FINK (1999). If the medium is locally homogeneous, G_{ss} can be approximated by $\delta(t)$ and this factor is neglected. In case of a strong heterogeneity in the vicinity of the source, as the free surface in seismology, G_{ss} exhibits secondary arrivals. They result in spurious contributions in the correlation. G_{ss} can nevertheless be neglected if we assume a homogeneous distribution of sources throughout the medium. In practice, when dealing with seismograms from earthquakes at different depths and in different environments, we expect that the spurious contributions cancel out with source averaging nonetheless their role in the eventual 'noise' in the reconstruction cannot be ruled out. The correlation of two signals given by the general form (8) does not have the dimension of a displacement. Depending on the quantity considered in practice, displacement or velocity for example, we expect to recognize the waveshape of the response from the time derivative, or the correlation itself.

Encouraged by the positive results obtained in acoustics, CAMPILLO and PAUL (2003) tested the feasability of the reconstruction with coda waves recorded in Mexico. The first striking observation is that long-range correlations of (multiply-scattered) coda waves can be measured between two stations. As intuitively expected, the correlation obtained for a single coda window does not exhibit any clear pulse. The surface wave part of the Green function is accurately reconstructed only after source averaging. The correlation of records from a single earthquake exhibits no visible arrivals associated with the Green function but, due to the limited time averaging performed with actual seismograms, only remnant fluctuations of the diffuse field. In the source averaging, the deterministic Green function emerges since it increases with N the number of time windows while the incoherent fluctuation varies only with the square root of N. Rayleigh and Love waves can be identified without ambiguity in the correlations, both from their speeds and from their polarizations. Furthermore, the average cross correlations between the different components of motions present the symmetries of the Green tensor. Depending on the couple of stations considered, it was observed that the correlations are symmetric or asymmetric in time, a feature that was interpreted by CAMPILLO and PAUL (2003) as the mark of the flow of diffuse energy from the source. This point will be discussed in section 7.

In the following we discuss how the Green function can emerge from the correlations for configurations close to the ones encountered in seismology. Two different approaches are presented. First, the reconstruction can be performed from the correlations of signals produced by a set of sources. The Green function will emerge through source averaging. This approach is independent of the regime of propagation given a sufficient distribution of sources. The presence of scattering is not required. Second, the reconstruction can be performed from a diffuse field, in theory even from a single source. Finally, it will be shown that the reconstruction can be performed with a limited number of sources if sufficiently strong scattering occurs.

6. Analogy with Time Reversal Experiments

Before entering into a detailed discussion of the results, let us propose a simple analogy that facilitates understanding the physical process leading to the reconstruction. DERODE *et al.* (2003) showed the relation between the reconstruction from correlations and time reversal experiments. In the original time reversal experiment (FINK, 1992), a series of active devices are used to collect the acoustic signals, time reverse them and re-emit them in the medium, forming a so-called time-reversed mirror. Because of time symmetry of the wave equation, the waves naturally back propagate to focus on the original source. This operation is similar to the concept of wave-equation migration developed in seismic exploration while in this case numerical propagation is achieved (e.g., TARANTOLA, 1999). The Time Reversal

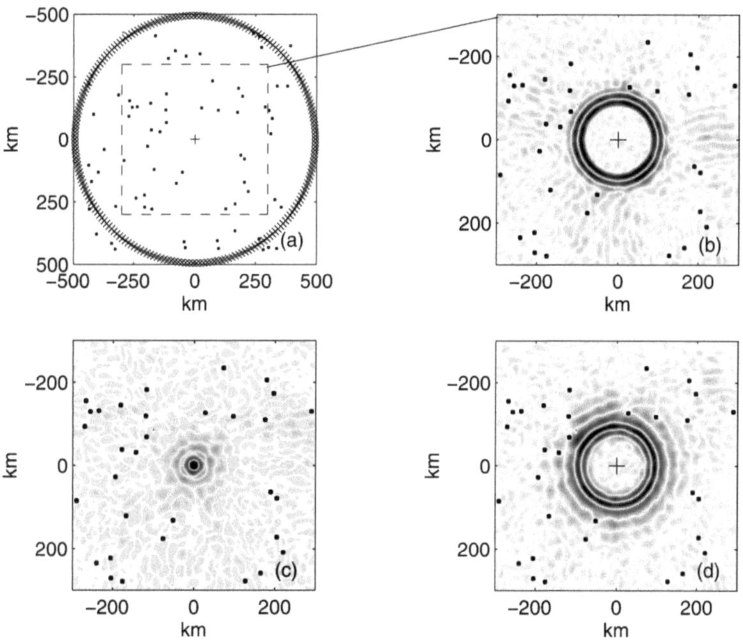

Figure 5
Numerical simulation of the reconstruction of the causal and anti-causal parts of the Green function from cross correlations. (a) Configuration of the numerical experiment. 1000 sources S ($\times$) surrounding the reference point A (+) are considered successively . The black dots indicate the point scatterers. (b) Snapshot of the cross correlation between the field in A with the field at location $(\mathbf{x}, \mathbf{y})$ after averaging over the sources S for correlation time −30 s. The weakly diffusive medium is characterized by the transport mean free path $l^* = 640$ km which is larger than the distance between the points where the correlations are computed. A converging wavefront is well-defined and constitutes the anti-causal part of the Green function. (c) Snapshot for correlation time $t = 0$ s : the wavefront is focused on A. (d) Snapshot for $t = 30$ s: the diverging wavefront corresponds to the causal part of the Green function (from PAUL *et al.*, 2005).

Mirror (TRM) technique was a breakthrough for practical applications in ultrasonics (e.g. FINK *et al.*, 2000) and marine acoustics (e.g., PARVULESCU, 1995; ROUX *et al.* 2004).

TRM can be understood easily with the Huyghens' principle. Imagine that a signal emitted by a source at $\mathbf{x}_a$ is recorded at every point $\mathbf{s}$ on a close surface Σ surrounding the source. For the sake of simplicity, we assume in the following that source functions are ideally impulsive as well as receiver responses. After time reversal and re-emission, the waves will naturally converge to the original source. The validity of this idea is independent of the medium in the vicinity of the point $\mathbf{x}_a$, which can be heterogeneous.

The relation of TRM and emergence of Green function from correlations is straightforward and based on reciprocity only. If the primary source is now at the point $\mathbf{s}$ on Σ, we can compute the cross correlation between the signals produced in $\mathbf{x}_a$ and another location $\mathbf{x}_b$. Because of reciprocity and by definition of correlation and convolution, the correlation $C(\mathbf{x}_a, \mathbf{x}_b)$ is strictly equal to the signal that would be produced if the source was in $\mathbf{x}_a$, recorded in $\mathbf{s}$ where it was time-reversed and re-emmitted to be finally recorded in $\mathbf{x}_b$. Reproducing this operation with a distribution of sources on Σ leads to the exact equivalent of a perfect time-reversal experiment. This analogy corresponds to visualizing the correlation between the field at a reference point and at surrounding points as a reconstructed physical field. Figure 5 gives a simple example that illustrates that reconstruction by correlation is not solely an artificial signal processing but results from physical wave propagation. The correlation map depicts the convergent (anti-causal solution: G(−t)) and divergent (causal G(t)) wave fronts. These two contributions are expected (LOBKIS and WEAVER, 2001) and observed in seismology (CAMPILLO and PAUL, 2003; PAUL *et al.*, 2005) and ultrasonics (e.g., MALCOLM *et al.*, 2004; LAROSE *et al.*, 2005).

Note that there is an implicit assumption in the time reversal and the use of the concept of the Green function. To consider the simplest case of acoustics, a signal, typically a particle velocity, cannot be directly time-reversed and re-emmitted as a pressure without a hypothesis about the impedance. Even though the proportionality of velocity and pressure, i.e., a constant impedance, is valid only for a homogeneous medium, it is a good approximation if the medium is locally homogeneous. If not, one must consider the local response at the source itself, that is the Green function *Gss*. We mentioned this point in the previous section and indicated that it can be reasonably ruled out by averaging over a source set since in every case *Gss* is dominated by the pulse at time 0.

The analogy allows to directly apply the results of the analysis of time reversal experiments to the reconstruction of the Green function by correlations. Most of the analysis was done for acoustic waves. They demonstrate that the identification of the Green function to source averaged field-field correlations is correct when the uncorrelated sources are distributed on a closed surface as for the perfect TRM, or if they are uniformly distributed in the volume (e.g., ROUX *et al.*, 2004; WEAVER and LOBKIS, 2004). SNIEDER (2004) used a stationary phase argument to demonstrate that the surface waves can be reconstructed by correlation when assuming an isotropic distribution of incoming scattered waves. WAPENAAR (2004) recently gave a formal demonstration of a generalization in the elastic case for a distribution of sources on a closed surface. These formal demonstrations cannot be directly applied to actual configurations in seismology since sources are not uniformly distributed, neither in the volume nor on a closed surface. In exploration seismics with a sufficiently large coverage of sources, the correlation method can be used to provide Green functions between receivers at depth (BAKULIN and CALVERT, 2004) in the same way as in laboratory.

7. Correlation and Diffusion Approximation

It was known for years from experiments with ultrasounds that time reversal focusing, or conversely correlation reconstruction can be achieved with a very limited mirror surface, owing the medium to be heterogeneous enough to produce a diffuse field (e.g., DERODE *et al.*, 1998; ROUX *et al.*, 2000). Following the analogy between source average correlation and TRM, the same argument holds for Green function reconstruction in seismology. First let us review a couple of theoretical results exclusively concerning diffuse waves. The point here is to isolate a single source. If multiple scattering occurs, the resulting wavefield evolves towards an equipartitioned state as discussed above. Under this condition, it is easy to show that the average correlation is proportional to the Green function (LOBKIS and WEAVER, 2001). The argument is based on a time average over a long lapse time and with the limitations of the equipartition principle on the role played by disorder in mixing the modes without changing the structure of the spectrum as already mentioned. VAN TIGGELEN (2003) used the diffusion approximation to derive an asymptotic relation between the field-field correlation and the ensemble averaged Green function of the medium, i.e., the response in a medium with the effective properties of the actual heterogeneous medium. The correlation of the fields at two points separated by $\mathbf{r}$ and located at $\mathbf{R}$ from the source can be written as

$$\begin{aligned} C_{\mathbf{R}}(\mathbf{r},\tau) &= \rho(\mathbf{R},\tau)\frac{\partial}{\partial\tau}[\langle G_B(\mathbf{r},\tau)\rangle - \langle G_B(\mathbf{r},-\tau)\rangle] \\ &\quad - 3\mathbf{J}(\mathbf{R},\tau)\cdot\nabla[\langle G_B(\mathbf{r},\tau)\rangle - \langle G_B(\mathbf{r},-\tau)\rangle] \end{aligned} \tag{10}$$

where the brackets $\langle\cdot\rangle$ denote an ensemble average, and G_B is the retarded causal Green function filtered in frequency band B. ρ is the diffuse energy density and $\mathbf{J}$ the energy current. With D the diffusion constant, $\mathbf{J}$ is given by the Fourier's law:

$$\mathbf{J}(\mathbf{R},t) = -D\nabla\rho(\mathbf{R},t). \tag{11}$$

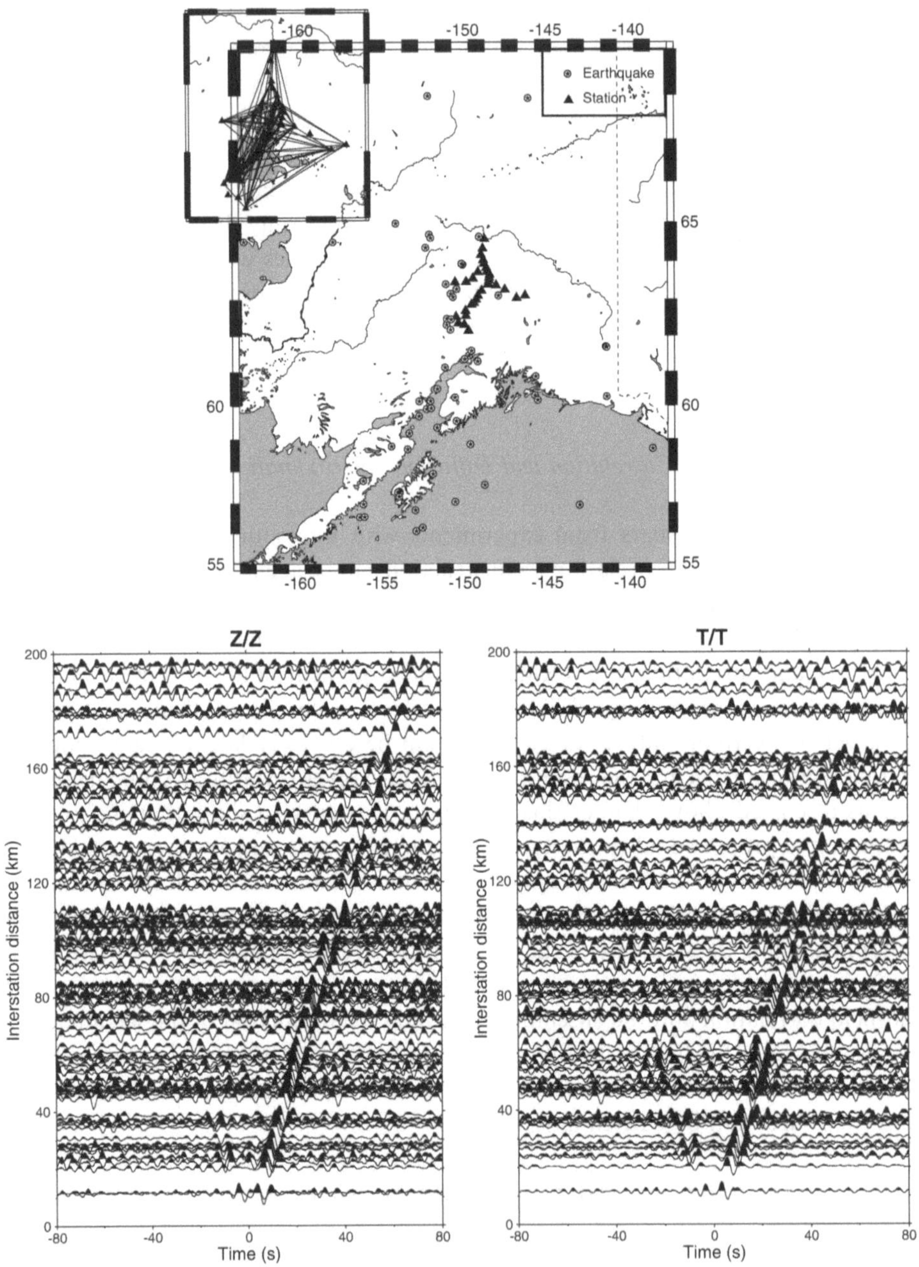

Figure 6

Top: Map of stations and earthquakes used in the analysis of Alaskan data. The paths between couples of stations for which average cross correlations have been computed are shown in the inset. Bottom: Average cross correlations as a function of interstation distance. The correlations have been computed for vertical (Z) components of motion (left), and transverse (T) (right).

The relation is valid for long times of correlation and ensemble averaged quantities. Indeed it is not directly relevant for seismological applications nonetheless it provides a useful relation between correlation of the diffuse field produced by a single source and the (time-symmetric) Green function.

These results on the correlation of diffuse waves provide another perspective on the reconstruction of Green function which is independent of the source distribution and which suggests that if the time of observation is long enough, a single source could be sufficient. To conclude this short summary on the theoretical approaches, we have two different arguments which can be invoked: Firstly, source averaging over an (almost) uniform volumic distribution of sources, or upon a uniform source distribution on an enclosing surface, and secondly infinite time averaging over a perfectly diffuse wavefield. In practical seismological applications neither of the two proper theoretical arguments can be used directly how ever both can be exploited simultaneously. Since the duration of the records is limited by the presence of noise and by absorption, averaging over a set of sources is required to expect the emergence of the Green function. It can be stated both ways: Multiple scattering (and the consequent equipartition) compensates the lack of sources or conversely source averaging compensates for the limited duration of the actual signals. DERODE *et al.* (2003b) and LAROSE *et al.* (2004) showed the role of multiple scattering in enhancing the efficiency of the reconstruction of the Green function with a limited number of sources and finite durations of recording, in conditions approximating those of seismology. Examples of numerical simulations of 2-D snapshots of correlations are presented by PAUL *et al.* (2005). They illustrate that even a discrete distribution of sources is enough to reconstruct the Green function as soon as multiple scattering is involved.

8. *Applications to Coda Waves*

CAMPILLO and PAUL (2003) reconstructed the surface wave part of the Green's functions between three pairs of stations in Mexico. To confirm these results in a different environment and to show that they are not dependent upon a particular azimuth of the path between the stations, PAUL *et al.* (2005) used the data from a temporary broadband experiment in Alaska (FERRIS *et al.*, 2003). Figure 6 (upper panel) shows the locations of stations and earthquakes. Correlations have been computed for every available pair of stations. The resulting paths span a large range of azimuths (Fig. 6). The records from about 100 regional earthquakes have been used. To obtain the source averaged correlations we first filtered the original broadband seismograms in the band 0.08–0.3 Hz. The horizontal components of the seismograms were then rotated assuming the inter station great circle path to be the radial direction. We used coda records starting 20 s after the arrival of the S wave and ending when the signal-to-noise ratio was smaller than 4. The cross correlation between coda records cannot be performed directly. In fact the exponential coda amplitude decay would result in a strong overweighting of the earliest part of the coda. In order to overcome this difficulty, we disregard the amplitudes completely and consider only one-bit signals. Since the phases are preserved when transforming the original data into 1-bit signals, this operation does not alter significantly the correlation of band-passed time series. CAMPILLO and PAUL (2003) checked that using 1-bit signals leads to the same results as a compensation for coda Q in successive time windows. This procedure has been successfully applied in time reversal experiments (DERODE *et al.*, 1999). The cross correlation of one-bit coda records at each pair of stations was computed for each earthquake, normalized to a maximum amplitude of one, and averaged over the entire set of events.

The resulting time-distance sections of the cross correlation of vertical with vertical and transverse with transverse components are also presented in Fig. 6. As for the Mexican data, the cross correlations present the symmetries of the Green tensor (PAUL *et al.*, 2005). The dominant phases are Rayleigh and Love waves as expected for the response at the surface of a source at the surface. The quality of the reconstruction is not enough to observe the body waves. Besides, it is possible that the coda at the period range we consider is dominated by surface waves and that the body waves are more difficult to reconstruct. Traces with a maximum amplitude at a negative time have been time-reversed to improve the identification at positive times and to attain a better view of the presence, or the absence, of symmetry in time. This experiment shows the anti-causal waves, which are expected from the symmetry of the wave equation to be valid solutions, but are not present in the signals produced by earthquakes or other actual sources in the real causal world.

We interpret the fact that causal and anti-causal Green functions have different amplitudes, i.e., time asymmetry, as the effect of the persistence of a net flux of energy from the source (Fig. 3), even in the late coda as it was discussed above. Since the sources are not evenly distributed, a preferential direction of transport remains after averaging. The data indicate that the time asymmetry is less pronounced when the correlation is performed in the later windows of the coda. This confirms that coda waves are behaving as multiply-scattered waves evolving towards equipartition and isotropy.

The relatively low frequency content (0.08–0.3 Hz) of the reconstructed signals has several possible causes. First of all the high frequency waves are attenuated more rapidly. It is therefore natural that they are more difficult to observe in noisy signals such as correlations. Furthermore the reconstruction at a given distance is more difficult for high frequency waves. It can be understood easily with a ray argument. Following a stationary phase argument (e.g., AKI and RICHARDS, 1980; ROUX *et al.*, 2004, and SNIEDER, 2004 for application to the reconstruction of Green function) the coherent constructive contributions of the direct wave part of the Green functions can be seen as the diffracted waves propagating in a limited 'endfire lobe' around the direction of the two stations. Indeed the width of this contributing region is proportional to the square root of the wavelength (e.g., SPETZLER and SNIEDER, 2004). Conversely, it is also decreasing with the distance, showing that it is naturally difficult to reconstruct high frequencies at large distances, i.e., a large number of wavelengths. It is also important to note that the signal-to-noise-ratio obtained with the correlation depends on the total duration of the signals which were correlated. Because of absorption, the available duration of coda is strongly decreasing with frequency varying between 0.1 to 5 Hz. This is an important limitation of the use of coda waves. The bandwidth of the signals is also important since an average upon frequency is also implicitly performed. Actually we know that the diffuse field remains correlated over a frequency interval called the Thouless frequency. The Thouless frequency can be written in our case as: $\omega_T \approx Vl/3R^2$ where V is the shear-wave velocity, l the mean free path and R the characteristic distance between the source and the pair of receivers. There is no advantage in frequency averaging when the bandwidth is small with respect to the Thouless frequency. In our example, ω_T is much smaller than the bandwidth and therefore our processing takes advantage of the independent information available over the entire frequency band.

9. *Field Correlation in the Seismic Noise*

In the previous section, we discussed the correlations between fields that are produced by deterministic sources such as earthquakes. We saw that, in our applications, the reconstruction of the Green function relies both on a source average over a large number of sources and on the (partial) isotropy of the field due to multiple scattering. It is therefore appealing to use the ambient noise although its actual origin is not fully known. The same principles must apply for diffuse coda waves from transient sources or for noise. The idea of using noise for exploration purposes is not new. Since 1957, Keiiti Aki proposed the use of noise to retrieve the propagation properties of the subsurface. CLAERBOUT (1968) also suggested the use of ambient noise to reconstruct the reflectivity profile. Later on, a similar concept of extraction of the response by correlation was successful in helioseismology (e.g., DUVALL *et al.*, 1993; RICKETT and CLAERBOUT, 1999, 2000).

Ambient noise has also been a long standing subject of interest in marine acoustics (KUPERMANN and INGENITO, 1980) and correlations of noise produced by the wavebreaking and ships have been used to reconstruct the acoustic response of the ocean (SABRA *et al.*, 2003). This calls to mind the fluctuation-dissipation theorem (e.g., KUBO, 1966) which states a relation between the Green function between two points and the correlation of the random fluctuations of the field at these two points. Originally, this relation was developed for thermal noise. An application in acoustics was shown by WEAVER and LOBKIS (2001) who successfully retrieved reflected paths in an aluminium block.

The thermal noise in the seismic frequency band is totally negligible with respect to the effects of ocean-continent coupling, wind and human activity and therefore the concept of fluctuation-dissipation does not seem to apply at the first order to what we call seismic noise. On the other hand, the property can be more robust and holds for a variety of wavefields. Indeed, there is no guarantee that the ambient seismic noise has the same behavior as thermal noise but we expect that the distribution of the sources of the seismic noise randomizes when averaged over sufficiently long times. Furthermore, the waves that compose the noise are randomized by scattering from heterogeneities and a form of equipartition is therefore expected.

AKI (1957) proposed the use of noise records from a seismic array to evaluate the phase velocity of the predominant surface wave, a technique often referred to as the SPAC method. The method is based on short distance correlation and azimuthal averaging. At short distance, the correlation in the frequency domain is expressed as a Bessel function of order zero, J_0, and a fit of the observed correlation makes it possible to measure locally the wave number assuming a dominant mode. The use of noise records to extract local propagation properties below the array has been successfully applied to investigate shallow layers, especially soft layers that could be responsible for seismic amplification during earthquakes (e.g., KUDO *et al.*, 2002). Note that there is a close relation between the SPAC method, where the identification of the Green function in the correlation is not explicit, and our reconstruction. Considering a single surface mode, the problem reduces to a 2-D scalar homogeneous problem. In this case (2D, SH) the Green function in the Fourier domain is:

$$G(x,y) = \frac{1}{4i\mu}[J_0(kr) - iY_0(kr)], \tag{12}$$

where Y denotes the Neuman function and μ the rigidity. Note that this expression gives the causal Green function: The Fourier transforms of imaginary and real part contribute equally in the positive times and cancel out exactly in negative times. The correlation ($J_0(kr)$) is proportional to the imaginary part of the Green function. It therefore gives a signal that is proportional to the causal Green function with an even contribution in the negative times, similar to what we described previously (e.g., Fig. 5). Indeed the assumptions under this reasoning, lateral homogeneity and single mode, cannot be put forward for long-range correlation. The canonical problem of a random distribution of plane waves within a homogeneous elastic medium was treated in 2-D and 3-D by SÁNCHEZ-SESMA and CAMPILLO (2006) who showed that equipartition between P and S is required for retrieving the exact Green function.

The use of long-range correlations for practical applications in seismology is recent. SHAPIRO and CAMPILLO (2004), SABRA *et al.* (2005), SHAPIRO *et al.* (2005) and PEDERSEN *et al.* (2005) demonstrated the validity of the approach. SHAPIRO and CAMPILLO (2004) performed group velocity measurements on reconstructed signals in the period range 10–125 s and for interstation distances from a few hundreds to several thousands of kilometers. SABRA *et al.* (2005) further described the emergence of the Green function. Nevertheless, the limits of application of the method in seismology are not fully understood since the origin of the noise itself is not precisely known in the different frequency bands.

In the low frequency domain ($f < 0.3$ Hz), the ambient noise seems widely dominated by the interaction of the ocean with the solid Earth (LONGUET-HIGGINS, 1950; FRIEDERICH *et al*, 1998; WEBB, 1998; RHIE and ROMANOWICZ, 2004; TANIMOTO, 2005; KEDAR and WEBB, 2005). The noise amplitude is strongly correlated with storm activity (e.g., BROMIRSKI, 2001; 2005) and even with climate change (GREVEMEYER *et al.*, 2000). At higher frequencies, the noise is produced locally by human activity and wind and, because of attenuation, cannot propagate over extended distances. In any case the noise is produced by surface sources which generate predominantly surface waves. It is therefore expected that the signals that can be extracted from noise records are also predominantly made of surface waves. Furthermore one must remember that the noise spectrum has marked peaks, and consequently the correlation (which spectrum is the product of the spectra of the records) is yet more strongly peaked. While the capability of reconstructing the body waves cannot be ruled out, it has not been convincingly demonstrated to date. In spite of the locations of the noise sources at the free surface, the scattering associated with the strong heterogeneity of the surface layers and with topography results in a coupling between surface waves and body waves. The efficiency of this process in the Earth is illustrated by the observation of the equipartition between the different modes of body and surface waves. It is therefore expected that body waves could be retrieved from correlation of records at closely located stations.

Even though only surface waves have been reconstructed from noise thus far, it is a valuable result since dispersive surface waves are widely used for imaging crustal and lithospheric shear velocity structures. SHAPIRO *et al.* (2005) presented the first example of the use of the ocean-generated noise to map Rayleigh wave group velocity. Here we present an example of this mapping. We use continuous noise records from 62 broadband stations in California. We selected 30 days from August to September 2004 during which no $M > 5.8$ earthquake occurred. The data were bandpass-filtered between 10 and 20 s and all pairs of crosscorrelations were computed. In most of the cases, we can clearly identify the fundamental Rayleigh wave. We computed the amplitude ratio between the emerging Rayleigh and the surrounding fluctuations. When the ratio is less than 4, the waveforms are rejected. This processing results in 785 group speed measurements at 18 s for paths larger than two wavelengths (Fig. 7). Velocity dispersion is measured by frequency-time analysis (LEVSHIN *et al.*, 1989; RITZWOLLER and LEVSHIN, 1998; SHAPIRO and SINGH, 1999). We finally applied a tomographic inversion (BARMIN *et al.*, 2001) to obtain a group speed map on a 28 km $\times$ 28 km grid across California.

The group velocity map (Fig. 8) shows the different geological features. The Sierra Nevada and the Peninsular Ranges, composed principally of Cretaceous

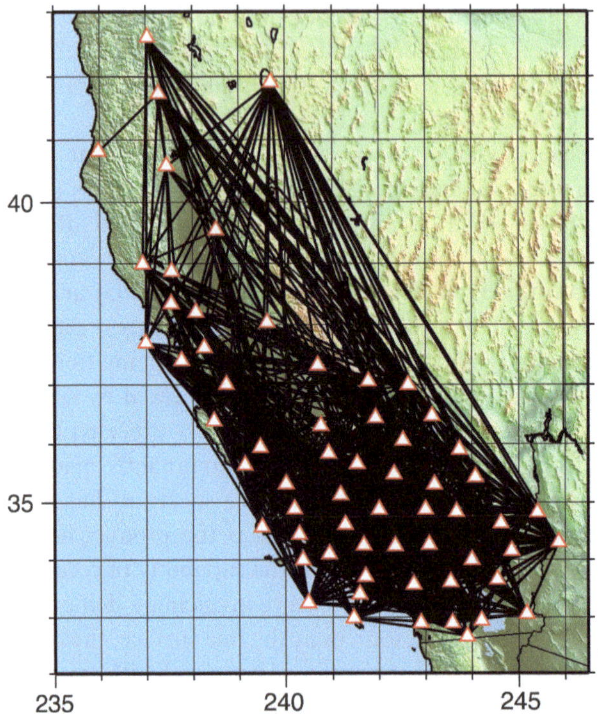

Figure 7
Paths where 18 s Rayleigh wave group speed measurements were obtained from cross correlations of ambient seismic noise. White triangles show locations of USArray stations used in this study.

granitic batholiths, are characterized by high group velocities. The group speeds are lower in the Great Basin and in the Mojave Desert, indicating that the middle crust in these areas is probably hotter and weaker than in the Sierra Nevada. Group speeds are low in the sedimentary mountain ranges, e.g., the Transverse Ranges, the southern part of the Coast Ranges, and the Diablo Range, while fast wave speeds are observed for the Salinian block. The signature at depth of the San Andreas fault is clearly visible in this area. The western wall is characterized by the high velocities of the plutonic rocks of the Salinian block while the low-speed eastern wall is associated with the sedimentary rocks of the Franciscan formation.

This example illustrates that the noise on Earth in the period band considered has the required properties for allowing the emergence of the Green function from averaging of correlation over long time series. Surprisingly, LAROSE *et al.* (2005b) succeeded in retrieving the Green function at stations distant of several wavelengths in a very different context. They used the high frequency noise recorded on the moon by a group of sensors installed during the Appolo 17 mission. Although the noise is of completely different origin on the Moon, namely heat-generated cracking, the

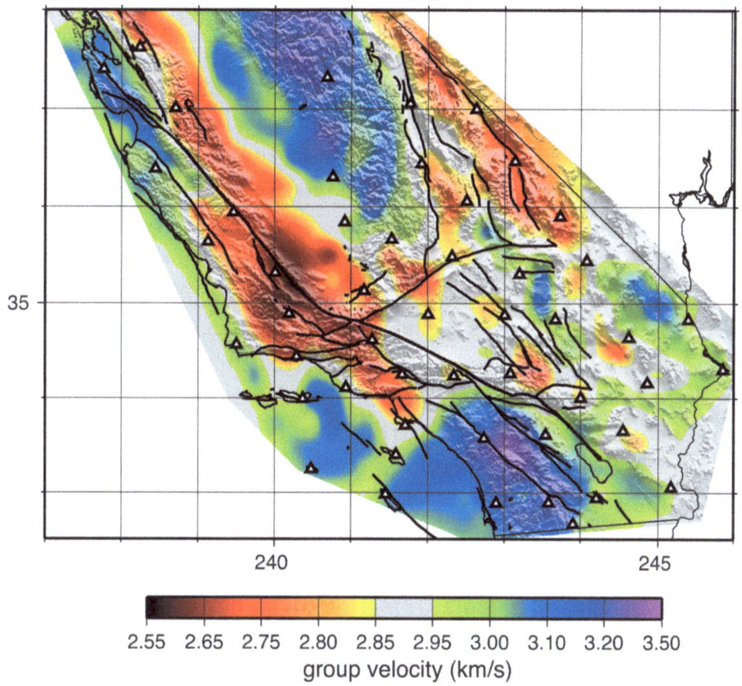

Figure 8
Rayleigh wave group speed map at 18 s period constructed by cross correlating 30 days of ambient noise between stations. Black solid lines show known active faults. White triangles show locations of the stations used in this study.

technique proved to be efficient to probe the shallow layers. It could be a valuable tool for planetary exploration.

10. Conclusion

We show that the average properties of the time decay of coda energy can be explained by the theory of radiative transfer assuming an uncoherent sum of contributions. This theory is valid only for intensity or total energy. Coda waves exhibit the transition from low-order scattering in the early coda to diffusion in the late coda. The stabilization of the ratio of S-to-P energies is a marker of the diffusive regime. It indicates that the waves are evolving toward equipartition, although equipartition is an asymptotic regime. The diffuse field keeps a significant preferential direction of flow even at extended lapse times.

Radiative transfer and diffusion neglect the phase effects in the combination of the contribution. We demonstrate the presence of mesoscopic effects that require consideration of the very nature of the waves, i.e., their property of producing interference effects. Weak localization can be observed with simple experiments in the field. It results in the so-called coherent backscattering enhancement at the source, a phenomenon characteristic of multiple scattering. This observation shows the importance of phase effects for fields that we used to consider to be 'random'. Since phase information is preserved, we used spatial correlation properties of the fields to infer deterministic propagation characteristics. Different approaches to the problem are related to the type of averaging performed to extract the deterministic Green function from the field fluctuations. Seismological applications rely on source averaging. The presence of scattering has an important positive effect by making the wavefield more isotropic. The results presented here concern coda waves produced by a set of earthquakes or long-time series of noise records for which we expect the source of the noise to randomize partially. In both cases, it is demonstrated that the field-field correlations contain, at least partially, the deterministic response of the Earth between the stations, offering the possibility of an empirical construction of seismograms based on the actual wave propagation in the complex Earth. They can be used for imaging the velocity structure of the Earth interior without waiting for earthquakes. As the measurements from the correlations are performed for paths between stations and for short periods, such methods are likely to contribute to improved resolution in tomographic imaging. Correlation of diffuse wavefields overall opens up a wide range of potential applications.

Acknowledgments

I am grateful to Keiiti Aki for his advice and encouragment over the years. The works presented here were accomplished in close collaboration with Ludovic Margerin, Nikolai Shapiro, Bart van Tiggelen, Anne Paul, Arnaud Derode, Eric Larose, Céline Lacombe and Laurent Stehly. I am indebted to Richard Weaver, Francisco Sánchez-Sesma, Roger Maynard, Philippe Roux and Haruo Sato for enlightening discussions. I thank Roel Snieder, Toshiro Tanimoto and Helle Pedersen for their careful reviews and their useful comments.

References

Abubakirov, I.R. and Gusev, A.A. (1990), *Estimation of scattering properties of lithosphere of Kamchatka based on Monte-Carlo simulations of record envelope of a near earthquake*, Phys. Earth Planet. Inter. *64*, 52–67.

Aki, K. (1957), *Space and time spectra of stationary stochastic waves with special reference to microtremors*, Bull. Earthq. Res. Inst. *35*, 415–456.

Aki, K. (1969), *Analysis of the seismic coda of local earthquakes as scattered waves*, J. Geophys. Res. *74*, 615–618.

AKI, K. (1980), *Scattering and attenuation of shear waves in the lithosphere*, J. Geophys. Res. *85*(B11), 6496–6504.

AKI, K. (1992), *Scattering conversions P to S versus S to P*, Bull. Seismol. Soc. Am. *82*, 1969–1972.

AKI, K. and CHOUET, B. (1975), *Origin of coda waves: source, attenuation, and scattering effects*, J. Geophys. Res. *80*, 3322–3342.

AKI, K. and RICHARDS, P. (1980), *Quantitative Seismology–Theory and Methods*, Vols. 1 and 2 (W. H. Freeman, San Francisco, 1980).

AKKERMANS, E., WOLF, P.E., MAYNARD, R., and MARET, G. (1988), *Theoretical study of the coherent backscattering of light by disordered media*, J. Phys. France *49*, 77–98.

BAKULIN, A. and CALVERT, R. (2004), *Virtual source: new method for imaging and 4-D below complex overburden*, SEG Technical Program Expanded Abstracts, pp. 2477–2480.

BARMIN, M.P., RITZWOLLER, M.H., and LEVSHIN, A.L. (2001), *A fast and reliable method for surface wave tomography*, Pure Appl. Geophys. *158*, 1351–1375.

BEAUDOIN, B.C., FUIS, G.S., MOONEY, W.D., NOKLEBERG, W., and CHRISTENSEN, N.I. (1992), *Thin, low-velocity crust beneath the southern Yukon Tanana terrane, east central Alaska: Results from Trans-Alaska crustal cransect refraction/wide-angle reflection data*, J. Geophys. Res. *97*, 1921–1942.

BROMIRSKI, P.D. (2001), *Vibrations from the perfect storm*, Geochem. Geophys. Geosyst. 2 Paper number 2000GC000119.

CAMPILLO, M. and PAUL, A. (2003), *Long-range correlations in the seismic coda*, Science *299*, 547–549.

CAMPILLO, M., MARGERIN, L., and SHAPIRO, N.M., *Seismic wave diffusion in the Earth lithosphere*. In *Waves in Complex Media* (J.P. Fouque, Ed.) (NATO ASI Series, Kluwer, 1999).

CHANDRASEKHAR, S., *Radiative Transfer* (Dover, New York, 1960).

CLAERBOUT, J.F. (1968), *Synthesis of a layered medium from its acoustic transmission response*, Geophysics *33*, 264–269.

DAINTY, A.M. and TOKSOZ, M.N. (1990), *Array analysis of seismic scattering*, Bull. Seismol. Soc. Am. *80*, 2248–2260.

DERODE, A., TOURIN, A., and FINK, M. (1999), *Ultrasonic pulse compression with one-bit time reversal through multiple scattering*, J. Appl. Phys. *85*, 6343–6352.

DERODE, A., TOURIN, A., and FINK, M. (1998), *Time reversal in multiply scattering media*, Ultrasonics *36*, 443–447.

DERODE, A., TANTER, M., SANDRIN, L., TOURIN, A., and FINK, M. (2001), *Numerical and experimental time-reversal of acoustic waves in random media*, J. Comput. Acoust. *9*(3), 991–998.

DERODE, A., LAROSE, E., TANTER, M., DE ROSNY, J., TOURIN, A., CAMPILLO, M., and FINK, M. (2003), *Recovering the Green's function from field-field correlations in an open scattering medium (L)*, J. Acoust. Soc. Am. *113*, 2973–2976.

DERODE, A., LAROSE, E., CAMPILLO, M., and FINK, M. (2003), *How to estimate the Green's function of a heterogeneous medium between two passive sensors? Application to acoustic waves*, Appl. Phys. Lett. *83*, 3054–3056.

DRAEGER, C. and FINK, M. (1999), *One-channel time-reversal in chaotic cavities: Theoretical limits*, J. Acoust. Soc. Am. *105*, 611–617.

DUVALL, T.L., JEFFERIES, S.M., HARVEY, J.W., and POMERANTZ, M.A. (1993), *Time distance helioseismology*, Nature *362*, 430–432.

FERRIS, A., ABERS, G.A., CHRISTENSEN, D.H., and VEENSTRA, E. (2003), *High resolution image of the subducted Pacific (?) plate beneath central Alaska, 50–150 km depth*, Earth. Planet. Sci. Lett. *214*, 575–588.

FINK, M. (1992), *Time reversal of ultrasonic fields- Part I: Basic principles*, IEEE Trans. Ultrason., Ferroelec., Freq. Contr. *39*(5), 555–566.

FINK, M., CASSEREAU, D., DERODE, A., PRADA, C., ROUX, P., TANTER, M., THOMAS, J-L., and WU, F. (2000), *Time-reversed acoustics*, Rep. Prog. Phys. *63*, 1933–1995.

FRIEDERICH, A., KRUGER, F., and KLINGE, K. (1998), *Ocean-generated microseismic noise located with the GRFO array*, J. Seismol. *2*, 47–64.

GREVEMEYER, I., HERBERT, R., and ESSEN, H.H. (2000), *Microseismological evidence for a changing wave climate in the northeast Atlantic Ocean*, Nature *408*, 349–352.

GUSEV, A. (1995), *Vertical profile of turbidity and coda Q*, Geophys. J. Int. *123*, 665–672.

Hennino, R., Tregoures, N., Shapiro, N.M., Margerin, L., Campillo, M., van Tiggelen, B.A., and Weaver, R.L. (2001), *Observation of equipartition of seismic waves*, Phys. Rev. Lett. *86*, 3447–3450.

Herraiz, M. and Espinoza, A.F. (1987), *Coda waves: A review*, Pure Appl. Geophys. *125*(4), 499–577.

Hoshiba, M. (1991), *Simulation of multiple scattered coda wave excitation based on the energy conservation law*, Phys. Earth Planet. Inter. *67*, 123–136.

Hoshiba, M. (1993), *Separation of scattering attenuation and intrinsic absorption in Japan using the multiple lapse time window analysis of full seismogram envelope*, J. Geophys. Res. *98*, 15809–15824.

Hoshiba, M. (1995), *Estimation of nonisotropic scattering in western Japan using coda wave envelopes: Application of a multiple nonisotropic scattering model*, J. Geophys. Res. *100*, 645–657.

Hoshiba, M., Rietbrock, A., Scherbaum, F., Nakahara, H., and Haberland, C. (2001), *Scattering attenuation and intrinsic absorption using uniform and depth dependent model: Application to full seismogram envelope recorded in Northern Chile*, J. Seismol. *5*, 157–179.

Kedar, S. and Webb, F.H. (2005), *The Ocean's Seismic Hum*, Science *307*, 682–683.

Kubo, R. (1966), *The fluctuation-dissipation theorem*, Rep. Prog. Phys. *29*, 255–284.

Kudo, K., Kanno, T., Okada, H., Zel, O., Erdik, M., Sasatani, T., Higashi, S., Takahashi, M., and Yoshida, K. (2002), *Site-specific issues for strong ground motions during the Kocaeli, Turkey, Earthquake of 17 August 1999, as inferred from array Observations of microtremors and aftershocks*, Bull. Seismol. Soc. Am. *92*, 448–465, doi: 10.1785/0120000812.

Kuperman, W.A. and Ingenito, F. (1980), *Spatial correlation of surface-generated noise in a stratified ocean*, J. Acoust. Soc. Am. *67*, 1988–1996.

Lacombe, C., Campillo, M., Paul, A., and Margerin, L. (2003), *Separation of intrinsic absorption and scattering attenuation from Lg coda decay in Central France*, Geophys. J. Int. *154*, 417–425.

Larose, E., Derode, A., Campillo, M., and Fink, M. (2004), *Imaging from one-bit correlations of wideband diffuse wavefields*, J. Appl. Phys. *95*, 8393–8399.

Larose, E., Margerin, L., Campillo, M., and van Tiggelen, B.A. (2004), *Weak localization of seismic waves*, Phys. Rev. Lett. *93*, 048501.

Larose, E., Derode, A., Clorennec, D., Margerin, L., and Campillo, M. (2005a), *Passive retrieval of Rayleigh wave in disordered elastic media*, submitted to Phys. Rev. E.

Larose, E., Khan, A., Nakamura, Y., and Campillo, M. (2005), *Lunar subsurface investigated from correlation of seismic noise*, Geophys. Res. Lett. doi: 10.1029/2005GL023518.

Levshin, A.L., Yanovskaya, T.B., Lander, A.V., Bukchin, B.G., Barmin, M.P., Ratnikova, L.I., and Its, E.N., *Recording, identification, and measurement of surface wave parameters*. In *Seismic Surface Waves in a Laterally Inhomogeneous Earth*, (ed. V. I. Keilis-Borok), (Kluwer Acad., Norwell, Mass 1989).

Lobkis, O.I. and Weaver, R.L. (2001), *On the emergence of the Green's function in the correlations of a diffuse field*, J. Acoust. Soc. Am. *110*, 3011–3017.

Malcolm, A.E., Scales, J., and van Tiggelen, B.A. (2004), *Extracting the Green function from diffuse, equipartitioned waves*, Phys. Rev. E *70*, 015601.

Margerin, L., *Introduction to radiative transfer of seismic waves*. In *Seismic Data Analysis with Global and Local Arrays* (A. Levander and G. Nolet eds.) (AGU Monograph Series, in press.)

Margerin L., Campillo, M., and van Tiggelen, B.A. (1998), *Radiative transfer and diffusion of waves in a layered medium: New insight on Coda Q*, Geophys. J. Int. *134*, 596–612.

Margerin, L., Campillo, M., Shapiro, N.M., and van Tiggelen, B.A. (1999), *Residence time of diffuse waves in the crust as a physical interpretation of coda Q: application to seismograms recorded in Mexico*, Geophys. J. Int. *138*, 343–352.

Margerin, L., Campillo, M., and van Tiggelen, B.A. (2000), *Multiple scattering of elastic waves*, J. Geophys. Res. *105*, 7873–7892.

Margerin L., van Tiggelen, B.A., and Campillo, M. (2001a), *Coherent back-scattering of acoustic waves in the near field*, Geophys. J. Int. *145*, 593–603.

Margerin, L., van Tiggelen, B.A., and Campillo, M. (2001b), *Effect of absorption on energy partition of elastic waves*, Bull. Seismol. Soc. Am. *91*, 624–627.

Mayeda, K. and Walter, W.R. (1996), *Moment, energy, stress drop, and source spectra of western United States earthquakes from regional coda envelopes*, J. Geophys. Res. *101*(B5), 11,19511,208.

Parvulescu, A. (1995), *Matched-signal (MESS) processing by the ocean*, J. Acoust. Soc. Am. *98*, 943–960.

PAUL, A., CAMPILLO, M., MARGERIN, L., LAROSE, E., and DERODE, A. (2005), *Empirical synthesis of time-asymmetrical Green functions from the correlation of coda waves*, J. Geophys. Res. *110*, B08302, doi:10.1029/2004JB003521.

PEDERSEN, H.A., KRÜGER, F., and SVEKALAPKO Seismic Tomography Working Group (2005), *Influence of the seismic noise characteristics on crustal tomography from noise correlations*, submitted to Geophysical Journal International.

PHILLIPS, W.S. and AKI, K. (1986), *Site amplification of coda waves from local earthquakes in Central California*, Bull. Seismol. Soc. Am. *76*(3), 627–648.

RICKETT, J.E. and CLAERBOUT, J.F. (1999), *Acoustic daylight imaging via spectral factorization; helioseismology and reservoir monitoring*. The Leading Edge *18*, 957–960.

RICKETT, J.E. and CLAERBOUT, J.F. (2000), *Calculation of the Sun's impulse response by multi-dimensional spectral factorization*, Solar Physics *192*, 203–210.

RHIE, J. and ROMANOWICZ, B. (2004), *Excitation of Earth's continuous free oscillation by atmosphere-ocean-seafloor coupling*, Nature *431*, 552–555.

RITZWOLLER, M.H. and LEVSHIN, A.L. (1998), *Eurasian surface wave tomography: Group velocities*, J. Geophys. Res. *103*, 4839–4878.

ROUX, P., DERODE, A., PEYRE, A., TOURIN, A., and FINK, M. (2000), *Acoustical imaging through a multiple scattering medium using a time-reversal mirror*, J. Acoust. Soc. Am. *107*(2), L7–L12.

ROUX, P., SONG, H.C., and KUPERMAN, W.A. (2003), *Time-reversal using ambient noise as a probe source*, J. Acoust. Soc. Am. *113*(4), 2218.

ROUX, P., KUPERMAN, W.A., HODGKISS, W.S., SONG, H.C., AKAL, T., and STEVENSON, M. (2004), *A nonreciprocal implementation of time reversal in the ocean*, J. Acoust. Soc. Am. *116*(2), 1009–1015

ROUX, P., SABRA, K.G., KUPERMAN, W.A., and ROUX, A. (2005), *Ambient noise cross correlation in free space: Theoretical approach*, J. Acoust. Soc. Am. *117*, 79–84.

RYZHIK, L.V., PAPANICOLAOU, G.C., and KELLER, J.B. (1996), *Transport equations for elastic and other waves in random media*, Wave Motion *24*, 327–370.

SABRA, K.G., ROUX, P., KUPERMAN, W.A., HODGKISS, W.S., SONG, H.C., AKAL, T., and STEVENSON, M. (2003), *Emergence of the time domain Green's function from ocean noise correlations*, J. Acoust. Soc. Am. *114*(4), 2462–2462.

SABRA, K.G., GERSTOFT, P., ROUX, P., KUPERMAN, W.A., and FEHLER, M.C. (2005), *Extracting time-domain Green's function estimates from ambient seismic noise*, Geophys. Res. Lett. *32*, L03310, doi:10.1029/2004GL021862.

SÁNCHEZ-SESMA, F.J. and CAMPILLO, M. (2006), *Retrieval of the Green function from cross correlation: The canonical elastic problem*, Bull. Seism. Soc. Am. in press.

SATO, H. (1995), *Formulation of the multiple non-isotropic scattering process in 3-D space on the basis of energy transport theory*, Geophys. J. Int. *121*, 523–531.

SATO, H. and FEHLER, M., *Seismic Wave Propagation and Scattering in the Heterogeneous Earth* (Springer-Verlag, New York, 1998).

SHAPIRO, N.M. and SINGH, S.K. (1999), *A systematic error in estimating surface-wave velocity dispersion curves and a procedure for its correction*, Bull. Seismol. Soc. Am. *89*, 1138–1142.

SHAPIRO, N.M., CAMPILLO, M., STEHLY, L., and RITZWOLLER, M. (2005), *High resolution surface wave tomography from ambient seismic noise*, Science *307*, 1615–1618.

SHAPIRO, N.M., CAMPILLO, M., MARGERIN, L., SINGH, S.K., KOSTOGLODOV, V., and PACHECO, J. (2000), *The energy partitioning and the diffusive character of the seismic coda*, Bull. Seismol. Soc. Am. *90*, 655–665.

SHAPIRO, N.M. and CAMPILLO, M. (2004), *Emergence of broadband Rayleigh waves from correlations of the ambient seismic noise*, Geophys. Res. Lett. *31*, L07614, doi:10.1029/2004GL019491.

SINGH, S. and HERRMANN, R.B. (1983), *Regionalization of crustal coda Q in the continental United States*, J. Geophys. Res. *88*, 527–538.

SPETZLER, J. and SNIEDER, R. (2004), *The Fresnel volume and transmitted waves*, Geophysics *69*, 653–663.

SNIEDER, R. (2004), *Extracting the Green's function from the correlation of coda waves: A derivation based on stationary phase*, Phys. Rev. E *69*, 046610.

TANTER M. (1999), *Application of time reversal to brain hyperthermia*, Ph.D Thesis, University Paris VII, France.

TANIMOTO, T. (2005), *The oceanic excitation hypothesis for the continuous oscillations of the Earth*, Geophys. J. Int. *160*, 276–288.

TOKSÖZ, M.N. (1964), *Microseisms and an attempted application to exploration*, Geophys. *29*(2), 154–177.

TURNER, J.A. (1998), *Scattering and diffusion of seismic waves*, Bull. Seismol. Soc. Am. *88*, 276–283.

VAN TIGGELEN, B.A. (2003), *Green function retrieval and time reversal in a disordered world*, Phys. Rev. Lett. *91*, 243904.

VAN TIGGELEN, B.A., MARGERIN, L., and CAMPILLO, M. (2001), *Coherent backscattering of elastic waves: Specific role of source, polarization, and near field*, J. Acoust. Soc. Am. *110*(3), 1291–1298.

WAPENAAR, K. (2004), *Retrieving the elastodynamic Green's function of an arbitrary inhomogeneous medium by cross correlation*, Phys. Rev. Lett. *93*, 254301.

WEAVER, R.L. (1982), *On diffuse waves in solid media*, J. Acoust. Soc. Am. *71*, 1608–1609.

WEAVER, R.L. (1990), *Diffusivity of ultrasound in polycrystals*, J. Mech. Phys. of Solids *38*, 55–86.

WEAVER, R.L. and LOBKIS, O.I. (2001), *Ultrasonics without a source: Thermal fluctuation correlation at MHz frequencies*, Phys. Rev. Lett. *87*, 134301–134304.

WEAVER, R.L. and LOBKIS, O.I. (2002), *On the emergence of the Green's function in the correlations of a diffuse field: Pulse-echo using thermal phonons*, Ultrasonics, *40*, 435–439.

WEAVER, R.L. and LOBKIS, O.I. (2004), *Diffuse fields in open systems and the emergence of the Greens function*, J. Acoust. Soc. Am. submitted.

WEBB, S.C. (1998), *Broadband seismology and the noise under the ocean*, Rev. Geophys. *36*, 105–142.

WU, F., THOMAS, J-L., and FINK, M. (1992), *Time reversal of ultrasonic fields- Part II: Experimental results*, IEEE Trans. Ultrason., Ferroelec., Freq. Contr., *39*(5), 567–578.

WU, R.S. (1985), *Multiple scattering and energy transfer of seismic waves – separation of scattering effect from intrinsic attenuation – I. Theoretical modeling*, Geophys. J. R. Astron. Soc. *82*, 57–80.

WU, R.S. and AKI, K. (1985), *Scattering characteristics of elastic waves by an elastic heterogeneity*, Geophysics *50*, 582–595, 10261–10273.

WU, R.S. and AKI, K. (1988a), *Multiple scattering and energy transfer of seismic waves – Separation of scattering effect from intrinsic attenuation. II. Application of the theory to Hindu- Kush region*, Pure Appl. Geophys. *128*, 49–80.

WU, R.S. and AKI, K., eds (1988b), *Scattering and attenuation of seismic waves*, Pure Appl. Geophys. *128*, 1–447.

WU, R.S. and AKI, K. eds (1990), *Scattering and attenuation of seismic waves*, Pure Appl. Geophys. *132*, 1–437.

ZENG, Y. (1991), *Compact solutions for multiple scattered wave energy in time domain*, Bull. Seismol. Soc. Am. *81*, 1022–1029.

(Received May 13, 2005; accepted September 16, 2005)
Published Online First: February 8, 2006

Pure appl. geophys. 163 (2006) 503–531
0033–4553/06/030503–29
DOI 10.1007/s00024-005-0027-5

Pure and Applied Geophysics

Modeling Full Seismogram Envelopes Using Radiative Transfer Theory with Born Scattering Coefficients

U. Wegler,[1] M. Korn,[1] and J. Przybilla[1]

Abstract—The equation of radiative transfer is used to model the transport of seismic energy in 2-D and 3-D acoustic random media. Monte-Carlo solutions of this equation using non-isotropic Born scattering coefficients are compared to three analytical solutions: Markov approximation, radiative transfer theory with isotropic scattering coefficients, and diffusion approximation. Additionally, we compare to finite differences solutions of the full wave equation in 2-D. We find a good correspondence of radiative transfer theory to Markov approximation for the case of multiple forward scattering. The comparison to radiative transfer theory with isotropic scattering coefficients, a model frequently used in data analysis, demonstrates that in the case of forward scattering the isotropic scattering model is not better than a diffusion approach. To compare radiative transfer theory with non-isotropic scattering coefficients to finite differences solutions of the full wave equation, the finite source duration and the bandpass filter process as well as the normalization of absolute amplitudes are explicitely taken into account. We find a good coincidence of both theories for scattering parameters, which are realistic for usual Earth crust. The theory correctly describes the unscattered direct wavefront, the envelope broadening caused by multiple forward scattering, as well as the late coda caused by multiple wide angle scattering. For strong scattering, which can be expected for very heterogeneous media such as strato volcanoes, the solutions of radiative transfer differ from the more complete solutions of the full wave equation.

Key words: Numerical comparison, radiative transfer, multiple scattering, attenuation, random medium, elastic wave propagation.

1. Introduction

In deterministic seismology it is assumed, that each inhomogeneity of the medium causes a travel time shift and/or special phases (e.g., reflections) in the observed seismograms. Using the observed travel times and/or reflections we may then analyze the parameters of the inhomogeneity such as its location and size. This method, however, is only successful for inhomogeneities larger than the station separation, the Fresnel zone, and the wavelength. In the Earth's crust, especially in strongly heterogeneous areas like island arcs and volcanoes there exist many small-scale heterogeneities, which cannot be described by deterministic models. Stochastic

[1]Institut für Geophysik und Geologie, Universität Leipzig, Germany. E-mail: uli@rz.uni-leipzig.de

models can be used to determine the most important statistical parameters of the small-scale inhomogeneities neglecting their exact locations. This idea of the statistical treatment of seismic wave propagation was first suggested by AKI (1969) and AKI and CHOUET (1975) for the coda.

For the description of seismic wave propagation in highly inhomogeneous random media two standard methods have been established. Firstly, Markov approximation (SATO, 1989; SCHERBAUM and SATO, 1991; SAITO *et al.*, 2002) and secondly, radiative transfer theory, which is also known as energy transport theory (WU, 1985; GUSEV and ABUBAKIROV, 1987; ZENG *et al.*, 1991; HOSHIBA, 1991; MARGERIN *et al.*, 1998). These theories model two different phenomena, which are caused by multiple scattering: Firstly, the development of a seismic coda and, secondly, the envelope-broadening. Markov approximation describes multiple forward scattering under the assumption of a wavelength considerably shorter than the correlation length of the medium heterogeneities. This theory can explain the delay of the maximum of energy compared to the direct S wave, which is observed in real data (SATO, 1989). FEHLER *et al.* (2000) confirmed that Markov approximation coincides with statistically averaged envelopes computed from finite differences solutions of the full wave equation. However, in this theory backscattering is neglected and as a result Markov approximation does not correctly model the late part of the seismogram (the coda).

The second method is the theory of radiative transfer. RYTOV *et al.* (1987) give a derivation of this approximation from the stochastic wave equation in the acoustic case. Radiative transfer takes into account both forward and backscattering. However, traditionally isotropic scattering is used. In that case the theory only correctly models the development of the coda, whereas the envelope broadening cannot be explained by isotropic scattering. GUSEV and ABUBAKIROV (1996) pointed out that using non-isotropic scattering coefficients, which take into account the strong forward scattering, radiative transfer can also simulate a broadening of envelopes. As numerical method a Monte-Carlo simulation (GUSEV and ABUBAKIROV, 1987; HOSHIBA, 1991, 1995; YOSHIMOTO, 2000) is used to model radiative transfer including non-isotropic scattering. Alternatively, SATO (1995) formulated an analytical approach to the problem. GUSEV and ABUBAKIROV (1996) compared numerical solutions of the equation of radiative transfer including nonisotropic scattering coefficients to diffusion approximation and to Markov approximation.

The theory of radiative transfer has the advantage that it can be generalized to more complicated problems: WEAVER (1990) and RYZHIK *et al.* (1996) derived an elastic equation of radiative transfer from the equation of motion. MARGERIN *et al.* (2000) and PRZYBILLA *et al.* (2006) presented a numerical Monte-Carlo method to solve this coupled elastic transport equation including non-isotropic scattering and conversions between P and S waves. A depth-dependent background velocity or a depth-dependent scattering coefficient can also be included in the theory (e.g., MARGERIN *et al.*, 1998; YOSHIMOTO, 2000; HOSHIBA *et al.*, 2001; LACOMBE *et al.*, 2003).

In this article we present a numerical comparison between five different theories for the computation of theoretical seismogram envelopes. Firstly, we compare radiative transfer theory using Born scattering coefficients to three simple analytical models, which are Markov approximation, diffusion approximation and radiative transfer theory with isotropic scattering coefficients. This comparison confirms results of the previous study by GUSEV and ABUBAKIROV (1996). Here, we pay special attention to a comparison between isotropic scattering, a model frequently used in data analysis, and the more realistic non-isotropic scattering. Moreover, we compare in detail Markov approximation to radiative transfer theory. The most complete solution to the problem is the ensemble averaged envelope computed from finite difference solutions of the full wave equation. FRANKEL and CLAYTON (1986) give an introduction to finite difference simulations of waves in random media. Due to the computational cost, this method is realized only in two-dimensional space. Comparing these complete solutions to Monte-Carlo solutions of radiative transfer theory with Born scattering coefficients confirms the broad validity range of the radiative transfer equation.

2. Radiative Transfer Theory

2.1. Basic Definitions

In this article we concentrate on the acoustic wave equation in an unbounded space with a homogeneous and isotropic random medium, which has fluctuations in velocity only, whereas density is assumed to be independent of location. The acoustic wave equation then is:

$$\frac{1}{v^2(\mathbf{r})}\frac{\partial^2}{\partial t^2}G(\mathbf{r},t) - \triangle G(\mathbf{r},t) = \delta(\mathbf{r})\delta(t)A_0. \tag{1}$$

Here we assume an impulsive and point-like source, where the constant A_0 is set to be equal to $1\,\mathrm{Nsm}^{-1}$ so that the Green's function G has a dimension of pressure $\mathrm{Pa} = Nm^{-2}$. In the stochastic approach we assume that the functions $v(\mathbf{r})$ and $G(\mathbf{r},t)$ are rapidly fluctuating in space. Therefore, it is impossible to record the full wave field $G(\mathbf{r},t)$ and it is also impossible to invert the complete medium parameters $v(\mathbf{r})$ from measured seismograms. Instead, we describe the propagation medium by its first two statistical moments, which are mean velocity and autocorrelation function of velocity. For the autocorrelation function we write the velocity as a sum of mean velocity v_0 and a fluctuating part (e.g., SATO and FEHLER, 1998, p. 12)

$$v(\mathbf{r}) = v_0[1 + \xi(\mathbf{r})]. \tag{2}$$

Then the autocorrelation function of the velocity fluctuations is defined as an ensemble average:

$$R_v(\mathbf{R},\mathbf{r}) := \left\langle \xi\left(\mathbf{R}+\frac{\mathbf{r}}{2}\right)\xi^*\left(\mathbf{R}-\frac{\mathbf{r}}{2}\right)\right\rangle \tag{3}$$

$$= R_v(|\mathbf{r}|). \tag{4}$$

Here $\mathbf{R}$ is the absolute location and $\mathbf{r}$ is the relative location. In equality (4) an isotropic and homogeneous random medium is assumed, which results in an autocorrelation function only depending on the absolute value $|\mathbf{r}|$ of the relative location. The power spectral density function is the spatial Fourier transform of the autocorrelation function:

$$R_v(\mathbf{R},\mathbf{k}) := \iiint_{-\infty}^{+\infty} R_v(\mathbf{R},\mathbf{r})\exp(-2\pi i\,\mathbf{r}\cdot\mathbf{k})d\mathbf{r} \tag{5}$$

$$= R_v(|\mathbf{k}|). \tag{6}$$

Equation (5) gives the general definition and equality (6) is valid in an isotropic and homogeneous random medium where the power spectral density function only depends on the absolute value $|\mathbf{k}|$ of the wave number.

Similarly, we also define the autocorrelation of the Green's function:

$$R_G(\mathbf{R},\mathbf{r},T,t) := \kappa_0\left\langle G\left(\mathbf{R}+\frac{\mathbf{r}}{2},T+\frac{t}{2}\right)G^*\left(\mathbf{R}-\frac{\mathbf{r}}{2},T-\frac{t}{2}\right)\right\rangle. \tag{7}$$

Here κ_0 is the ensemble averaged compressibility with a dimension of Pa^{-1}. This factor is introduced, so that the dimension of $R_G(\mathbf{R},\mathbf{r},T,t)$ corresponds to energy per volume ($\mathrm{J/m^3}$) and $R_G(\mathbf{R},\mathbf{r}=0,T,t=0)$ can be associated with energy density. In contrast to the autocorrelation function of the velocity fluctuations there is an additional time dependency. Moreover, one cannot simply drop the dependency on the absolute coordinates $\mathbf{R}$ and T, because the wavefield is neither homogeneous nor stationary. Applying a spatial Fourier transform with respect to relative location $\mathbf{r}$ and a temporal Fourier transform with respect to relative time t, one obtains a quantity, which is the analogue of the power spectral density function:

$$R_G(\mathbf{R},\mathbf{k},T,f) := \iiiint_{-\infty}^{+\infty} R_G(\mathbf{R},\mathbf{r},T,t)\exp[-2\pi i(\mathbf{k}\cdot\mathbf{r}+ft)]dt\,d\mathbf{r}. \tag{8}$$

This quantity is referred to as the four-dimensional Wigner transformation of the Green's function $G(\mathbf{R},T)$. It is a fundamental quantity in radiative transfer theory (e.g., RYZHIK *et al.*, 1996; SHENG, 1995, pp. 115–120). The dependencies on the parameters $\mathbf{R}$ and T describe the properties of the envelope. $\mathbf{k}$ and f, on the contrary, refer to properties of the underlying wave and are called internal wavenumber and internal frequency, respectively. The dimension of Wigner distribution $R_G(\mathbf{R},\mathbf{k},T,f)$ is energy per unit volume, per unit wavenumber volume element and per unit frequency range ($\mathrm{J/(m^3m^{-3}Hz)}$). It can be associated with energy density in phase space which is also called wave number resolved energy density.

In radiative transfer theory we use the additional fact that internal frequency f and absolute value of internal wave number $|\mathbf{k}|$ are connected by a dispersion relation. In the simplest case, it is assumed that waves in the heterogeneous medium travel with the same velocity as in the homogeneous medium. Then, we can express wavenumber resolved energy density by another quantity, which is called direction resolved energy density $I(\mathbf{R}, \hat{\mathbf{k}}, T, f)$ (RYTOV *et al.*, 1987, p. 149):

$$R_G(\mathbf{R}, \mathbf{k}, T, f) = \delta\left(|k| - |\frac{f}{v_0}|\right) I(\mathbf{R}, \hat{\mathbf{k}}, T, f) \frac{1}{k^2}. \tag{9}$$

Quantity $I(\mathbf{R}, \hat{\mathbf{k}}, T, f)$ is a function of the direction of $\hat{\mathbf{k}}$ only, where $\hat{\mathbf{k}} = \mathbf{k}/|\mathbf{k}|$ is a unit vector in the same direction as $\mathbf{k}$. The dimension of I is $\mathrm{Jm^{-3}Hz^{-1}}$.

Summing up over all directions, the local energy density $E_G(\mathbf{R}, T, f)$ of the Green's function G results:

$$E_G(\mathbf{R}, T, f) := \iint I(\mathbf{R}, \hat{\mathbf{k}}, T, f) d\hat{\mathbf{k}} \tag{10}$$

$$= \kappa_0 \int_{-\infty}^{+\infty} \left\langle G\left(\mathbf{R}, T + \frac{t}{2}\right) G^*\left(\mathbf{R}, T - \frac{t}{2}\right) \right\rangle \exp(-2\pi i f t)\, dt. \tag{11}$$

Here, $d\hat{\mathbf{k}}$ indicates an integration over the unit sphere. The dimension of E_G is $\mathrm{Jm^{-3}Hz^{-1}}$ (energy per unit volume and per unit frequency range). Equation (11) indicates that $E_G(\mathbf{R}, T, f)$ can be expressed as the one-dimensional (temporal) Wigner transformation of the Green's function $G(\mathbf{R}, T)$. This quantity can be associated with a bandpass-filtered and squared velocity seismogram.

Finally, integrating over space the spectral energy $W_G(f)$ is obtained:

$$W_G(T, f) := \iiint_{-\infty}^{+\infty} E_G(\mathbf{R}, T, f) d\mathbf{R}. \tag{12}$$

In the case of zero dissipation this quantity is constant in time and corresponds to the spectral source energy $W_G(T, f) = W_G(f)$ with a dimension of $\mathrm{JHz^{-1}}$.

2.2. *Equation of Radiative Transfer*

If scattering is weak, direction resolved energy density obeys the following transport equation:

$$\frac{1}{v_0}\partial_T I(\mathbf{R}, \hat{\mathbf{k}}, T, f) + \hat{\mathbf{k}} \cdot \mathrm{grad}_R I(\mathbf{R}, \hat{\mathbf{k}}, T, f) = -g^0(f) I(\mathbf{R}, \hat{\mathbf{k}}, T, f) + \frac{1}{4\pi}\int g(\hat{\mathbf{k}} \cdot \hat{\mathbf{k}}', f) I(\mathbf{R}, \hat{\mathbf{k}}', T, f) d\hat{\mathbf{k}}', \tag{13}$$

which is known as equation of radiative transfer. Here $d\hat{\mathbf{k}}'$ indicates an integration over the unit sphere. This equation can be interpreted as a local conservation of energy in phase space (ISHIMARU, 1997; MARGERIN *et al.*, 1998) and corresponds to the Boltzmann equation of kinetic gas theory. After its phenomenological derivation based on the conservation of energy (CHANDRASEKHAR, 1960), equation (13) was also derived directly from the wave equation (e.g., RYTOV *et al.*, 1987, pp. 141–152). For equation 13 to be valid the scattering mean free path l^0 is required to be considerably larger than the wave length λ (e.g., VAN TIGGELEN, 1999):

$$l^0 > \frac{\lambda}{2\pi}. \tag{14}$$

In the simplest approach assuming that scattering is not too strong in forward direction, we can use the Born scattering coefficients as scattering coefficients in equation (13) (RYTOV *et al.*, 1987, p. 150). The validity range for the use of Born scattering coefficients is given by (SATO and FEHLER, 1998, p. 121):

$$l^0 > a. \tag{15}$$

Here a is the correlation length of the random medium. Condition (15) is stricter than condition (14), because strong scattering always requires a correlation length larger than or equal to the wavelength $(a \geq \frac{1}{2\pi}\lambda)$. The connection between the angular-dependent scattering coefficient $g(\hat{\mathbf{k}}\cdot\hat{\mathbf{k}}', f)$ and the total scattering coefficient $g^0(f)$ in equation (13) is given by:

$$g^0(f) = \frac{1}{4\pi}\int g(\hat{\mathbf{k}}\cdot\hat{\mathbf{k}}', f)d\hat{\mathbf{k}}', \tag{16}$$

where $d\hat{\mathbf{k}}'$ indicates an integration over the unit sphere. Another important quantity in radiative transfer theory is the transport scattering coefficient $g^*(f)$, which is also called momentum transfer scattering coefficient. In the definition of this quantity forward scattering is excluded by an additional factor $[1 - \cos(\hat{\mathbf{k}}\cdot\hat{\mathbf{k}}')]$ before integration:

$$g^*(f) = \frac{1}{4\pi}\int g(\hat{\mathbf{k}}\cdot\hat{\mathbf{k}}', f)[1 - \cos(\hat{\mathbf{k}}\cdot\hat{\mathbf{k}}')]d\hat{\mathbf{k}}'. \tag{17}$$

The inverse quantities have a dimension of length and are referred to as (total) scattering mean free path $l^0(f) = 1/g^0(f)$ and transport scattering mean free path $l^*(f) = 1/g^*(f)$, respectively. The angular-dependent, total, and transport Born scattering coefficients for the 3-D and 2-D acoustic case are given in Appendix A.

2.3. Numerical Solution of the Equation of Radiative Transfer Using a Monte-Carlo Simulation

To solve the equation of radiative transfer numerically we follow the method of GUSEV and ABUBAKIROV (1996) andYOSHIMOTO (2000). This Monte-Carlo simulation uses the similarity of equation (13) to the Boltzmann equation of kinetic gas theory and simulates the transport of energy through the heterogeneous medium by the transport of particles. At the source each particle, representing a small wave package of unit energy, is initialized with a position at the origin and with a velocity in random direction (isotropic source). During propagation the particle is characterized by the time it already has traveled, its position, and its current direction of propagation. In each time step the particle moves a distance Δx. This constant step size takes the numerical value of the n-th part of the scattering mean free path ($\Delta x = l^0/n$), where n controls the accuracy of the Monte-Carlo simulation. Empirically, we found that a value of n around 5 is a good compromise between accuracy and computational cost. We use the (total) Born scattering coefficients in three and two dimensions g^0_{3D} and g^0_{2D}, respectively, which are summarized in Appendix A. In the second process, which is scattering, the direction of the particle changes, but its position is constant. Scattering does not occur during each time step, but a random process decides whether scattering occurs or not. For each time step a random variable P_1 is computed with equal distribution between 0 and 1. If $P_1 < \Delta x/l^0 = 1/n$, scattering takes place.

So far, this algorithm is identical to the method of YOSHIMOTO (2000). Additionally, non-isotropic scattering has to be implemented. To do so, firstly, the scattering angles are computed in a local coordinate system, where the z axis points in the direction of the current direction of propagation. Using the rotational symmetry of the problem the azimuth angle is chosen as $\varphi = 2\pi P_2$, where P_2 is an equally distributed random number between 0 and 1. The scattering incidence angle ϑ, on the contrary, is not equally distributed, but its probability distribution function is used to simulate the non-isotropic scattering according to the Born scattering coefficients. Here, the rejection method (PRESS *et al.*, 1996, p. 281) is used: Firstly, an guessed scattering incidence angle is computed as in the isotropic scattering case using $\vartheta = \arccos(1 - 2P_3)$, where P_3 is equally distributed in [0,1]. Then, in a second step to simulate the non-isotropic scattering, another random variable P_4 is computed. If $g(\vartheta)/g(0) < P_4$, ϑ is accepted as scattering incidence angle, otherwise the computer code returns to the first step and computes angle ϑ again using a new random number P_3. Step one (guessing) and step two (rejecting) are repeated until one scattering angle ϑ is accepted. The rejection method becomes inefficient for strong forward scattering. In that case the constant comparison function $g(0)$, which is the value of the scattering coefficient in forward direction, differs significantly from $g(\vartheta)$ resulting in many wrong guesses of the scattering angle. After selecting the new direction of propagation in the local coordinate system, an Euler rotation is used to rotate into absolute coordinates.

At each time step, it is checked whether the particle is located near the receiver. Using the fact that the medium is isotropic and stationary, all receivers at distance R from the source can be used simultaneously. Therefore, the receiver volume is a shell of a sphere with thickness $\Delta r = l^0/n$. All particles, which are in a certain time step within the receiver volume, are added up. For a large number of particles we obtain the local energy density as a function of time and source-receiver distance by normalizing with the total number of particles and with the receiver volume.

3. Comparison to Analytical Solutions in 2-D and 3-D

Firstly, we compare radiative transfer theory with Born scattering coefficients to three analytical solutions: Markov approximation, radiative transfer theory with isotropic scattering coefficients, and diffusion theory. More numerical comparisons concerning radiative transfer theory with non-isotropic scattering coefficients can be found in GUSEV and ABUBAKIROV (1996).

3.1. Markov Approximation

Markov approximation is used to describe the envelope broadening caused by multiple forward scattering (SATO, 1989; SCHERBAUM and SATO, 1991). It can be derived from wave theory under the condition that wavelength is much shorter than the scale of medium heterogeneity (SATO, 1989). GUSEV and ABUBAKIROV (1996) compared the peak delay predicted by Markov approximation to numerical peak delays computed from radiative transfer theory and found good coincidence. In the case of a Gaussian random medium an analytical solution is known for local energy density $E(R, T, f)$, which is independent of frequency f. Neglecting the "wandering effect" which is caused by phase fluctuations (e.g., SATO and FEHLER, 1998, p. 247) the formula for the 3-D case is (SAITO *et al.*, 2002):

$$E^{3D}_{mark}(R,T) = \frac{1}{4\pi R^2} H\left[T - \frac{R}{v_0}\right] \frac{\pi^2}{2v_0 t_m} \sum_{n=1}^{\infty} (-1)^{n+1} n^2 \exp\left[-\frac{\pi^2 n^2}{4t_m}\left(T - \frac{R}{v_0}\right)\right]. \quad (18)$$

Here R is source receiver distance, T is time after source time, v_0 is mean velocity, and $H(x)$ is the Heaviside-(step-) function. $t_m = \sqrt{\pi}\epsilon^2 R^2 (2av_0)^{-1}$ is called the characteristic time, where ϵ^2 and a are variance and correlation length of a Gaussian random medium.

In 2-D the formula is (FEHLER *et al.*, 2000):

$$E^{2D}_{mark}(R,T) = \frac{1}{2\pi R} \int_{-\infty}^{+\infty} \sqrt{\frac{2\sqrt{2\pi i F t_m}}{\sin\left[2\sqrt{2\pi i F t_m}\right]}} \exp\left[-2\pi i F\left(T - \frac{R}{v_0}\right)\right] dF. \quad (19)$$

Here t_m is the same quantity as in the 3-D case and integration over F is done using a numerical Fourier transform.

The characteristic time t_m can also be expressed using the transport mean free path l^*. For the 3-D Gaussian random medium this connection is:

$$t_m = \frac{R^2}{4 v_0 l^*_{3D}}, \quad (20)$$

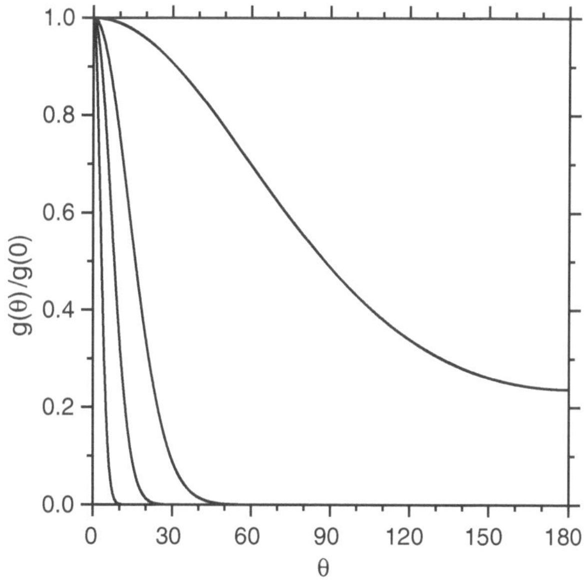

Figure 1
Angular dependent scattering coefficient $g(\vartheta, f)$ normalized by scattering coefficient in forward direction $g(0, f)$ for 2-D and 3-D Gaussian random medium as a function of scattering angle ϑ. Each of the four curves corresponds to a different value of frequency f, where from the top to the bottom most curve $2\pi a/\lambda = 1.2$, 6, 12, and 30, respectively. (The angular dependency for 2-D and 3-D media is identical).

where we used the high frequency approximation (equation (A17) in Appendix A) for l^*_{3D}. This equation indicates that the only information on the random medium, which is used in Markov approximation, is its transport mean free path.

For the numerical solution of the equation of radiative transfer with Born scattering coefficients we used the values of $2\pi a/\lambda = 1.2, 6, 12$, and 30. The corresponding angular dependencies of the Born scattering coefficient are shown in Figure 1 for a 3-D Gaussian random medium. For the case of $2\pi a/\lambda = 1.2$ forward scattering is weak and we do not expect Markov approximation to be valid. Figure 2 shows a comparison of 2-D Markov approximation to 2-D radiative transfer theory for the case of $2\pi a/\lambda = 6$ (moderate forward scattering). The used random medium in Figure 2 is of Gaussian type with mean velocity $v_0 = 1$ km s^{-1}, standard deviation $\epsilon = 5\%$ and autocorrelation length $a = 1$ km. The internal angular frequency of the wave package is $\omega = 6$ Hz. The scattering mean free path and the transport mean free path for these parameters are $l^0 \approx 6.2$ km and $l^* \approx 220$km, respectively. The figure shows energy density curves for the two source receiver distances of $R = 74$ km (roughly one third of a transport mean free path) and $R = 2200$ km (roughly ten transport mean free paths). The curve for radiative transfer theory with Born scattering coefficients shows a concentration of energy density shortly after the first onset for the case of the small source receiver distance. Markov approximation agrees well with radiative transfer theory within this maximum

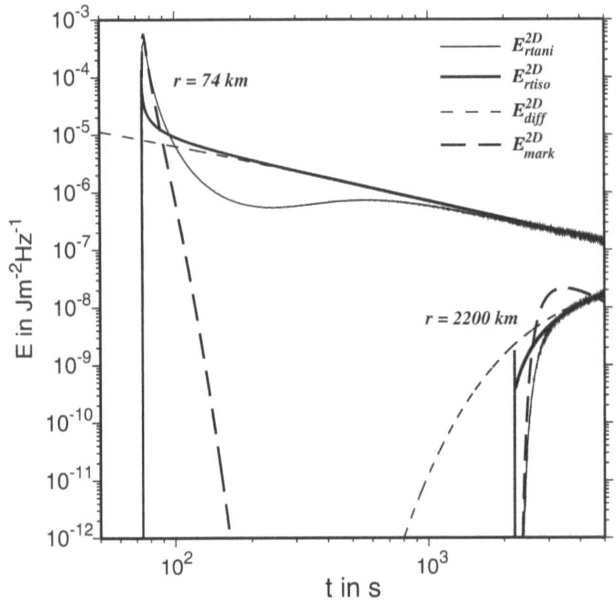

Figure 2
Numerical solutions of the 2-D equation of radiative transfer using anisotropic scattering coefficients for $2\pi a/\lambda = 6$ (thin, continuous curves) in comparison to three analytical approximations: 2-D radiative transfer with isotropic scattering coefficients (thick, continuous curves), 2-D diffusion (thin, dashed curves), and 2-D Markov approximation (thick, dashed curves). Source receiver distances are R = 74 km ($\approx l^*/3$) and R = 2200 km ($\approx 10l^*$).

of energy. The late coda has only little energy compared to this maximum. Figure 2 shows that Markov approximation and radiative transfer theory with Born scattering coefficients do not coincide for the late coda. The discrepancy between both theories becomes important for times larger than about 100 s to 200 s for the case of the source receiver distance of $R = 74$ km. This disagreement of the two theories is due to the fact that in this regime backscattering becomes important, which is neglected in Markov approximation. As a rough approximation we can say that Markov approximation is valid for times smaller than the transport mean free time l^*/v_0 which equals 220 s in Figure 2. In the large source receiver distance of $R = 2200$ km Markov approximation and radiative transfer theory disagree in general. Again this indicates the failure of Markov approximation caused by wide angle scattering. However, we note that very close to the first onset Markov approximation is still the best analytical approximation even at large distances.

Figure 3 is similar to Figure 2 but shows the 3-D case. Using the same parameters for the random medium as in 2-D ($v_0 = 1$ km s^{-1}, $\epsilon = 5\%$, $a = 1$ km, and $\omega = 6$ Hz) in three-dimensional space the scattering mean free path and the transport mean free path are $l^0 \approx 6.2$ km and $l^* \approx 110$ km, respectively. Qualitatively, we observe no important differences between the 2-D and the 3-D case.

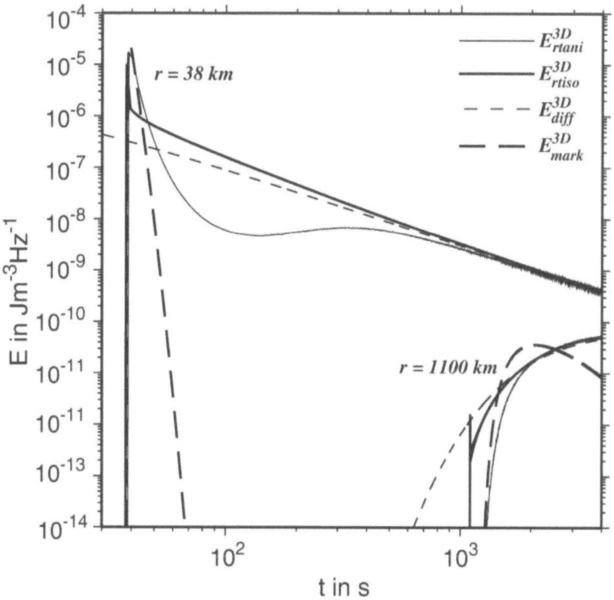

Figure 3
Same as figure 2, but for 3-D case. Source receiver distances are $R = 38$ km ($\approx l^*/3$) and $R = 1100$ km ($\approx 10l^*$).

Figures 2 and 3 used the case of moderate forward scattering ($2\pi a/\lambda = 6$) and showed energy density curves for times as large as thousands of seconds. For increasing values of $2\pi a/\lambda$ forward scattering gets more pronounced (Fig. 1) and the numerical Monte-Carlo simulation becomes increasingly more time consuming. Therefore, Figure 4 shows shorter time windows but uses the larger values of $2\pi a/\lambda = 12$ and 30. In contrast to Figures 2 and 3, which use a double logarithmic representation, Figure 4 is in linear scale which gives more details near the maximum of energy. The figure presents numerical solutions of the 3-D equation of radiative transfer using anisotropic scattering coefficients in comparison to 3-D Markov solutions. The used random medium is the same as before (Gaussian type, $v_0 = 1$ km s^{-1}, $\epsilon = 5\%$, a = 1 km). The internal angular frequencies of the wave package are $\omega = 12$ Hz and 30 Hz, respectively. The transport mean free path for these parameters is $l^* \approx 110$ km independent of frequency. The scattering mean free path is $l^0 \approx 1.6$ km at $\omega = 12$ Hz and $l^0 \approx 0.25$ km at $\omega = 30$ Hz. We note that in the case of $\omega = 30$ Hz the formal condition for the validity of the Born approximation in equation (15) is violated. In Figure 4A we use a source receiver distance of $R = 11$ km (roughly one tenth of the transport mean free path). The curve for radiative transfer with $\omega = 30$ Hz is almost identical to the Markov approximation. The curve for radiative transfer with $\omega = 12$ Hz, on the contrary, differs from the Markov approximation. It shows more energy near the direct onset and has a smaller

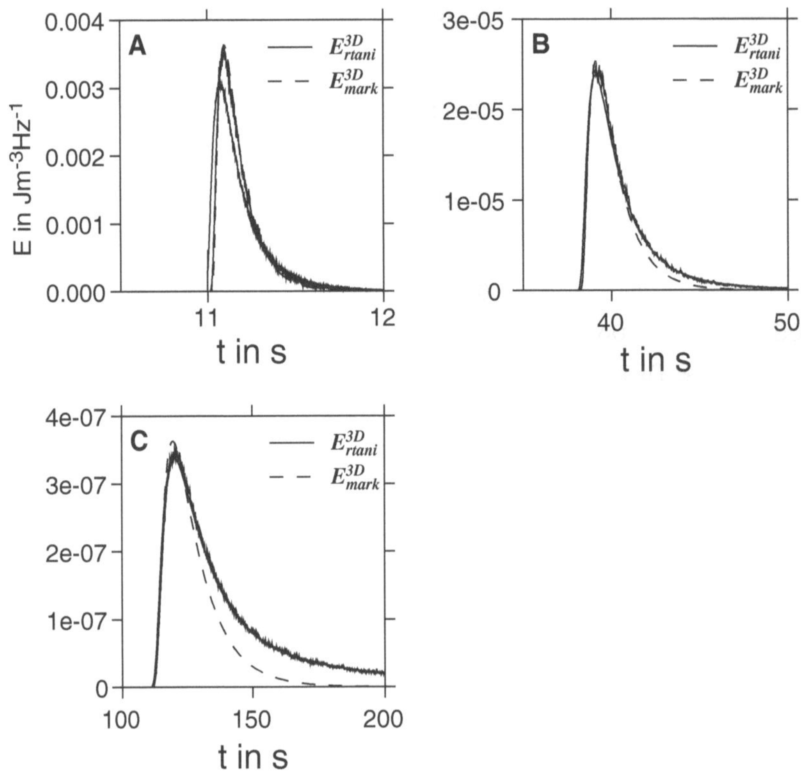

Figure 4

Solutions of the 3-D equation of radiative transfer using Born scattering coefficients (solid) in comparison to 3-D Markov approximation (dashed). For radiative transfer two curves for $\omega = 12$ Hz and 30 Hz are shown, whereas Markov approximation is independent of frequency. In A the source receiver distance is R = 11 km ($\approx l^*/10$). The curve for radiative transfer with $\omega = 30$ Hz is almost identical to Markov approximation. The curve for radiative transfer with $\omega = 12$ Hz is the one which has the smaller maximum. In B R = 38 km ($\approx l^*/3$). In C R = 110 km ($\approx 1l^*$). In B and C the curves of radiative transfer for ω = 30Hz and $\omega = 12$ Hz, respectively, are almost identical and cannot be distinguished in the figure.

maximum of energy. We interpret this discrepancy as follows: The scattering mean free path for $\omega = 12$ Hz is $l^0 \approx 1.6$ km. In a source receiver distance of only 11 km some energy of the unscattered wavefield remains in the curves for radiative transfer theory. Markov approximation, on the contrary, assumes multiple (forward) scattering and neglects the direct waves. Therefore, Markov approximation also fails to model small distances and early times. In Figure 4B the source receiver distance is 38 km (roughly one third of the transport mean free path). In this case both curves of radiative transfer for $\omega = 12$ Hz and 30 Hz, respectively, are very similar and cannot be distinguished in the figure. Moreover, they show excellent agreement with Markov approximation for early times. For late times some deviation develops, which is caused by backscattering as in Figure 3. Finally, in

Figure 4C the source receiver distance is 110 km (roughly one transport mean free path). Again the two curves of radiative transfer for $\omega = 30$ Hz and $\omega = 12$ Hz, respectively, cannot be distinguished in the figure. The discrepancy between radiative transfer and Markov approximation increases in comparison to Figure 4B, because backscattering becomes more important in the longer distance.

In summary our numerical comparison shows that Markov approximation without wandering effect gives identical results to radiative transfer theory with Born scattering coefficients under the condition that the time T after the source time is required to be considerably larger than the mean free time l^0/v_0 and to be much smaller than the transport mean free time l^*/v_0.

$$\frac{l^0}{v_0} \ll T \ll \frac{l^*}{v_0}. \tag{21}$$

This condition is slightly different from the normally used condition for Markov approximation which requires that $2\pi a/\lambda$ should be large. However, we found that Markov approximation is a good approximation using a value of $2\pi a/\lambda = 12$ in Figure 4B. In a shorter distance, on the contrary, using the same value for $2\pi a/\lambda$ Markov approximation is not a good approximation (Fig. 4A). Therefore, we believe that the condition in equation (21) is more general than the condition that $2\pi a/\lambda$ should be large. Of course, equation (21) implicitly also requires strong forward scattering. The wavelength λ of the seismic waves has to be substantially less than the correlation length a of the random medium to obtain a mean free path l^0 much smaller than the transport mean free path l^*:

$$\frac{a}{\lambda} \gg \frac{1}{2\pi} \Leftrightarrow l^0 \ll l^*. \tag{22}$$

The use of Born scattering coefficients in radiative transfer theory is questionable for very large frequencies f, because the scattering mean free path $l^0(f)$ decreases with increasing frequency (e.g., Appendix A, equation (A8)) and finally violates the condition in equation (15). However, in our numerical simulations we obtained useful results also for the case of very strong forward scattering ($2\pi a/\lambda = 30$) outside of the validity range of the Born approximation. Especially, we found good coincidence of radiative transfer theory with Markov approximation for this range. We speculate that this observation is related to the fact that Markov approximation can be expressed using the transport scattering coefficient (equation (20)). This transport scattering coefficient is independent of the degree of forward scattering. It might be estimated accurately by the Born approximation even outside of the formal range of validity of the Born approximation, because the most critical range of forward scattering is excluded in the computation of g^*.

Taking the wandering effect into account in Markov approximation results in deteriorated coincidence with radiative transfer theory. This indicates that travel-time fluctuations are neglected in radiative transfer theory, and as a result this theory describes the broadening of individual seismograms rather than the correct statistical ensemble average. We speculate that radiative transfer becomes wrong for high frequencies in the sense that it does not include the wandering effect. In a recent paper PRZYBILLA *et al.*, (2006) suggest correcting radiative transfer by using an additional convolution with the wandering effect of Markov approximation. After correction these authors compare their results to numerical computation of statistically averaged seismogram envelopes of finite differences calculations and find good agreement between both methods even outside of the validity range of the Born approximation.

3.2. Isotropic Scattering Approximation

In a next step we compare radiative transfer theory using isotropic scattering coefficients with radiative transfer theory using anisotropic Born scattering coefficients. The analytic solution for the 3-D case assuming isotropic scattering is (ZENG *et al.*, 1991; SATO and FEHLER, 1998, pp. 175–182):

$$E^{3D}_{RTiso}(R,T,f) = \frac{1}{iR}\int\int_{-\infty}^{+\infty} \frac{K E_{<G>}(K,F,f)}{1 - v_0 g^*_{3D}(f) E_{<G>}(K,F,f)} \exp[2\pi i(FT + KR)]dF\, dK. \tag{23}$$

$g^*_{3D}(f)$ is the transport scattering coefficient as a function of internal frequency f, which can be computed from the power spectral density of the velocity fluctuations according to equation A14 given in Appendix A. The two integrations over external frequency F and external wave number K are done using a numerical two-dimensional Fourier transform. $E_{<G>}(K,F,f)$ is the local energy density of the mean field in frequency-wavenumber domain given by the following formula:

$$E_{<G>}(K,F,f) = \frac{1}{v_0}\frac{1}{2\pi K}\arctan\frac{2\pi K}{g^*_{3D}(f) - \frac{2\pi i F}{v_0}}. \tag{24}$$

In 2-D the formula for isotropic scattering is (SATO, 1993; SATO and FEHLER, 1998, p. 185):

$$\begin{aligned} E^{2D}_{RTiso}(R,T,f) = {} & \frac{1}{2\pi R}\delta(v_0 T - R)\exp(-g^*_{2D}(f)R) \\ & + \frac{g^*_{2D}(f)}{2\pi\sqrt{v_0^2T^2 - R^2}}\exp\left[g^*_{2D}(f)\left(\sqrt{v_0^2T^2 - R^2} - v_0 t\right)\right]H(v_0T - R). \end{aligned} \tag{25}$$

Here $H(x)$ is the Heaviside-step-function. Solutions (23) and (25) can be used, if the wave number of the seismic waves is considerably larger than the correlation length a of the random medium, because in that case scattering is isotropic. In this regime, which is known as Rayleigh regime, scattering coefficient, angular dependent scattering coefficient, and transport scattering coefficient are equal.

$$\frac{a}{\lambda} \ll \frac{1}{2\pi} \Leftrightarrow g(\vartheta, \varphi) = g^0 = g^* \tag{26}$$

However, for Rayleigh scattering in 3-D media we also have to expect a frequency dependency of the form $g^*_{3D} \sim f^4$. Thus far, to the knowledge of the authors, such a frequency dependency of g^*_{3D} was never observed for real seismological data. This indicates that for the seismological data interpretation the Rayleigh regime is not of relevance, but we have to consider anisotropic (forward) scattering. Since isotropic scattering is much simpler, the question arises, can we approximate the anisotropic scattering coefficient by an effective isotropic scattering coefficient. Using solutions (23) and (25) for isotropic scattering as an approximation for anisotropic scattering, it is important to use as the effective isotropic scattering coefficient the transport scattering coefficient g^* rather than the scattering coefficient g^0. This guarantees that the curves for radiative transfer theory E_{RTiso} and E_{RTani} using either isotropic or anisotropic scattering, approach each other in the diffusion limit (Figs. 2 and 3).

The result is disappointing. Although we use only moderate forward scattering ($2\pi a/\lambda = 6$) in Figures 2 and 3, the isotropic scattering model E_{RTiso} approaches the anisotropic scattering model E_{RTani} only for late times. For example, for a source receiver distance of $R = 74$ km (roughly one third of a transport mean free path) the anisotropic scattering model and the isotropic scattering model approach each other for $T > 1000$ s. This range can be described by the condition that time T should be larger than the transport mean free time:

$$T \gg \frac{l^*}{v_0} \tag{27}$$

This corresponds to the validity range of the diffusion solution and indicates that using the isotropic scattering model as an approximation for anisotropic scattering gives no better results than a simple diffusion approach. Therefore, in contradiction to the common practice, for seismological data interpretation the radiative transfer theory with anisotropic scattering coefficients rather than isotropic scattering approximations should be used.

The sag occurring in the curves for anisotropic radiative transfer theory at short distances (Figs. 2 and 3) depends on the degree of forward scattering. With increasing value of $2\pi a/\lambda$ the sag becomes deeper. Using random media with von Karman rather than Gaussian autocorrelation functions, the sag diminishes for the same value of $2\pi a/\lambda$, because forward scattering is less pronounced in such media.

3.3. *Diffusion Approximation*

Finally, we compare radiative transfer theory with Born scattering coefficients to the diffusion solution, which can be used for strong multiple scattering. This topic is only briefly touched on, because considerable literature already exists on comparisons of radiative transfer theory and diffusion approximation (e.g., GUSEV and ABUBAKIROV, 1996; MARGERIN *et al.*, 1998, 2000; WEGLER, 2003, 2005). The diffusion solution for an unbounded space is (e.g., DAINTY and TOKSÖZ, 1981):

$$E_{\text{diff}}(R,T,f) = (4\pi DT)^{-\frac{p}{2}} \exp\left(-\frac{R^2}{4DT}\right) \tag{28}$$

p is the dimension and equals 2 or 3, $D = D(f)$ is diffusivity and can be computed from the transport mean free path:

$$D(f) = \frac{v_0}{p} l^*(f). \tag{29}$$

Figures 2 and 3 show a comparison of radiative transfer theory using anisotropic Born scattering coefficients to the diffusion model. As expected, we observe a good coincidence of both theories for late times ($T \gg l^*/v_0$).

In summary, we find that radiative transfer theory E_{RTani} using anisotropic Born scattering coefficients agrees with Markov approximation in the regime of multiple forward scattering. Additionally, E_{RTani} agrees with the diffusion approximation for late times. Therefore, the theory is able to simultaneously explain the effects of envelope broadening as well as the coda generation, which confirms the results of GUSEV and ABUBAKIROV (1996). We expect that radiative transfer theory using anisotropic Born scattering coefficients is suitable to explain entire seismogram envelopes from the first arrival to the late coda. To test this, in the next section we compare radiative transfer theory using anisotropic Born scattering coefficients with local energy density E computed from finite difference simulations of the full wave equation.

4. *Comparison to Finite-difference Calculations of the Full Wave Equation in 2-D*

4.1. *Effects of Finite Source Duration and of Filtering*

In most previous studies data were compared to local energy density E assuming an impulsive source. FEHLER *et al.* (2000) compared finite difference solutions of the wave equation to analytical solutions of Markov approximation. These authors demonstrated that convolving the Markov envelopes with the squared source time function gives better results than using the pure envelopes. In real data we have to consider two effects. The source process is of finite duration and, additionally, a bandpass filter is applied before comparing the data to the theoretical curves. In our finite differences simulation we take into account both effects: The connection between pressure $p(\mathbf{R}, T)$ and the Green's function $G(\mathbf{R}, T)$ is given by:

$$p(\mathbf{R}, T) = G(\mathbf{R}, T) * s(T) * B(T) \tag{30}$$

Here $s(T)$ is a point-like source with finite source duration, and $B(T)$ is the impulse response of a bandpass filter. Using Fourier transforms the autocorrelation function of the pressure field can be written as:

$$\left\langle p\left(\mathbf{R}, T+\frac{t}{2}\right)p^*\left(\mathbf{R}, T-\frac{t}{2}\right)\right\rangle = \int_{-\infty}^{+\infty}\int_{-\infty}^{+\infty}\left\langle G\left(\mathbf{R}, f+\frac{F}{2}\right)G^*\left(\mathbf{R}, f-\frac{F}{2}\right)\right\rangle \times s\left(f+\frac{F}{2}\right)s^*\left(f-\frac{F}{2}\right)B\left(f+\frac{F}{2}\right)B^*\left(f-\frac{F}{2}\right) \times \exp[2\pi i(FT+ft)]df\,dF, \tag{31}$$

where we use internal or carrier frequency f and modulation or external frequency F. This indicates that in the most general case one has to compute the local energy density $E_G(R, T, f)$ for several frequencies f. However, assuming that $B(T)$ is a narrow bandpass filter with center frequency f_0 and bandwidth σ, the problem simplifies to the following equation:

$$\langle|p(\mathbf{R}, T)|^2\rangle \approx E_G(R, T, f_0) * |s(T) * B(T)|^2 \tag{32}$$

$$\approx |s(f_0)|^2 E_G(R, T, f_0) * |B(T)|^2. \tag{33}$$

To derive equation (32) it is assumed that $E_G(R, T, f)$ is constant for f in the interval $[f_0 - \sigma, f_0 + \sigma]$. Additionally, we assume that modulation frequency F is always smaller than carrier frequency f, (which might be violated near the delta-like onset of the direct wave front). Equation (32) corresponds to the formula applied by FEHLER *et al.* (2000). Under the additional assumption that the source spectrum $s(f)$ is constant in $[f_0 - \sigma, f_0 + \sigma]$ equation (32) reduces to equation (33). Using a general result of filter theory (e.g., PAPOULIS, 1977, p. 18) the following equation results from equation (33):

$$\int_{-\infty}^{+\infty}\langle|p(\mathbf{R}, T)|^2\rangle dT = |s(f_0)|^2\left[\int_{-\infty}^{+\infty}E(R, T, f_0)dT\right]\left[\int_{-\infty}^{+\infty}|B(T)|^2 dT\right]. \tag{34}$$

Therefore, to guarantee the conservation of local energy density in the filter process we use the following normalizations in our comparison of radiative transfer to finite difference calculations:

$$|s(f_0)| = 1 \tag{35}$$

$$\int_{-\infty}^{+\infty}|B(T)|^2 dT = 1. \tag{36}$$

4.2. *Normalization of Local Energy Density*

In addition to the finite duration of the seismic source process and of the impulse response of the bandpass filter, we also have to address the normalization factor: In radiative transfer theory the spectral source energy $W_G(f)$ commonly is set equal to one, whereas in the wave equation a unit delta impulse is used as a source. To compare the two theories one must use the correct normalization of local energy density. To find this factor, firstly, the local energy density $E^{2D}_{G_0}(\mathbf{R}, T, f)$ of the Green's function in homogeneous two-dimensional space is computed:

$$E^{2D}_{G_0}(\mathbf{R}, T, f) = \kappa_0 \int_{-\infty}^{+\infty} G_0(\mathbf{R}, f + \frac{F}{2})\, G_0^*(\mathbf{R}, f - \frac{F}{2}) \exp(2\pi i F T) dF. \quad (37)$$

Here G_0 is the Green's function of the 2-D wave equation in frequency domain, which equals [e.g. MORSE and FESHBACH, 1953, p. 891]:

$$G_0^{2D}(R, f) = -\frac{i}{4} H_0^{(2)}\left(\frac{2\pi R}{v_0} f\right) \quad (38)$$

$H_0^{(2)}(x)$ is the Hankel function of second kind and zero order (ABRAMOWITZ and STEGUN, 1970, p. 358). Using a far field approximation $(\frac{R}{v_0}, T \gg \frac{1}{f})$ we get:

$$E^{2D}_{G_0}(\mathbf{R}, T, f) = \frac{v_0}{16\pi^2 R f} \kappa_0 \delta\left(T - \frac{R}{v_0}\right) \quad (39)$$

$$W^{2D}_{G_0}(T, f) = \frac{v_0^2}{8\pi f} \kappa_0. \quad (40)$$

In the homogeneous space with zero dissipation the spectral energy $W^{2D}_{G_0}(T, f)$ of the unit impulse is independent of time T and can be interpreted as spectral source energy $(W_{G_0}(T, f) = W_{G_0}(f))$. This value is also the spectral source energy in the case of an inhomogeneous medium $(W_{G_0}(f) = W_G(f))$. Therefore, it can be used as a normalization factor to compare radiative transfer theory, where usually $W_G(f)$ is set to 1, and full wave theory, where a unit delta impulse is used as a source.

The same analysis in 3-D yields:

$$E^{3D}_{G_0}(\mathbf{R}, T, f) = \frac{1}{16\pi^2 R^2} \kappa_0 \delta\left(T - \frac{R}{v_0}\right) \quad (41)$$

$$W^{3D}_{G_0}(f) = \frac{v_0}{4\pi} \kappa_0 \quad (42)$$

4.3. Computation of FD Envelopes

To compute the acoustic wavefield in a random medium we use finite difference solutions of the full wave equation in two-dimensional space. The size of the grid is 8100 times 8100 points with a grid spacing of 20 m resulting in a model size of 162 km times 162 km. We use a Gaussian autocorrelation function with a mean velocity of $v_0 = 1$ km s^{-1}. As standard deviations of the velocity fluctuations we use $\epsilon = 0.05$ and $\epsilon = 0.15$ to simulate moderate and strong fluctuations, respectively. The correlation length takes the values of a = 1.0 km and a = 0.2 km, respectively. With a central frequency of the source signal at $f_0 = 1.0$ Hz, these two correlation lengths correspond to the two cases of wide angle scattering $2\pi a/\lambda \approx 1.2$ and forward scattering $2\pi a/\lambda \approx 6$. The angular dependency of the Born scattering coefficient $g_{2D}(\vartheta, f)$ for these two cases is shown in Figure 1. The source is located in the center of the grid, whereas the receivers are placed on circles with a radius of 10, 20, ..., 70 km, respectively. 72 receivers for each distance are used for spatial averaging. At the border of the grid we use absorbing boundary conditions. However, we consider only times before the first theoretical arrival of any reflected wave from the boundary. Therefore the choice of the boundary condition is not important in our analysis. The configuration of the numerical experiment is shown in Figure 5. As a source function a Küpper wavelet is used:

$$s(t) = \begin{cases} 2f_0\{\sin(2\pi t f_0) - \frac{1}{2}\sin(4\pi t f_0)\} & 0 \quad < t \quad < \frac{1}{f_0} \\ 0 & \text{else} \end{cases} \tag{43}$$

which obeys the normalization in equation (35). After computing the wavefield for 5 realizations, each simulated seismogram is bandpass filtered with a Gaussian bandpass filter with normalization according to equation (36) and center frequency f_0 chosen to be identical to the center frequency f_0 of the Küpper wavelet. Then each trace is squared and the ensemble average is computed by taking the mean value over 5 realizations, each consisting of 72 seismograms with the same source receiver distance but 72 different azimuths. This spatial averaging over a circle of constant

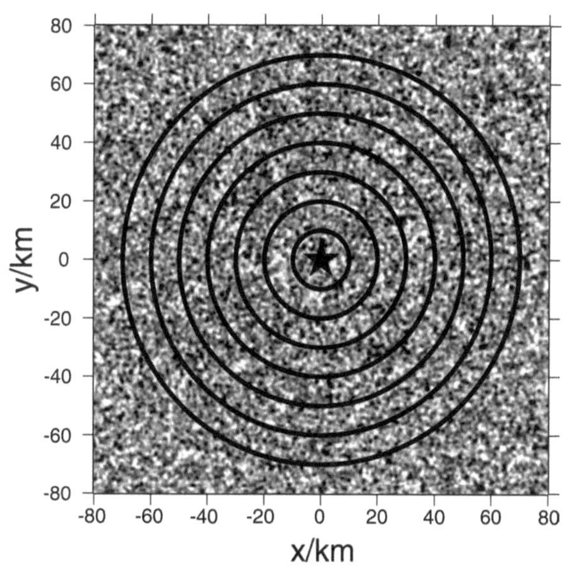

Figure 5
Finite difference model with 162 km times 162 km grid. A 2-D Gaussian random medium with correlation length a = 1 km is shown. The source is located in the center and the receivers are distributed along circles with radii 10 km, 20 km,..., 70 km. Each circle contains 72 equally spaced receivers, which are used for spatial averaging for each distance.

distance is not identical to averaging over 72 different realizations, because some neighboring seismograms may not be statistically independent. However, it proved to be very efficient, so that 5 different realizations of the random medium were sufficient to compute smooth envelopes. On the contrary, using ensemble averaging only, FEHLER *et al.* (2000) needed 100 realizations to obtain smooth envelopes indicating a much larger computational cost. Finally, the envelopes are normalized with the spectral source energy of the delta impulse according to equation (40).

For the numerical solutions of the equation of radiative transfer, firstly, the Küpper wavelet source function $s(t)$ in equation (43) is filtered with a normalized Gaussian bandpass using a center frequency $f_0 = 1.0$ Hz. Then, according to equation (32), the result is squared and used as a filter for the simulated envelope Green's functions $E^{2D}_{RTani}(\mathbf{R}, T, f_0)$. Therefore, in fact a comparison between $\langle |p(\mathbf{R}, T)|^2 \rangle_{FD}$ and $\langle |p(\mathbf{R}, T)|^2 \rangle_{RTani}$ near internal frequency f_0 is done. We can consider it as a low passed version of the desired comparison between $E^{2D}_{FD}(\mathbf{R}, T, f_0)$ and $E^{2D}_{RTani}(\mathbf{R}, T, f_0)$.

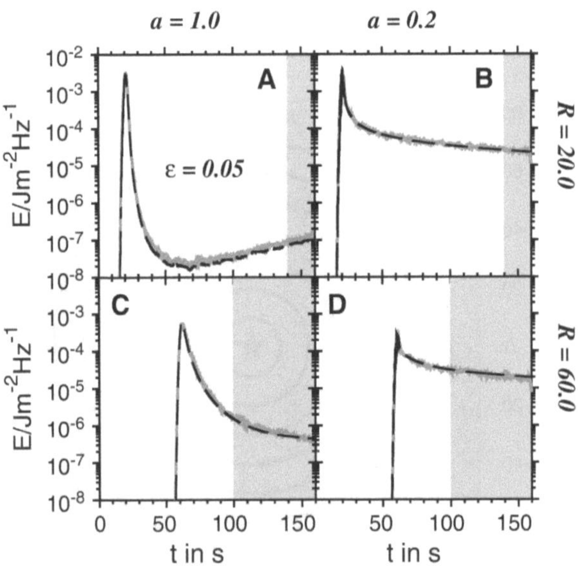

Figure 6
Numerical solutions of the 2-D equation of radiative transfer using anisotropic scattering coefficients (dashed, black) in comparison to numerical solutions of the full wave equation (continuous, grey) for a standard deviation of velocity fluctuations $\epsilon = 5\%$. The grey area indicates times, where the wavefield of the finite difference calculations might be influenced by the model boundaries. We use a velocity of $v_0 = 1.0$ km s^{-1}, which results in a wavelength of $\lambda = 1$ km in the undisturbed medium. Source receiver distances are $R = 20$ km (top) and $R = 60$ km (bottom). Correlation lengths are $a = 1.0$ km, which results in $l^0 \approx 5.7$ km and $l^* \approx 220$ km (left), and $a = 0.2$ km, $l^0 \approx 24$ km, $l^* \approx 38$ km (right).

4.4. Results of Comparison

Figures 6 and 7 summarize the results of the numerical comparison between radiative transfer theory using Born scattering coefficients and finite difference simulations of the full acoustic wave equation in two-dimensional space. In general, we find a good agreement between the two results. In detail, they coincide for the case of wide-angle scattering ($2\pi a/\lambda \approx 1.2$) as well as for forward scattering ($2\pi a/\lambda \approx 6$). Moreover, using anisotropic radiative transfer theory the first onset, the envelope broadening, as well as the late coda of the seismogram envelopes are computed correctly. This indicates that radiative transfer is suitable to explain full seismogram envelopes including direct waves as well as multiple scattering of arbitrary order. In the case of a small standard deviation such as $\epsilon = 5\%$, we could not find discrepancies between radiative transfer and full wave theory. In the case of large fluctuations such as $\epsilon = 15\%$, there appears some disagreement between the two theories (Fig. 7D) at large distances. This indicates a failure of the used transfer theory. We compare this failure of radiative transfer to the theoretical predictions of equation (14) and equation (15). Firstly, according to equation (14) we expect a failure of radiative transfer for small values of $2\pi l^0/\lambda$. In our numerical comparison we use the values of $2\pi l^0/\lambda \approx 36$ (Fig. 6AC), 150 (Fig. 6BD), 4.0 (Fig. 7AC), and 17 (Fig. 7BD). This indicates that the condition in equation (14) is never violated in our simulations. Moreover, the smallest value of $2\pi l^0/\lambda \approx 4.0$ occurs in Figure 7AC, where we

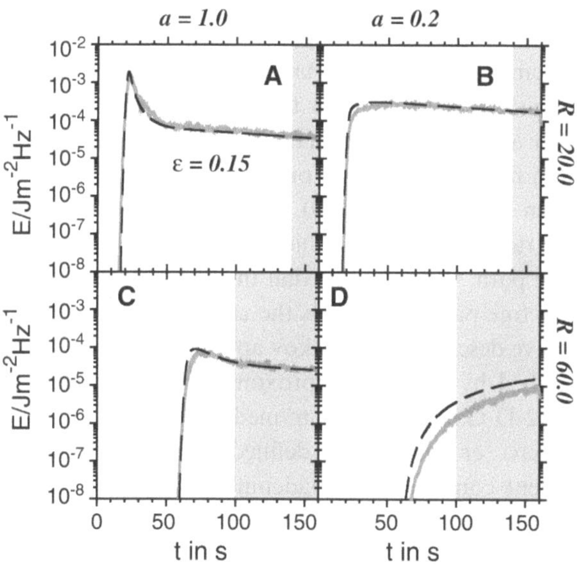

Figure 7
Same as Figure 6, but for velocity fluctuations with $\epsilon = 15\%$. In the left column $l^0 \approx 0.63$ km, $l^* \approx 24$ km and in the right column $l^0 \approx 2.7$ km, $l^* \approx 4.2$ km.

numerically find good coincidence of radiative transfer envelopes and finite differences envelopes. The observed failure of radiative transfer in Figure 7D corresponds to a larger value of $2\pi l^0/\lambda \approx 17$ and, therefore, cannot be explained by a violation of equation (14). Secondly, we compare to the condition for the validity of the use of Born scattering coefficients in equation (15). Here, in our numerical simulation we use the values of $l^0/a \approx 5.7$ (Fig. 6AC), 120 (Fig. 6BD), 0.63 (Fig. 7AC), and 14 (Fig. 7BD). We find that equation (15) is violated in Figure 7AC, where we numerically find good coincidence of radiative transfer and finite differences. The observed deviation of both theories, on the contrary occurs in Figure 7D, where the value of l^0/a is larger than in Figure 7AC. This indicates that neither equation (14) nor equation (15) can explain our numerical findings. We come back to this point in the discussion section.

5. Discussion

Several previous studies showed that the observed broadening near the direct S-wave onset can be used to extract information about small-scale heterogeneities of the propagation medium. Other studies used the late coda to obtain such information. Radiative transfer theory with Born scattering coefficients is a modeling method that can describe both effects, envelope broadening and coda generation, in one single theory. Therefore, in the inversion procedure we can use the stricter constraint that the theoretical envelopes fit the observed ones for all times to reveal information on small-scale heterogeneities.

Here we like to emphasize the importance of the transport mean free path l^*. It is well known that this quantity controls the coda generation at late times. On the other hand, we can also express the characteristic time of Markov approximation using the transport mean free path. For the 3-D Gaussian random medium this connection is given by equation (20). This equation indicates that the only information on the random medium, which is used in Markov approximation, is its transport mean free path. We conclude that in the case of strong forward scattering the transport mean free path l^* controls the early part of the seismogram envelope near the direct S wave described by Markov approximation as well as the late part of the envelope described by diffusion approximation. A relation similar to equation (20) also holds for 2-D Gaussian random media. For the case of von Karman type random media SATO *et al.* (2004) defined a "broadening coefficient." This broadening coefficient controls the broadening of Markov envelopes and again is proportional to the transport scattering coefficient. We conclude that seismogram envelopes contain considerable information on the transport mean free path l^*, which in the case of a Gaussian random medium is up to a constant factor identical to ϵ^2/a. Therefore, in inversion of real data we expect a good resolution for this quantity.

The degree of forward scattering only slightly influences the seismogram envelope. This influence only becomes apparent for times T near $T \approx l^*/v_0$ where neither Markov nor diffusion theory can be used. This range corresponds to the sag of the envelopes in Figures 2 and 3, where the sag becomes more pronounced with increasing value of $2\pi a/\lambda$. The sag can only be observed for source receiver distances smaller than the transport mean free path. In the case of real data we expect it to be even more difficult to observe this sag, because additional energy from Moho reflections exists arriving at such intermediate times. We conclude that seismogram envelopes contain only little information on the degree of forward scattering. Therefore, in inversion of real data we expect poor resolution for the scattering mean free path l^0 and poor resolution for $2\pi a/\lambda$. We can turn this negative result into a positive one: In the forward modeling for the inversion of real seismological data it is sufficient to use some moderate forward scattering (e.g., $2\pi a/\lambda \approx 6$). Using such degree of anisotropic scattering will produce much more realistic seismogram envelopes than the isotropic scattering model and can be done with reasonable computational effort using a Monte-Carlo method. The use of very strong forward scattering (e.g., $2\pi a/\lambda \approx 30$), which requires large computational effort in the Monte-Carlo method, is not necessary, because it changes the seismogram envelope only slightly in comparison to moderate forward scattering. Using such a fixed value of forward scattering does of course not allow an inversion for the value of $2\pi a/\lambda$, although on the other hand we can accurately measure the transport scattering coefficient with reasonable computational effort. The transport scattering coefficient contains all information on the medium, if it is known for a large frequency band. For example, in the acoustic case, taking the derivative of equation (A14) we obtain an equation connecting the power spectrum of the velocity fluctuations $R_v(k)$ to the derivative with respect to frequency of the transport scattering coefficient:

$$R_v(k) = \frac{4\pi}{k^3}\frac{d}{dk} g^*_{3D}\left(\frac{k}{2}\right). \tag{44}$$

Instead of using radiative transfer theory with anisotropic scattering coefficients, SAITO *et al.* (2003) and SATO *et al.* (2004) suggested the use of "hybrid models" to combine Markov approximation with the isotropic scattering model. The advantage of such models is that an analytical solution can be found for special cases (e.g., scattering in full space). On the other hand, currently the theoretical background of such models is not yet well established and their derivation is phenomenologically only. Comparing it to radiative transfer theory with Born scattering coefficients we expect a failure of such hybrid models at intermediate times near $T \approx l^*/v_0$, where neither Markov nor diffusion approximation can be used. At such intermediate times radiative transfer theory with Born scattering exhibits a frequency-dependent sag of energy density, which cannot be modelled using hybrid models. This is in agreement with SATO *et al.* (2004), who compared their 2-D hybrid model to 2-D finite difference envelopes. These authors found the hybrid model to be reliable at early and at late times, but they found some deviations from the finite difference envelopes at intermediate times.

The most common model for real data inversion is the isotropic scattering model. Strictly, it can only be applied for times considerably larger than the transport mean free time ($T \geq 5l^*/v_0$). Additionally, we should not use times near the direct wave onset ($T > R/v_0$). However, in the inversion process it turned out that trade-off problems exist using the late coda only. Therefore, additional to the late coda, LACOMBE *et al.* (2003) used information on direct waves. Also an inversion strategy based on "multiple lapse time window analysis" (FEHLER *et al.*, 1992) uses information of an early time window. In this inversion method a smoothing over time windows is applied which makes the difference between isotropic and the more realistic anisotropic scattering smaller. It appears that such methods also yield realistic estimations of the transport scattering coefficient. The advantage of taking anisotropic scattering into account seems to be that more detailed information of the observed envelopes can be used for the inversion.

Although radiative transfer theory with Born scattering coefficients is a quite general theory, it can only be used for weak scattering. We observed a difference between radiative transfer theory and solutions of the full wave equation which might be caused by different effects. Firstly, in the heterogeneous medium velocity is a function of frequency, which can be approximated by the frequency-independent mean velocity v_0 only in the case of weak fluctuations. Additionally, for strong fluctuations the scattering coefficients in the equation of radiative transfer can no longer be approximated by the Born scattering coefficients. Instead effects like resonant scattering (e.g., LAGENDIJK and VAN TIGGELEN, 1996) might become important, where the energy is caught within one scatterer for a certain time. If scattering becomes even stronger, the regime of Anderson localization is reached (e.g., VAN TIGGELEN, 1999) and the equation of radiative transfer cannot be used at all. In three-dimensional space the condition for the occurrence of such strong scattering effects is given by the value of l^0/λ (equation (14)). However, in two-dimensional space Anderson localization is also present for small fluctuations, but can be observed only for large time and distance scales (e.g., VAN TIGGELEN, 1999). Numerically, we find that the value of $2\pi l^0/\lambda$ used in equation (14) does not seem to be so important. We observed that radiative transfer is valid for values of $2\pi l^0/\lambda$ as small as 4.0, but invalid for the larger value of $2\pi l^0/\lambda = 17$. Based on numerical simulations we suggest another criterion for the validity of radiative transfer with Born scattering coefficients similar to equation (14), which uses the transport mean free path l^* instead of the scattering mean free path l^0:

$$l^* \gg \frac{\lambda}{2\pi}. \tag{45}$$

Using the value of $2\pi l^*/\lambda$, we may summarize our comparison of radiative transfer envelopes to finite difference envelopes as follows: We observe deviations of both theories for small values of l^*/λ such as $l^*/\lambda \approx 4.2$, but we observe good coincidence for large values of l^*/λ such as $l^*/\lambda \approx 24, 38$, and 200. However, we note that the condition in equation (45) is purely based on numerical comparison, however a good theory for this finding is missing. We can compare these values of $2\pi l^*/\lambda$ to expected values in real seismological experiments: For the Earth's crust we estimate $2\pi l^*/\lambda \approx 180$, assuming a transport mean free path of 100 km (e.g., SATO and FEHLER, 1998), a frequency of 1 Hz and a mean shear-wave velocity of 3.3 km/s. Comparing this to our numerical experiments, we conclude that radiative transfer theory is well suited to model seismic wave propagation in the Earth's crust. For the case of strato volcanoes, on the contrary, we estimate $2\pi l^*/\lambda \approx 0.63$ assuming a transport mean free path of $l^* \approx 0.15$ km (WEGLER and LÜHR, 2001; WEGLER, 2003), a frequency of 1 Hz and a mean shear-wave velocity of 1.5 km/s. Clearly, comparing this value to our numerical findings, it is questionable as to whether we can use radiative transfer theory with Born scattering coefficients for seismic wave propagation in strato volcanoes. Recently, AKI and FERRAZZINI (2000) and FRIEDRICH and WEGLER (2005) suggested that Anderson localization might be important in volcano seismology.

Due to computation time limitations, finite difference computations were only done for the two-dimensional case. To test more realistic models we have to use three-dimensional models. Additionally, elastic waves rather than acoustic waves should be used. MARGERIN *et al.* (2000) and PRZYBILLA *et al.* (2006) showed that the elastic equation of radiative transfer can also be solved with a Monte-Carlo method. Moreover, for the comparison with data a layered random medium including effects such as the crust mantle boundary must be included in the theory. Such depth-dependent phenomena can also be described within the theory of radiative transfer (e.g., MARGERIN *et al.*, 1998; YOSHIMOTO, 2000; HOSHIBA *et al.*, 2001; LACOMBE *et al.*, 2003).

6. *Conclusion*

We compared five methods for the synthesis of seismogram envelopes. Three analytical models (Markov approximation, radiative transfer theory with isotropic scattering coefficients, and diffusion approximation) and two numerical models (radiative transfer theory with non-isotropic scattering coefficients and finite differences solutions of the full wave equation) were used. All analytical solutions have some severe limitations: Markov approximation cannot be used for late times, radiative transfer theory with isotropic scattering coefficients and diffusion approximation give the wrong results for early times. Radiative transfer theory with Born scattering coefficients seems to be a good compromise: it correctly models full seismogram envelopes from the first onset to the multiple scattered late coda and the numerical solution is much faster than the computation of statistically averaged envelopes of finite difference solutions of the full wave equation. The restriction that the fluctuations are not allowed to be large does not seem to hinder the modeling of wave propagation in usual Earth crust, but might be problematic in volcano seismology.

Appendix A: Acoustic Born Scattering Coefficients

The acoustic angular-dependent Born scattering coefficients $g(\vartheta, f)$ used in radiative transfer theory are summarized for the 2-D and 3-D case:

$$g_{3D}(\vartheta, f) = \frac{k^4}{\pi} R_v(2k \sin\frac{\vartheta}{2}), \tag{A1}$$

$$g_{2D}(\vartheta, f) = k^3 R_v(2k \sin\frac{\vartheta}{2}). \tag{A2}$$

Here $k = 2\pi f / v_0$ and $R_v(k)$ is the power spectral density function of the velocity fluctuations, which in our simulations is of Gaussian type:

$$R_v^{3D}(k) = \epsilon^2 \sqrt{\pi^3} a^3 \exp\left(-\frac{k^2 a^2}{4}\right), \tag{A3}$$

$$R_v^{2D}(k) = \epsilon^2 \pi a^2 \exp\left(-\frac{k^2 a^2}{4}\right). \tag{A4}$$

From these the total Born scattering coefficients can be computed:

$$g_{3D}^0(f) = \frac{k^2}{2\pi} \int_0^{2k} R_v(k') k' dk' \tag{A5}$$

$$= \sqrt{\pi} \, \frac{\epsilon^2}{a} \, (ka)^2 [1 - \exp(-a^2 k^2)] \tag{A6}$$

$$\approx \sqrt{\pi} \, \frac{\epsilon^2}{a} \, (ka)^4 \quad 2\pi a / \lambda \ll 1 \tag{A7}$$

$$\approx \sqrt{\pi} \, \frac{\epsilon^2}{a} \, (ka)^2 \quad 2\pi a / \lambda \gg 1 \tag{A8}$$

$$g_{2D}^0(f) = \frac{k^2}{\pi} \int_0^{2k} R_v(k') \left[1 - \left(\frac{k'}{2k}\right)^2\right]^{-\frac{1}{2}} dk' \tag{A9}$$

$$= \pi \, \frac{\epsilon^2}{a} \, (ka)^3 I_0 (\frac{(ak)^2}{2}) exp(- \frac{(ak)^2}{2}) \tag{A10}$$

$$\approx \pi \, \frac{\epsilon^2}{a} \, (ka)^3 \quad 2\pi a / \lambda \ll 1 \tag{A11}$$

$$\approx \sqrt{\pi} \, \frac{\epsilon^2}{a} \, (ka)^2 \quad 2\pi a / \lambda \gg 1 \tag{A12}$$

Equations (A5) and (A9) are for arbitrary and equations (A6) and A(10) are for the special case of a Gaussian power spectral density of velocity fluctuations. I_0 is the modified Bessel function of first kind and zero order (e.g., ABRAMOWITZ and STEGUN, 1970, p. 374), which is computed after PRESS *et al.*, (1996).

Finally, we also use the transport scattering coefficients:

$$g^*_{3D}(f) = \frac{1}{4\pi}\int_0^{2k} R_v(k')k'^3 dk' \tag{A14}$$

$$= 2\sqrt{\pi}\,\frac{\epsilon^2}{a}\left[1 - (a^2k^2 + 1)\exp\left(-a^2k^2\right)\right] \tag{A15}$$

$$\approx \sqrt{\pi}\,\frac{\epsilon^2}{a}\,(ka)^4 \quad 2\pi a/\lambda \ll 1 \tag{A16}$$

$$\approx 2\sqrt{\pi}\,\frac{\epsilon^2}{a} \quad 2\pi a/\lambda \gg 1 \tag{A17}$$

$$g^*_{2D}(f) = \frac{1}{2\pi}\int_0^{2k} R_v(k')k'^2\left[1 - \left(\frac{k'}{2k}\right)^2\right]^{-\frac{1}{2}} dk' \tag{A18}$$

$$= \pi\,\frac{\epsilon^2}{a}\,(ak)^3\, M\!\left(\frac{3}{2}, 2, -a^2k^2\right) \tag{A19}$$

$$\approx \pi\,\frac{\epsilon^2}{a}\,(ak)^3 \quad 2\pi a/\lambda \ll 1 \tag{A20}$$

$$\approx \sqrt{\pi}\,\frac{\epsilon^2}{a} \quad 2\pi a/\lambda \gg 1 \tag{A21}$$

where again equations (A14) and (A18) are for arbitrary and equations (A15) and (A19) are for the special case of a Gaussian power spectral density of velocity fluctuations. $M(a, b, z)$ is the Kummer function as defined in ABRAMOWITZ and STEGUN (1970, p. 504), which is computed numerically after NARDIN *et al.*, (1992).

Acknowledgements

We are grateful to H. SATO and Y. ZENG for their helpful reviews.

REFERENCES

ABRAMOWITZ, M. and STEGUN, I.A., *Handbook of Mathematical Functions*, 9 ed. (Dover Publications, New York, 1970).

AKI, K. (1969), *Analysis of the seismic coda of local earthquakes as scattered waves*, J. Geophys. Res. *74*, 615–631.

AKI, K. and CHOUET, B. (1975), *Origin of coda waves: Source, attenuation, and scattering effects*, J. Geophys. Res. *80*, 3322–3342.

AKI, K. and FERRAZZINI, V. (2000), *Seismic monitoring and modeling of an active volcano for prediction*, J. Geophys. Res. *105*, 16,617–16,640.

Chandrasekhar, S., *Radiative Transfer* (Dover Publications, New York, 1960).
Dainty, A.M. and Toksöz, M.N. (1981), *Seismic codas on the earth and the moon: A comparison*, Phys. Earth Planet. Inter. *26*, 250–260.
Fehler, M., Hoshiba, M., Sato, H., and Obara, K. (1992), *Separation of scattering and intrinsic attenuation for the Kanto-Tokai region, Japan, using measurements of S-wave energy versus hypocentral distance*, Geophys. J. Int. *108*, 787–800.
Fehler, M., Sato, H., and Huang, L.-J. (2000), *Envelope broadening of outgoing waves in 2-D random media: A comparison between the Markov approximation and numerical simulations*, Bull. Seismol. Soc. Am. *90*, 914–928.
Frankel, A. and Clayton, R.W. (1986), *Finite difference simulation of seismic scattering: Implications for the propagation of short-period seismic waves in the crust and models of crustal heterogeneity*, J. Geophys. Res. *91*, 6465–6489.
Friedrich, C. and Wegler, U. (2005), *Localization of seismic coda at Merapi volcano (Indonesia)*, Geophys. Res. Lett. *32*, L14,312, doi:10.1029/2005GL023,111.
Gusev, A.A. and Abubakirov, I.R. (1987), *Monte-Carlo simulation of record envelope of a near earthquake*, Phys. Earth Planet. Inter. *49*, 30–36.
Gusev, A.A., and Abubakirov, I.R. (1996), *Simulated envelopes of non-isotropically scattered body waves as compared to observed ones: Another manifestation of fractal heterogeneity*, Geophys. J. Int. *127*, 49–60.
Hoshiba, M. (1991), *Simulation of multiple-scattered coda wave excitation based on the energy conservation law*, Phys. Earth Planet. Inter. *67*, 123–136.
Hoshiba, M. (1995), *Estimation of non-isotropic scattering in western Japan using coda wave envelopes: Application of a multiple nonisotropic scattering model*, J. Geophys. Res. *100*, 645–657.
Hoshiba, M., Rietbrock, A., Scherbaum, F., Nakahara, H., and Haberland, C. (2001), *Scattering attenuation and intrinsic absorption using uniform and depth-dependent model — Application to full seismogram envelope recorded in northern Chile*, J. Seismology *5*, 157–179.
Ishimaru, A., *Wave Propagation and Scattering in Random Media*, (IEEE Press and Oxford University Press, New York, 1997), first published by Academic Press 1978.
Lacombe, C., Campillo, M., Paul, A., and Margerin, L. (2003), *Separation of intrinsic absorption and scattering attenuation from Lg coda decay in central France using acoustic radiative transfer theory*, Geophys. J. Int. *154*, 417–425.
Lagendijk, A. and van Tiggelen, B.A. (1996), *Resonant multiple scattering of light*, Phys. Rep. *270*, 143–216.
Margerin, L., Campillo, M., and van Tiggelen, B.A. (1998), *Radiative transfer and diffusion of waves in a layered medium: New insight into coda Q*, Geophys. J. Int. *134*, 596–612.
Margerin, L., Campillo, M., and van Tiggelen, B. (2002), *Monte-Carlo simulation of multiple scattering of elastic waves*, J. Geophys. Res. *105*, 7873–7892.
Morse, P.M. and Feshbach, H., *Methods of Theoretical Physics*, vols. 1 and 2 (McGraw-Hill, New York, 1953).
Nardin, M., Perger, W.F., and Bhalla, A. (1992), *Algorithm 707, solutions to the confluent hypergeometric function*, ACM Transact. Math. Software *18*, 345–349.
Papoulis, A., *Signal Analysis* (McGraw-Hill, New York, 1977).
Press, W.H., Flannery, B.P., Teukolsky, S.A., and Vetterling, W.T. *Numerical Recipes in Fortran 77: The Art of Scientific Computing*, 2 ed. (Cambridge University Press, Cambridge, 1996).
Przybilla, J., Korn, M., and Wegler, U. (2006), *Radiative transfer of elastic waves versus finite difference simulations in 2-D random media*, J. Geophys. Res, in press.
Rytov, S.M., Kravtsov, Y.A., and Tatarskii, V.I., *Principles of Statistical Radiophysics, vol. 4, Wave Propagation Through Random Media* (Springer-Verlag, Berlin, 1987).
Ryzhik, L., Papanicolaou, G., and Keller, J.B. (1996), *Transport equations for elastic and other waves in random media*, Wave Motion *24*, 327–370.
Saito, T., Sato, H., and Ohtake, M. (2002), *Envelope broadening of spherically outgoing waves in three-dimensional random media having power-law spectra*, J. Geophys. Res. *107*, 10.1029.
Saito, T., Sato, H., Fehler, M., and Ohtake, M. (2003) *Simulating the envelope of scalar waves in 2-D random media having power-law spectra of velocity fluctuations*, Bull. Seismol. Soc. Am. *93*, 240–252.

SATO, H. (1989), *Broadening of seismogram envelopes in the randomly inhomogeneous lithosphere based on the parabolic approximation: Southeastern Honshu, Japan*, J. Geophys. Res. *94*, 17,735–17,747.
SATO, H. (1993), *Energy transportation in one- and two-dimensional scattering media: Analytic solutions of the multiple isotropic scattering model*, Geophys. J. Int. *112*, 141–146.
SATO, H. (1995), *Formulation of the multiple non-isotropic scattering process in 3-D space on the basis of energy transport theory*, Geophys. J. Int. *121*, 523–531.
SATO, H. and FEHLER, M.C., *Seismic Wave Propagation and Scattering in the Heterogeneous Earth* (Springer-Verlag, New York, 1998).
SATO, H., FEHLER, M., and SAITO, T. (2004), *Hybrid synthesis of scalar wave envelopes in two-dimensional random media having rich short-wavelength spectra*, J. Geophys. Res. *109*, doi: 10.1029/2003JB002,673.
SCHERBAUM, F. and SATO, H. (1991), *Inversion of full seismogram envelopes based on the parabolic approximation: Estimation of randomness and attenuation in southeast Honshu, Japan*, J. Geophys. Res. *96*, 2223–2232.
SHENG, P., *Introduction to Wave Scattering, Localization, and Mesoscopic Phenomena*, (Academic Press, San Diego, 1995).
VAN TIGGELEN, B.A., *Localization of waves. In Diffuse Waves in Complex Media* ed. J.-P. Fouque, pp. 1–60 (Kluwer Academic Publishers, Dortrecht, The Netherlands, 1999).
WEAVER, R.L. (1990), *Diffusivity of ultrasound in polycrystals*, J. Mech. Phys. Solids *38*, 55–86.
WEGLER, U. (2003), *Analysis of multiple scattering at Vesuvius volcano, Italy, using data of the TomoVes active seismic experiment*, J. Volcanol. Geotherm. Res. *128*, 45–63.
WEGLER, U. (2005), *Diffusion of seismic waves in layered media: Boundary conditions and analytical solutions*, Geophys. J. Int. *163*, 1123–1135.
WEGLER, U., and LÜHR, B.-G. (2001), *Scattering behaviour at Merapi volcano (Java) revealed from an active seismic experiment*, Geophys. J. Int. *145*, 579–592.
WU, R. (1985), *Multiple scattering and energy transfer of seismic waves; separation of scattering effect from intrinsic attenuation; I theoretical modelling*, Geophys. J. R. astr. Soc. *82*, 57–80.
YOSHIMOTO, K. (2000), *Monte-Carlo simulation of seismogram envelopes in scattering media*, J. Geophys. Res. *105*, 6153–6161.
ZENG, Y., SU, F., and AKI, K. (1991), *Scattering wave energy propagation in a random isotropic scattering medium: 1. Theory*, J. Geophys. Res. *96*, 607–619.

(Received April 5, 2005; accepted September 9, 2005)

Pure appl. geophys. 163 (2006) 533–548
0033–4553/06/030533–16
DOI 10.1007/s00024-005-0025-7

Pure and Applied Geophysics

Scattered Surface Wave Energy in the Seismic Coda

YUEHUA ZENG[1]

Abstract—One of the many important contributions that Aki has made to seismology pertains to the origin of coda waves (AKI, 1969; AKI and CHOUET, 1975). In this paper, I revisit Aki's original idea of the role of scattered surface waves in the seismic coda. Based on the radiative transfer theory, I developed a new set of scattered wave energy equations by including scattered surface waves and body wave to surface wave scattering conversions. The work is an extended study of ZENG *et al.* (1991), ZENG (1993) and SATO (1994a) on multiple isotropic-scattering, and may shed new insight into the seismic coda wave interpretation. The scattering equations are solved numerically by first discretizing the model at regular grids and then solving the linear integral equations iteratively. The results show that scattered wave energy can be well approximated by body-wave to body wave scattering at earlier arrival times and short distances. At long distances from the source, scattered surface waves dominate scattered body waves at surface stations. Since surface waves are 2-D propagating waves, their scattered energies should in theory follow a common decay curve. The observed common decay trends on seismic coda of local earthquake recordings particular at long lapse times suggest that perhaps later seismic codas are dominated by scattered surface waves. When efficient body wave to surface wave conversion mechanisms are present in the shallow crustal layers, such as soft sediment layers, the scattered surface waves dominate the seismic coda at even early arrival times for shallow sources and at later arrival times for deeper events.

Key words: Seismic coda, scattering waves, surface waves, body waves.

1. Introduction

Over the past three decades, seismologists have recognized the importance of lateral inhomogeneity of the earth at various scales. New instrumentation and computer technology developments have provided new opportunities for better observations and theoretical and numerical interpretations, particularly at high frequencies. An example of the early pioneering studies on high frequency wave propagation is Aki's work on the origin of the coda waves (AKI, 1969). In this paper, Aki attempted to interpret the coda waves by surface wave scattering. The pervasive

[1]US Geological Survey, Denver Federal Center, MS 966, Box 25046, Denver, CO 80225, U.S.A.
E-mail: zeng@usgs.gov

observation of coda wave envelope decays at close-in distances to the sources has convinced him that the dominant contributions to coda were from scattered body waves (AKI and CHOUET, 1975). Following this work, many new theories have been proposed to explain the seismic scattering waves. For example, WU (1985) has introduced the stationary energy transport theory to study multiple seismic scattering. ZENG *et al.* (1991) has formulated the energy transport theory in the time-domain with analytical solutions and provided a unified approach to the energy transport approach (WU, 1985) and the ray theoretical approach (AKI and CHOUET, 1975; SATO, 1977; KOPNICHEV, 1977; GAO *et al.*, 1983a,b). ZENG (1993) and SATO (1994a) later proposed a multiple-scattering theory that includes a P-and S-waves scattering conversion to examine the evolving energy budget between P and S waves. These theories consider only isotropic scattering. SATO (1994b) extended the theory to consider nonisotropic scattering. Numerical solutions to the nonisotropic scattering (e.g., GUSEV and ABUBAKIROV, 1996; HOSHIBA, 1995; MARGERIN *et al.*, 1998, 2000) indicated that the observed broadening of the direct wave envelopes were from strong forward scattering.

With the mainstream acceptance of the body wave scattering origin of the seismic coda, questions still remain as to how scattered surface waves contribute to the seismic coda. Despite that, much progress has been made in the study of surface wave scattering in recent years. For examples, SNIEDER (1986 and 1988) has studied the effect of topographic variation and buried inclusions on surface wave scattering. SATO and NOHECHI (2001) developed the envelope formulation of long-period Rayleigh waves using an isotropic scattering model and compared their theoretical calculations with observations. As indicated by SNIEDER (2002) there are two main problems: One involves the coupling of surface waves to body waves and the second involves the multiple scattering of surface waves. These problems are well defined but difficult to solve numerically. In this paper, I will extend the scattering theory to include scattered surface waves and surface-wave and body-wave scattering conversions under the framework of radiative transfer theory. Instead of taking the exact solution for the scattering of converted surface waves from body waves, a heuristic propagator will be developed so that a realistic numerical calculation could be made. By discretizing the model onto regular grids, I will formulate the scattering equations for each discrete scatterer and solve those scattering equations iteratively. The results will be carefully examined on the contribution of scattered surface waves and body waves to the seismic coda. Based on the study, new interpretation and physical insight to the coda wave propagations and observations will be proposed.

2. *Theory*

Consider acoustic waves propagating in a 3-D elastic medium with unperturbed velocity v and randomly distributed scatterers which will generate incoherent scattered waves upon an incident wave arrival. Following ZENG *et al.* (1991), the scattered wave energy can be written as

$$E(\mathbf{r},t) = E_{in}\left(t - \frac{|\mathbf{r} - \mathbf{r}_0|}{v}\right)\frac{e^{-\eta|\mathbf{r}-\mathbf{r}_0|}}{4\pi|\mathbf{r} - \mathbf{r}_0|^2} + \int_V \eta_s\, E\left(\mathbf{r}_1,t - \frac{|\mathbf{r}_1 - \mathbf{r}|}{v}\right)\frac{e^{-\eta|\mathbf{r}_1-\mathbf{r}|}}{4\pi|\mathbf{r}_1 - \mathbf{r}|^2}\,dV_1 \quad (1)$$

where $E(\mathbf{r}, t)$ is the scattered wave energy at receiver $\mathbf{r}$, E_{in} is the incident wave energy originated from $\mathbf{r}_0$, η_s is the scattering coefficient, η_s is the absorption coefficient, and $\eta = \eta_s + \eta_i$ Clearly the scattered wave energy at the receiver is given by incident wave energy and all the scattered wave energies integrated over all possible scatterers at $\mathbf{r}_1$ in volume V.

The above theory could be extended to include body-wave and surface-wave scattering and their scattering conversions. We write the body-wave scattering energy with contribution from the surface-wave scattering conversion as

$$\begin{aligned}E_b(\mathbf{R},t) =& E_{in}\left(t - \frac{|\mathbf{R} - \mathbf{R}_0|}{v_b}\right)\frac{\gamma_b e^{-\eta^b|\mathbf{R}-\mathbf{R}_0|}}{4\pi|\mathbf{R} - \mathbf{R}_0|^2}\\ &+ \int_{V_1} \eta_s\left[E_b\left(\mathbf{R}_1, t - \frac{|\mathbf{R}_1 - \mathbf{R}|}{v_b}\right) + E_c\left(\mathbf{r}_1, z_1, t - \frac{|\mathbf{R}_1 - \mathbf{R}|}{v_b}\right)\right]\frac{\gamma_b e^{-\eta^b|\mathbf{R}_1-\mathbf{R}|}}{4\pi|\mathbf{R}_1 - \mathbf{R}|^2}dV_1\end{aligned} \quad (2a)$$

where E_b is the body-wave scattering energy, E_c is the surface-wave scattering energy, $\mathbf{R} = (\mathbf{r}, z)$, $\mathbf{R}_0$ is the source location of incident wave, γ_b is the partition factor of the source energy to body waves, v_b is the body wave velocity, η_s is the scattering coefficient and η^b is the body wave total attenuation coefficient, $\mathbf{R}_1$ is the integration variable for all possible scatterers in volume V_1. For mathematic convenience, we redefined $\mathbf{r} = (x, y)$ instead of (x, y, z) in equation (1).

For surface-wave scattering energy, we can write

$$\begin{aligned}E_c(\mathbf{r}, z, t) =& E_{in}\left(t - \frac{|\mathbf{r} - \mathbf{r}_0|}{v_c}\right)\gamma_c G_c(\mathbf{r}, \mathbf{r}_0, z, z_0)e^{-\eta^c|\mathbf{r}-\mathbf{r}_0|}\\ &+ \int_{V_1} \eta_s\left[E_b\left(\mathbf{R}_1, t - \frac{|\mathbf{r}_1 - \mathbf{r}|}{v_c}\right)\right.\\ &\left.+ E_c\left(\mathbf{r}_1, z_1, t - \frac{|\mathbf{r}_1 - \mathbf{r}|}{v_c}\right)\right]\gamma_c G_c(\mathbf{r}, \mathbf{r}_1, z, z_1)e^{-\eta^c|\mathbf{r}_1-\mathbf{r}|}\,dV_1\end{aligned} \quad (2b)$$

where G_c describes the geometrical spreading of surface-wave energy and its amplitude distribution with depth, γ_c is the partition factor of the source energy to surface waves, v_c is the surface wave velocity and η^c is the surface wave total attenuation coefficient. The partition factor γ_c, which describes the energy conversion of body wave to surface wave, depends on the near-surface velocity structure of the earth. For example, a soft surface layer will tend to trap a large amount of body waves thus has a larger value of γ_c. The value of γ_c depends on the source or scatterer depth. But in our current analysis, we assume γ_c is common for all source and scatterer depths. For function G_c, there are analytical expressions for the Green's functions in the far field (AKI and RICHARDS, 1980). However, a general analytical expression does not exist for a general inhomogeneous medium. Thus a good approximation of G_c is vital for our numerical simulation. Any valid approximation of G_c should include a geometrical spreading $1/r$ (surface-wave energy) at the far field and an exponential decay of amplitude with depth. Figure 1 shows an example of the exponential decay of surface-wave amplitude with depth (AKI and RICHARDS, 1980). This depth dependent amplitude decay is also a function of frequency. With consideration of those characteristics, I assume G_c has the form

$$G_c(\mathbf{r}, \mathbf{r}_0, z, z_0) = \frac{|\mathbf{r} - \mathbf{r}_0|}{2\pi[|\mathbf{r} - \mathbf{r}_0|^2 + c^2(z_0 + z)^2]} \frac{e^{-(Z+Z_0)/a}}{a} \tag{3}$$

where r_o and z_o describe the horizontal and depth location of the scatterer, a is frequency dependent, c is a distance and depth ratio parameter that determines how

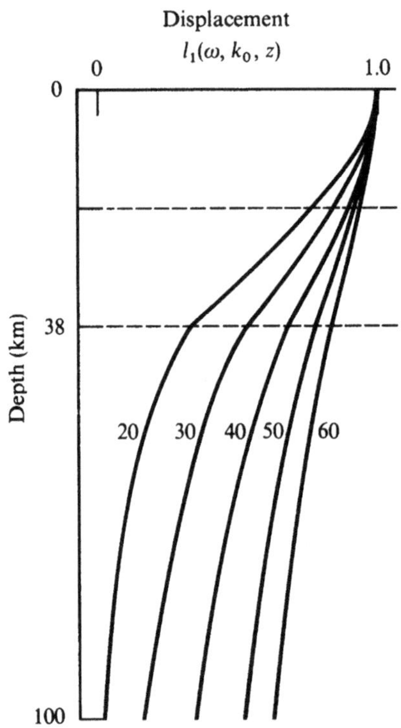

Figure 1
Examples of the amplitude dependences of Love waves with depth at various periods from AKI and RICHARDS (1980). The amplitudes are normalized to unit at z = 0.

surface waves are developed away from the scattering point source. By carefully examining the fundamental Rayleigh waves with periods ranging from 1 to 20 seconds (e.g., in CHEN, 1993; AKI and RICHARDS, 1980), I found that a ranges from 2 to 40 km. In equation (3), G_c reaches its maximum value at the horizontal distance of c times the sum of the source and receiver depth, namely, $|\mathbf{r}-\mathbf{r}_o| = c(z_o + z)$. This provides a critical distance constraint for the development of surface waves. According to AKI and RICHARDS (1980), equation (6.71), p. 239–240, the ratio between the horizontal distance and the sum of the source and receiver depth must be at least five for many rock materials in order for the Rayleigh pulse to be well developed. In our following numerical modeling, we assume c = 5.

For the scattering wave energy at a point scatterer, the sum of the scattered energy flux partitioned to body wave and to surface wave should be equal to the total scattered energy flux at the point. Thus, we have

$$\int_{\Sigma_1} \frac{\gamma_b E_s(\mathbf{R}_0, t - |\mathbf{R}-\mathbf{R}_0|/v_b)}{4\pi|\mathbf{R}-\mathbf{R}_0|^2} v_b\, d\Sigma + \int_{\Sigma_2} \gamma_c E_s\left(\mathbf{R}_0, t - \frac{|\mathbf{r}-\mathbf{r}_0|}{v_c}\right) G_c(\mathbf{r}, \mathbf{r}_0, z, z_0) v_c\, d\Sigma = \int_{\Sigma_1} \frac{E_s(\mathbf{R}_0, t - |\mathbf{R}-\mathbf{R}_0|/v_b)}{4\pi|\mathbf{R}-\mathbf{R}_0|^2} v_b\, d\Sigma, \tag{4}$$

where E_s is the scattered energy at $\mathbf{R}_0$. In this integration equation, the first term on the left-side of the equation is the scattered body-wave energy flux integrated over a spherical interface $\sum$ of radius $|\mathbf{R}-\mathbf{R}_0|$ centered at $\mathbf{R}_0$. The second term is the scattered surface-wave energy flux integrated over a cylindrical surface of radius $|\mathbf{r}-\mathbf{r}_0|$ centered at $\mathbf{r}_0$. The sum of the two terms gives the total scattered energy flux, which is equivalent to the total scattered body-wave energy flux assuming no scattering conversion to surface waves. Equation (4) can be reduced to

$$\gamma_b v_b + \int_0^\infty \gamma_c 2\pi|\mathbf{r}-\mathbf{r}_0| G_c(r, r_0, z, z_0) v_c\, dz = v_b. \tag{5}$$

Substituting the expression for G_c in equation (3), we get

$$\gamma_b \approx 1 - \frac{\gamma_c(|\mathbf{R}-\mathbf{R}_0|v_c/v_b)^2 v_c/v_b}{(|\mathbf{R}-\mathbf{R}_0|v_c/v_b)^2 + c^2 z_0^2} e^{-z_0/a}, \tag{6}$$

where we replace $|\mathbf{r}-\mathbf{r}_0|$ by an identity value of $|\mathbf{R}-\mathbf{R}_0|v_c/v_b$ Equation (6) shows that γ_b is a function of distance and depth. At distance near a scatterer, $\gamma_b = 1$, suggesting that near the scatterer, scattered body waves dominate the scattered wavefield. Surface wave energy develops when these waves reach the surface. The amount of scattered wave energy partitioning into surface wave increases as waves propagate away from the scatterer.

3. Energy Conservation

Equation (2) provides the basic formulations for solving the scattered surface-wave and body-wave energies and their scattering conversion. Equation (3) and (6) provide the analytical function form for the surface wave energy propagator and the constraint for the scattering partitioning parameters between the body and surface waves. In this section, I will further confirm that this system of equations satisfies the fundamental law of energy conservation.

Integrating the terms in both sides of equation (2a) over space, we find the total energy for the body can be written as

$$\begin{aligned}\int_V E_b(\mathbf{R},t)dV &= \int_V E_{in}\left(t-\frac{|\mathbf{R}-\mathbf{R}_0|}{v_b}\right)\frac{\gamma_b e^{-\eta^b|\mathbf{R}-\mathbf{R}_0|}}{4\pi|\mathbf{R}-\mathbf{R}_0|^2}dV \\ &\quad + \int_{V_1}\int_V \eta_s\left[E_b\left(\mathbf{R}_1,t-\frac{|\mathbf{R}_1-\mathbf{R}|}{v_b}\right)\right. \\ &\quad \left.+E_c\left(\mathbf{r}_1,z_1,t-\frac{|\mathbf{R}_1-\mathbf{R}|}{v_b}\right)\right]\frac{\gamma_b e^{-\eta^b|\mathbf{R}_1-\mathbf{R}|}}{4\pi|R_1-R|^2}dVdV_1 \\ &= \int_0^t\{E_{in}(t-\tau)+\eta_s\int_{V_1}[E_b(\mathbf{R}_1,t-\tau)+E_c(\mathbf{r}_1,z_1,t-\tau)]dV_1\}\gamma_b v_b e^{-\eta^b v_b\tau}d\tau.\end{aligned} \tag{7a}$$

The total energy for the surface waves can be found as

$$\begin{aligned}\int_V E_c(\mathbf{r},z,t)dV &= \int_V E_{in}\left(t-\frac{|\mathbf{r}-\mathbf{r}_0|}{v_c}\right)\gamma_c G_c(\mathbf{r},\mathbf{r}_0,z,z_0)e^{-\eta^c|\mathbf{r}-\mathbf{r}_0|}dV \\ &\quad + \int_{V_1}\int_V \eta_s\left[E_b\left(\mathbf{R}_1,t-\frac{|\mathbf{r}_1-\mathbf{r}|}{v_c}\right)\right. \\ &\quad \left.+E_c\left(\mathbf{r}_1,z_1,t-\frac{|\mathbf{r}_1-\mathbf{r}|}{v_c}\right)\right]\gamma_c G_c(\mathbf{r},\mathbf{r}_1,z,z_1)e^{-\eta^c|\mathbf{r}_1-\mathbf{r}|}dVdV_1 \\ &= \int_0^t\{E_{in}(t-\tau)+\eta_s\int_{V_1}[E_b(\mathbf{R}_1,t-\tau) \\ &\quad + E_c(\mathbf{r}_1,z_1,t-\tau)]\,dV_1\}(1-\gamma_b)v_b e^{-\eta^b v_b\tau}d\tau.\end{aligned} \tag{7b}$$

In order to derive equation (7b), we have used the result in equation (5). If we further assume $E(t)=\int_V[E_b(\mathbf{R},t)+E_c(\mathbf{r},z,t)]\,dV$ and $\eta^c v_c=\eta^b v_b$, the sums of (7a) and (7b) can be reduced to

$$E(t) = \int_0^t [E_{in}(t-\tau) + \eta_S E(t-\tau)] v_b e^{-\eta^b v_b \tau} \, d\tau. \tag{8}$$

Taking a Fourier Transfer to both sides of equation (8), we obtain

$$E(\omega) = [E_{in}(\omega) + \eta_s E(\omega)] v_b / (i\omega + \eta^b v_b).$$

Assuming there is no intrinsic attenuation, then we have $\eta_s = \eta^b$. The above equation can be reduced to

$$E(\omega) = E_{in}(\omega) v_b / (i\omega)$$

or

$$E(t) = \int_0^t E_{in}(\tau) v_b \, d\tau. \tag{9}$$

Equation (9) indicates that if the duration of the incidence wave is less than t, E(t) will be a constant value. Thus the total energy of the scattered body waves and surface waves satisfies the energy conservation law.

4. *Numerical Examples and Discussion*

Although we have simplified the Green's function form for the scattered surface wave energy propagator, equations (2a) and (2b) are difficult to solve analytically in a closed form solution. One approach is to solve the problem using a Monte-Carlo simulation scheme. For practical purposes, I instead discretize the above equation at uniformly distributed grid points with constant spacing between each point, and solve the equation iteratively following the approach of ZENG (1991). I choose a $110 \times 110 \times 110$ km finite half-space model with about 10^5 scatterers uniformly distributed inside. To test the numerical method, I first compute only body wave to body-wave scattering by taking $\gamma_c = 0$ in (4). Then I calculate the same scattering energy distribution using the analytical solution of ZENG *et al.* (1991) and compare the numerical calculations. I assume $\eta_s = 0.02$ km^{-1}, $\eta_i = 0.02$ km^{-1}, $v_b = 3$ km/sec source depth of 5 km and stations located on the free surface. Figure 2 shows the numerical solution in comparison with the analytical solution of ZENG *et al.* (1991). In this figure, each curve starts with a direct energy pulse arrival followed by its coda wave energy decay with time. The scattering wave energy decay curves are obtained at different source-receiver distances indicated by their direct energy pulse arriving times. Red curves are for the analytical solutions and blue curves are for the numerical calculations. The two sets of solutions agree quite well with each other. A discrepancy between the analytical and numerical solutions occurred at later arriving

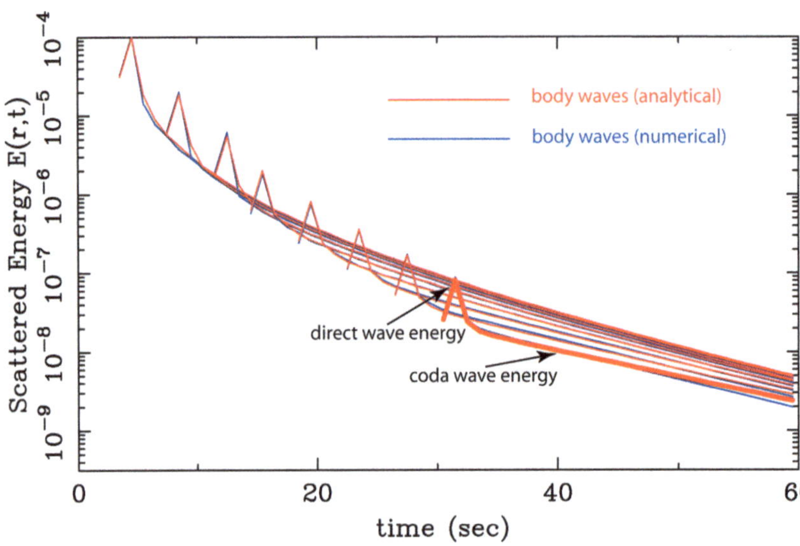

Figure 2
Comparison between the numerical solution of equation (2) with only body-wave scattering and the analytical solution of ZENG *et al.* (1991). The source is located at a depth of 5 km and stations located on the free surface. For the scattering parameters, $\eta_s = 0.02\ \text{km}^{-1}$ and $\eta_i = 0.02\ \text{km}^{-1}$. For other model parameters, $v_b = 3$ km/sec and $\gamma_c = 0.0$.

coda for the receivers near the boundary of the numerical model. Clearly this discrepancy is expected since those receivers in the numerical model do not receive multiple-scattered energies from the other side of the boundary in contrast to the receivers in the unbounded analytical model.

With the confidence from the above comparison of numerical simulations, we compute an example with a = 4.0 km, $v_b = 3$ km/sec, $v_c = 2.8$ km/sec, and $\gamma_c = 0.8$ (Fig. 3a). The value for a is chosen correspondent to a wave frequency of about 0.5–1.5 Hz. In this figure, red lines indicate the total energies, blue lines represent the body-wave energies and green lines are the surface-wave energies. Several interesting observations can be made from this figure. For the direct wave energies, the direct surface wave energies in green start to exceed the direct body wave energies in blue at distances over 50 km. This could be attributed to the effect of difference in geometrical spreading between direct surface waves and direct body waves. The slower spreading rate of the surface wave energy at $1/r$ in comparison with that for the body wave energy at $1/r^2$ has eventually lead to the direct surface wave energy dominating over that of the body wave energy. By comparing the scattered body wave and surface energies, we find that at long distances scattered surface-wave energy dominates over scattered body-wave energy at surface stations. This may have important implications for coda wave interpretation. Since AKI and CHOUET (1975), seismic coda waves have been largely interpreted as scattered shear waves in the

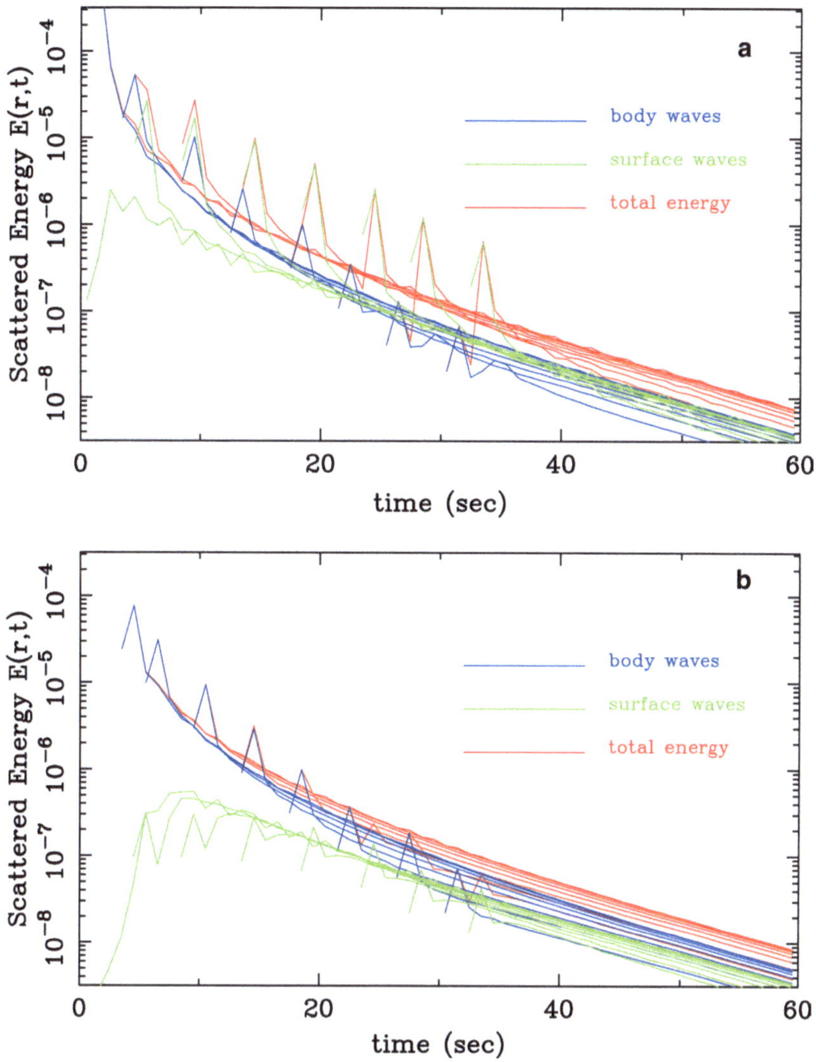

Figure 3

(a) Comparison between total scattered energies (red), scattered body-wave energies (blue) and scattered surface-wave energies (green) calculated using equations (2a) and (2b) at different receiver locations. The source is located at a depth of 5 km and stations located on the free surface. For the other scattering model parameters, $\eta_s = 0.02\ m^{-1}$, $\eta_i = 0.02\ km^{-1}$, $a = 4.0$ km, $v_b = 3$ km/sec, $v_c = 2.8$ km/sec, and $\gamma_c = 0.8$. (b) Same as in Figure 3a but with a source depth of 15 km.

crust. AKI (1992) has shown that the scattered energy conversion from P to S waves is about $(\alpha/\beta)^4$ times more efficient that that from S to P waves, where α is the P-wave velocity and β is the S-wave velocity, leading to the S waves dominating the coda generated from local earthquakes. This is later confirmed by numerical simulations based on the theory of scattered P-and S-wave energy in a random scattering medium by ZENG (1993). However, those studies have not considered the scattering conversion between body waves and surface waves. Although the result of our current study supports the hypothesis that S-wave scattered energies dominates for most part of the seismic coda waves, the later arrivals of the seismic coda waves are perhaps dominated by scattered surface wave energies.

In examining Figure 3a, we also find that the scattered surface wave energies follow a common decay curve. As the surface wave energies dominate the coda, the same should be true for the later arriving coda waves. This is consistent with the observation made by AKI (1980) that the envelopes of coda waves share a common decay shape independent of the source-receiver locations, especially at later arrival times. Based on this observation, a coda normalization method to correct for source and site power factors was proposed and widely used in coda wave analysis (e.g., WU and AKI, 1988; FEHLER *et al.*, 1992; MAYEDA *et al.*, 1992 and others). For a uniformly distributed random scattering medium, the scattered energy shape depends on the source-receiver distance (Fig. 2). This model is not supported by the observation for a strong scattering case. The scattering model with a depth-dependent scattering coefficient provides an alternative explanation for the observations (ZENG, 1991; HOSHIBA, 1994). I favor the scattered surface-wave energy model because it is a natural consequence of scattering wave conversions and the slower geometrical decay of the surface waves that lead to the scattered surface waves dominate the later seismic coda. Physically, surface waves are 2-D propagating waves. SATO (1993) has shown theoretically that for a 2-D random medium, coda waves generated from random scattering waves follow a common decay curve despite the station distance from the primary wave source. This was also illustrated by 2-D finite difference simulations for wave propagation in a random scattering medium (FRANKEL and CLAYTON, 1986; FEHLER *et al.*, 2000). Physically the spatial distribution of 2-D scattering wave energy evolves quickly into a diffusion-like solution. In a 3-D case, the process is further complicated by the added degree of freedom in space. To further test the result, I simulate a case with a source located at a greater depth. Figure 3b shows the results with the same scattering model parameters as for results in Figure 3a except it incorporates a deeper source at a depth of 15 km. As we would expect, the direct surface wave energy is considerably smaller compared to the body-wave energy at the distances considered for this model. At the close-in stations, the scattered surface-wave energies gradually catch up with the scattered body-wave energies at later arrival times. At the distant stations, the scattered surface-wave energies start to dominate over body-wave energies even at early arriving times following the direct wave arrivals. As the scattered surface wave energies dominate the seismic coda, the general coda energy decay starts to follow a common decay curve at all stations at large lapse time windows. For a depth-dependent intrinsic Q where Q decreases away from the near surface, we expect a faster decay for the scattered surface waves than that of the scattered body waves since the scattered body waves transverse deeper portions of the crust. Thus a depth-dependent Q may tend to inhibit the dominance of scattered surface waves in the later coda at very high frequency.

Figures 4a and 4b show the simulated scattered wave energy fields using the same model parameters and primary source locations as for results in Figures 3a and 3b, respectively, except a smaller partition factor γ_c from body-wave energy to surface-wave energy. Obviously, the energy conversions between body waves and surface waves are less efficient. The scattered body-wave energies dominate the coda waves at all times. Studies of site amplification estimation at stations less contaminated by deep alluvial structure found that the site amplification factor estimated using coda waves is in good agreement with that from direct S waves (e.g., SU *et al.*, 1996). On the other hand, within deep sedimentary basins or extremely soft soil sites, like the bay mud sites in the San Francisco Bay area, site responses estimated using coda and direct S waves do not agree with each other (MARGHERITI *et al.*, 1994; BONILLA *et al.*, 1997). In examining the content of the waves, they have found that those extremely soft sites or deep alluvial sites contain significant trapped waves in the coda. In other words, those soft sites have a more efficient mechanism to convert body waves to surface waves, resulting in trapped surface waves that dominate the seismic coda. For sites without the soft layer, the high frequency coda waves more likely consist of scattered S waves. As a result, the site responses estimated using coda at those sites closely agree with that estimated using direct S waves.

Another important effect on scattering conversion between surface waves and body waves is the frequency. The examples of frequency-dependent surface wave amplitude distributions with depth in Figure 1 show that the lower the frequency, the deeper the surface wave amplitudes extend. The parameter that simulates this frequency-dependent effect is the factor a in equation (3). The larger the factor a is, the lower the frequency for the scattered surface waves. For the fundamental Rayleigh waves of period ranging from 1 to 5 seconds, a ranges from 2 to 10 km (CHEN, 1993; AKI and RICHARDS, 1980). We simulate this frequency-dependent effect in Figures 5a and 5b using the same model parameters and primary source locations as for results in Figures 3a and 3b, respectively, except we apply a larger frequency-dependent parameter a of 10 km. Clearly as frequency decreases, the primary surface-wave amplitude increases for deep sources, and the conversion rate from body to surface wave also increases. This is also expected as the surface wave amplitude extends to a greater depth, more deep scatterers will participate in the scattering conversion between surface waves and body waves. By comparing body-wave energies in Figures 3, 4 and 5, we find that they are approximately the same at near source distance. This suggests that the scattered body-wave energy can be well approximated by body wave to body-wave scattering at near source distance. Unlike the presence of a fixed ratio for S/P wave energy in body-wave scattering, there is no fixed ratio of surface/body-wave energy even in the long lapse time limit, because the two are different types of propagating waves with surface wave propagating along a 2-D surface and decaying exponentially with depth.

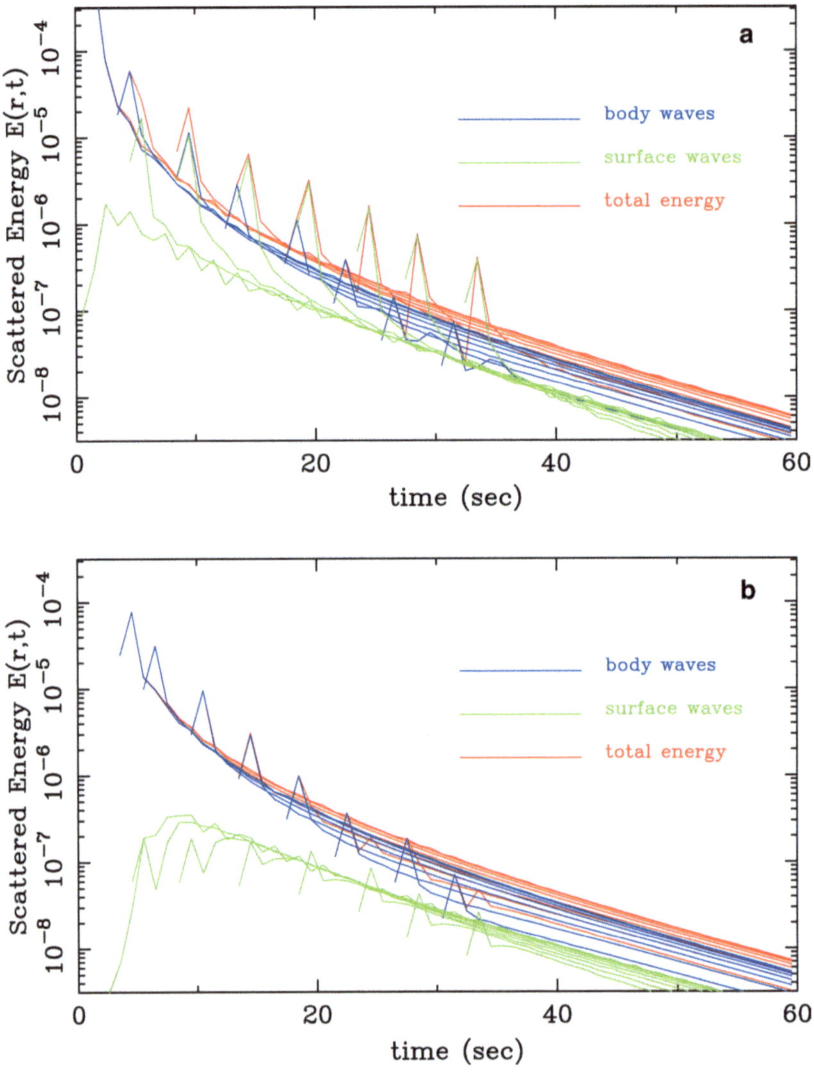

Figure 4
(a) Same as in Figure 3a but with $\gamma_c = 0.5$. (b) Same as in Figure 3b but with $\gamma_c = 0.5$.

5. Conclusion and Remarks

I have extended the scattered wave energy theory for body waves to include scattered surface waves and body wave to surface-wave scattering conversions based on the radiative transfer theory. By starting from those fundamental scattering wave equations, I demonstrated analytically that these scattering wave equations satisfy

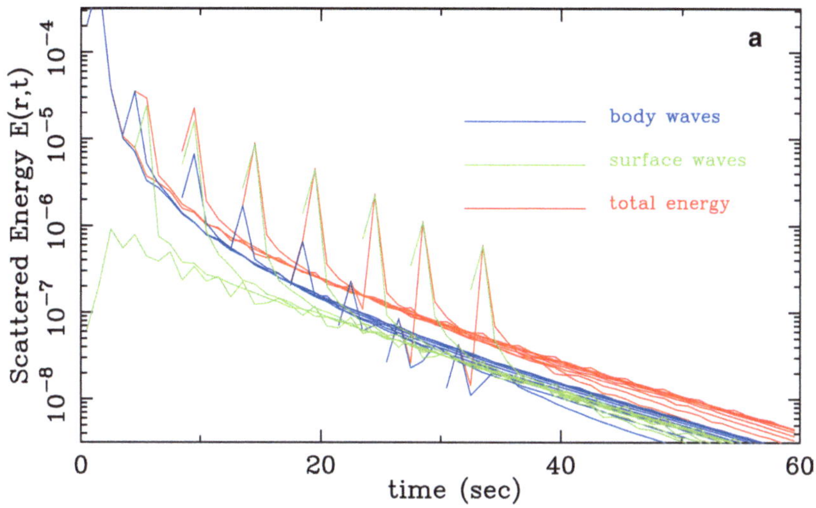

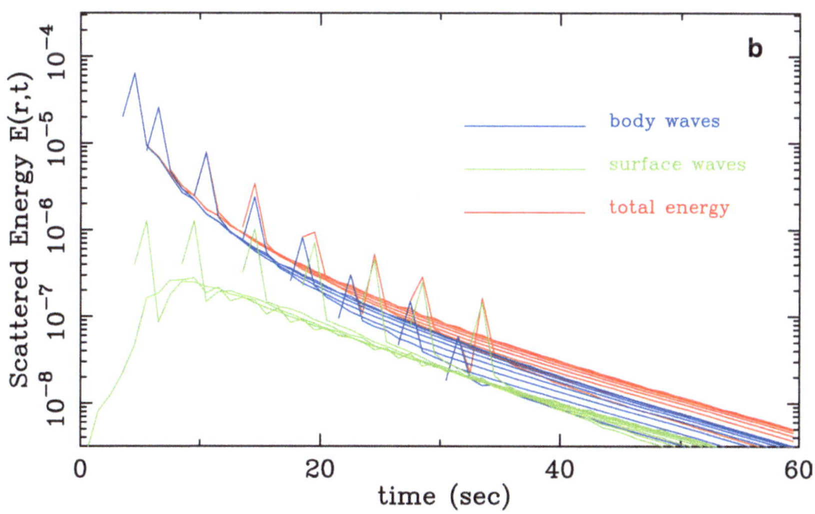

Figure 5
(a) Same as in Figure 3a except a = 10 km. (b) Same as in Figure 3b except a = 10 km.

the energy conservation. I have developed a simple analytical form for the scattered surface-wave energy propagator that takes into consideration the far-field surface geometrical spreading and the exponential decay of the surface-wave amplitude with depth. Based on this simple heuristic solution, efficient numerical simulations have been performed to examine how the physical parameters that characterize the random scattering medium influence scattered surface-wave energy and the energy conversion between scattered surface wave and body wave.

The numerical simulation results show that the scattered wave energy can be well approximated by body to body-wave scattering at earlier arrival times and at short distances. As distance from the source increases, scattered surface waves become important. The scattered surface waves will dominate even earlier coda when an efficient body wave to surface-wave conversion mechanism is present in the shallow layers, for example, in the soft sediment layer. Since surface waves are basically 2-D propagation waves, theoretically their scattered energies should follow the same constant decay curve (SATO, 1993). The observed common decay trends in seismic coda of local earthquake recordings suggest that later seismic codas could be dominated by scattered surface waves. Thus our result provides a useful physical explanation for the widely used coda normalization scheme based on the observation. The numerical simulation also indicates that shallow sources are more effective in converting scattered body waves to scattered surface waves than that for deep sources. This could provide a useful theoretical basis for discrimination between earthquakes and other explosive types of sources such as nuclear explosions. Differences in shallow velocity structures will strongly influence the energy conversion rate between body wave and surface wave. As a result, the content of later seismic coda will differ. The observed difference between site responses estimated using coda and direct S waves may be a consequence of domination of scattered surface waves in the coda. For lower frequencies, surface-wave amplitude increases for deep sources, and the conversion rate from body to surface wave also increases, assuming the same scatterer density distribution. This could provide additional insight on the frequency dependence of Q, particularly in areas of active tectonics.

Still there are phenomena not explained in this study; for example the primary wave pulses broadening effect and the depth-dependent scatterer distribution. Our future studies should extend equations (2a) and (2b) to non-isotropic scattering analysis following SATO (1994b). We should also consider depth-dependent model parameters to understand the effect of depth-dependent scattering, intrinsic attenuation and seismic wave velocity and the role of scattered surface waves to the observed common seismic coda decay.

Acknowledgements

I thank Haruo Sato, Michael Korn, Art Frankel, Mark Petersen, and David Wald for their helpful reviews of the manuscript. I thank Feng Su for her many constructive suggestions. I am also grateful to Haruo Sato and Michel Campillo for their helpful discussions during the poster section of the 2004 AGU meeting. Considerable research work was done by the author while a faculty member at the Seismological Laboratory, University of Nevada at Reno.

REFERENCES

AKI, K. (1969), *Analysis of seismic coda of local earthquakes as scattered waves*, J. Geophys. Res. *74*, 615–631.

AKI, K. (1992), *Scattering conversions P to S vs. S to P*, Bull. Seismol. Soc. Am. *82*, 1969–1972.

AKI, K. and CHOUET, B. (1975), *Origin of coda waves: Source, attenuation and scattering effects*, J. Geophys. Res. *80*, 3322–3342.

AKI, K. (1980), *Attenuation of shear-waves in the lithosphere for frequencies from 0.05 to 25 Hz*, Phys. Earth Planet. Inter. *21*, 50–60.

AKI, K. and RICHARDS, P. G., *Quantitative Seismology: Theory and Methods* (W. H. Freeman, San Francisco 1980).

BONILLA, L.F., STEIDL, J.H., LINDLEY, G.T., TUMARKIN, A.G., and ARCHULETA, R. J. (1997), *Site amplification in the San Fernando Valley, California: Variability of site-effect estimation using the S-wave, coda and H/V methods*, Bull. Seismol. Soc. Am. *87*, 710–730.

CHEN, X. F. (1993), *A systematic and efficient method of computing normal modes for multi-layered half space*, Geophys. J. Int. *115*, 391–409.

FEHLER, M., HOSHIBA, M., SATO, H., and OBARA, K. (1992), *Separation of scattering and intrinsic attenuation for the Kanto-Tokai Region, Japan, using measurements of S-wave energy versus hypocentral distance*, Geophys. J. Int. *108*, 787–800.

FEHLER, M., SATO, H., and HUANG, L-J, (2000), *Envelope broadening of outgoing waves in 2-D random media: A comparison between the Markov approximation and numerical simulations*, Bull. Seismol. Soc. Am. *90*, 914–928.

FRANKEL, A. and CLAYTON, R. W. (1986), *Finite difference simulations of seismic scattering: Implications for the propagation of short-period seismic waves in the crust and models of crustal heterogeneity*, J. Geophys. Res. *91*, 6465–6489.

GAO, L. S., LEE, L. C., BISWAS, N. N., and AKI, K. (1983a), *Comparison of the effects between single and multiple scattering on coda waves for local earthquakes*, Bull. Seismol. Soc. Am. *73*, 377–390.

GAO, L. S., BISWAS, N. N., LEE, L. C., and AKI, K. (1983b), *Effects of multiple scattering on coda waves in a three-dimensional medium*, Pure Appl. Geophys. *121*, 3–15.

GUSEV, A. A. and ABUBAKIROV, I. R. (1996), *Simulated envelopes of nonisotropically scattered waves as compared to observed ones: New evidence for fractal heterogeneity*, Geophys. J. Int. *127*, 49–60.

HOSHIBA, M. (1994), *Simulation of coda wave envelope in depth-dependent scattering and absorption structure*, Geophys. Res. Lett. *21*, 2853–2856.

HOSHIBA, M. (1995), *Estimation of nonisotropic scattering in western Japan using coda wave envelopes: Application of a multiple nonisotropic scattering model*, J. Geophys. Res. *100*, 645–657.

KOPNICHEV, Y. F. (1977), *The role of multiple scattering in the formation of a seismogram's tail*, Izv. Acad. Sci., USSR, Phys. Solid Earth *13*, 394–398.

MARGHERITI, L., WENNERBERG, L., and BOATWRIGHT, J. (1994), *A comparison of coda and S-wave spectral ratio estimates of site response in the southern San Francisco Bay area*, Bull. Seismol. Soc. Am. *84*, 1815–1830.

MARGERIN, L., CAMPILLO, M., and VAN TIGGELEN, B. A. (1998), *Radiative transfer and diffusion of waves in a layered medium: New insight into coda Q*, Geophys. J. Int. *134*, 596–612.

MARGERIN, L., CAMPILLO, M., and VAN TIGGELEN, B. A. (2000), *Monte-Carlo simulation of multiple scattering of elastic waves*, J. Geophy. Res. *105*, 7873–7892.

MAYEDA, K., KOYANAGI, S., HOSHIBA, M., AKI, K., and ZENG, Y. (1992), *A comparative study of scattering, intrinsic and coda 1/Q for Hawaii, Long Valley and Central California between 1.5 and 15.0 Hz*, J. Geophy. Res. *97*, 6643–6660.

SATO, H. (1977), *Energy propagation including scattering effects, single isotropic scattering approximation*, J. Phys. Earth *25*, 27–41.

SATO, H. (1993), *Energy Transportation in one- and two-dimensional scattering media: Analytic solutions of the multiple isotropic scattering model*, Geophys. J. Int. *112*, 141–146.

SATO, H. (1994a), *Multiple isotropic scattering model including P-S conversions for the seismogram envelope formation*, Geophys. J. Int. *117*, 487–494.

SATO, H. (1994b), *Formulation of the multiple non isotropic scattering process in 2-D space on the basis of energy transport theory*, Geophys. J. Int. *117*, 727–732.

SATO, H. and NOHECHI, M. (2001), *Envelope formation of long-period Rayleigh waves in vertical component seismograms: Single isotropic scattering model*, J. Geophys. Res. *106*, 6589–6594.

SNIEDER, R. (1986), *The influence of topography on the propagation and scattering of surface waves*, Phys. Earth Planet. Interiors *44*, 226–241.

SNIEDER, R. (1988), *The optical theorem for surface waves and the relation with surface wave attenuation*, Geophys. J. *95*, 293–302.

SNIEDER, R., *Scattering of surface waves. In Scattering and Inverse Scattering in Pure and Applied Science* (eds. Pike, R. and P. Sabatier) (Academic Press, San Diego 2002) pp. 562–577.

SU, F., ANDERSON, J., BRUNE, J., and ZENG, Y. (1996), *A comparison of direct S-wave and coda wave site amplification determined from aftershocks of the Little Skull Mountain earthquake*, Bull. Seismol. Soc. Am. *86*, 1006–1018.

WU, R. S. (1985), *Multiple scattering and energy transfer of seismic waves — separation of scattering effect from intrinsic attenuation, I. Theoretical modeling*, Geophys. J. *82*, 57–80.

WU, R. S. and AKI, K. (1988), *Multiple scattering and energy transfer of seismic waves — Separation of scattering effect for intrinsic attenuation, II. Application of the theory to Hindu Kush Region*, Pure Appl. Geophys. 128, 49–80.

ZENG, Y., SU, F., and AKI, K. (1991), *Scattered wave energy propagation in random isotropic scattering medium - Part I: Theory*, J. Geophy. Res. *96*, 607–619.

ZENG, Y., *Deterministic and stochastic modelling of the high frequency seismic wave generation and propagation in the lithosphere*, Ph.D. Thesis (University of Southern California, Los Angeles, California 1991).

ZENG, Y. (1993), *Theory of scattered P-and S-wave energy in a random isotropic scattering medium*, Bull. Seismol. Soc. Am. *83*, 1264–1276.

(Received April 21, 2005; accepted September 29, 2005)

Pure appl. geophys. 163 (2006) 549–566
0033–4553/06/030549–18
DOI 10.1007/s00024-005-0031-9

Pure and Applied Geophysics

Constituents of Vertical-component Coda Waves at Long Periods

TAKUTO MAEDA,[1] HARUO SATO,[1] and MASAKAZU OHTAKE[2]

Abstract—AKI (1969) first modeled coda waves of a local earthquake as a superposition of scattered surface waves. This paper attempts to clarify the constituents of surface-wave coda at long periods at very long lapse times. For a large earthquake of magnitude 7 or larger, vertical component oscillation in periods from 90 to 180 s persists for more than 20 hours from the earthquake origin time. Although the early portion of the coda envelope is successfully modeled by assuming incoherent scattered Rayleigh waves by heterogeneities distributed all over the Earth, the later potion of the observed coda envelope (roughly later than 35,000 s) has systematically larger amplitude than theoretical prediction. To clarify the cause of this discrepancy, we studied the constituents of vertical-component seismograms of three large earthquakes recorded by the F-net in Japan using the *f-k* power spectral analysis. We found that the direct and scattered fundamental-mode Rayleigh waves of velocity about 3.7 km/s are dominant in the earlier part of each envelope. It justifies the use of a scattering model of the fundamental Rayleigh waves for synthesizing the envelope. At lapse times later than 20,000 s–35,000 s, higher modes with phase velocities around 20 km/s become dominant. The transition time to the dominance of higher modes is found to become earlier for a deeper focus earthquake. The small coda attenuation factor from $(1.90 \pm 0.23) \times 10^{-3}$ to $(2.38 \pm 0.32) \times 10^{-3}$ estimated from later coda envelopes recorded at IRIS stations distributed worldwide also agrees with the attenuation factor of spheroidal modes according to PREM. We may interpret that higher mode waves are uniformly distributed at large lapse time due to large velocity dispersion and/or scattering and they dominate over the fundamental mode waves because of smaller attenuation in the lower mantle. The coda attenuation measurement proposed by Aki is found to be useful even for long periods and at very large lapse times.

Key words: Rayleigh waves, coda, heterogeneity, seismic array, scattering, attenuation, global seismology.

1. Introduction

Following direct-wave arrivals in seismograms of local earthquakes, we observe wave trains of which the amplitude decreases smoothly with increasing lapse time. AKI (1969) first modeled these coda waves as a superposition of surface waves scattered by distributed random inhomogeneities in the earth medium. Subsequently, AKI and CHOUET (1975) modeled coda waves as a superposition of single-backscattered waves, and concluded that coda waves of period longer than 1 s are

[1]Department of Geophysics, Graduate School of Science, Tohoku University, Sendai, Japan
[2]Tohoku University, Sendai, Japan

mainly composed by scattered surface waves from their weak attenuation in the Tsukuba region (Japan). KOPNICHEV (1975) also proposed a single isotropic scattering of surface waves for the synthesis of coda envelopes. Array analysis of coda waves gave evidence that coda waves are composed of randomly scattered waves (*e.g.*, SCHERBAUM *et al.*, 1991; DAINTY and TOKSÖZ, 1990; MENKE *et al.*, 1990). Recent development has come from the field of radiative transfer theory (*e.g.*, SATO and FEHLER, 1998). In this paper, we attempt to examine the origin of coda waves in 90–180 s, which is an extraordinary long-period band compared to previous studies on coda waves.

Figure 1 shows a vertical component velocity seismogram of the 1999 Izmit earthquake (M_W = 7.6) for the period band of 90–180 s recorded at station YAK (Δ = 58.92°) belonging to the IRIS/GSN. The bottom and top trace are velocity waveform and its Mean-Squared (MS) envelope in the logarithmic scale representation which corresponds to the energy density of seismic waves, respectively. Multiple arrivals of direct Rayleigh waves are clearly observed as multiple peaks in the MS envelope, as denoted by indices from R1 to R5. In addition to these signals, we observe several characteristic features in the coda decay curve. First, the wave train has significant amplitude between multiple arrivals of Rayleigh waves. The amplitude between phases R1 and R2 is about three orders of magnitude smaller than the direct Rayleigh-wave amplitudes, but it is considerably larger than the usual noise level denoted by a horizontal arrow in Figure 1. Second, the coda oscillation lasts as long as 70,000 s ($\sim$20 hours) in lapse time measured from the origin time of the earthquake. Third, there seems to exist a bend in the envelope decay at around 35,000 s into the lapse time.

To understand the nature of surface wave propagation through the inhomogeneous Earth structure, there have been many theoretical developments based on the Born approximation and its extensions (SNIEDER, 1986; FRIEDERICH *et al.*, 1993; SPETZLER *et al.*, 2002) mainly focusing on the phase change due to scattering. However, these studies paid little attention to the amplitude of scattered surface waves, especially at long lapse time.

On the other hand, we have developed a model in our previous studies (SATO and NOHECHI, 2001; MAEDA *et al.*, 2003) to explain the whole MS envelope by transplantation of the radiative transfer theory for body waves in the Euclidean space to the single-scattering process of fundamental-mode Rayleigh waves on the spherical surface of the Earth. It successfully explained the observed envelope of vertical component seismogram up to about 35,000 s in lapse time. Applying the model to observed seismograms, they estimated the total scattering coefficient $g_0 \approx 2 \times 10^{-6}\,\mathrm{km}^{-1}$, which is a phenomenological parameter that stochastically characterizes the strength of scattering of the crust and upper mantle. However, the above single scattering model cannot explain coda amplitude observed at lapse times longer than 35,000 s. The observed MS amplitudes are systematically larger than the theoretical MS envelopes predicted by the single scattering model. This discrepancy

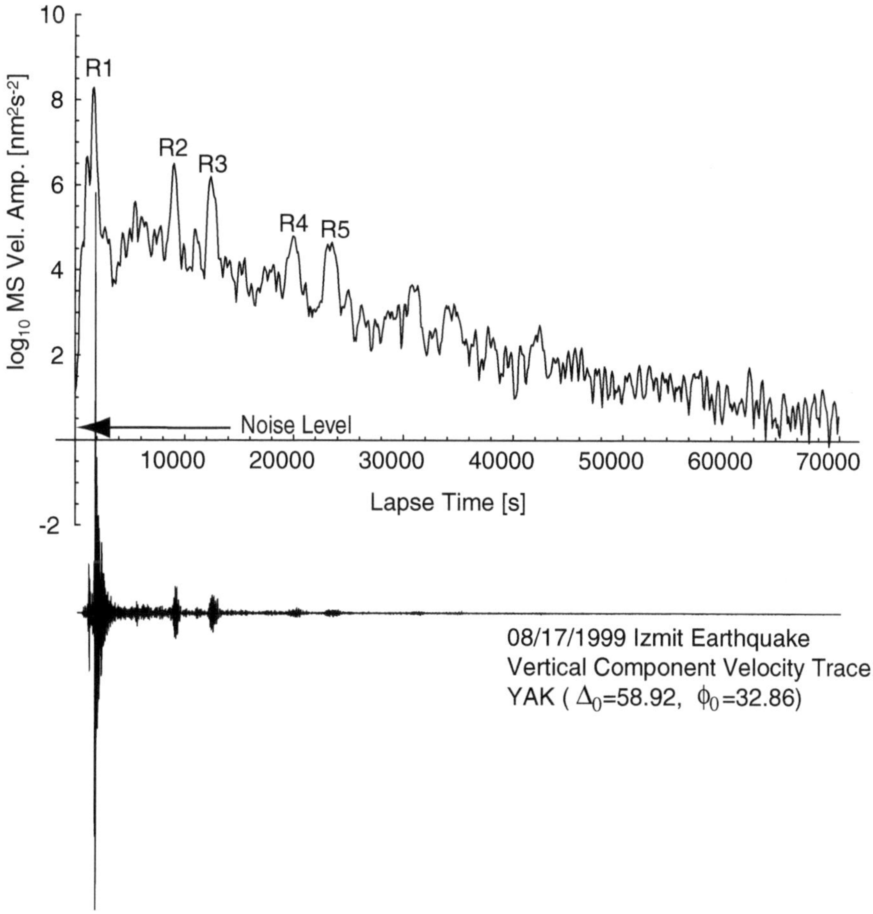

Figure 1
(Bottom) Vertical-component velocity seismogram of the 1999 Izmit earthquake (M_w = 7.4) for the period band 90–180 s recorded at station YAK. (Top) Logarithmic plot of its MS envelope. Multiple arrivals of fundamental-mode Rayleigh waves are denoted as R1 to R5.

increases with increasing lapse time even if multiple scattering effects are considered (SATO and NISHINO, 2002). The departure of theoretical envelope from observed envelope suggests a possibility that there might exist a contribution of other types of waves, especially for the later coda. This is similar to what is observed in local and regional coda studies.

In this paper we perform a frequency-wavenumber (*f-k*) power spectral analysis of seismograms of large earthquakes recorded by the F-net broadband seismic network in Japan to clarify the constituents of the Rayleigh-wave coda. Then we measure the attenuation factor from the later coda observed at globally distributed

IRIS broadband stations. Comparing the phase velocity obtained by *f-**k*** power spectral analysis and attenuation factor obtained by the coda measurement with those predicted by PREM, we investigate the constituents of vertical-component seismograms at long periods.

2. Frequency-wavenumber Power-spectral Analysis

2.1. Method

We measure constituents of vertical-component seismograms at long periods by estimating the arrival directions and slowness (or phase velocity) using the sliding window *f-**k*** power spectral analysis (ROST and THOMAS, 2003). It is useful for discriminating predominant modes having different phase velocities at each lapse time. For the estimate of the power spectrum, we use Capon's method (CAPON, 1969; JOHNSON and DUDGEON, 1993) to minimize the contamination of the array response to the estimated power. We estimate the *f-**k*** power spectrum for the period of 128 s, which is the center of the period band (90–180 s) of this study. The time window length is 512 s and the sliding step is 64 s. In each time window, we estimate the *f-**k*** power spectrum in slowness space, where the estimate of slowness ranges from 0.02 s/km to 0.5 s/km with grid spacing of 0.005 s/km and 2° in azimuthal angle.

2.2. Data

We investigate three large earthquakes of different locations and mechanisms: (A) southeastern off Taiwan earthquake of M_W = 7.4 (May 3, 1998, Lat. = 22.37°N, Long. = 125.53°E, Depth = 22.9 km); (B) Izmit, Turkey earthquake of M_W = 7.6 (August 17, 1999, Lat. = 41.01°N, Long. = 29.97°E, Depth = 17.0 km); and (C) Hindu-Kush earthquake of M_W = 7.6 (March 3, 2002, Lat. = 36.57°N, Long. = 70.42°E, Depth = 228.5 km). Figures 2(a) and 2(c) show event locations and source mechanisms from Harvard's Centroid Moment Tensor (CMT) solution, respectively. These events commonly have very long coda duration of more than 20 hours in lapse time. In the following, these three events are referred to as the Off-Taiwan, Izmit, and Hindu-Kush earthquakes, respectively.

We select vertical-component velocity seismograms of 1 s sampling (LHZ component) from the F-net in Japan operated by NIED (FUKUYAMA *et al.*, 1996). Data with considerable noise or those including records of local earthquakes are excluded. Though F-net stations are distributed all over Japan, we select 12 stations equipped with STS-1 seismometers located in the central part of Japan to avoid the influence of Earth curvature (see Fig. 2b). The average station separation of this array and the maximum aperture of the array are about 150 km and 700 km, respectively. Waveform data are used for the *f-**k*** power spectrum estimate after correcting for the instrumental response.

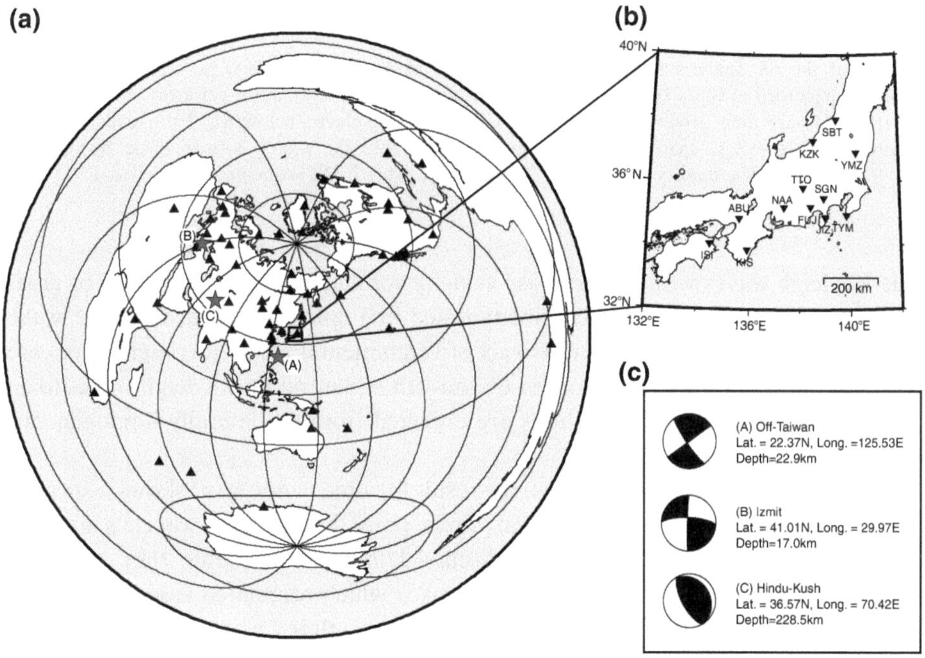

Figure 2
(a) Plots of earthquake epicenters (stars) used for the analysis and IRIS stations (triangles) used for the coda decay analysis. (b) F-net array configuration used for the *f-k* power spectral analysis. (c) Source mechanisms of three events.

2.3. Results

Figure 3 shows estimated *f-k* power spectra for the case of the Izmit earthquake. The top panel (a) shows a vertical component velocity trace recorded at the TTO station that is located at the center of the F-net array. The lower panels (b) show density plots of estimated *f-k* power spectra in slowness space for time windows denoted by gray hatches in the velocity trace in (c). In each panel, the estimated spectrum is normalized by the maximum value of the spectrum. Time window (i) includes the arrival of direct Rayleigh waves (R1). Estimated *f-k* power spectrum has a strong peak in the northwest direction, in good agreement with the back azimuth of the epicenter from the array, which is represented by a white line. The spectral peak shows a phase velocity of 4 km/s, which is also in good agreement with the average phase velocity of the fundamental mode Rayleigh wave at this period. These results demonstrate that the *f-k* power spectral analysis can correctly detect the arrival of Rayleigh waves of the fundamental mode. The *f-k* power spectrum for time window (iii) also has a strong peak. However, the arrival direction is opposite, and its phase velocity is about 7 km/s. This observation can be interpreted as an arrival of higher-

▶

Figure 3
Examples of the *f-k* analysis for the Izmit earthquake at a period of 128 s. (a) Vertical-component seismogram recorded at the TTO station (F-net) located at the center of the F-net array. Hatched areas from (i) to (iv) are time windows used for the *f-k* analysis displayed below. (b) Estimated *f-k* power spectrum in the slowness space for each time window, where the power is normalized by its peak amplitude. The density scale of the *f-k* power spectrum in dB is shown at the bottom.

mode Rayleigh waves with larger phase velocity coming along the major arc of great circle path denoted as X in the velocity trace in Figure 3(a). Indeed, these wave packets appearing between direct arrivals of fundamental mode Rayleigh waves are known as mantle X-waves that consist of 2nd–5th higher mode Rayleigh waves (*e.g.*, TANIMOTO, 1987). These observations are expected from the laterally homogeneous Earth structure.

In contrast to the above two cases, the result for time window (ii) shows a spectral peak strongly deviated from the great circle path. Since these wavelets are not attributed to other earthquakes as is discussed in a later section, they are most probably interpreted as scattered Rayleigh waves which originated from topographical changes and/or medium heterogeneities. Many scattered wave packets that have oblique arrival directions are observed between multiple arrivals of fundamental and higher-mode Rayleigh waves for all the three events. The existence of these scattered waves allows the use of a scattering model to describe the envelope as a first approximation, at least for the earlier part of the MS envelope.

However, in the later portion of seismogram, spectral peaks tend to have a higher phase velocity. Time windows (iv) in Figure 3(b) show the *f-k* power spectrum at 41,024 s in lapse time as an example. A clear coherent peak with the phase velocity around 20 km/s appears near the center of the slowness space. This phase velocity is much higher than that of mantle X-waves. Such a spectral peak with a large phase velocity is observed frequently in the later portion of coda waves for all three earthquakes.

Figure 4 shows a temporal change of the dominant phase velocity for the Izmit earthquake. Performing the sliding window *f-k* power spectral analysis, we plot a radial section of the normalized *f-k* power spectrum containing the origin and the peak spectrum in the slowness space for each time window, where bright color indicates stronger power spectrum. Multiple peaks of about 4 km/s in phase velocity appear at lapse times earlier than about 35,000 s. They correspond to multiple arrivals of fundamental mode Rayleigh waves. The estimated phase velocity agrees well with the dispersion curve of the fundamental mode predicted by PREM (DZIEWONSKI and ANDERSON, 1981). As lapse time increases, the dominant phase velocity becomes higher than that of the fundamental mode. Alternatively, continuous spectral peaks with phase velocities around 20 km/s as shown in Figure 3(d) often appear at lapse times exceeding than about 35,000 s. Hereafter we

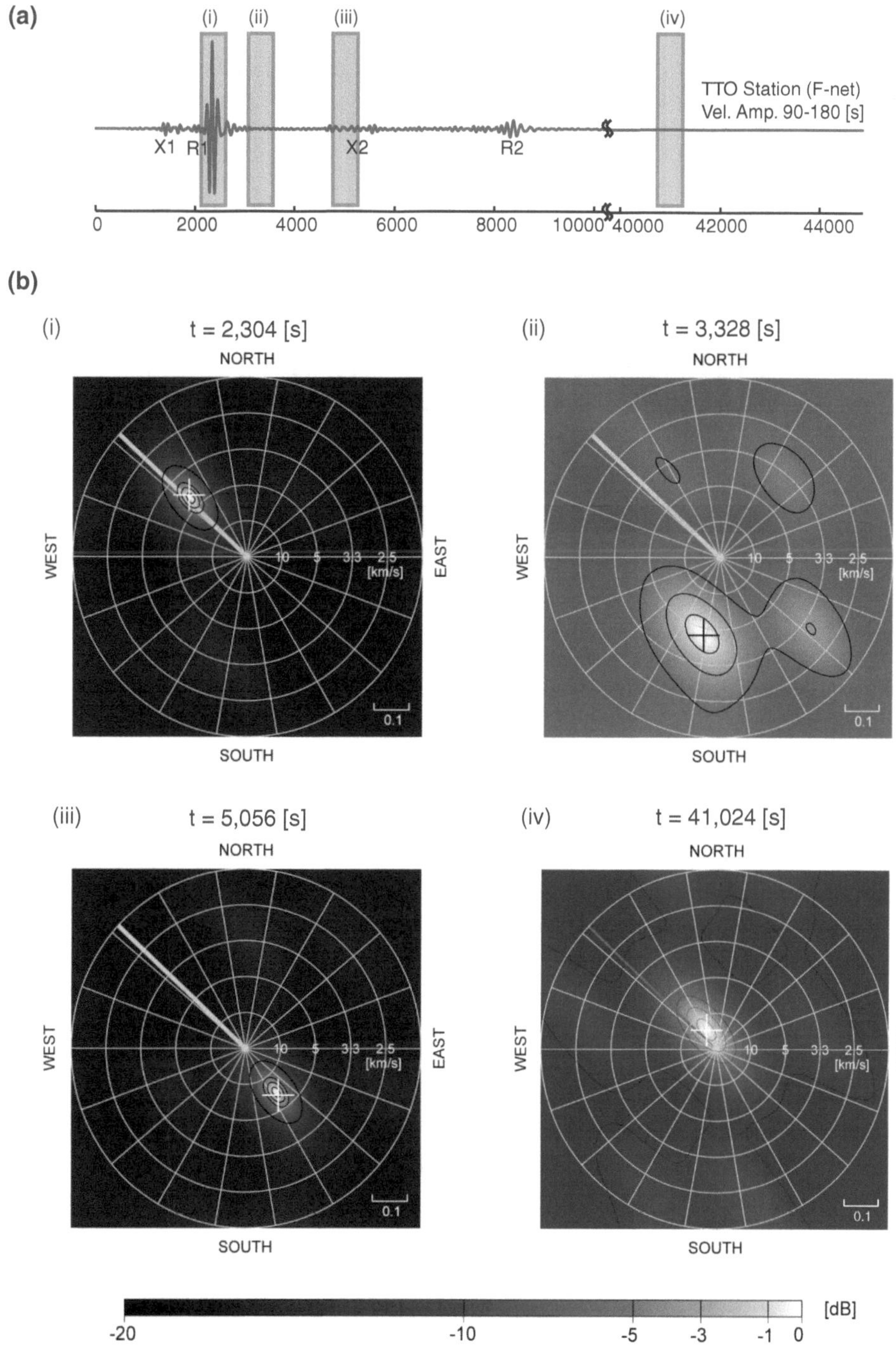
(a)
(i)
(ii)
(iii)
(iv)
TTO Station (F-net)
Vel. Amp. 90-180 [s]
X1
R1
X2
R2
0
2000
4000
6000
8000
10000
40000
42000
44000
(b)
(i) t = 2,304 [s]
(ii) t = 3,328 [s]
(iii) t = 5,056 [s]
(iv) t = 41,024 [s]
NORTH
SOUTH
WEST
EAST
10
5
3.3
2.5
[km/s]
0.1
[dB]
-20
-10
-5
-3
-1
0

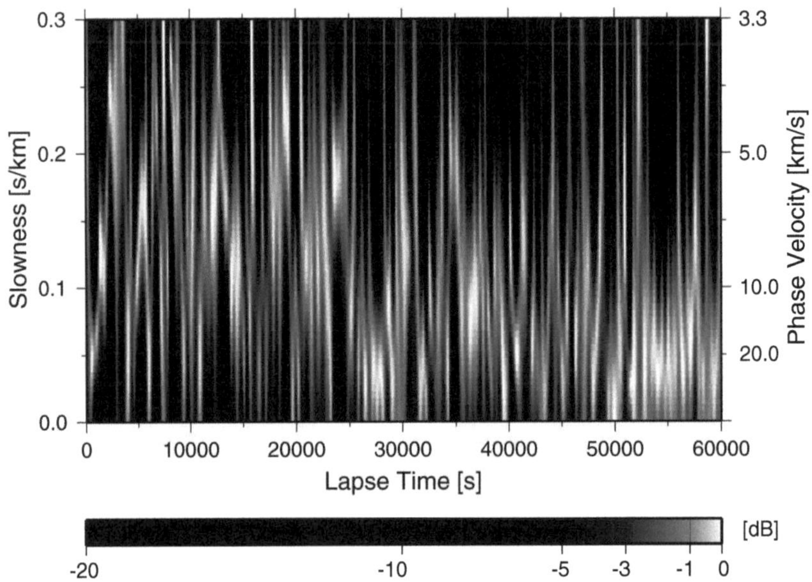

Figure 4
Temporal change in the dominant phase velocity estimated from the *f*-***k*** power spectral analysis for the Izmit earthquake at a period of 128 s. Normalized *f*-***k*** power spectrum along the radial line containing the maximum power in slowness space is plotted for every 512 s. Density scale in dB is given at the bottom.

refer to the time when the dominant mode changes as transition time T_0. In the case of the Izmit earthquake, $T_0^{\mathrm{Izmit}} \approx 35,000$ s.

Figure 5 displays dispersion curves of phase velocity predicted by PREM for periods from 50 s to 200 s for the fundamental mode to the 15th higher mode spheroidal oscillations. Large phase velocities mean that the constituents of the later coda are higher-mode spheroidal oscillations. Phase velocity of 20 km/s means that the degrees of modes are around 10th in this period band.

Figures 6(a) and 6(b) show temporal changes of the dominant phase velocity for the Off-Taiwan and Hindu-Kush earthquakes. Because of larger phase velocities, the dominance of higher modes in the later portion of the coda is common to all three events. However, the transition time $T_0^{\mathrm{Hindu}} \approx 20,000$ s from the fundamental mode to the higher mode is relatively earlier for the Hindu-Kush event, which has a deeper focus than the other two events. For the Off-Taiwan event with shallow focal depth, the transition time $T_0^{\mathrm{Taiwan}} \approx 35,000$ s is comparable to that of the Izmit earthquake.

3. *Coda Attenuation and Synthesis of the Entire Envelope*

Observed MS envelopes decay smoothly with increasing lapse time irrespective of observation sites as shown in Figure 7, which is a superimposition of envelopes of the

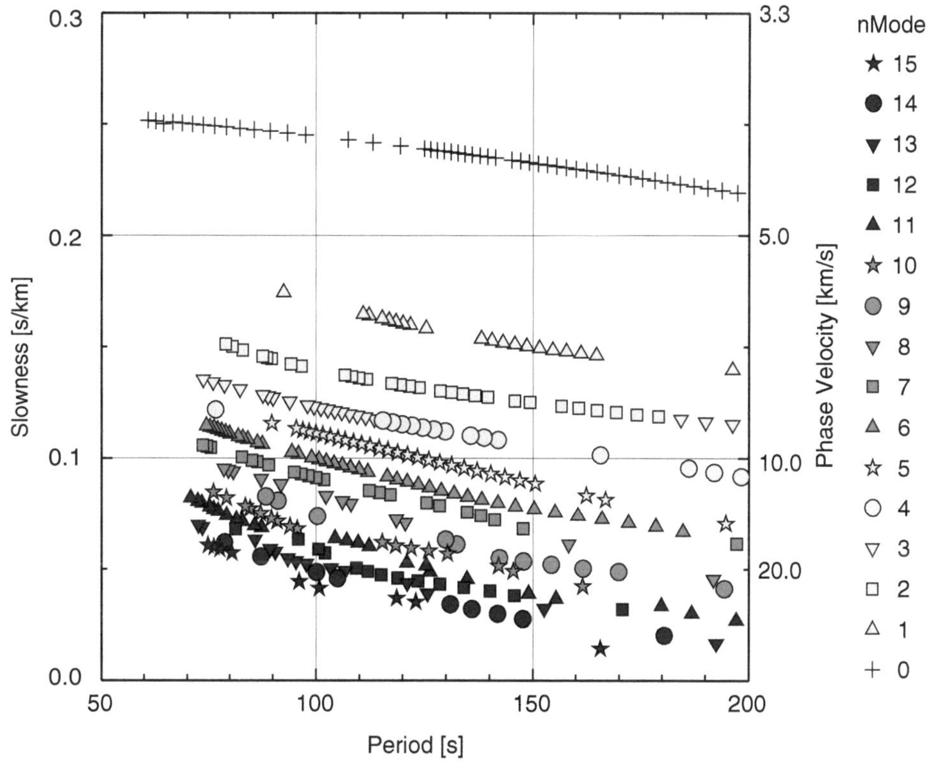

Figure 5
Dispersion curves of phase velocities according to PREM. Fundamental mode and fifteen higher modes of spheroidal oscillations are marked by different symbols.

Izmit earthquake observed worldwide. Therefore, the decay rate of later coda amplitudes could be a characteristic parameter to describe the Earth's heterogeneity and inelasticity since higher mode waves spread over the Earth due to large dispersion of group velocity and/or scattering due to medium inhomogeneities at considerable lapse time. In this section we extract further information about the degree of modes relative to the attenuation structure of the Earth.

If later coda waves are uniformly distributed throughout the Earth, their energy density E^C can be written by using an exponential decay as

$$E^C(t) = A_C e^{-Q_c^{-1}\omega t}, \tag{1}$$

where Q_c^{-1} is the attenuation factor of coda waves. Different from conventional coda decay, (see AKI and CHOUET, 1975) any geometrical decay factor is not introduced for the evaluation of Q_c^{-1} since we have assumed a uniform distribution of coda waves worldwide. We estimate Q_c^{-1} by applying the least-squares method to logarithmic MS

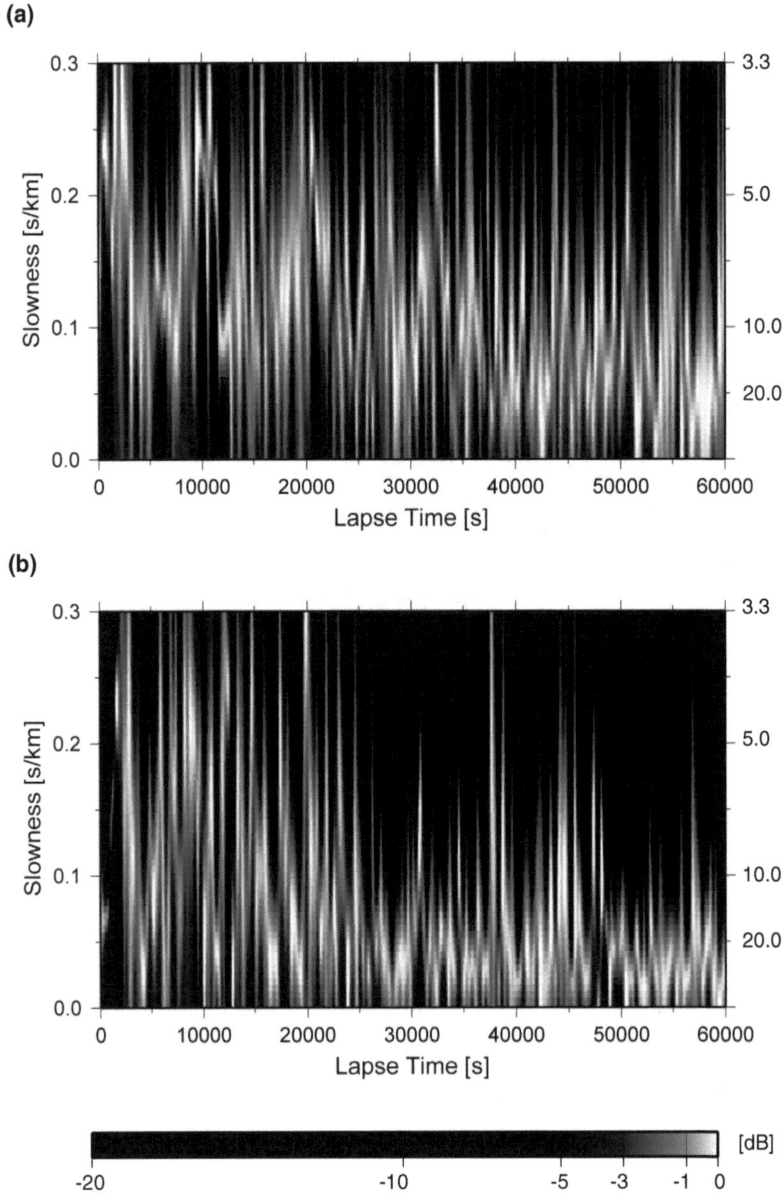

Figure 6
Temporal change in the dominant phase velocity estimated from *f*-***k*** spectral analysis at a period of 128 s for (a) Off-Taiwan event and (b) Hindu-Kush event (see Fig. 5).

envelopes recorded at 61 stations for 35,000–60,000 s in lapse time for the period band of 90–180 s. Data contaminated by noise and other earthquakes are excluded. The stations used for the coda attenuation analysis are plotted by triangles in Figure 2(a).

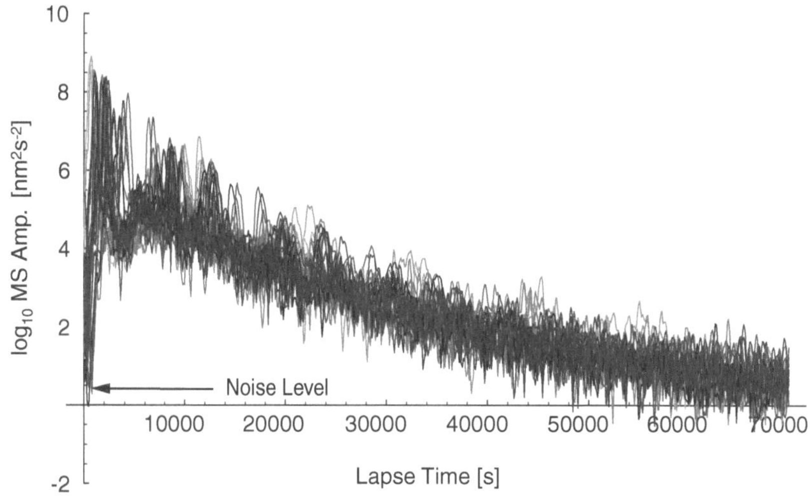

Figure 7
Gathered logarithmic plots of MS envelopes recorded at 24 IRIS stations for periods from 90 s to 180 s for the Izmit earthquake.

The average and the standard deviations of Q_c^{-1} are estimated as $(2.38 \pm 0.30) \times 10^{-3}$ for the Off-Taiwan earthquake, $(2.38 \pm 0.32) \times 10^{-3}$ for the Izmit earthquake, and $(1.90 \pm 0.23) \times 10^{-3}$ for the Hindu-Kush earthquake, respectively. Two shallow strike-slip events (Off-Taiwan and Izmit earthquakes) have the same coda attenuation factor, while a deeper event (Hindu-Kush) has a smaller coda attenuation factor, though within the measurement error.

In Figure 8 the frequency dependence of intrinsic attenuation Q^{-1} for the spheroidal modes given by PREM is plotted by different symbols. A dashed line and a gray-hatched area show the mean and $\pm 1\sigma$ region of estimated Q_c^{-1} for the case of the Izmit earthquake. The estimated Q_c^{-1} corresponds to the intrinsic attenuation Q^{-1} of PREM for higher modes of orders 9th to 14th. These modes are in good agreement with those estimated from the *f*-***k*** analysis. The excitation of higher mode oscillation at the source is generally smaller than that of the fundamental mode. However, the fundamental mode waves rapidly decrease due to the larger attenuation as illustrated in Figure 9. Fundamental Rayleigh waves in periods around 100 s are highly sensitive to the structure down to 200 km from the Earth's surface. Eigenfunctions penetrate deeper in the Earth as the degree of mode becomes higher; the 10th higher mode in the period band of this research has sensitivity to the entire mantle (DAHLEN and TROMP, 1998). Therefore, the later coda waves are sensitive to the structure of the deep mantle. The difference between the coda attenuation factor estimated for shallow events and that for a deep focus event might be caused by the difference in higher mode excitation depending on the focal depth and mechanism.

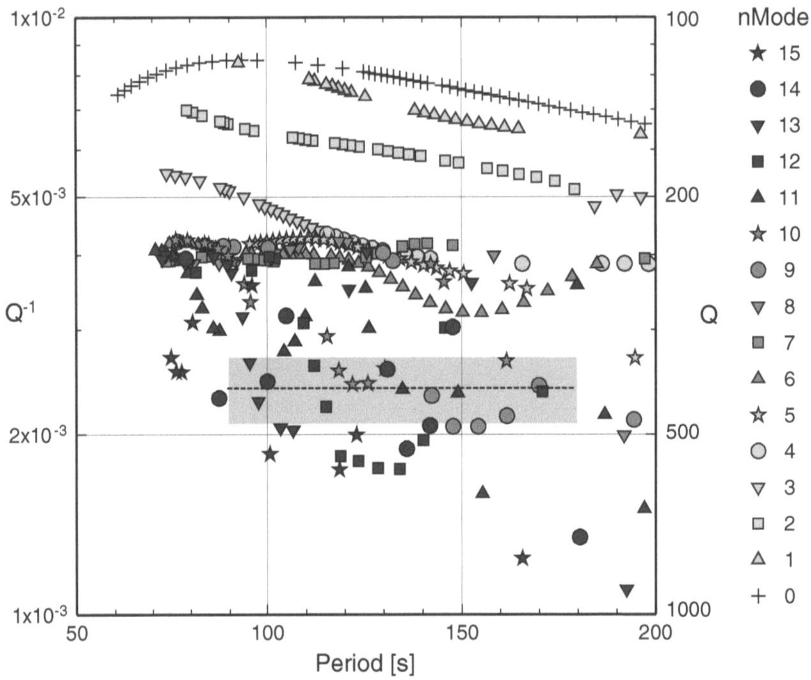

Figure 8
Dispersion curves of Q^{-1} of spheroidal modes given by PREM. Thick dotted line and gray hatched area show the mean Q_c^{-1} value and $\pm 1\sigma$ range estimated from the later-coda analyses of vertical component seismograms of three earthquakes recorded at 61 IRIS stations for the case of the Izmit earthquake.

According to the above weakly-decaying coda envelope, we model the total energy density of vertical oscillation $E(\Delta, \phi, t)$ as

$$E(\Delta, \phi, t) = E^{F0}(\Delta, \phi, t) + E^{FS}(\Delta, \phi, t) + E^{C}(t), \tag{2}$$

where Δ is epicentral distance and ϕ is azimuth. Function $E^{F0}(\Delta, \phi, t)$ is the energy density of the multiple direct arrivals of the fundamental mode Rayleigh waves, $E^{FS}(\Delta, \phi, t)$ is scattered energy density of the fundamental mode, and $E^{C}(t)$ denotes uniformly distributed energy density of higher modes.

We estimate the energy density and total attenuation from the early envelope to synthesize $E^{F0}(\Delta, \phi, t)$ and $E^{FS}(\Delta, \phi, t)$ based on the single isotropic scattering model for a nonisotropic source radiation following the procedure of SATO and NOHECHI (1999) for the case of the Izmit earthquake. For R1 to R4 phases, energy density $E^{F0}(\Delta, \phi, t)$ is estimated by taking the time integration. The integration for each phase is performed for a time window in which the MS amplitude is larger than 10% of the maximum MS amplitude of each phase. The time window length increases with later arrivals because of dispersion. Total attenuation factor Q^{-1} is estimated by the average of the ratio of energy of R1 to energy of R3, and that of R2 to R4. The

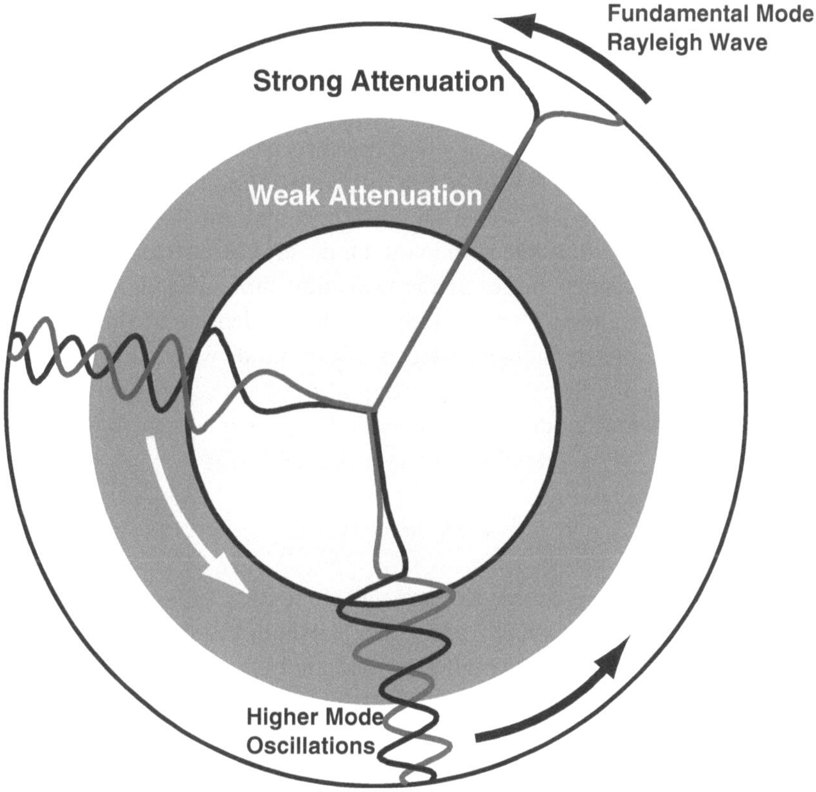

Figure 9
Cartoon illustration of the propagation of fundamental-mode and higher-mode oscillations and the depth-dependent attenuation structure.

total attenuation factor Q^{-1} includes both intrinsic and scattering attenuation. Then we estimate the total radiated energy W from time-integrated energy by correcting the total attenuation, geometrical spreading, and the radiation pattern of the source. A strike-slip source is assumed for the source radiation of fundamental-mode Rayleigh waves in the case of the Izmit earthquake. Averaging for all stations, we finally derive $W = 6.28 \times 10^{15}$ nm^2 km^2 s^{-2} and $Q^{-1} = 7.82 \times 10^{-3}$. The estimation of the attenuation factor of the fundamental mode Rayleigh wave is supported by a good agreement with Q^{-1} of the fundamental surface wave from PREM and a classic report by KANAMORI (1970). By using the value of total attenuation of the fundamental-mode Rayleigh waves Q^{-1} and the total energy radiation from the source W, we can synthesize the single scattered MS envelope for various values of total scattering coefficient g_0. Using a grid search, we estimate $g_0 = 2.0 \times 10^{-6}$ km^{-1} for the case of the Izmit earthquake.

Figure 10 shows a comparison between the observed MS envelope and the synthetic model at station YAK (thin solid curve). A thin solid curve shows the observed MS envelope. A thick solid curve shows the envelope predicted by the single isotropic scattering model $E^{F0} + E^{H}$ for fundamental-mode Rayleigh waves, where the time width of multiple direct arrivals in E^{F0} increases with lapse time due to dispersion of the group velocity. The gray straight curve shows E^{C} for the later coda according to (1), where scale factor A_C is allowed to be free since the excitation of higher modes is unknown. As is shown, the single scattering model and the uniform exponential decay well explain the early envelope and the later coda envelope, respectively. We can see that the dominant mode in MS envelope changes from fundamental mode to higher modes at the transition time around $T_0 \approx 35,000$ s in lapse time. After the transition time, predicted energy density of the fundamental Rayleigh waves is completely hidden under the higher mode waves. Earlier transition time for the case of the Hindu-kush earthquake indicates relatively diminished excitation of fundamental-mode Rayleigh waves from such a deep source.

4. Discussions

In this paper we analyzed seismogram envelopes for lapse times as long as ~20 hours. Since many aftershocks and other earthquakes occurred in that time window, some of them might have affected our analysis. Therefore, we examined earthquakes that occurred within 20 hours from the origin time of each event

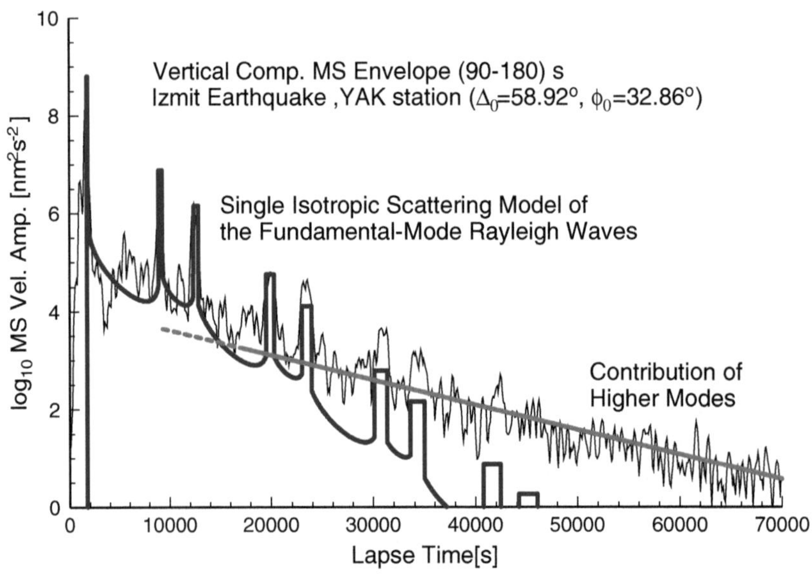

Figure 10
Observed MS envelope (thin black curve) at the YAK station (IRIS) and the best-fit single scattering MS envelope $E^{0} + E^{FS}$ (thick black curve), and the best-fit uniform exponential decay function for higher mode oscillation E^{H} (thick gray line) for the case of the Izmit earthquake.

according to the NEIC catalogue. Figure 11 displays the epicenter distribution and M-T diagram of those earthquakes for each large event we studied. They include both aftershocks in the vicinity of each mainshock epicenter and other small earthquakes that occurred globally. The largest magnitude was $M = 5.2$ for the Izmit and the Hindu-Kush events, and $M = 5.1$ for the Off-Taiwan event. We examined observed amplitudes of $M = 5$ events that occurred globally from 1998 to 2000 to clarify the contribution of these earthquakes to the MS envelopes of large earthquakes. We found that the maximum MS amplitude of a typical $M = 5$ event for the period band of 90–180 s did not exceed the typical noise level 10^1 nm^2/s^2, irrespective of source-receiver distance. Therefore, we conclude that these small earthquakes do not affect the coda attenuation analysis in this period band. Since the *f*-***k*** power spectrum detects a coherent wave packet with maximum amplitude in each time window, such smaller earthquakes (aftershocks and other earthquakes) do not affect our estimate of *f*-***k*** spectral analysis.

In this study we only considered the vertical-component Rayleigh waves, and assumed only Rayleigh-to-Rayleigh wave scattering of the fundamental mode. In general, however, the scattering processes in inhomogeneous mediums also include mode conversion. By incorporating horizontal components, we may clarify the contribution of mode-converted waves in the later coda waves. However, the noise level is generally higher in horizontal components than that in vertical components and thus made it difficult to perform the array analysis of our horizontal component data. We will address this issue in future work.

5. *Conclusions*

We found a dominance of spheroidal higher mode oscillation, especially in the late coda of vertical component seismograms at long periods from *f*-***k*** power spectral analysis and coda attenuation measurements.

By performing an *f*-***k*** power spectral analysis of three large earthquakes, we examined the constituents of the vertical-component seismogram up to 60,000 s in lapse time for 90–180 s period band. Though direct and scattered Rayleigh waves of the fundamental mode are found in early envelopes, higher modes with phase velocities around 20 km/s become dominant at lapse times later than 20,000–35,000 s. The degrees of these modes are estimated from 9 to 13 by comparing to PREM.

From the coda decay analysis, the attenuation factor of later coda waves (Q_c^{-1}) ranges between $(1.90 \pm 0.23) \times 10^{-3}$ and $(2.38 \pm 0.30) \times 10^{-3}$, depending on the focal depth. These values are almost three times smaller than that of the fundamental mode. These decay rates are in agreement with those of higher mode spheroidal oscillations estimated by the *f*-***k*** power spectral analysis. This coincidence also supports the dominance of higher modes in the later portion of coda waves, which

(a) Off-Taiwan

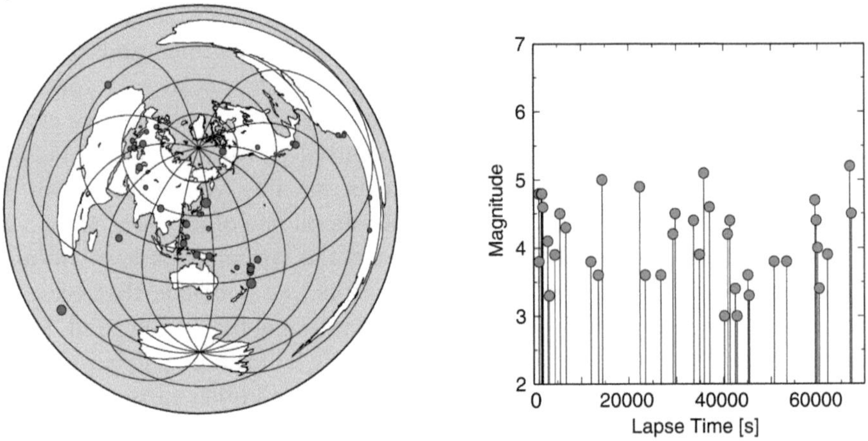

(b) Izmit

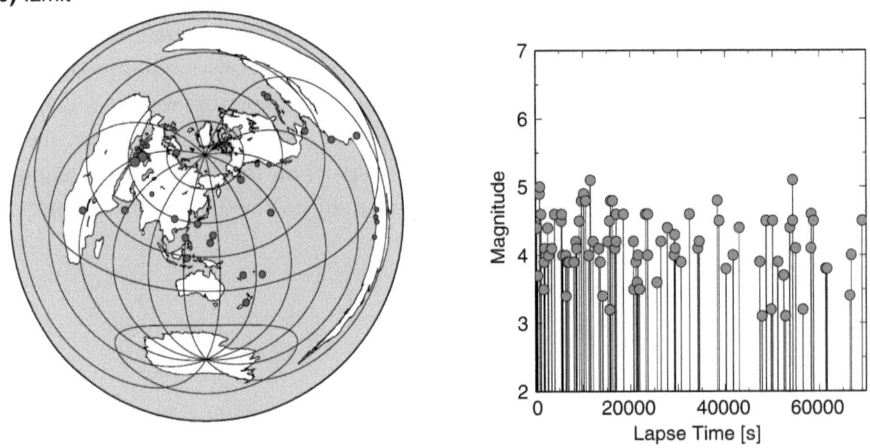

(c) Hindu-Kush

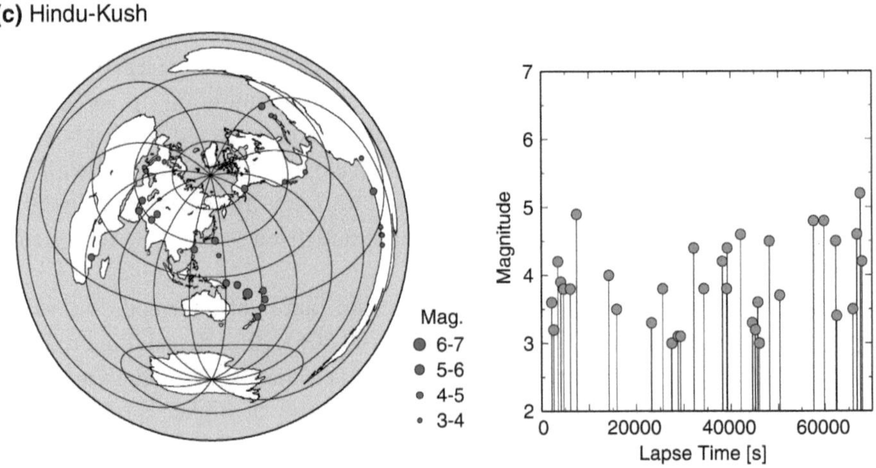

◀

Figure 11
Seismicity and M-T diagram within the 70,000 s from the earthquake origin time: (a) Off-Taiwan, (b) Izmit, and (c) Hindu-Kush earthquake. Events with M ≥ 3 are taken from the NEIC catalog.

was not considered in past studies. The entire MS envelope can be well represented by a summation of the power of scattered fundamental mode Rayleigh waves with strong attenuation and the power of uniformly distributed higher mode oscillation with weak attenuation.

Our results suggest that analyses of the coda at long periods have great potential for inverting medium inhomogeneities in the entire mantle because these coda waves may contain accumulated scattering effect of medium inhomogeneities. We note that this coda attenuation at long periods is a stable observable quantity describing the Earth's inhomogeneity and attenuation characteristics. The coda attenuation measurement first proposed by Aki works well even for long periods at very long lapse times.

Acknowledgements

The authors thank Editor William H. K. Lee and two reviewers, Niren Biswas and Kevin Mayeda, for their careful reviews and valuable comments. We are grateful to the NIED (F-net) and the IRIS consortium that provided us with broadband waveform data. TM is supported by Grant-in-Aid for JSPS Research fellow (No. 07346) and TM and HS are supported by the Earthquake Research Institute cooperative research program (No. 2003-B-04). This study is partially supported by the 21st century COE program "Advanced Science and Technology Center for the Dynamic Earth" in Tohoku University.

References

Aki, K. (1969), *Analysis of seismic coda of local earthquakes as scattered waves*, J. Geophys. Res. *74*, 615–631.

Aki, K. and Chouet, B. (1975), *Origin of coda waves: Source, attenuation, and scattering effects*, J. Geophys. Res. *80*, 3322–3342.

Capon, J. (1969), *High-resolution frequency-wavenumber spectrum analysis*, Proc. IEEE, *57*, 1408–1418.

Dahlen, F.A. and Tromp, J., *Theoretical Global Seismology* (Princeton University Press, Princeton, New Jersey, 1998).

Dainty, A. and Toksöz, M.N. (1990), *Array analysis of seismic scattering*, Bull. Seismol. Soc. Am. *80*, 2242–2260.

Dziewonski, A.M. and Anderson, D.L. (1981), *Preliminary reference Earth model*, Phys. Earth Planet. Interiors *25*, 97–356.

Friederich, W., Wielandt, E., and Stange, S. (1993), *Multiple forward scattering of surface waves: Comparison with an exact solution and Born single-scattering methods*, Geophys. J. Int. *112*, 264–275.

FUKUYAMA, E., ISHIDA, M., HORI, S., SEKIGUCHI, S., and WATADA, S. (1996), *Broadband seismic observation conducted under the FREESIA project*, Report of the National Research Institute for Earth Science and Disaster Prevention *57*, 23–31.

JOHNSON, D.H. and DUDGEON, D.E., *Array Signal Processing: Concepts and Techniques* (Prentice Hall Signal Processing Series, Prentice Hall, New Jersey, 1993).

KANAMORI, H. (1970), *Synthesis of long-period surface waves and its application to earthquake source studies—Kuril islands earthquake of October 13, 1963*, J. Geophys. Res. *75*, 5011–5027.

KOPNICHEV, Y.F. (1975), *A model of generation of the tail of the seismogram*, Dokl. Akad. Nauk SSSR *222*, 333–335.

MAEDA, T., SATO, H., and Ohtake, M. (2003), *Synthesis of Rayleigh-wave envelope on the spherical Earth: Analytic solution of the single isotropic-scattering model for a circular source radiation*, Geophys. Res. Lett. *30*(6), 1286, doi:10.1029/2002GL016629.

MENKE, W., LEERNER-LAM, A.L., DUBENDORFF, B., and PACHECO, J. (1990), *Polarization and coherence of 5 to 30 Hz seismic wavefields at a hard-rock site and their relevance to velocity heterogeneities in the crust*, Bull. Seismol. Soc. Am. *80*, 430–449.

RITZWOLLER, M.H., SHAPIRO, N.M., BARMIN, M.P., and LEVSHIN, A.L. (2002), *Global surface wave diffraction tomography*, J. Geophys. Res. *107* (B12), 2335, doi:10.1029/2002JB001777.

ROST, S. and THOMAS, C. (2003), *Array seismology: Methods and applications*, Rev. Geophys. *40*(3), 1008, doi:10.1029/2000RG000100.

SATO, H. and FEHLER, M., *Seismic Wave Propagation and Scattering in the Heterogeneous Earth* (AIP Press/Springer Verlag, New York, 1998).

SATO, H. and NISHINO, M. (2002), *Multiple isotropic-scattering model on the spherical Earth for the synthesis of Rayleigh-wave envelopes*, J. Geophys. Res. *107*(B12), 2343, doi:10.1029/2001JB000915.

SATO, H. and NOHECHI, M. (2001), *Envelope formation of long-period Rayleigh waves in vertical component seismograms: Single isotropic scattering model*, J. Geophys. Res. *106*, 6589–6594.

SCHERBAUM, F., GILLARD, D., and DEICHMANN, N. (1991), *Slowness power spectrum analysis of the coda composition of two microearthquake clusters in northern Switzerland*, Phys. Earth. Planet. Interiors *67*, 137–161.

SNIEDER, R. (1986), *3-D linearized scattering of surface waves and a formalism for surface wave holography*, Geophys. J. R. astr. Soc. *84*, 581–605.

SPETZLER, J., TRAMPERT, J., and SNIEDER, R. (2002), *The effect of scattering in surface wave tomography*, Geophys. J. Int. *149*, 755–767.

TANIMOTO, T. (1987), *The three-dimensional shear wave structure in the mantle by overtone waveform inversion — I. Radial seismogram inversion*, Geophys. J. R. astr. Soc. *89*, 713–740.

(Received April 27, 2005; accepted September 24, 2005)

Pure appl. geophys. 163 (2006) 567–600
0033–4553/06/030567–33
DOI 10.1007/s00024-005-0034-6

Pure and Applied Geophysics

Temporal Changes of Shallow Seismic Velocity Around the Karadere-Düzce Branch of the North Anatolian Fault and Strong Ground Motion

ZHIGANG PENG,[1,2] and YEHUDA BEN-ZION[1]

Abstract—We analyze temporal variations of seismic velocity along the Karadere-Düzce branch of the north Anatolian fault using seismograms generated by repeating earthquake clusters in the aftershock zones of the 1999 M_w7.4 İzmit and M_w7.1 Düzce earthquakes. The analysis employs 36 sets of highly repeating earthquakes, each containing 4–18 events. The events in each cluster are relocated by detailed multi-step analysis and are likely to rupture approximately the same fault patch at different times. The decay rates of the repeating events in individual clusters are compatible with the Omori's law for the decay rate of regional aftershocks. A sliding window waveform cross-correlation technique is used to measure travel time differences and evolving decorrelation in waveforms generated by each set of the repeating events. We find clear step-like delays in the direct *S* and early *S*-coda waves (sharp seismic velocity reduction) immediately after the Düzce main shock, followed by gradual logarithmic-type recoveries. A gradual increase of seismic velocities is also observed before the Düzce main shock, probably reflecting post-seismic recovery from the earlier İzmit main shock. The temporal behavior is similar at each station for clusters at various source locations, indicating that the temporal changes of material properties occur in the top most portion of the crust. The effects are most prominent at stations situated in the immediate vicinity of the recently ruptured fault zones, and generally decrease with normal distance from the fault. A strong correlation between the co-seismic delays and intensities of the strong ground motion generated by the Düzce main shock implies that the radiated seismic waves produced the velocity reductions in the shallow material.

Key words: Seismic velocity changes, crustal structure, scattering, coda waves, repeating earthquakes, strong ground motion.

1. Introduction

Laboratory experiments and theory indicate that the elastic properties of crustal rocks evolve when the stress-strain conditions are beyond those associated with the linear elastic regime (e.g., MOGI, 1962; SCHOLZ, 1968; JAEGER and COOK, 1976;

[1]Department of Earth Sciences, University of Southern California, Los Angeles, CA, 90089-0740, USA
[2]Department of Earth and Space Sciences, University of California, Los Angeles, CA, 90095-1567, USA E-mail: zpeng@ess.ucla.edu

LOCKNER *et al.*, 1992; LYAKHOVSKY *et al.*, 1997; TEN CATE *et al.*, 2000; HAMIEL *et al.*, 2004; JOHNSON and SUTIN, 2005). Since the stress-strain fields exceed the linear elastic regime at some regions during the occurrence of earthquakes, such changes should be, in principle, measurable through seismological observations. Observing temporal variations of material properties in the seismogenic zone has been a long-sought goal of the geophysics community. In the past, three seismological approaches have been applied to measure temporal changes of crustal properties: travel time of *P* and *S* arrivals or *Vp*/*Vs* ratio (e.g., WHITCOMB *et al.*, 1973), seismic anisotropy (e.g., GUPTA, 1973; CRAMPIN *et al.*, 1990, 1991; CRAMPIN and GAO, 2005), and analysis of coda wave (e.g., AKI and CHOUET, 1975; AKI, 1985; JIN and AKI, 1986). Some studies claimed to observe fairly large temporal changes (e.g., more than 10%) using natural earthquake sources (e.g., GUPTA, 1973; Jin and AKI, 1986; CRAMPIN *et al.*, 1990). However, these claims were not convincing, mainly due to uncertainties associated with earthquake locations and origin times (e.g., MCEVILLY and JOHNSON, 1974; RYALL and SAVAGE, 1974; KANAMORI and FUIS, 1976; ASTER *et al.*, 1990, 1991; BEROZA *et al.*, 1995; HELLWEG *et al.*, 1995; ANTOLIK *et al.*, 1996; LIU *et al.*, 2004, 2005; PENG and BEN-ZION, 2005).

Recent results from controlled source studies have shown considerably smaller but detectable temporal changes (on the order of a few percent) associated with the occurrence of major earthquakes (e.g., MATSUMOTO *et al.*, 2001; VIDALE and LI, 2003; IKUTA and YAMAOKA, 2004). One advantage of controlled source studies is that the space-time coordinates of the sources are known more accurately than those associated with natural earthquakes. However, studies based on controlled sources have several disadvantages including expensive cost, inadequate temporal sampling and poor depth penetration. The latter point is especially important for studies concerned with the physics of earthquakes and faults.

An effective way of using natural earthquake sources for monitoring changes of material properties is to employ sets of events that have very similar locations and focal mechanisms. These events, referred to as repeating earthquakes, generate highly similar waveforms at recording stations and are associated with nearly identical propagation paths to the stations (e.g., POUPINET *et al.*, 1984; NADEAU *et al.*, 1994; SCHAFF *et al.*, 1998). Recent relocation techniques based on waveform cross correlation (e.g., SCHAFF *et al.*, 2004) and the double-difference algorithm (e.g., WALDHAUSER and ELLSWORTH, 2000) improved significantly the determination of relative earthquake locations. This has allowed the identification of many sets of repeating earthquake clusters, mainly along the central and northern sections of the San Andreas fault system (e.g., SCHAFF *et al.*, 1998, 2002; RUBIN, 2002; WALDHAUSER *et al.*, 2004; PENG *et al.*, 2005), and along portions of subduction zones in Japan (e.g., IGARASHI *et al.*, 2003).

Using repeating earthquakes, BOKELMANN and HARJES (2000) found ~2% decrease in shear-wave anisotropy associated with a fluid injection experiment at the 9 km deep KTB borehole in Germany. NIU *et al.* (2003) detected systematic temporal variations in early *S*-coda waves from repeating earthquakes along the Parkfield segment of the San Andreas fault. They interpreted the observation as resulting from migration of seismic scatterers due to a transient aseismic event. SCHAFF and BEROZA (2004) and RUBINSTEIN and BEROZA (2004a,b) reported clear co-seismic reduction of seismic velocity followed by logarithmic post-seismic recovery in the aftershock zones of the 1984 M6.2 Morgan Hill earthquake, the 1989 M6.9 Loma Prieta earthquake, and its largest aftershock, the 1990 M5.4 Chittenden earthquake. They suggested that the changes reflect evolution in the microcrack density in the top-most portion of the crust, initiated by the strong ground motions of the main shocks. Using repeatable seismic sources around volcanic regions, such as repeating earthquakes (YAMAWAKI *et al.*, 2004) and gas bubbles (GRET *et al.*, 2005), clear reductions of waveform cross-correlation coefficients have been found in *P*, *S* and coda waves, indicating rapid changes in the scattering properties associated with volcanic activity. Recently, repeating ultrasonic coda waves have been used to determine the nonlinear dependence of the seismic velocity in granite (SNIEDER *et al.*, 2002). Correlation analysis of the coda waves can also be used to infer relative source separation between repeating earthquakes (SNIEDER and VRIJLANDT, 2005).

PENG and BEN-ZION (2004) performed a systematic analysis of crustal anisotropy along the Karadere-Düzce branch of the North Anatolian fault (NAF), which ruptured during the 1999 M_w7.4 İzmit and M_w7.1 Düzce earthquakes. They found ~1 km broad zones around the surface traces of the İzmit and Düzce main shocks with enhanced anisotropy associated with fault-parallel cracks and/or structural fabric. The belt of anisotropic rock is confined primarily to the top ~3 km of the crust and surrounds ~100 m wide zone of highly damaged rock acting as trapping structure for seismic fault zone waves (BEN-ZION *et al.*, 2003). Using repeating earthquake clusters, PENG and BEN-ZION (2005) investigated the fine-scale spatial and temporal variations of crustal anisotropy in the same area. Large apparent co-seismic changes (up to 30%) of shear-wave splitting delay times are observed across the time of the Düzce main shock at stations near the epicentral region. However, the changes can be mostly explained by spatial variations of the ray paths due to the changing seismicity, rather than changes in the properties of the anisotropic medium. Splitting parameters measured using repeating earthquake clusters in the area indicate at most 2% changes in delay times associated with the occurrence of the Düzce main shock.

In this study, we use the early *S*-coda waves generated by 36 repeating earthquake sequences identified by PENG and BEN-ZION (2005) to document seismic velocity changes around the time of the Düzce earthquake. The occurrence of the İzmit earthquake three months before the Düzce main shock activated many repeating earthquake sequences in the area. This provides an important data set that can be used to constrain possible temporal changes during the pre-, co-, and post-seismic periods of the Düzce main shock. As mentioned above, our related high-resolution study of seismic anisotropy with the same data showed only small co-seismic changes, and demonstrated the lack of robust changes of anisotropy properties either before or after the Düzce main shock. In contrast, the high resolution analysis of the present work, based on travel-time changes of the *S* and early *S*-coda waves, reveals clear co- and post-seismic changes of seismic velocities manifested, respectively, as sharp velocity reductions followed by gradual (logarithmic-type) recoveries. The results also show small gradual increases of seismic velocities before the Düzce main shock, which are likely the post-seismic recoveries from the earlier İzmit main shock. The effects are observed generally at all stations, with larger amplitudes at stations in the immediate vicinity of the damaged fault zone layer of the İzmit and Düzce rupture zones.

In the following sections 2 and 3, we describe the overall data set and properties of the repeating earthquake clusters. In section 4 we use a sliding window waveform cross-correlation technique (e.g., POUPINET *et al.*, 1984; NIU *et al.*, 2003) to measure travel-time differences and evolving decorrelation in waveforms generated by the different sets of repeating earthquakes. In section 5 we analyze the evolution of the observed temporal changes, constrain the possible regions where the temporal changes occur, and investigate possible relations between the temporal changes and the strong ground motions produced by the Düzce main shock. The implications of the results to seismic properties of the crust are discussed in section 6.

2. Data

2.1. The Seismic Experiment

The analysis employs seismic waveform data sets recorded by a temporary 10-station PASSCAL seismic network (Fig. 1a) along and around the Karadere-Düzce branch of the NAF (SEEBER *et al.*, 2000; BEN-ZION *et al.*, 2003). The seismometers were deployed a week after the August 17, 1999, M_w7.4 İzmit earthquake. All 10 stations had REFTEK recorders and three-component L22 short-period sensors with a sampling frequency of 100 Hz. In addition, 8 stations (not including MO and GE) had three-component force-balance accelerometers (FBA). The deployment location was fortunate in that three month later, the November 12, 1999, M_w7.1 Düzce earthquake occurred in the same area. Our temporary seismic network straddled the rupture zones of both main shocks and recorded about 26,000 earthquakes during its six-month operational period.

The "standard" locations for the 26,000 earthquakes, used in our previous studies, were obtained with the code HYPOINVERSE (KLEIN, 1978) and station corrections. The horizontal errors of those locations are less than 1 km around the center of the network and 1–2 km near the margins. The vertical errors are somewhat

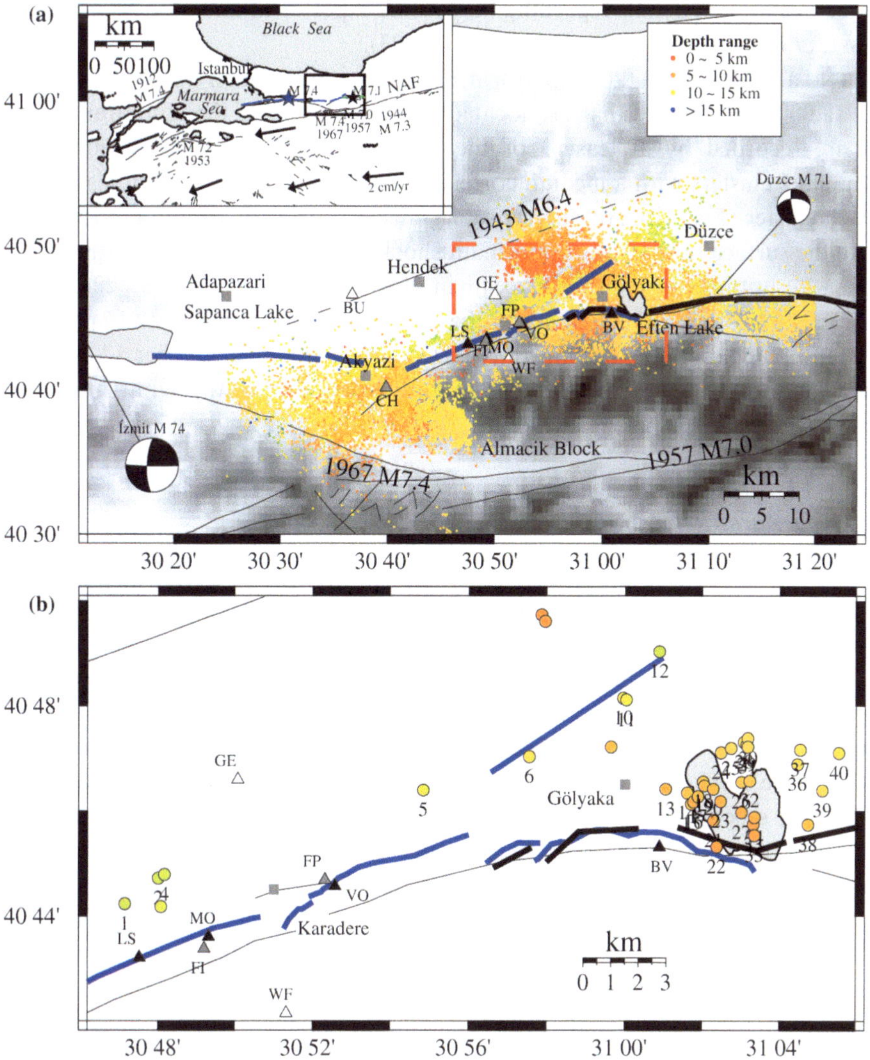

Figure 1

(a) Hypocentral distribution of ~18000 earthquakes recorded by the PASSCAL seismic experiment along the Karadere-Düzce branch of the NAF. Aftershock locations are marked by small dots with colors denoting different depth ranges. Shaded background indicates topography with white being low and dark being high. The surface ruptures of the İzmit and Düzce earthquakes are indicated with thick blue and black lines, respectively. Dark thin lines associated with earthquake information denote faults that were active during recent ruptures. Other dark thin lines are geologically inferred fault traces. Stations within, near and outside the fault zone are shaded with dark, gray and white triangles, respectively. Gray squares denote locations of nearby cities. The area bounded by the red dashed lines is shown in (b). The inset illustrates the tectonic environment in northwestern Turkey with the box corresponding to our study area. Arrow vectors represent plate deformation rate (REILINGER *et al.*, 1997) from GPS data. Modified from PENG and BEN-ZION (2005). (b) Hypocentral locations of 40 repeating event clusters in the study area. The ID is assigned to each cluster sequentially according to the centroid longitude of that cluster. The 36 clusters that are used in this study are marked by the cluster ID. Other symbols and notations are the same as in (a).

greater. Additional details regarding the seismic experiment and data set are given by SEEBER *et al.* (2000) and BEN-ZION *et al.* (2003).

PENG and BEN-ZION (2005) performed cross correlations on waveforms generated by ~18,000 earthquakes that are located within ~20 km of the temporary network and are recorded by at least 3 of the 10 stations. They identified 40 repeating earthquake clusters with a median correlation coefficient ≥ 0.95 and at least 5 events in each cluster (Fig. 1b). Among those, 28 clusters occurred around the Eften Lake, where the surface trace of the İzmit main shock terminated. Three months later, this segment reruptured during the Düzce main shock (e.g., HARTLEB *et al.*, 2002). The 28 clusters in the Eften Lake area are in the depth range of 5–10 km, and appear to be on a fault plane dipping ~65° north (UTKUCU *et al.*, 2003). The other 12 clusters are located along the Karadere segment that ruptured during the İzmit main shock. Most of these clusters are in the depth range of 10–15 km.

2.2. Hypocentral Relocation

We relocate earthquakes in each cluster using the double-difference technique (WALDHAUSER and ELLSWORTH, 2000). The technique incorporates both the catalog travel-time data and relative P and S arrival times from waveform cross correlation. The correlation measurement is obtained by a two-step procedure. We first use a 1-s time window to align P and S arrivals to a reference trace. To minimize the effect of temporal changes in the P- and S- coda waves on earthquake relocation, we then use a 0.32-s time window starting 0.12 s before the P arrival, and a 0.64-s time window starting 0.2 s before the S arrival, to compute relative travel times with subsample precision. These choices produce narrow time windows centered on the direct P and S arrivals. After relocation, the source dimensions for most earthquakes in each cluster almost overlap each other, indicating that they rupture approximately the same fault patch repeatedly (Fig. 2). However, we find a small percentage of events that are relatively far from other members of a cluster.

The events used in the analysis have magnitudes smaller than $M = 3$ (ZHU, 2003). To improve further the similarity of the employed event locations, we estimate the earthquake source dimensions using the analytical scaling relation for a circular crack in an infinite Poisson solid (e.g., KANAMORI and ANDERSON, 1975; BEN-ZION, 2003)

$$r^3 = (7/16)(P_0/\Delta\varepsilon), \tag{1}$$

where r is the crack radius, P_0 is the scalar potency of the event (given by the integral of the final slip distribution over the failure area), and $\Delta\varepsilon$ is the static strain drop of the event (equal to the ratio of a nominal stress drop divided by a nominal rigidity). To implement (1), we assume that each event sustains a strain drop of 10^{-4}

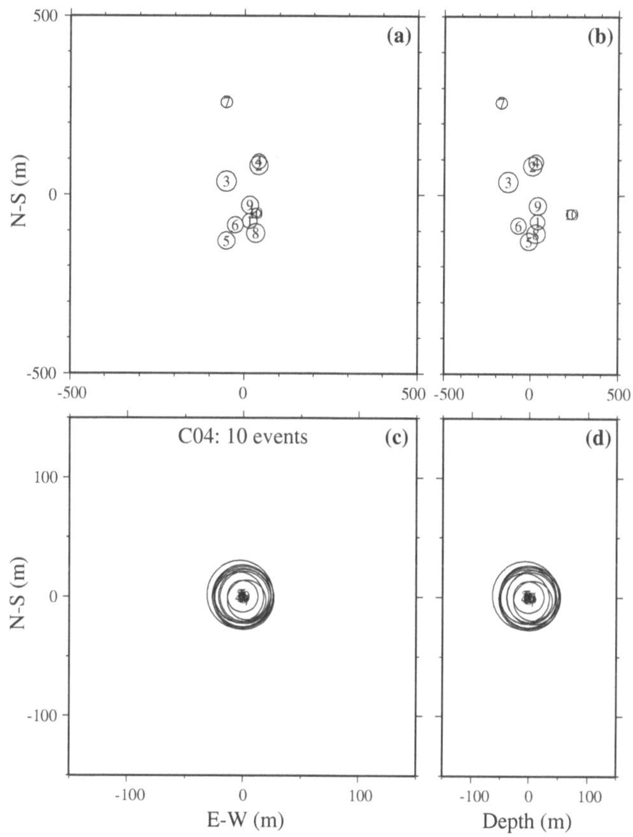

Figure 2
Map view (a) and cross section (b) of the relative locations of the 10 events in cluster C04 before relocation. Map view (c) and cross section (d) of the same 10 events after relocation. The zero coordinates correspond to the centroid location of the cluster. The circle size scales with the source radius, which is estimated based on the event magnitude (ZHU, 2003) and the potency-magnitude relation of BEN-ZION and ZHU (2002).

(KASAHARA, 1981) and employ the empirical potency-magnitude relation of BEN-ZION and ZHU (2002) for earthquakes with magnitudes $M < 3$,

$$\log P_0 = 1.00\,M - 4.72. \tag{2}$$

To ensure that only clusters of repeating events that are very close in space are used in the study of temporal changes (GOT and COUTANT, 1997), we discard an individual event if its source area does not overlap 50% with any other member in that cluster (WALDHAUSER and ELLSWORTH, 2002). This removes 25 events out of 326 events in the 40 clusters. Most of the removed events occurred during the first two weeks of the deployment when only a few of the 10 stations were installed. Consequently, their hypocenter locations are generally less well constrained. In addition, we require that the duration of the repeating clusters is more than 10 days and that each cluster has at least four events. Clusters C03, C07, C08, and C09 do not satisfy these criteria. Ultimately, we are left with 36 clusters having 292 highly repeating events that are used for further detailed analysis.

3. Occurrence Pattern of Repeating Clusters

The remaining 36 repeating clusters in the aftershock zones of the İzmit and Düzce earthquakes are associated with sequences of failures on the same (or nearly the same) small fault patches. As shown in Fig. 3a, the recurrence intervals *Tr* of most repeating event clusters exhibit clear changes at the time of the Düzce main shock. In general, the recurrence intervals become gradually longer as the time of the Düzce main shock is approached (and the elapsed time from the earlier İzmit main shock increases), decrease sharply at the time of the Düzce main shock, and then gradually increase again after the Düzce main shock. The occurrence of the Düzce main shock affected most strongly the nearby 28 clusters in the Eften Lake area, and less so the 8 clusters that are further away along the Karadere segment.

As shown in Figure 3b, the inverse of the recurrence intervals *Tr* is approximately proportional to the elapsed times *t* since the Düzce main shock for clusters around the Eften Lake area on a log-log scale. To ensure the data quality, we select 16 out of the 28 clusters that have at least five events after the Düzce main shock, and at least a 0.8 range in the logarithm of the recurrence interval range. Since the number of repeating events in a given time interval is inversely proportional to *Tr*, each cluster of events plotted in Figure 3b is compatible individually with the empirical $1/t$ Omori law for aftershock decay rate (OMORI, 1895). However, it is important to note that the data shown in Figure 3b, covering a narrow range of time intervals, can be fitted equally well by exponential and other functions (e.g., KISSLINGER, 1996; BEN-ZION and LYAKHOVSKY, 2006). Similar patterns have been observed in the aftershock zones of the 1989 Loma Prieta earthquake (SCHAFF *et al.*, 1998) and the 1984 Morgan Hill earthquake (PENG *et al.*, 2005). The decay rates of repeating events following the Düzce main shock (and the previous İzmit event) can be explained by a variety of relaxation processes, including stress corrosion (DAS and SCHOLZ, 1981), rate- and state-dependent friction (DIETERICH, 1994), dislocation creep (ZÖLLER *et al.*, 2005) and damage rheology (BEN-ZION and LYAKHOVSKY, 2006).

The increasing rates of many repeating events at the time of the Düzce main shock provide a dense temporal sampling immediately after the Düzce event, which is very useful for a detailed study of post-seismic effects. The repeating events that were activated by the earlier İzmit main shock provide important data that can be used to study possible pre- and co-seismic changes associated with the Düzce main shock. These effects are examined in the following sections.

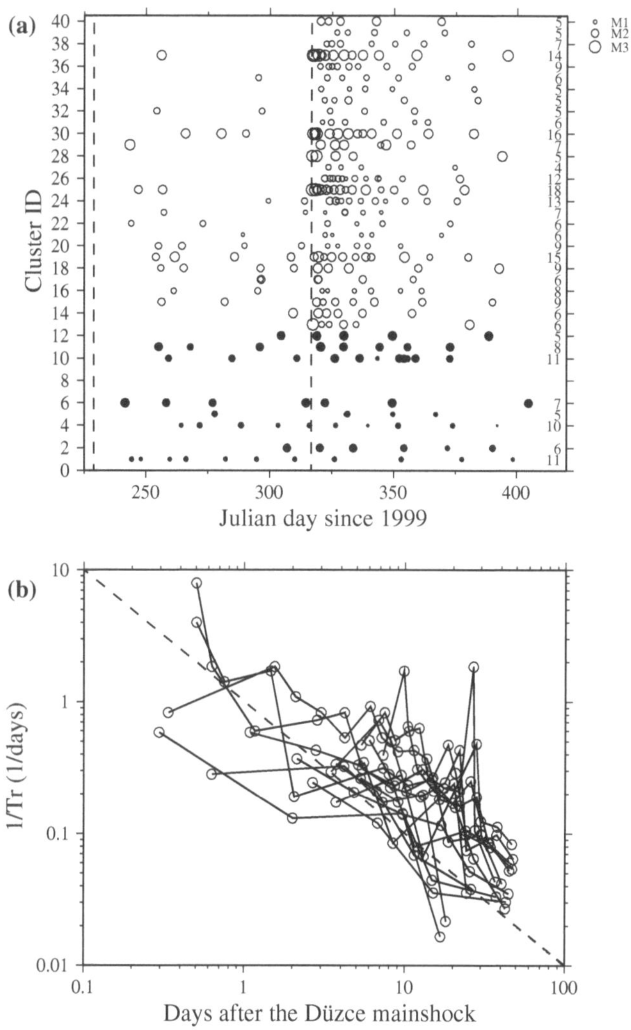

Figure 3
(a) Occurrence times (Julian day since 1999) of the 36 repeating earthquake clusters. The vertical dash lines mark the origin time of the İzmit and Düzce earthquakes. Open and filled circles mark the clusters around the Eften Lake and Karadere segment, respectively. The number of events in each cluster is marked on the right side of the figure. (b) The inverse of recurrence intervals $1/Tr$ as a function of their elapsed times t for 16 clusters near the Eften Lake that have at least 5 events after the Düzce main shock, and a range of logarithmic recurrence interval of at least 0.8. The dash line denotes the $1/t$ OMORI law for aftershock decay rates.

4. Analysis Procedure

Before performing detailed analysis, we conduct the following preprocessing steps. We examine three-component data that are generated by 292 microearthquakes in 36 repeating clusters and recorded by the L22 short-period sensors. Strongly clipped seismograms are removed. The remaining seismograms are detrended, shifted to have zero mean, and bandpass filtered at 1–20 Hz using a two-pass 4-pole Butterworth filter from subroutines in the Seismic Analysis Code (GOLDSTEIN *et al.*, 2003). The bandpass filter is applied to enhance signals and remove possible incoherent noise at higher frequencies.

Next, we select for each cluster a reference seismogram that all other seismograms are compared with. We are mostly interested in temporal changes induced by the Düzce main shock, so we choose reference seismograms that are generated distant in time from the main shock, so that the influence of the Düzce main shock is small. Since our 10-station seismic network was not complete in the first several weeks of the deployment, the reference seismogram for each station is chosen to be either the last trace within each sequence, or the second-to-last if the signal-to-noise ratio for the last trace is less than 3.

The *P*- and *S*-wave arrival times are manually picked on each reference trace. The other seismograms of the repeating events in that cluster are aligned at 1 s with the manually picked *P* arrival on the reference trace. This is done by cross-correlating a 1-s-long window starting 0.2 s before the automatically picked *P* arrival on these traces (Fig. 4a). Because of this alignment, the delay times obtained from the analysis are relative to the *P* arrival of the reference trace and reflect changes in the later *S* (and *S*-coda) minus *P* time. Thus, we are unable to measure temporal changes in the *P* and early *P*-coda waves, and the obtained temporal changes from the *S* and *S*-coda waves are probably lower bounds. However, this procedure removes possible errors in the absolute timing of a seismogram (e.g., RUBINSTEIN and BEROZA, 2004a).

A sliding window waveform cross-correlation technique (NIU *et al.*, 2003) is used to measure the travel-time differences and evolving decorrelation in waveforms of the repeating earthquakes in each cluster. The delay time $\tau(t)$ and decorrelation index $D(t)$ (defined as 1 minus the maximum cross-correlation value) between the reference and subsequent seismograms are computed using a 0.5-s time window, sliding forward at 0.01 s interval through 9 s after the *P* arrival (Fig. 4). To obtain a subsample accuracy, we interpolate the waveforms from 100 samples to 10,000 samples per second. A cosine taper is applied to each window with 10% of the entire width to reduce the Gibbs' phenomenon.

Figure 4 shows clear changes of both delay time $\tau(t)$ and decorrelation index $D(t)$ for the vertical-component seismograms generated by cluster C04 and recorded at station VO. The $\tau(t)$ values at 3–4 s in the early *S*-coda waves for the earthquakes that occur before and after the Düzce main shock differ by up to 10 ms. Since the temporal evolution is most prominent in the early *S*-coda waves, we compute the

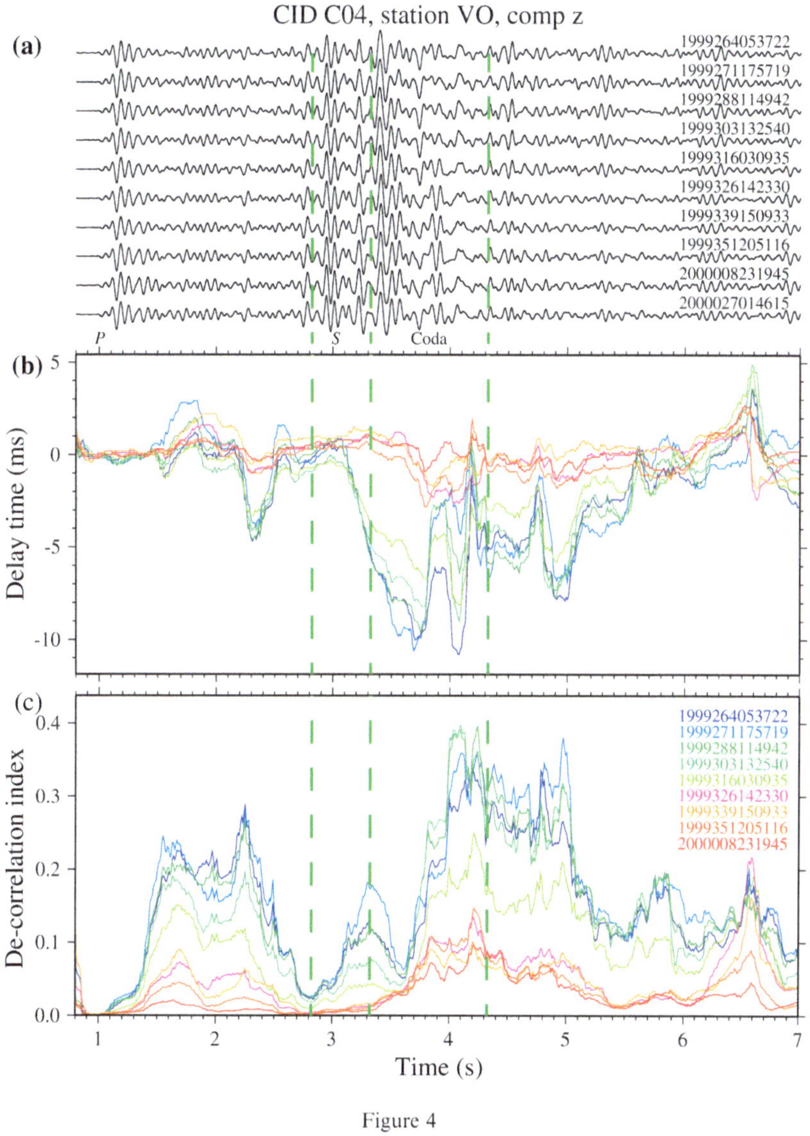

Figure 4

(a) Vertical-component seismograms generated by the 10 events in cluster C04 and recorded at station VO. Seismograms are aligned by *P* wave arrivals at 1 s. Calculated (b) delay time $\tau(t)$ and (c) decorrelation index $D(t)$. The vertical dashed lines denote the 1.0-s time window for the early *S*-coda waves, and the 0.5-s time window around the *S* arrival that are used to compute the median delay times. The event ID numbers are marked on the right side of panels (a) and (c), and consist of 4 digits year, 3-digits Julian day, 2-digits hour, 2-digits minute, and 2-digits seconds of the earthquake occurrence time.

median of $\tau(t)$ and $D(t)$ functions over a 1.0 s time window starting 0.3 s after the direct *S* arrival, and use these values in the later analysis. Slightly different time windows may result in different median values of $\tau(t)$ and $D(t)$. However, the general patterns and overall conclusions that are based on statistics of hundreds of measurements remain essentially the same. To ensure the data quality, we keep only measurements that have a median $D(t)$ smaller than 0.3 and a signal-to-noise ratio larger than 3.

5. *Results*

5.1. Systematic Behavior of Temporal Changes in Delay Times

Figure 5 shows the median $\tau(t)$ and $D(t)$ values measured from the vertical-component seismograms for cluster C04 at 10 stations. The $\tau(t)$ values increase at the time of the Düzce main shock for all stations other than BV and CH, and the $D(t)$ values generally decrease. Since all the seismograms are compared with the last trace after the Düzce main shock, a decrease in $D(t)$ value (increase in correlation coefficient) is consistent with an increase in $\tau(t)$ value across the occurrence time of the Düzce main shock. Station VO that was installed inside the surface rupture zone of the İzmit earthquake has the largest coseismic change in median $\tau(t)$ and $D(t)$ values. Figure 6 displays the median $\tau(t)$ changes for 36 clusters at the fault zone station VO and a nearby station FP that is about 400 m from the İzmit rupture zone. The occurrence of the Düzce main shock produces distinct signals that are shown at both stations for 8 clusters along the Karadere segment (Figs. 6a,b), and 28 clusters around the Eften Lake area (Figs. 6c,d). However, the co-seismic changes at the fault zone station VO are about 50% larger than at station FP. We note that the median $\tau(t)$ values right after the Düzce main shock are larger for the 28 clusters around the Eften Lake area than those for the 8 clusters along the Karadere segment, possibly due to denser samplings of (larger) delays immediately following the main shock by the 28 clusters that are reactivated by the Düzce event. However, the temporal evolutions of delay times have similar shapes for different clusters with various source locations. This observation indicates that the temporal changes in $\tau(t)$ are most likely originated from the near-station structures, instead of the near-source regions or the propagation paths. We also note that, in some cases, changes with the same sense as those induced by the Düzce main shock begin before the Düzce event, and hence may be interpreted as precursory. However, these changes are associated with large scatter and require additional analysis that will be the subject of a future work.

Next, we merge the $\tau(t)$ measurements from the vertical-component seismograms of all clusters at each station (Fig. 7). Since our instruments worked in a trigger mode and most events are small, this results in a different number of recorded events at the different stations. In addition, only results with median $D(t)$ smaller than 0.3 and signal-to-noise ratio larger than 3 are kept. The final number of events and clusters analyzed for each station are indicated at the top of each panel in Figure 7. Most

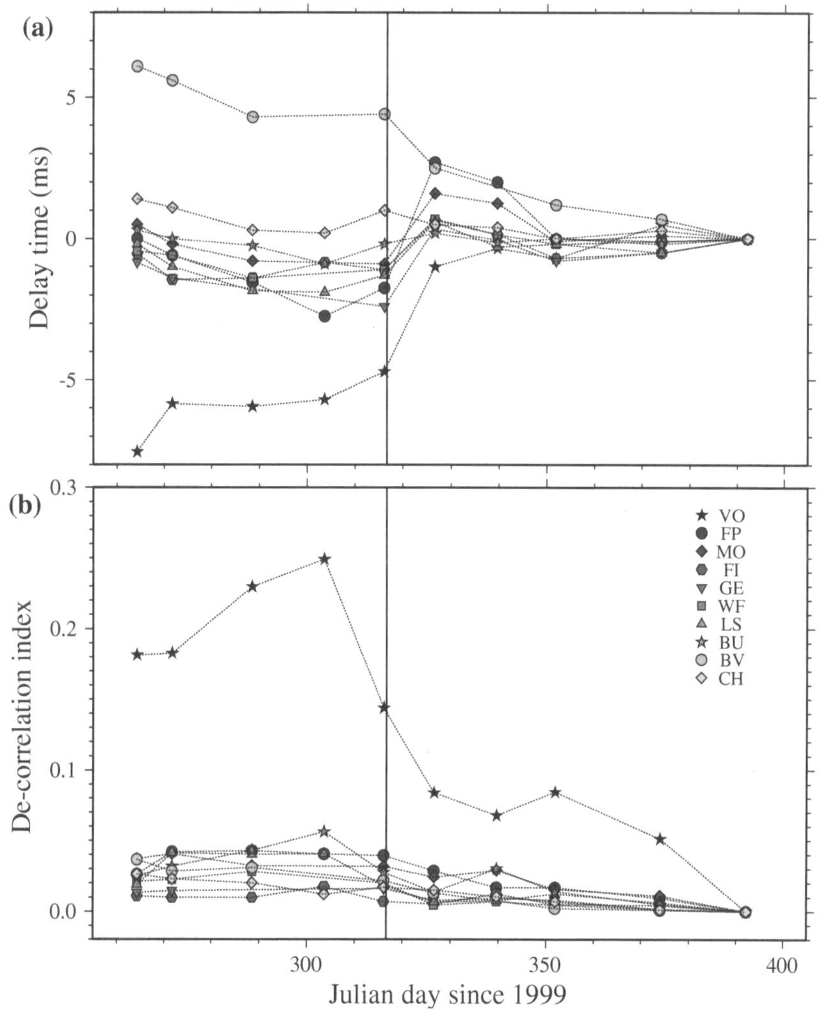

Figure 5
(a) Median delay time $\tau(t)$ and (b) decorrelation index $D(t)$ for the early S-coda waves plotted against the earthquake occurrence times for the vertical-component seismograms recorded at the 10 stations and generated by the earthquakes in cluster C04. The vertical line denotes the occurrence time of the Düzce main shock.

stations show clear increases in the median $\tau(t)$ values across the time of the Düzce main shock. An increase of $\tau(t)$ corresponds to a drop in the seismic velocity that is induced by the main shock. As noted before, the strength of the co-seismic increase of $\tau(t)$ is different among the 10 stations. In general, the fault zone station VO has the largest (up to 20 ms) co-seismic increase in $\tau(t)$ in the vertical component seismograms. The corresponding changes for stations FP, MO, FI, GE, WF, and

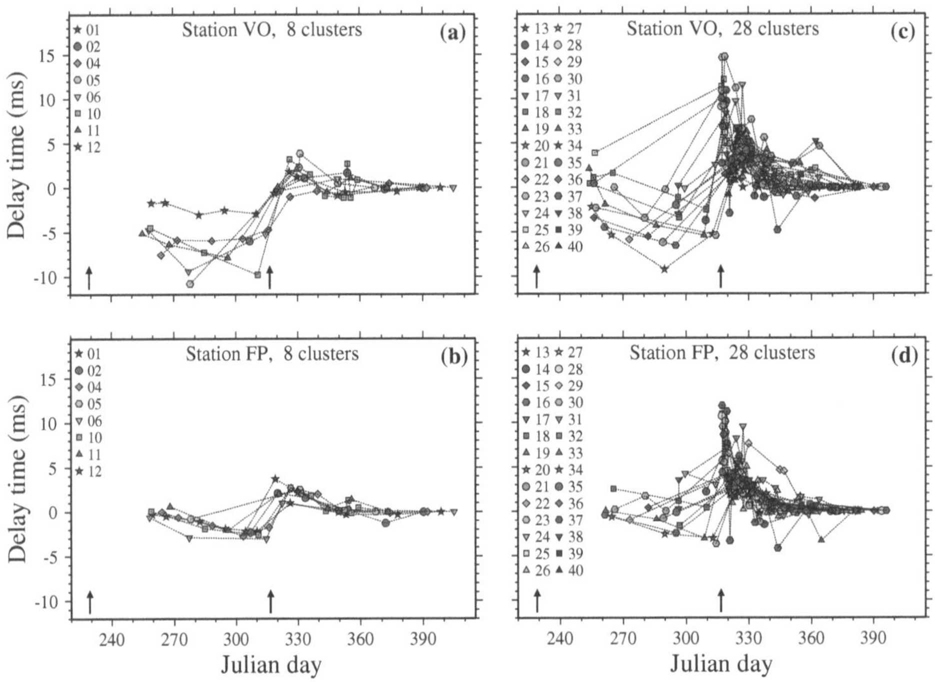

Figure 6
Median $\tau(t)$ for the early *S*-coda waves plotted against the earthquake occurrence times for the vertical-component seismograms recorded at stations VO (a) and FP (b) for 8 clusters along the Karadere segment. Median $\tau(t)$ values for 28 clusters around the Eften Lake area are shown in (c) and (d) for stations VO and FP, respectively. The $\tau(t)$ value for each cluster is denoted by a different symbol. The two vertical arrows mark the occurrence times of the İzmit and Düzce main shock.

LS are about 10–15 ms. Relatively small changes (around 5 ms) are observed for stations BU and BV, and station CH shows little or no observable co-seismic changes.

In addition to the clear co-seismic increases of delay times, we also observe that the $\tau(t)$ values for most stations gradually recover towards the levels that existed before the Düzce main shock (Fig. 8). This indicates that the rock damage heals (and the seismic velocity increases) in the post-main shock period. As shown in Figure 8, the decrease of $\tau(t)$ is approximately linear with logarithmic time. We fit the data by a least-squares procedure using a logarithmic relation

$$\tau(t) = a - b\log_{10}(t/\mathrm{day}), \tag{3}$$

where a is the delay time in ms at 1 day after the main shock, b is a measure of the decreasing rate of the median $\tau(t)$ and $t/\mathrm{day} \geq 0.1$. Because $\tau(t)$ diverges as t approaches 0 or infinity, this equation is valid for finite (relatively small) positive times. The threshold time of 0.1 day is chosen for all stations because it is about the

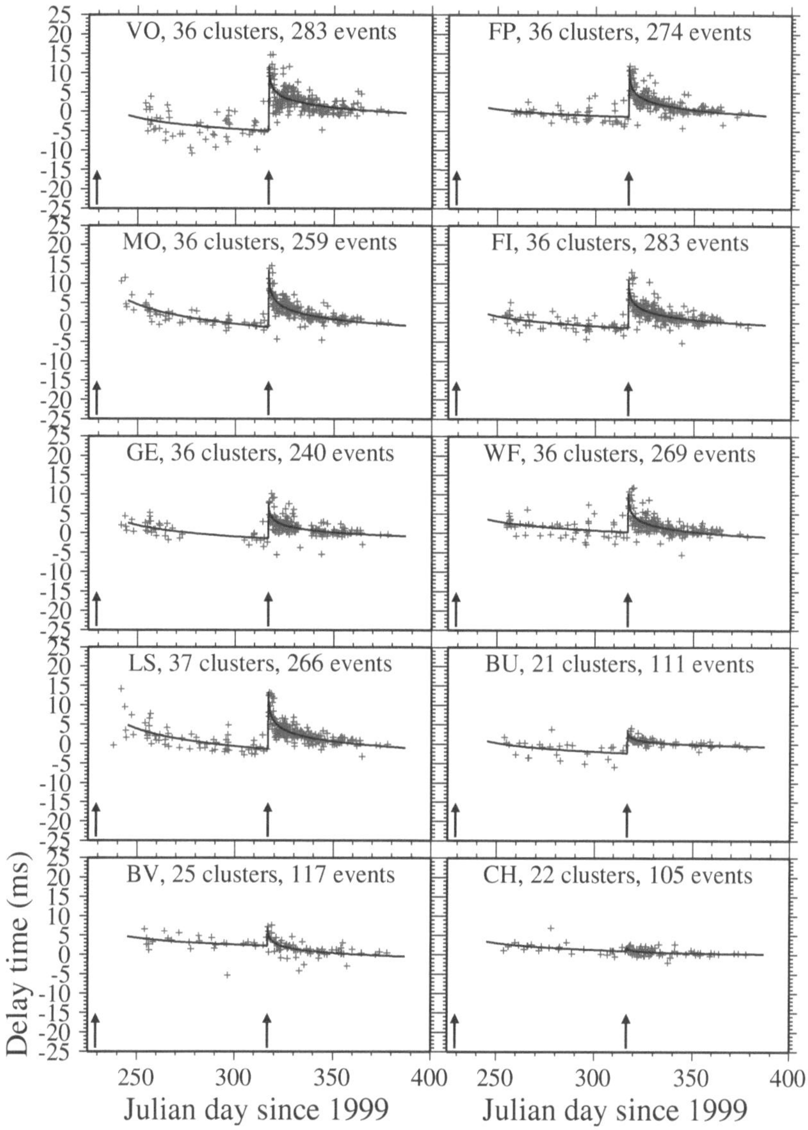

Figure 7
Median $\tau(t)$ for the early *S*-coda waves plotted against the earthquake occurrence times for the vertical-component seismograms recorded at the 10 stations. The two vertical arrows in each panel mark the occurrence times of the İzmit and Düzce main shock. The solid line in each panel denotes a least-squares fit to the data for each station using two logarithmic decay relations (Equation (3)) separately for the repeating events that occur between the İzmit and Düzce main shocks, and the events following the Düzce main shock.

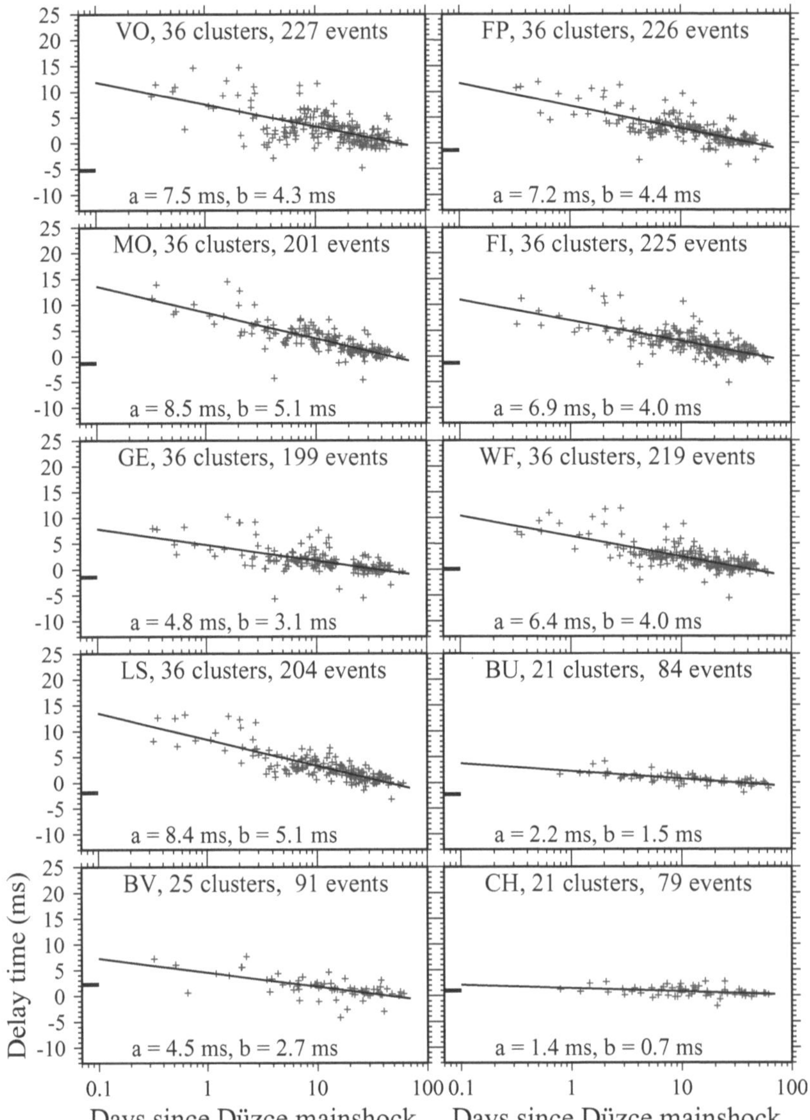

Figure 8

Median $\tau(t)$ for the early *S*-coda waves plotted against the logarithmic elapsed times after the Düzce main shock for the vertical-component seismograms recorded at the 10 stations. The solid line in each panel denotes a least-squares fit to the data for each station using the logarithmic decay relation (Equation (3)). The short horizontal bars on the left indicate the $\tau(t)$ levels before the Düzce main shock. The name of each station, total number of clusters and events for that station are marked at the top of each panel. The obtained values of *a* and *b* are indicated at the bottom of each panel.

earliest time when repeating earthquakes identified in this study start after the Düzce main shock. The parameter a is associated with initial conditions, while the parameter b represents a healing rate. For example, in the damage rheology model of LYAKHOVSKY *et al.* (1997, 2005), the fitting parameters a and b correspond, respectively, to the initial damage level and the damage healing rate. The obtained values of a and b for the time period following the Düzce main shock are indicated in the panels of Figure 8. After three months, the delays for most stations return close to the levels that existed before the Düzce main shock. However, this does not hold for the fault zone station VO which experienced the largest co-seismic change and is likely to have longer recovery time and larger permanent deformation than the other stations.

Since our temporary network straddled the rupture zones of both the İzmit and Düzce main shocks, it recorded many clusters that were activated right after the İzmit main shock. These clusters can be used to obtain information on the temporal changes associated with the İzmit event. Unfortunately, the network was deployed one week after the İzmit main shock and some stations that were installed in the first two weeks were later removed. Thus, we cannot constrain the temporal changes in the co- and early post-seismic period of the İzmit event. However, most stations show a gradual decrease in $\tau(t)$ after the İzmit main shock that is similar to the healing process observed more clearly after the Düzce main shock (Fig. 8). As shown in Figure 7, the healing process after the İzmit event was interrupted abruptly by the occurrence of the Düzce main shock three month later. Similar patterns have been observed for the M7.3 1992 Landers and M7.1 1999 Hector Mine earthquakes (VIDALE and LI, 2003), and for the M6.9 1989 Loma Prieta earthquake and its largest aftershock, the M5.4 1990 Chittenden earthquake (RUBINSTEIN and BEROZA, 2004b).

We use the same logarithmic functional form (Equation (3)) to fit the data observed between the İzmit and Düzce main shocks, as well as after the Düzce main shock. Due to lack of data in the early post-İzmit period, the parameters a and b are probably not well constrained for that period. Nevertheless, the procedure allows us to estimate the level of delay time right before the Düzce main shock, and the co-seismic increase in delay time at each station associated with the Düzce event. Specifically, the pre-Düzce-main shock level is evaluated using the fitting parameters obtained for the post-İzmit data at the time of the Düzce main shock. The level of delay time right after the Düzce main shock is computed at 0.1 day after the main shock using the fitting parameters obtained for the post-Düzce data. The differences between the pre- and post-Düzce levels represent the co-seismic increases in delay times at the various stations (Table 1).

Similar patterns of median $\tau(t)$ values for the early *S*-coda waves are shown, with larger amplitudes, for the north component seismograms at each station (Fig. 9), indicating that our observations are robust. Analysis of the east components produces very similar patterns to those associated with the north components. These results are not shown in a separate figure but are included in a later summary

Table 1

Station parameters, observed co-seismic delays, and peak ground accelerations during the Düzce main shock

Station	Latitude	Longitude	Distance to rupture (km)	Co-seismic delays (ms)			PGA (g)		
				Z	N	E	Z	N	E
VO	40.7429	30.8761	0.00	16.81	30.65	26.67	0.20	0.89	0.56
FP	40.7448	30.8720	0.41	12.83	17.41	17.88	0.10	0.15	0.14
MO	40.7267	30.8220	0.00	14.63	15.54	15.05	N/A	N/A	N/A
FI	40.7229	30.8220	0.34	12.29	14.55	17.08	0.10	0.14	0.27
GE	40.7448	30.8720	4.86	9.23	7.67	9.16	N/A	N/A	N/A
WF	40.7025	30.8553	3.49	9.99	13.73	15.24	0.06	0.16	0.12
LS	40.7201	30.7922	0.00	14.62	19.92	21.90	0.05	0.10	0.13
BU	40.7774	30.6128	11.34	6.00	8.47	9.02	0.02	0.03	0.05
BV	40.7552	31.0149	1.00	5.00	6.72	4.21	0.07	0.08	0.11
CH	40.6698	30.6655	1.42	1.12	7.04	6.82	0.02	0.03	0.05

Station VO, MO, and LS are inside the surface rupture of the İzmit earthquake. Station BV is about 1 km from a 2 m scarp formed during the Düzce earthquake (SEEBER *et al.*, 2000). Station FP is about 410 m from fault zone station VO. Station FI is about 340 m from fault zone station MO. The distances to the Karadere segment for the remaining stations (GE, WF, and CH) are computed by projecting to a line that is centered at station MO with a strike of 70°.

Figures 11 and 13. We note that the observed co-seismic changes for the horizontal components are larger than those of the vertical component (Figs. 7, 9; Table 1). This is probably because the horizontal seismograms contain more (and hence are more sensitive to changes of) *S*-wave energy than the vertical seismograms. In addition, the horizontal seismograms are more sensitive to site amplifications and other local near-station structures than the vertical ones (e.g., CASTRO *et al.*, 1997). As illustrated in Figure 4, we also measure changes in delay time for the direct *S* arrival using the median of $\tau(t)$ with a 0.5 s window starting 0.2 s before the picked *S* arrival. A summary of results associated with such analysis of the direct *S* waves is given in Figure 10. The patterns are similar to those observed for the early *S*-coda waves (Fig. 7), although the amplitudes in Figure 10 for the direct *S* waves are smaller that those associated with the *S*-coda waves.

Figure 11a gives the healing rates following the Düzce main shock based on analysis of all three components of motion at the ten different stations versus the corresponding co-Düzce time delays. Figure 11b plots the post-Düzce healing rates at all stations and components of motion versus the corresponding post-İzmit healing rates. The results of Figure 11a show that the post-Düzce healing rates are strongly correlated with the co-seismic delays. This indicates that the function that governs the recovery process is nonlinear, i.e., the healing rate depends strongly on the damage level, as in the damage model of LYAKHOVSKY *et al.* (1997, 2005). For most stations, a similar time interval of about three months is needed for near-complete post-Düzce recovery. The results at the fault zone station VO deviate from

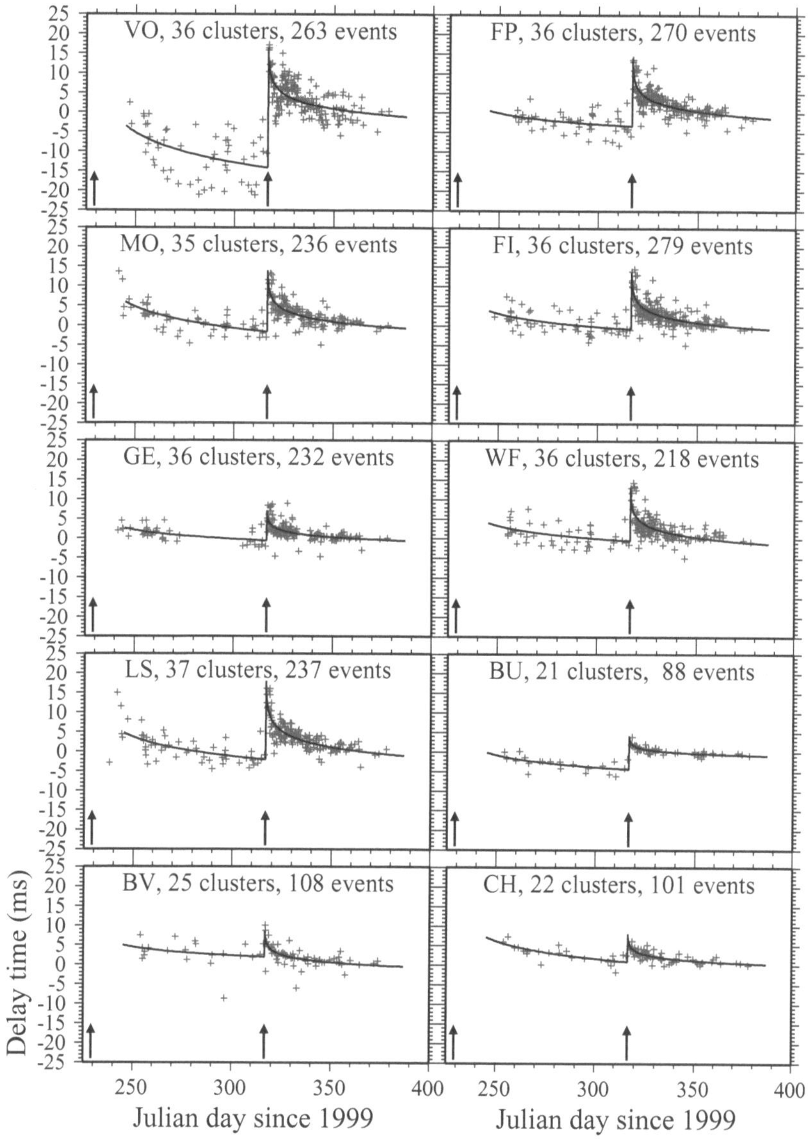

Figure 9
Median $\tau(t)$ for the early *S*-coda waves plotted against the earthquake occurrence times for the north-component seismograms recorded at the 10 stations. Other symbols and notations are the same in Figure 7.

this trend and indicate, as mentioned before, a longer recovery process and possible larger permanent change. The results of Figure 11b show that the post-Düzce healing rates correlate well with, and are about half of, the post-İzmit healing rate. Since the

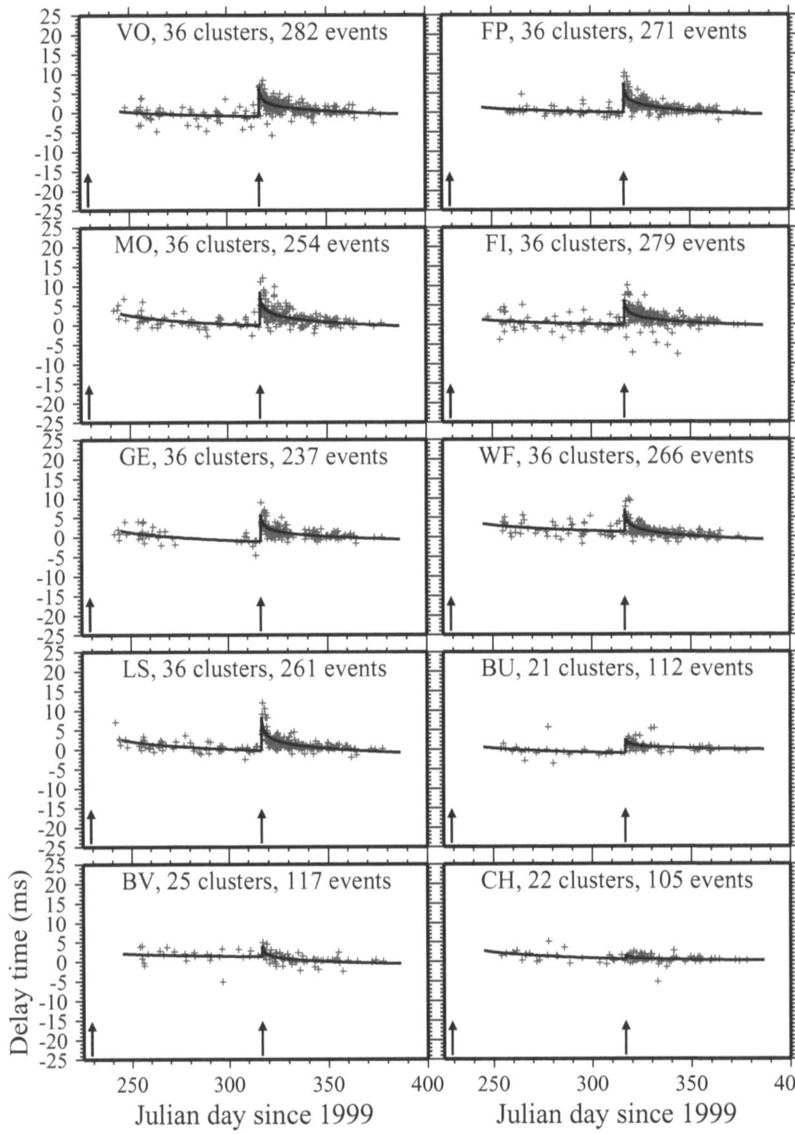

Figure 10
Median $\tau(t)$ for the *S* waves plotted against the earthquake occurrence times for the vertical-component seismograms recorded at the 10 stations. Other symbols and notations are the same in Figure 7.

healing rates depend strongly on the co-seismic delays, this suggests that the co-seismic delays for the İzmit main shock could be twice as large as those generated by the Düzce main shock.

The observed correlations lend support to the robustness of the measurements and indicate that the results reflect characteristics of the various sites. As will be

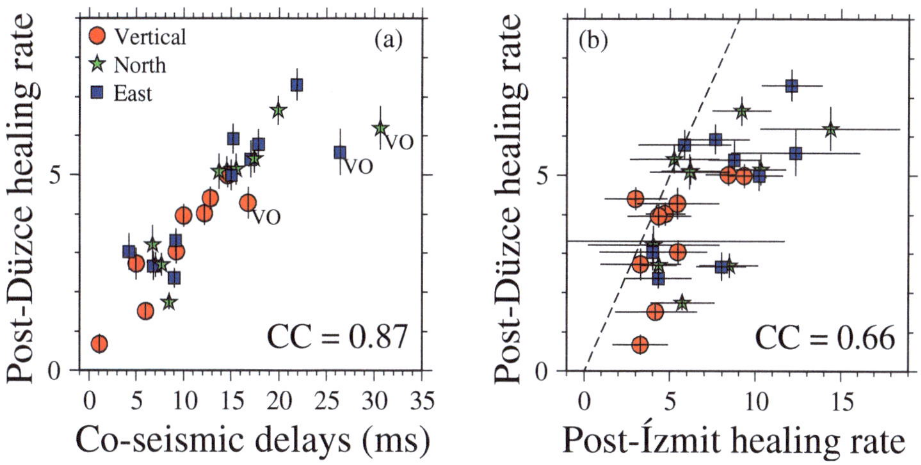

Figure 11
(a) The healing rates (ms per decade change in time) after the Düzce main shock plotted versus the co-seismic delays (ms) for all three components of motions at the 10 stations. Red circles, green stars, and blue squares denote results for the vertical, north, and east components, respectively. The bar on top of each symbol marks the standard error of each measurement. The correlation coefficient (CC) between the two values is marked in the bottom right of the figure. The results for station VO are marked by its station name. (b) The healing rates after the Düzce main shock plotted versus the healing rates after the İzmit main shock. The dashed line denotes a 1-1 relationship.

discussed below, the co-seismic delays strongly depend on the peak ground acceleration generated by the Düzce main shock. Since the İzmit event had about three times as large moment release as the Düzce event, it is plausible that the damage levels created during the İzmit main shock are twice as large as those generated during the Düzce main shock. However, the lack of recordings during the co-seismic and early post-seismic period of the İzmit main shock prevents us from reaching more definitive conclusions.

5.2. Possible Regions of Temporal Changes

The observed temporal evolution of $\tau(t)$ in the *S*-coda waves can be caused by systematic shifts in the hypocentral locations, and/or temporal changes of material properties near the source regions, along the source-receiver paths, or near the receivers. In this section, we attempt to constrain the factors that contribute to the observed temporal changes.

It has been noted that up to 0.1 s delay changes in later *S*-coda waves can be solely generated by event separation over a distance of about 500 m (Got and Coutant, 1997). However, it is unlikely that our observed temporal changes of $\tau(t)$ are caused mainly by changes in source locations. This is because the detailed relocation analysis discussed in section 2.2 enables us to collocate the employed events in each cluster to ~10 m precision (Fig. 2), thereby removing most of the effects from source shifts. In addition, we found that temporal changes for clusters having considerably different locations are similar to each other at a given station (Fig. 6). This indicates again that source shifts are not the primary cause for the observed temporal changes. The same observation can be used to rule out velocity changes near the source regions as the primary cause. We also observe (Fig. 5) that temporal changes of waveforms in a given cluster are not the same at nearby stations (e.g., stations VO and FP). If the contributions to $\tau(t)$ come mostly from the source regions, we would expect to see the opposite. Thus, neither shifts in hypocentral locations nor near-source velocity changes are the dominant causes of the observed $\tau(t)$ in the *S*-coda waves.

It is difficult to constrain the precise fractions of changes contributed by source-receiver path and near-station site effects. The most likely candidate regions that produce the observed $\tau(t)$ are the heavily-damaged top few hundred meters of the crust and the recently ruptured fault zone sections. If the seismic velocities were reduced over much of the seismogenic zone, we would expect to observe the largest $\tau(t)$ at the most distant stations. However, the most distant station BU has the smallest $\tau(t)$ among all stations. Station BV is very close to the rupture zone of the Düzce main shock and right above the clusters around the Eften Lake. However, the $\tau(t)$ values at station BV are the second smallest among all stations. The observed changes of $\tau(t)$ are largest at the fault zone stations VO, MO, and LS; however, the changes do not increase with increasing propagation distance along the rupture zone (Fig. 5). We also observe that stations GE and WF that are clearly outside the fault zone have significant delays.

We have shown that the temporal changes are similar at most stations for clusters with quite different source locations (Fig. 6). We also observe clear differences in the strength of the co-seismic increase in $\tau(t)$ for stations that are close to each other. For example, station VO that was deployed inside the surface rupture zone of the İzmit earthquake along the Karadere segment has the largest (up to 30 ms, Table 1) co-seismic increase in $\tau(t)$. The changes for station FP ~400 m away from station VO are somewhat smaller, on the order of 10–20 ms. Finally, we found that the post-Düzce healing rate correlates well with the post-İzmit healing rate, suggesting that the responses at a given site to the two main shocks are similar. The results indicate that most of the temporal changes are associated with shallow site effects that are largest for the fault zone stations.

5.3. Relations of Temporal Changes and Düzce Main shock Ground Motion

In section 5.2, we have argued that the near-station structure plays a dominant role in the observed temporal delays in the early *S*-coda waves. Here we show that the ground motion generated by the seismic waves radiated from the Düzce main shock are most likely the cause of the observed co-seismic changes of delay times. All stations other than MO and GE had three-component FBAs, collocated with the L22

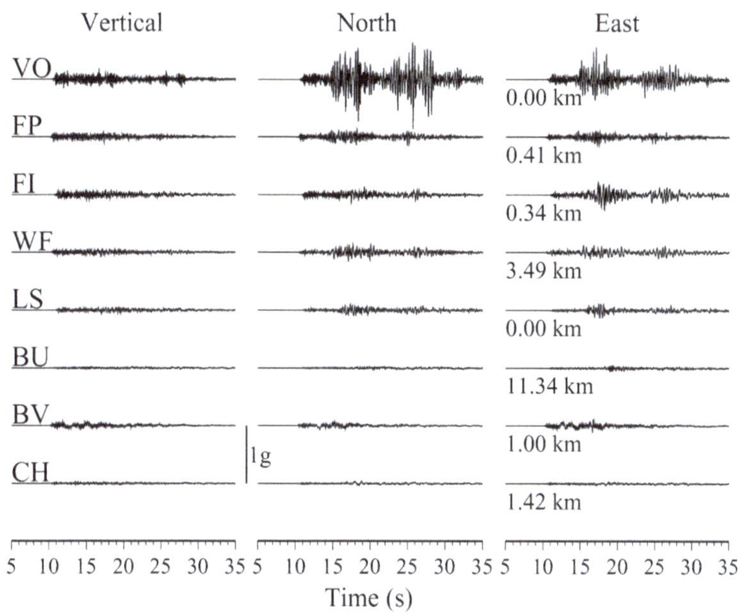

Figure 12
Three-component strong ground motion seismograms for the Düzce main shock recorded at 8 stations (not including MO and GE). All seismograms are plotted using a fixed scale of 1 g. The value under each east-component seismogram denotes the normal distance of each station to the recently ruptured fault zone.

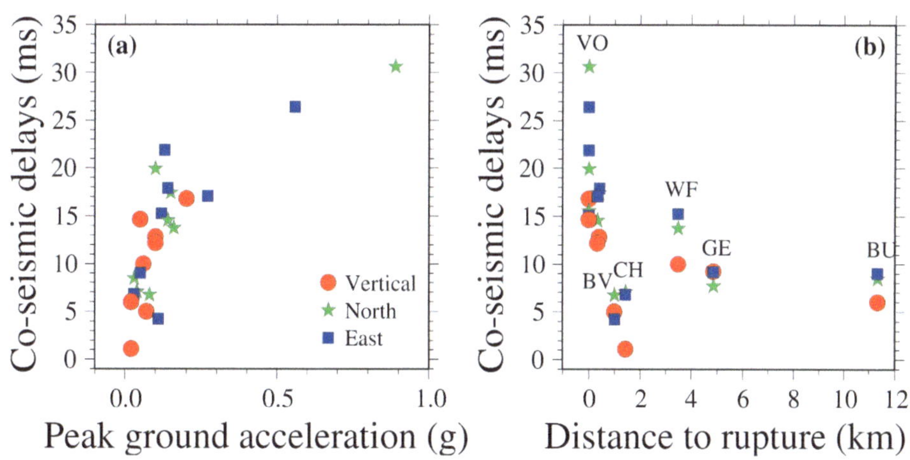

Figure 13
(a) The co-seismic delays plotted against the peak ground accelerations for eight stations (not including MO and GE). Red circles, green stars, and blue squares denote results for the vertical, north, and east components, respectively. (b) The co-seismic delays plotted versus the normal distances to the recently ruptured fault zone for the 10 stations. The results for stations VO, BV, CH, WF, GE, and BU are marked by their station names on the top of the corresponding symbols.

short-period sensors, that recorded strong ground motions generated by the M_w7.1 Düzce main shock (Fig. 12). The peak ground acceleration (PGA) parameters at the different stations are listed in Table 1. Figure 13a compares the PGA values with the observed co-seismic delays in the early *S*-coda waves. The results show a clear correlation between the observed delays and observed PGA values. Similar correlations were observed around the 1989 Loma Prieta earthquake (RUBINSTEIN and BEROZA, 2004a,b) and the 2004 Parkfield earthquake (RUBINSTEIN and BEROZA, 2005).

Figure 13b plots the co-seismic delay caused by the Düzce main shock at each station against its normal distance from the recently ruptured fault trace (D_r). Except for station BV which is close to the rupture zone of the Düzce main shock, the D_r values for the other nine stations are measured against the Karadere segment that ruptured during the İzmit earthquake. The co-seismic delays show an inverse correlation with the D_r values, except for stations BV and CH. A comparison of the results in Figures 13a and 13b indicates that the PGA values, observed time delays, and D_r values, all are generally correlated. Station VO which is inside the Karadere segment recorded up to 0.9 g during the Düzce main shock, and has up to 30 ms co-seismic delays. In comparison, peak accelerations recorded at station FP ~400 m away from station VO is 0.16 g and the largest co-seismic delay there is ~18 ms. However, station BV that is close to the western terminus of the Düzce rupture zone recorded only 0.1 g during the Düzce main shock and has small co-seismic delays. These low values suggest that station BV is probably located on a competent rock site near the end of the Düzce rupture zone. This is consistent with the relatively low triggers at that station during the experiment (SEEBER *et al.*, 2000). Stations MO and LS which are inside the Karadere segment have similar ground motions and co-seismic delays to those at nearby stations FI, GE, and WF. The small PGA and co-seismic delays at station CH could be explained by the relatively large distance from the Düzce epicenter. The different responses in fault zone stations VO, MO and LS indicate strong variations of motion amplification along the strike of the fault.

6. Discussion and Conclusions

We constructed a data set consisting of 36 clusters of highly repeating earthquakes in the aftershock zones of the 1999 M_w7.4 İzmit and M_w7.1 Düzce earthquakes. The decay rates of repeating events in individual clusters following the Düzce main shock are compatible with the $1/t$ Omori law for aftershock decay rates (OMORI, 1895), although other functions can also fit the limited available data (e.g., KISSLINGER, 1996; BEN-ZION and LYAKHOVSKY, 2006). Using waveforms generated by the repeating earthquake clusters, we identified systematic changes in travel times of the *S* and early *S*-coda waves relative to the *P* waves. We interpret the travel-time changes as temporal variations of seismic velocity in the shallow crust around the

Karadere-Düzce branch of the NAF that ruptured during the two main shocks. The co-seismic delay times at the various stations correlate well with the corresponding strong ground motions recorded during the Düzce main shock, indicating that shaking-induced damage is the most likely mechanism for the observed temporal changes. The amplitudes of delay times after the Düzce main shock follow a logarithmic decay, showing a post-main shock recovery of the shallow rock damage. Our results are compatible with recent observations of abrupt velocity reductions followed by post-seismic recoveries associated with several medium to large ($M > 5$) earthquakes, including the Loma Prieta earthquake and its largest aftershock (RUBINSTEIN and BEROZA, 2004a,b), the Morgan Hill earthquake (SCHAFF and BEROZA, 2004), the Hector Mine earthquake (VIDALE and LI, 2003), and the 2004 Parkfield earthquake (RUBINSTEIN and BEROZA, 2005).

The largest co-seismic velocity decrease is observed at station VO that was deployed inside the rupture zone of the 1999 İzmit earthquake and recorded ~0.9 g ground acceleration during the Düzce main shock. The observed co-seismic changes and ground motions generated by the Düzce main shock at the different stations generally decrease with normal distance from the rupture zone. As summarized by BEN-ZION and SAMMIS (2003), large fault zones are characterized by zones of damaged rock with intense microcracks and lower seismic velocity than the surrounding rocks. Such zones are expected to produce large motion amplification (BEN-ZION and AKI, 1990), as observed here and the studies of CORMIER and SPUDICH (1984), SPUDICH and OLSEN (2001), SEEBER *et al.* (2000), and ROVELLI *et al.* (2002). Recent systematic analyses of fault zone trapped waves (e.g., BEN-ZION *et al.*, 2003; PENG *et al.*, 2003; LEWIS *et al.*, 2005) and near-fault crustal anisotropy (e.g., COCHRAN *et al.*, 2003; BONESS and ZOBACK, 2004; PENG and BEN-ZION, 2004; LIU *et al.*, 2004) associated with several large earthquakes (including the İzmit and Düzce events) showed that the belts of damaged fault zone rock are confined primarily to the top 3–4 km of the crust. Seismic waves generated by sources outside such shallow low velocity fault zone layers can be significantly amplified inside the damaged fault zone layers (e.g., BEN-ZION *et al.*, 2003; FOHRMANN *et al.*, 2004). Owing to the nonlinearity of damage evolution (e.g., LYAKHOVSKY *et al.*, 1997), it is easier to produce additional damage in a rock volume that is already damaged. The above factors contribute to the observed large ground motions and co-seismic delays at the fault zone stations. Our results also show strong variations of fault zone amplification along strike, compatible with our previous analyses of fault zone trapped waves and anisotropy.

The observation that stations clearly outside the fault zone also have clear changes in delay times implies that the rock damage induced by the İzmit and Düzce main shocks is not limited to the immediate vicinity of the fault zone. The co-seismic and post-main shock patterns generated by repeating earthquake clusters of different depth are very similar, indicating that the changes of material properties occur in the shallow crust. The observation of similar patterns following the occurrence of smaller earthquakes (RUBINSTEIN and BEROZA, 2004a,b, 2005) suggest strongly that the damage occurred in the very shallow, heavily damaged, surface layer of the crust.

While our observations and other studies suggest that the dominant changes occur near stations and are likely to be shallow, we do not rule out small temporal changes near the source or along the propagation path. The 8 clusters along the Karadere segment and the 28 clusters around the Eften Lake in Figure 6 exhibit similar trends in delays; however, there are slight variations in the results. Although most of the variations may be explained by the differences in occurrence times that are modulated by the Düzce main shock, the co-seismic delays and decay rates after the Düzce main shock for the 8 clusters along the Karadere segment appear to be smaller than those associated with the 28 clusters around the Eften Lake. Since the latter 28 clusters around the Eften Lake are located on the north-dipping faults that failed during the Düzce main shock, the difference may include changes near the source or along the propagation path. However, the delays at station BV that is within 1 km of the Düzce rupture zone are very small for all clusters and we do not find systematic relations between the delays and the hypocentral depths of the clusters. These observations indicate that changes near the stations play the dominant role in explaining the general behavior of the time-dependent effects in our data.

Some properties of the regions generating the time-dependent effects may be inferred by examining the relation between the portions of the seismograms showing the largest effects and a clear reference phase like the direct *S* arrival. In the example of Figure 4, the largest delays are observed in the early *S*-coda waves and similar results were obtained by NIU *et al.* (2003) for the Parkfield segment of the San Andreas fault. SCHAFF and BEROZA (2004) found that the delays are peaked at around 2–3 s after the *S* arrival and then drop back for the aftershock zone of the Loma Prieta earthquake. We performed a systematic check for all 36 clusters at all 10 stations, but did not find any clear relation between the largest delays and the times after the *S* arrival. We speculate that although the velocity reductions occur mostly near the surface, the distribution of the regions generating the changes is not uniform. As indicated before, regions with higher initial damage are expected to produce larger time-dependent effects. It is likely that the fault zone stations (especially VO) pick some changes associated with the fault zone layer producing trapped waves, but the observed temporal changes are widespread and indicate overall a shallow regional origin associated with a near-surface layer.

Laboratory studies have shown that rocks, soils, and other granular materials exhibit universal nonlinear characteristics in response to dynamic deformation (e.g., JOHNSON and SUTIN, 2005). These characteristics include a decrease in elastic modulus or wave speed and increased dissipation with increasing input strain, which are generally termed "nonlinear fast dynamics" (NFD). Another important behavior termed "slow dynamics" (SD) is a logarithmic recovery of material properties over 10^3–10^4 s after dynamic excitation (e.g., TEN CATE and SHANKLAND, 1996; TEN CATE *et al.*, 2000). In the last 15 years, wide-spread nonlinear strong ground motion effects were reported at sediment sites (e.g., CHIN and AKI, 1991; FIELD *et al.*, 1997; AGUIRRE and IRIKURA, 1997; PAVLENKO and IRIKURA, 2002, 2003; BOUCHON *et al.*, 2004). The validity of some of those observations has been questioned by WENNERBERG (1996) and O'CONNELL (1999). Our observations of co-seismic changes of seismic velocities followed by logarithmic recovery, along with those others (VIDALE and LI, 2003; SCHAFF and BEROZA, 2004; RUBINSTEIN and BEROZA, 2004a,b), provide additional support that nonlinearity is wide-spread in the shallow crust during strong ground motion of moderate to large earthquakes. However, it is important to note that the time scale of the logarithmic recovery observed in the above lab studies is on the order of hours to days, rather than the months to years scales associated with this and other field studies. Longer recovery times than those observed in the dynamic deformation experiments are implied by the quasi-static experiments associated with rate- and state-dependent friction (e.g., DIETERICH, 1978; BEELER *et al.*, 1994; DIETERICH and KILGORE, 1996; MARONE, 1998). The differences in the recovery time scales probably reflect differences in the mechanisms operating during the dynamically-driven lab experiments, quasi-static lab experiments, and the yet slower quasi-static process in the field, as well as (potentially large) differences in the initial damage and frequency content of the waves.

Because the earliest repeating earthquakes that were used in this study were at least 0.1 days (2.4 hours) after the Düzce main shock, the co-seismic delays were estimated from the differences between pre-Düzce delay times and post-Düzce delay times at 0.1 day after the main shock. Since the changes follow a logarithmic recovery with time, we expect larger changes in time right after the main shock than those in latter times. Thus, the obtained co-seismic delays in this study are lower bounds. A systematic search for repeating earthquakes in the early aftershock period (e.g., PENG and VIDALE, 2004) could provide useful information on the co-seismic changes in material properties of the shallow crust, and how they evolve within the initial minutes to hours after strong shaking. However, it is extremely difficult to obtain reliable measurements in the noisy coda waves generated by main shocks and large aftershocks.

In a previous analysis of shear-wave splitting using the same data set (PENG and BEN-ZION, 2005), we found only small (2% or less) co-seismic changes of anisotropy parameters at the time of the Düzce main shock, and no appreciable post-main shock recoveries (or pre-main shock precursory evolution). In this study we found clear travel-time changes in *S* and early *S*-coda waves (up to 30 ms). The obtained values correspond to fractional slowness (inverse of velocity) changes of up to 3%, assuming that the decay is accumulated within the 1-s early *S*-coda waves window. This is similar to the values obtained in other studies (e.g., VIDALE and LI, 2003; SCHAFF and BEROZA, 2004). We cannot compare directly the percentage of changes obtained from the earlier shear-wave splitting analysis (PENG and BEN-ZION, 2005) and the early *S*-coda waves of this study. However, it is clear that measuring travel-time changes in coda waves by the sliding window waveform cross-correlation technique generally provides a better way of detecting possible temporal changes than analysis of shear-wave splitting. This is probably because of the following three reasons. First, shear-wave splitting is sensitive to a dominant subset of cracks with given orientation, whereas coda waves are sensitive to the entire crack population. Second, shear-wave splitting delay times are obtained from the direct *S* waves, which are typically less sensitive to possible temporal changes in the medium than the *S*-coda waves with longer travel time and propagation distance (e.g., SNIEDER, 2006). Third, standard shear-wave splitting analysis typically requires clear *S* waves that are recorded on three-component instruments, similar wave forms between fast and slow *S* waves, and earthquake sources within the shear-wave window (NUTTLI, 1961). Sometimes strict criteria are applied to obtain a "high quality" shear-wave splitting measurement (e.g., MATCHAM *et al.*, 2000; PENG and BEN-ZION, 2004, 2005). These requirements prevent many seismograms from being used in the anisotropy analysis. On the other hand, the sliding window waveform cross-correlation technique can be applied to single-component seismograms only (e.g., SCHAFF and BEROZA, 2004), and for earthquakes at various distances from the recording stations. Thus, although shear-wave splitting is a useful tool for analyzing structural anisotropy in the crust and upper mantle (e.g., LEARY *et al.*, 1990; SILVER, 1996; SAVAGE, 1999, and references therein), it appears that measuring travel-time changes in the *S*-coda waves has significant advantages over shear-wave splitting analysis in terms of identifying temporal changes of crustal properties.

We note that the observations discussed in this work and our previous studies on the north Anatolian fault and other large faults and rupture zones (BEN-ZION *et al.*, 2003; PENG *et al.*, 2003; LIU *et al.*, 2004, 2005; PENG and BEN-ZION, 2004, 2005; LEWIS *et al.*, 2005), as well as other related works based on trapped waves, anisotropy and coda waves (e.g., VIDALE and LI, 2003; SCHAFF and BEROZA, 2004; RUBINSTEIN and BEROZA, 2004a,b) reflect only properties of the shallow (e.g., ~3–4 km) or very shallow (e.g., ~200 m) top sections of the crust. Imaging the key mechanical structure of faults at seismogenic depths, where earthquakes nucleate and the bulk of co-seismic slip occurs, remains a significant observational challenge. Fault zone head waves refracting along material discontinuity interfaces in the structure of large faults provide a promising approach for resolving important fault zone properties throughout the seismogenic zone (BEN-ZION *et al.*, 1992; MCGUIRE and BEN-ZION, 2005).

Acknowledgments

We are grateful to John Armbruster, David Okaya, Naside Ozer and Nano Seeber for essential help with the fieldwork, and to PASSCAL for providing the equipment and technical support for the experiment. We thank Fenglin Niu for sharing his sliding window cross-correlation code, and Elizabeth Cochran, Joan Gomberg, Yunfeng Liu, and Paul Silver for useful discussions. The manuscript benefited from valuable comments by referees Justin Rubinstein and Roel Snieder and Editor Willie Lee. All figures were generated using GMT (WESSEL and SMITH, 1998). The study was supported by the National Science Foundation (Grant EAR0003401).

REFERENCES

AGUIRRE, J. and IRIKURA, K. (1997), *Nonlinearity, liquefaction, and velocity variation of soft soil layers in Port Island, Kobe, during the Hyogoken-Nanbu earthquake*, Bull. Seismol. Soc. Am. *87*, 1244–1258.

AKI, K. (1985), *Theory of earthquake prediction with special reference to monitoring of the quality factor of lithosphere by the coda method*, Earthq. Pred. Res. *3*, 219–230.

AKI, K. and CHOUET, L.B. (1975), *Origin of coda waves: Source, attenuation, and scattering effects*, J. Geophys. Res. *80*, 3322–3342.

ANTOLIK, M., NADEAU, R.M., ASTER, R.C., and MCEVILLY, T.V. (1996), *Differential analysis of coda Q using similar microearthquakes in seismic gaps, Part 2: Application to seismograms recorded by the Parkfield High Resolution Seismic Network*, Bull. Seismol. Soc. Am. *86*, 890–910.

ASTER, R.C., SHEARER, P.M., and BERGER, J. (1990), *Quantitative measurements of shear-wave polarizations at the Anza seismic network, southern California: Implications for shear-wave splitting and earthquake prediction*, J. Geophys. Res. *95*, 12449–12473.

ASTER, R.C., SHEARER, P.M., and BERGER, J. (1991), *Comments on Quantitative measurements of shear-wave polarization at the Anza seismic network, Southern California: implications for shear-wave splitting and earthquake prediction by ASTER, R.C., Shearer, P.M. and Berger J. – Reply*, J. Geophys. Res. *96*, 6415–6419.

BEELER, N., TULLIS T.E., and WEEKS, J.D. (1994), *The roles of time and displacement in the evolution effect in rock friction*, Geophys. Res. Lett. *21*, 1987–1990.

BEN-ZION, Y., *Appendix 2. Key formulas in earthquake seismology*. In *The IASPEI International Handbook of Earthquake and Engineering Seismology, Part B* (eds. Lee, W.H.K., Kanamori, H., Jennings, P.C., and Kisslinger C.) (Academic Press, London 2003) pp. 1857–1875.

BEN-ZION, Y. and AKI, K. (1990), *Seismic radiation from an SH line source in a laterally heterogeneous planar fault zone*, Bull. Seismol. Soc. Am. *80*, 971–994.

BEN-ZION, Y., KATZ, S., and LEARY, P. (1992), *Joint inversion of fault zone head waves and direct P arrivals for crustal structure near major faults*, J. Geophys. Res. *97*, 1943–1951.

BEN-ZION, Y. and LYAKHOVSKY, V. (2006), *Analysis of aftershocks in a lithospheric model with seismogenic zone governed by damage rheology*, Geophys. J. Int., 164, doi: 10.1111/j.1365-246x.2006.02878.x.

BEN-ZION, Y., PENG, Z., OKAYA, D., SEEBER, L., ARMBRUSTER, J.G., OZER, N., MICHAEL, A.J., BARIS, S., and AKTAR, M. (2003), *A shallow fault zone structure illuminated by trapped waves in the Karadere-Düzce branch of the north Anatolian Fault, western Turkey*, Geophys. J. Int. *152*, 699–717.

BEN-ZION, Y. and SAMMIS, C.G. (2003), *Characterization of fault zones*, Pure Appl. Geophys. *160*, 677–715.

BEN-ZION, Y. and ZHU, L. (2002), *Potency-magnitude scaling relations for southern California earthquakes with $1.0 < M_L < 7.0$*, Geophys. J. Int. *148*, F1–F5.

BEROZA, G.C., COLE, A.T., and ELLSWORTH, W.L. (1995), *Stability of coda wave attenuation during the Loma Prieta, California, earthquake sequence*, J. Geophys. Res. *100*, 3977–3987.

BOKELMANN, G.H.R. and HARJES, H.P. (2000), *Evidence for temporal variation of seismic velocity within the upper continental crust*, J. Geophys. Res. *105*, 23879–23894.

BONESS, N.L. and ZOBACK, M.D. (2004), *Stress-induced seismic velocity anisotropy and physical properties in the SAFOD Pilot Hole in Parkfield, CA*, Geophys. Res. Lett. *31*, L15S17, doi:10.1029/2004GL019020.

BOUCHON, M., KARABULUT, H., DIETRICH, M., BOUIN, M., and AKTAR, M. (2004), *Effect of fault zone on strong ground motion*, EOS Trans. AGU *85(47)*, Fall Meet. Suppl., Abstract S22B-05.

CASTRO, R.R., MUCCIARELLI, M., PACOR, F., and PETRUNGARO, C. (1997), *S-wave site-response estimates using horizontal-to-vertical spectral ratios*, Bull. Seismol. Soc. Am. *87*(1), 256–260, 1997.

CHIN, B. and AKI, K. (1991), *Simultaneous study of the source, path, and site effects on strong ground motion during the 1989 Loma Prieta earthquake: A preliminary result on pervasive nonlinear site effects*, Bull. Seismol. Soc. Am. *81*, 1859–1884.

COCHRAN, E.S., VIDALE, J.E., and LI, Y.-G. (2003), *Near-fault anisotropy following the Hector Mine earthquake*, J. Geophys. Res. *108*(B9), 2436, doi:10.1029/2002JB002352.

CORMIER, V.F. and SPUDICH, P. (1984), *Amplification of ground motion and waveform complexities in fault zones: Examples from the San Andreas and the Calaveras faults*, Geophys. J. R. Astr. Soc. *79*, 135–152.

CRAMPIN, S. and GAO, Y. (2005), *Comment on "Systematic Analysis of Shear-Wave Splitting in the Aftershock Zone of the 1999 Chi-Chi, Taiwan, Earthquake: Shallow Crustal Anisotropy and Lack of Precursory Changes" by LIU, Y. Teng, T.-L., BEN-ZION, Y.*, Bull. Seismol. Soc. Am. *95*, 354–360.

CRAMPIN, S., BOOTH, D.C., EVANS, R., PEACOCK, S., and FLETCHER, J.B. (1990), *Change in shear-wave splitting at Anza near the time of the North Palm Springs earthquake*, J. Geophys. Res. *95*, 11197–11212.

CRAMPIN, S., BOOTH, D.C., EVANS, R., PEACOCK, S., and FLETCHER, J.B. (1991), *Comment on "Quantitative measurements of shear-wave polarizations at the Anza seismic network, Southern California: Implications for shear-wave splitting and earthquake prediction" by ASTER, R.C., Shearer, P.M., and Berger, J.*, J. Geophys. Res. *96*, 6403–6414.

DAS, S. and SCHOLZ, C.H. (1981), *Theory of time-dependent rupture in the earth*, J. Geophys. Res. *86*, 6039–6051.

DIETERICH, J.H. (1978), *Time-dependent friction and the mechanics of stick-slip*, Pure Appl. Geophys. *116*, 790–806.

DIETERICH, J.H. (1994), *A constitutive law for rate of earthquake production and its application to earthquake clustering*, J. Geophys. Res. *99*, 2601–2618.

DIETERICH, J.H. and KILGORE, B.D. (1996), *Imaging surface contacts; power law contact distributions and contact stresses in quarts, calcite, glass, and acrylic plastic*, Tectonophysics *256*, 219–239.

FIELD, E.H., JOHNSON, P.A., BERESNEV, I.A., and ZENG, Y., (1997), *Nonlinear ground-motion amplification by sediments during the 1994 Northridge earthquake*, Nature *390*, 599–602.

FOHRMANN, M., JAHNKE, G., IGEL, H., and BEN-ZION, Y. (2004), *Guided waves from sources out side faults: An indication of shallow trapping structure?*, Pure Appl. Geophys. *161*, 2125–2137.

GRET, A., SNIEDER, R., ASTER, R.C., and Kyle, P.R. (2005), *Monitoring rapid temporal change in a volcano with coda wave interferometry*, Geophys. Res. Lett. *32*, L06304, doi:10.1029/2004GL021143.

GOLDSTEIN, P., DODGE, D., FIRPO, M., and MINNER, L., *SAC2000: Signal processing and analysis tools for seismologists and engineers*. In *The IASPEI International Handbook of Earthquake and Engineering Seismology, Part B* (eds. Lee, W.H.K., KANAMORI, H., Jennings, P.C., and KISSLINGER C.) (Academic Press, London 2003), Chapter 85.5.

GOT, J. L., and COUTANT, O. (1997), *Anisotropic scattering inferred from travel-time delay analysis in Kilauea volcano, Hawaii, earthquake coda waves*, J. Geophys. Res. *102*, 8397–8410.

GUPTA, I.N. (1973), *Premonitory variations in S-wave velocity anisotropy before earthquakes in Nevada*, Science *182*, 1129–1132.

HAMIEL, Y., LIU, Y., LYAKHOVSKY, V., BEN-ZION, Y., and LOCKNER, D. (2004), *A Visco-elastic damage model with applications to stable and unstable fracturing*, Geophys. J. Int. *159*, 1155–1165, doi: 10.1111/j.1365-246X.2004.02452.x.

HARTLEB, R.D., DOLAN, J.F., AKYÜZ, H.S., DAWSON, T.E., TUCKER, A.Z., YERLI, B., ROCKWELL, T.K., TORAMAN, E., ÇAKIR, Z., DIKBAS, A., and ALTUNEL, E. (2002), *Surface rupture and slip distribution along the Karadere segment of the 17 August 1999 İzmit and the western section of the 12 November 1999 Düzce, Turkey, earthquakes*, Bull. Seismol. Soc. Am. *92*(1), 67–78.

HELLWEG, M., SPUDICH, P., FLETCHER, J. and BAKER, L.M. (1995), *Stability of coda Q in the region of Parkfield, California: The view from the USGS Parkfield dense seismograph array*, J. Geophys. Res. *100*, 2089–2102.

IGARASHI, T., MATSUZAWA, T. and HASEGAWA, A. (2003), *Repeating earthquakes and interplate aseismic slip in the northeastern Japan subduction zone*. J. Geophys. Res. *108*(B5), 2249, doi:10.1029/2002JB001920.

IKUTA, R. and YAMAOKA, K. (2004), *Temporal variation in the shear-wave anisotropy detected using the Accurately Controlled Routinely Operated Signal System (ACROSS)*, J. Geophys. Res. *109*, B09305, doi:10.1029/2003JB002901.

JAEGER, J.C. and COOK, N.G.W., *Fundamentals of Rock Mechanics* (Chapman and Hall, New York 1976).

JIN, A. and AKI, K. (1986), *Temporal change in coda Q before the Tangshan earthquake of 1976 and the Haicheng earthquake of 1975*, J. Geophys. Res. *91*, 665–673.

JOHNSON, P. and SUTIN, A. (2005), *Slow dynamics and anomalous nonlinear fast dynamics in diverse solids*, J. Acoust. Soc. Am. *117*, 124–130.

KANAMORI, H. and ANDERSON, D.L. (1975), *Theoretical basis of some empirical relations in seismology*, Bull. Seismol. Soc. Am. *65*, 1073–1095.

KANAMORI, H. and FUIS, G. (1976), *Variation of P-wave velocity before and after the Galway Lake earthquake (M_L = 5.2) and the Goat Mountain earthquakes (M_L = 4.7, 4.7), 1975, in the Mojave desert, California*, Bull. Seismol. Soc. Am. *66*, 2027–2037.

KASAHARA, K., *Earthquake Mechanics* (Cambridge University Press, New York 1981).

KLEIN, F.W. (1978), *Hypocenter location program HYPOINVERSE*, USGS, Open File Report, 78–694.

KISSLINGER, C. (1996), *Aftershocks and fault-zone properties*, Adv. Geophys. *38*, 1–36.

LEARY, P.C., CRAMPIN, S., and MCEVILLY, T.V. (1990), *Seismic fracture anisotropy in the Earth's crust: An overview*, J. Geophys. Res. *95*(7), 11105–11114.

LEWIS, M.A, PENG, Z., BEN-ZION, Y., and VERNON, F.L. (2005), *Shallow seismic trapping structure in the San Jacinto fault zone near Anza, California*, Geophys. J. Int. *162*(3), 867–881, doi: 10.1111/j.1365-246X.2005.02684.

LIU, Y., TENG, T.-L., and BEN-ZION, Y. (2004), *Systematic analysis of shear-wave splitting in the aftershock zone of the 1999 Chi-Chi earthquake: Shallow crustal anisotropy and lack of precursory variations*, Bull. Seismol. Soc. Am. *94*, 2330–2347.

LIU, Y., BEN-ZION, Y., and TENG T.-L. (2005), *Reply to Comment by Crampin and Gao on "Systematic Analysis of Shear-Wave Splitting in the Aftershock Zone of the Chi-Chi, Taiwan, Earthquake: Shallow Crustal Anisotropy and Lack of Precursory Changes," by LIU, Y. Teng, T.-L., and BEN-*ZION, Y., Bull. Seismol. Soc. Am. *95*, 361–366.

LOCKNER, D. A., BYERLEE, J.D., KUKSENKO, V., PONOMAREV, A., and SIDORIN, A. (1992), *Observations of quasi-static fault growth from acoustic emissions*. In *Fault Mechanics and Transport Properties of Rocks*, International Geophysics Series (eds. Evans, B., and Wong, T.-f.) (Academic Press, San Diego, 1992) 51, 3–31.

LYAKHOVSKY, V., BEN-ZION, Y., and AGNON, A. (1997), *Distributed damage, faulting and friction*, J. Geophys. Res. *102*, 27635–27649.

LYAKHOVSKY, V., BEN-ZION Y., and Agnon, A. (2005), *A visco-elastic damage rheology and rate- and state-dependent friction*, Geophys. J. Int. *161*, 179–190, doi: 10.1111/j.1365-246X.2005.02583.x.

MARONE, C. (1998), *Laboratory-derived frictional laws and their application to seismic faulting*, Annu. Rev. Earth Planet. Sci. *26*, 643–696.

MATCHAM, I., SAVAGE, M.K., and GLEDHILL, K.R. (2000), *Distribution of seismic anisotropy in the subduction zone beneath the Wellington region, New Zealand*, Geophys. J. Int. *140*, 1–10.

MATSUMOTO, S., OBARA, K., YOSHIMOTO, T., SAITO, A., ITO, A., and HASEGAWA, A. (2001), *Temporal change in P-wave scatterer distribution associated with the M6.1 earthquake near Iwate volcano, northeASTERn* Japan, Geophys. J. Int. *145*, 48–58.

MCGUIRE, J. and BEN-ZION, Y. (2005), *High-resolution imaging of the Bear Valley section of the San Andreas Fault at seismogenic depths with fault-zone head waves and relocated seismicity*, Geophys. J. Int. *163*, 152–164, doi: 10.1111/j.1365-246X.2005.02703.x.

MOGI, K. (1962), *Study of the elastic shocks caused by the fracture of heterogeneous materials and its relation to earthquake phenomena*, Bull. Earthquake Res. Inst. Tokyo Univ. *40*, 125–173.

MCEVILLY, T.V. and JOHNSON, L.R. (1974), *Stability of P and S velocities from central California quarry blasts*, Bull. Seismol. Soc. Am. *64*, 343–353.

NADEAU, R.M., ANTOLIK, M., JOHNSON, P.A., FOXALL, W., and MCEVILLY, T.V. (1994), *Seismological studies at Parkfield III: Microearthquake clusters in the study of fault-zone dynamics*, Bull. Seismol. Soc. Am. *84*, 247–263.

NIU, F., SILVER, P.G., NADEAU, R.M., and MCEVILLY, T.V. (2003), *Stress-induced migration of seismic scatterers associated with the 1993 Parkfield aseismic transient event*, Nature *426*, 544–548.

NUTTLI, O. (1961), *The effect of earth's surface on the S-wave particle motion*, Bull. Seismol. Soc. Am. *51*, 237–246.

OMORI, F. (1895), *On the aftershocks of earthquakes*, J. Coll. Sci. Imp. Univ. Tokyo. *7*, 111–200.

O'CONNELL, D.R.H. (1999), *Replication of apparent nonlinear seismic response with linear wave propagation models*, Science *283*, 2045–2050.

PAVLENKO, O.V. and IRIKURA, K. (2002), *Changes in shear moduli of liquefied and nonliquefied soils during the 1995 Kobe earthquake and its aftershocks at three vertical-array sites*, Bull. Seismol. Soc. Am. *92*, 1952–1969.

PAVLENKO, O.V. and IRIKURA, K. (2003), *Estimation of nonlinear time-dependent soil behavior in strong ground notion based on vertical array data*, Pure Appl. Geophys. *160*, 2365–2379.

PENG, Z., BEN-ZION, Y., MICHAEL, A.J., and ZHU, L. (2003), *Quantitative analysis of fault zone waves in the rupture zone of the Landers, 1992, California earthquake: Evidence for a shallow trapping structure*, Geophys. J. Int. *155*, 1021–1041.

PENG, Z. and BEN-ZION, Y. (2004), *Systematic analysis of crustal anisotropy along the Karadere-Düzce branch of the north Anatolian fault*, Geophys. J. Int. *159*, 253–274, doi: 10.1111/j.1365-246X.2004.02379.x.

PENG, Z. and BEN-ZION, Y. (2005), *Spatio-temporal variations of crustal anisotropy from similar events in aftershocks of the 1999 M7.4 İzmit and M7.1 Düzce, Turkey, earthquake sequences*, Geophys. J. Int. *160*(3), 1027–1043, doi: 10.1111/j.1365-246X.2005.02569.x.

PENG, Z., VIDALE, J.E., MARONE, C., and RUBIN, A. (2005), *Systemic variations in recurrence interval and moment of repeating aftershocks*, Geophys. Res. Lett. *32*(15), L15301, doi: 10.1029/2005GL022626.

PENG, Z. and VIDALE, J.E. (2004), *Early aftershock decay rates of the M6 Parkfield earthquake*, EOS Trans. AGU, *85*(47), Fall Meet. Suppl., Abstract S51C-0170X.

POUPINET, G., ELLSWORTH, W.L., and FRECHET, J. (1984), *Monitoring velocity variations in the crust using earthquake doublets: An application to the Calaveras Fault, California*, J. Geophys. Res. *89*, 5719–5731.

ROVELLI, A., CASERTA, A., MARRA, F., and RUGGIERO, V. (2002), *Can seismic waves be trapped inside an inactive fault zone? The case study of Nocera Umbra, Central Italy*, Bull. Seismol. Soc. Am. *92*, 2217–2232.

ROUMELIOTI, Z. and BERESNEV, I.A. (2003), *Stochastic finite-fault modeling of ground motions from the 1999 Chi-chi, Taiwan, earthquake: Application to rock and soil sites with implications for nonlinear site response*, Bull. Seismol. Soc. Am. *93*, 1691–1702.

REILINGER, R.E., MCCLUSKY, S.C., ORAL, M.B., KING, R.W., TOKSOZ, M.N., BARKA, A.A., KINIK, I., LENK, O., and SANLI, I. (1997), *Global Positioning System measurements of the present day crustal movements in the Arabia-Africa-Eurasia plate collision zone*, J. Geophys. Res. *102*, 9983–9999.

RUBIN, A.M. (2002), *Using repeating earthquakes to correct high-precision earthquake catalogs for time-dependent station delays*, Bull. Seismol. Soc. Am. *92*, 1647–1659.

RUBINSTEIN, J.L. and BEROZA, G.C. (2004a), *Evidence for widespread nonlinear strong ground motion in the M_w6.9 Loma Prieta earthquake*, Bull. Seismol. Soc. Am. *94*(5), 1595–1608.

RUBINSTEIN, J.L. and BEROZA, G.C. (2004b), *Nonlinear strong ground motion in the Ml5.4 Chittenden earthquake: Evidence that pre-existing damage increases susceptibility to further damage*, Geophys. Res. Lett. *31*, L23614, doi: 10.1029/2004GL021357.

RUBINSTEIN, J.L. and BEROZA, G.C. (2005). *Depth constraints on nonlinear strong ground motion*, Geophys. Res. Lett. *32*, L14313, doi:10.1029/2005GL023189.

RYALL, A. and SAVAGE, W. (1974), *S wave splitting: Key to earthquake prediction?* Bull. Seismol. Soc. Am. *64*, 1943–1951.

SAVAGE, M.K. (1999), *Seismic anisotropy and mantle deformation: What have we learned from shear-wave splitting*, Rev. Geophys. *37*, 65–106.

SCHAFF, D.P., BEROZA, G.C., and SHAW, B.E. (1998), *Postseismic response of repeating aftershocks*, Geophys. Res. Lett. *25*(24), 4549–4552.

SCHAFF, D.P., BOKELMANN, G.H.R., BEROZA, G.C., WALDHAUSER, F., and ELLSWORTH, W.L. (2002), *High-resolution image of Calaveras fault seismicity*, J. Geophys. Res. *107*(B9), 2186, doi:10.1029/2001JB000633.

SCHAFF, D. P., BOKELMANN, G.H.R., ELLSWORTH, W.L., ZANZERKIA, E., WALDHAUSER, F., and BEROZA, G.C. (2004), *Optimizing correlation techniques for improved earthquake location*, Bull. Seismol. Soc. Am. *94*, 705–721.

SCHAFF, D.P. and BEROZA, G.C. (2004), Coseismic and postseismic velocity changes measured by repeating earthquakes, J. Geophys. Res. *109*, B10302, doi:10.1029/2004JB003011.

SCHOLZ, C.H. (1968), *Microfracturing and the inelastic deformation of rock in compression*, J. Geophys. Res. *73*, 1117–1432.

SEEBER, L., ARMBRUSTER, J.G., OZER, N., AKTAR, M., BARIS, S., OKAYA, D., BEN-ZION, Y., and FIELD, E., *The 1999 earthquake sequence along the North Anatolia transform at the juncture between the two main ruptures*. In *The 1999 İzmit and Düzce Earthquakes: Preliminary Results* (eds. Barka *et al.*) (Istanbul Technical University, Turkey 2000) pp. 209–223.

SILVER, P.G. (1996), *Seismic anisotropy beneath the continents: probing the depths of geology*, Annu. Rev. Earth Planet. Sci. *24*, 385–432.

SNIEDER, R., (2006), *The Theory of Coda Wave Interferometry*, this volume.

SNIEDER, R. and VRIJLANDT, M. (2005), *Constraining the source separation with coda wave interferometry: Theory and application to earthquake doublets in the Hayward fault, California*, J. Geophys. Res. *110*, B04301, doi:10.1029/2004JB003317.

SNIEDER, R., GRET, A., DOUMA, H., and SCALES, J. (2002), *Coda wave interferometry for estimating nonlinear behavior in seismic velocity*, Science *295*, 2253–2255.

SPUDICH, P. and OLSEN, K.B. (2001), *Fault zone amplified waves as a possible seismic hazard along the Calaveras fault in central California*, Geophys. Res. Lett. *28*, 2533–2536.

TEN CATE, J.A. and SHANKLAND, T. (1996), *Slow dynamics in the nonlinear elastic response of Berea sandstone*, Geophys. Res. Lett. *23*, 3019–3022.

TEN CATE, J.A., SMITH, E., and GUYER, R.A. (2000), *Universal slow dynamics in granular solids*, Phys. Rev. Lett. *85*, 1020–1023.

UTKUCU, M., NALBANT, S.S., MCCLOSKEY, J., STEACY, S., and ALPTEKIN, O. (2003), *Slip distribution and stress changes associated with the 1999 November 12, Düzce (Turkey) earthquake (M_w = 7.1)*, Geophys. J. Int. *153*, 229–241.

VIDALE, J.E. and LI, Y.-G. (2003), *Damage to the shallow Landers fault from the nearby Hector Mine earthquake*, Nature *421*, 524–526.

WALDHAUSER, F. and ELLSWORTH W.L. (2000), *A double-difference earthquake location algorithm: Method and application to the northern Hayward fault, California*, Bull. Seismol. Soc. Am. *90*, 1353–1368.

WALDHAUSER, F. and ELLSWORTH, W.L. (2002), *Fault structure and mechanics of the Hayward fault, California, from double-difference earthquake locations*, J. Geophys. Res. *107*(B3), 2054, doi:10.1029/2000JB000084.

WALDHAUSER, F., ELLSWORTH, W.L., SCHAFF, D.P., and COLE, A. (2004), *Streaks, multiplets, and holes: High-resolution spatio-temporal behavior of Parkfield seismicity*, Geophys. Res. Lett. *31*, L18608, doi:10.1029/2004GL020649.

WESSEL, P. and SMITH, W.H.F. (1998), *New, improved version of the Generic Mapping Tools released*, EOS Trans. Amer. Geophys. Union *79*, 579.

WENNERBERG, L. (1996), *Comment on "Simultaneous study on the source, path, and site effects on strong ground motion during the 1989 Loma Prieta earthquake: A preliminary result on pervasive nonlinear site effects" by Chin, B.-H., and Aki, K.*, Bull. Seismol. Soc. Am. *86*, 259–267.

WHITCOMB, J.H., GARMANY, J.E. and ANDERSON, D.L. (1973), *Earthquake prediction: Variation of seismic velocities before the San Francisco earthquake*, Science *180*, 632–635.

YAMAWAKI, T., NISHIMURA, T., and HAMAGUCHI, H. (2004), *Temporal change of seismic structure around Iwate volcano inferred from waveform correlation analysis of similar earthquakes*, Geophys. Res. Lett. *31*, L24616, doi:10.1029/2004GL021103.

ZHU, X. (2003), *Management of large earthquake data using the antelope relational database and seismicity analysis of the 1999 Turkey earthquake sequences*, M.S. Thesis, 96 pp., University of Southern California, Los Angeles, California.

Zöller, G., Hainzl, S., Holschneider, M., and Ben-Zion, Y. (2005), *Aftershocks resulting from creeping sections in a heterogeneous fault*, Geophys. Res. Lett. *32*, L03308, doi:10.1029/2004GL021871.

(Received April 5, 2005, accepted September 9, 2005)

Pure appl. geophys. 163 (2006) 601–616
0033–4553/06/030601–16
DOI 10.1007/s00024-005-0024-8

Pure and Applied Geophysics

Crustal Heterogeneity in the Source Region of the 2004 Mid Niigata Prefecture Earthquake: Inversion Analysis of Coda Envelopes

K. Nishigami[1]

Abstract—The 2004 Mid Niigata Prefecture earthquake (M_{JMA} 6.8) and its aftershock sequences generated complicated, i.e., several conjugate fault planes in their source region. In order to understand the generating process of these earthquakes, we estimated a 3-D distribution of relative scattering coefficients in the source region. The large slip area during the main shock rupture seems to be bounded by strong heterogeneous zones with larger scattering coefficients. Hypocenters of the main shock and major large aftershocks with M 5-6 classes tend to be located close to stronger scattering areas. We found that one of these strong heterogeneities already existed before the occurrence of the M 5.9 aftershock on November 8. We suppose that heterogeneous structures in the source region of this earthquake sequence affected the initiation and growth of ruptures of the main shock and major large aftershocks.

Key words: 2004 Mid Niigata Prefecture earthquake, crustal heterogeneity, scattering coefficient, inversion analysis, coda envelope, rupture process.

1. Introduction

The M_{JMA} 6.8 Mid Niigata Prefecture earthquake occurred in central Japan on Oct. 23, 2004 (Fig. 1). This earthquake was accompanied by many large aftershocks with M 5–6 classes and these earthquake sequences formed a complicated distribution of hypocenters, i.e., several conjugate reverse-fault planes. In order to understand the generating process of these earthquakes, several analyses were made based on temporary seismic observations. For example, Shibutani *et al.* (2005) delineated the conjugate fault planes clearly by a joint hypocenter determination (JHD) method. Kato *et al.* (2005) estimated a 3-D velocity structure using a double-difference tomography and showed that the main shock fault plane is located along the northwest-dipping velocity boundary, which suggests the reactivation of the past normal-fault plane, in the crustal stretching age before ~3.5 Ma (e.g., Sato, 1994), as

[1]Disaster Prevention Research Institute, Kyoto University, Gokasho, Uji, Kyoto, 611-0011, Japan.
E-mail: nishigam@eqh.dpri.kyoto-u.ac.jp

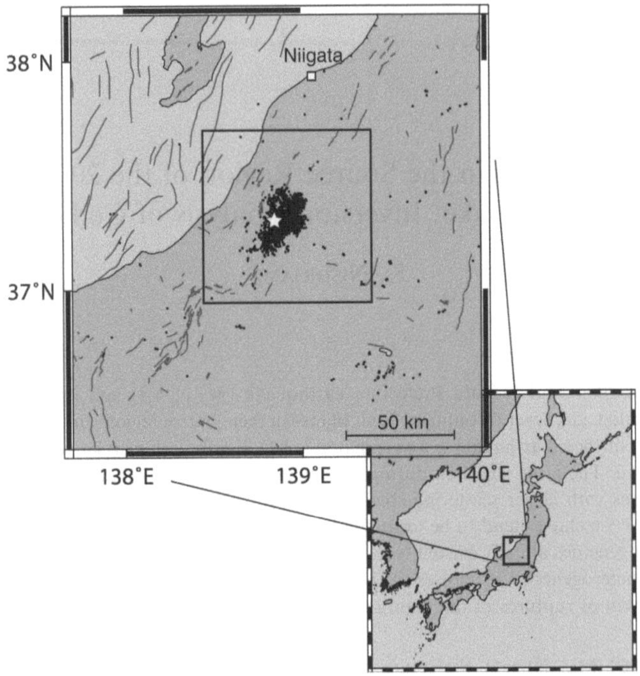

Figure 1
Map showing the analysis area in central Japan. A star and the dots show the Mid Niigata prefecture earthquake (M_{JMA} 6.8) on Oct. 23, 2004, and its aftershocks from Oct. 23 to Nov. 8 (SHIBUTANI *et al.*, 2005), respectively. Solid line segments represent active faults in this region.

a reverse faulting. This earthquake sequence occurred in the large strain rate, i.e., greater than 0.1 ppm/y contraction, zone detected by the GPS measurement, which is called the Niigata-Kobe tectonic zone (SAGIYA *et al.*, 2000), and it is also important to find some structural properties related to such active tectonics. In the present study, we estimate a 3-D distribution of relative scattering coefficients by an inversion of coda envelopes of aftershocks, aiming at elucidating the small-scale heterogeneities related to the generating process of these earthquakes.

The Earth's medium contains heterogeneous structure with various scale length as summarized by WU and AKI (1998). The heterogeneity with scale length greater than the seismic wavelength mainly causes a fluctuation of travel times and amplitudes of the seismic waves. These data observed by a seismographic network can be inverted to estimate a 3-D distribution of seismic wave velocities (e.g., AKI and LEE, 1976). Detailed images of the velocity structure in the lithosphere have been extensively studied and for example, ZHAO *et al.* (2000) revealed the correlation between the generation of large inland earthquakes and heterogeneous structures in the crust, such as low velocity anomalies. On the other hand, heterogeneities with scale length smaller than the seismic wavelength cause strong scattering of the seismic waves, such that we observe an excitation of coda waves. Analyzing the coda waves has a merit to extract small-scale heterogeneities more sensitive to the dimension of seismic sources and the related structures, and also to extract medium properties in a much wider area than analyzing the direct waves. Analysis methods to estimate the statistical or deterministic properties of coda have been extensively developed, as described in SATO and FEHLER (1998).

In the present study we use the inversion analysis of coda waves observed by local or regional seismographic networks. Compared with the conventional array analyses which detect coherent wavetrains and estimate their propagation and generation properties (e.g., NEIDELL and TANER, 1971; CAPON, 1969), this method treats incoherent wave data and tries to locate the secondary sources of scattered wave energies. REVENAUGH (1995) analyzed teleseismic P coda recorded by the Southern California Seismic Network with a Kirchhoff coda migration stacking and detected strong scattering related to the subducted slab at 50–200 km depths. NISHIGAMI (1991) presented an inversion method of coda energy fluctuation observed by a station network to estimate a 3-D distribution of relative scattering coefficients. NISHIGAMI (2000) applied this method to the Northern California Seismic Network and showed the segmentation structure along the San Andreas fault system. ASANO and HASEGAWA (2004) revised this inversion analysis to estimate the absolute scattering coefficients and showed scattering properties along the earthquake source fault of the M_{JMA} 7.3 Western Tottori earthquake in 2000. Inversion analysis of coda envelopes is considered to be one of the effective approaches to detect detailed small-scale heterogeneities in the source region.

2. Data and Analysis

In the present study we use the analysis method presented by NISHIGAMI (1991). We assume spherical source radiation, single isotropic scattering, S-to-S wave conversion for coda, and constant attenuation property in the medium. If the distribution of scatterers is random and uniform in space, the coda energy decays smoothly and it is expressed as exponential at large lapse times (SATO, 1977). We consider that the fluctuation of coda energy observed in the real seismograms is most effectively realized by nonuniform distribution of scattering coefficients in space, under the above assumptions. We can estimate the 3-D distribution of scattering coefficients by deterministic analyses such as inverting the coda energy fluctuation with the use of a large data set (NISHIGAMI, 1991).

In this study we used seven seismic stations in and nearby the source region of the 2004 Mid Niigata Prefecture earthquake (Fig. 2), i.e., 3 Hi-net stations (NGOH, KWNH, YNTH) of the National Research Institute for Earth Science and Disaster Prevention (NIED), one station (HIROKA) of the Japan Meteorological Agency

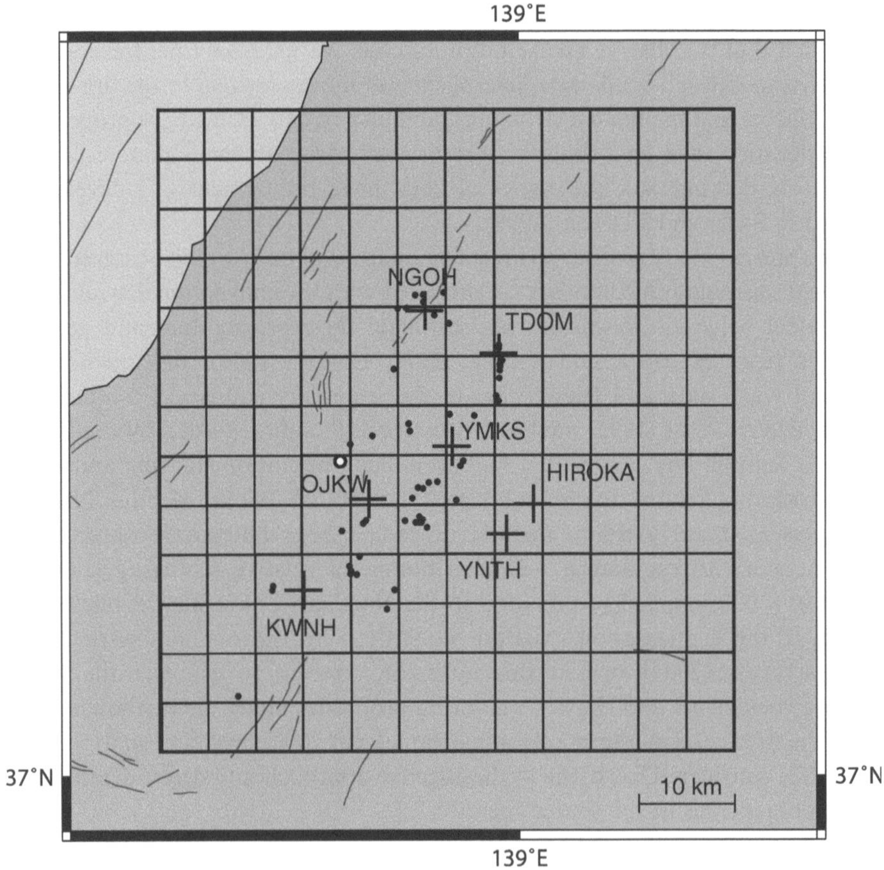

Figure 2
Map showing the block assignment and a distribution of events (dots) and stations (crosses) used in the analysis. An open circle indicates the event whose waveform is shown in Figure 3. Solid line segments represent active faults.

(JMA), and three temporary stations (TODM, OJKW, YMKS) operated by the joint observation of Kyoto and Kyushu Universities (SHIBUTANI, *et al.*, 2005). All are the telemetered stations and the seismometers are velocity-type transducers with a natural frequency of 1 Hz. We chose 70 earthquakes ($2.5 \le M \le 4.5$) which occurred on Nov. 1 to 7, considering their spatial distribution and good signal-to-noise ratio. Especially to avoid the contamination of small successive events in the coda, we used earthquakes more than one week after the main shock occurrence and carefully checked the waveforms. We took the analysis area, 60 km × 65 km in horizontal and 0–60 km in depth, effectively covering the distribution of stations and events selected, and the analysis area was divided into 1,872 small blocks with one side of 5 km in both horizontal and depth as shown in Figure 2.

Figure 3 shows an example of the wave data processing to evaluate the "coda energy residuals", i.e., observational data in an inversion, for an earthquake on Nov. 4 (*M* 2.5) recorded at HIROKA (JMA) station. Coda envelope (c) is calculated from the observed waveform (a) after the direct S wave, by bandpass filtering with 7–15 Hz (b), and taking the running mean with 0.5 s window length after squaring and correcting the geometrical spreading effect. We used only the vertical-component seismograms. We assume a single isotropic scattering and the coda energy thus evaluated is approximated by an exponential decay at lapse times of coda greater than about two times the S wave travel time (SATO, 1977). This approximation is shown by a straight line in the logarithmic coda energy in Figure 3(d). The deviation of coda energy (e) from the average decay curve is averaged at every 0.5 s and finally

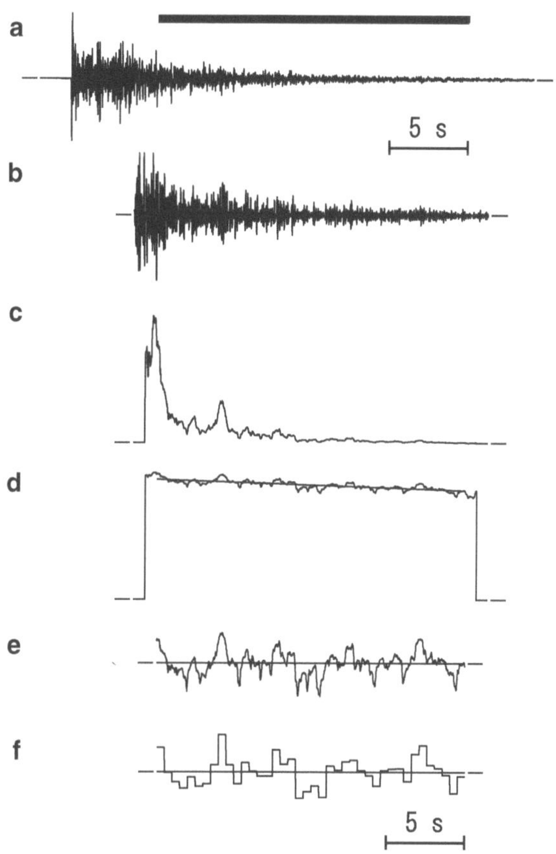

Figure 3
An example of the wave data processing for an earthquake shown by an open circle in Figure 2, recorded at HIROKA; (a) Original waveform with a thick bar indicating the final analysis interval, (b) bandpass filtered waveform, (c) running mean of squared waveform, (d) logarithmic trace with a straight line showing its average decay, (e) logarithmic deviation from the average coda decay, and (f) logarithm of coda energy residuals averaged at every 0.5 s. Time axes are common for all the traces.

the coda energy residuals (f) are obtained. Each of the averaged residuals at respective lapse times is the observational data. In the actual analysis, we took the beginning time of the analysis interval as earlier as possible than two times of S arrivals, by confirming that fitting in Figure 3(d) does not fail, in order to increase the resolution near the source region. The resultant beginning times were approximately 1.5 to 2.0 times the S travel times. The maximum lapse time of the analysis interval was restricted to 20 s measured from the event origin time to prevent the effect of a multiple scattering, which tends to dominate over a single scattering in the late coda (SATO, 1988). Coda energy residuals were thus obtained from 382 seismograms.

The number of observational data is 7,835 and that of the unknown parameters, i.e., relative scattering coefficient in each block, is 1,872. We solved the observational equation by a recursive stochastic inversion (e.g., TARANTOLA, 1987) in the same way as in NISHIGAMI (1991). The initial values in the inversion were taken to be zero perturbation for all the blocks, and the damping factor, which is defined as the ratio of *a priori* variances of data to those of parameters, was taken to be 0.1 to suppress the solutions with large negative perturbations, which may cause negative scattering coefficient. In constructing the observation equation, we assumed a layered velocity structure, i.e., three layers with the velocity and thickness, 2.5 km/s and 3 km, 3.5 km/s and 13 km, and 3.8 km/s and 16 km, respectively on a half space with a velocity of 4.5 km/s, referring to the 1-D velocity model estimated from the aftershocks data (SHIBUTANI *et al.*, 2005).

In the present analysis, solutions fitting the single isotropic scattering model are extracted from the observed coda energy residuals. Other data contained in the coda fluctuation, such as ground noise or site effects at certain stations and contamination of small events in the coda and so on, may cause some noises for the solutions. These noises do not fit the model and their effects can be reduced, i.e., the former kind of noise is reduced by using a large number of stations, and the latter one is reduced by analyzing a large number of events because the contamination of small events is rare and random in time.

3. *Results*

The result of inversion is shown in Figure 4. Solutions with diagonal elements of the resolution matrix, which is called the "resolution" hereafter, greater than 0.2 are

▶

Figure 4
Distribution of (a) relative scattering coefficients, (b) diagonal elements of the resolution matrix, and (c) result of the checkerboard resolution test, estimated for blocks at depths from 0–5 km to 15–20 km or to 35–40 km. Solutions with resolution greater than 0.2 are plotted in (a) and (c) and there are no plots deeper than 15–20 km. Solutions indicated by open and solid symbols in (a) and (c) represent stronger and weaker scattering than the average in the analysis area, respectively.

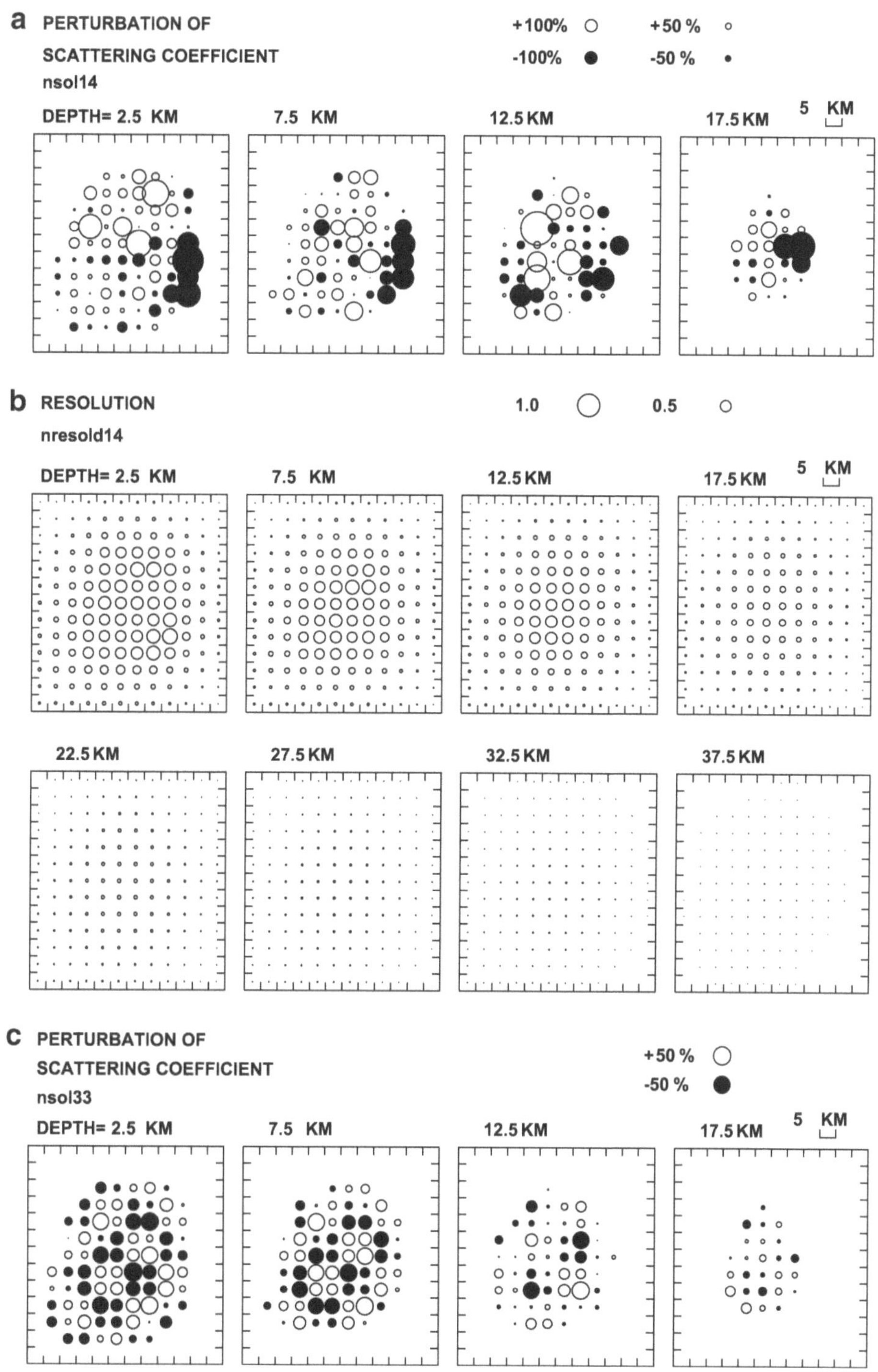
a PERTURBATION OF
SCATTERING COEFFICIENT
nsol14
+100%
+50 %
-100%
-50 %
DEPTH= 2.5 KM
7.5 KM
12.5 KM
17.5 KM
5 KM
b RESOLUTION
nresold14
1.0
0.5
DEPTH= 2.5 KM
7.5 KM
12.5 KM
17.5 KM
5 KM
22.5 KM
27.5 KM
32.5 KM
37.5 KM
c PERTURBATION OF
SCATTERING COEFFICIENT
nsol33
+50 %
-50 %
DEPTH= 2.5 KM
7.5 KM
12.5 KM
17.5 KM
5 KM

plotted at depths from 0–5 km to 15–20 km. There are no plots for the depths below 15–20 km. Solutions denoted by open and solid symbols represent stronger and weaker scattering than the average in the analysis area, respectively. Figure 4(c) shows a result of the checkerboard (CB) resolution test with resolutions greater than 0.2. The CB pattern assumed for the model contains a perturbation of 50% alternating at every 10 km intervals in horizontal and depth. The recovery of the CB pattern is good for solutions with resolution greater than 0.5 and also the polarity of the perturbation is true for solutions with resolution greater than 0.2. The solution shown in Figure 4, therefore, seems to be reliable in the aftershock area with depths approximately shallower than 15 km.

To see the scatterer distribution in more detail, we shifted the block assignment to the east, north, and both directions, respectively, with a half-block interval, and inverted the same data set. The solutions are superimposed in Figure 5. At each depth layer from 0–5 km to 10–15 km, the solutions with resolution greater than 0.2 are plotted, with aftershocks (light gray dots) determined by JHD method (SHIBUTANI *et al.*, 2005), and the main shock (a cross with no. 1) and major large aftershocks with *M* 5–6 classes (crosses with nos. 2–10) within each depth range. Distribution of stronger scatterers denoted by open circles in Figure 5 shows clearer clustering at several places at every depth than that in Figure 4, and therefore Figure 5 is more relevant to extracting the scattering properties in the source region.

Major active faults in this region are represented by red solid lines in Figure 5 (RESEARCH GROUP FOR ACTIVE FAULTS OF JAPAN, 1991). We can recognize a distribution of stronger scatterers at the shallow depth, 0–5 km, along the active faults northwest of the aftershock area. The earthquake fault did not appear on the ground for the main shock and large aftershocks sequences. Field surveys immediately after the main shock revealed the flexure displacements at several points along the Obiro and the Muikamachi faults and these two active faults were estimated to be the possible surface extension of the earthquake source fault (ACTIVE FAULT RESEARCH CENTER, 2004; NAGOYA UNIVERSITY *et al.*, 2005). KATO *et al.* (2005) also showed that the main shock fault plane extends to the surface trace of the Obiro and Muikamachi faults from their relocated distribution of aftershocks by double-difference tomography. Rough locations of these two active faults are indicated by red broken lines in Figure 5, referring to the Geographical Survey Institute (2004). We can find stronger scatterers northwest of the Obiro fault at the shallow depth of 0–5 km.

There exists a large area with weaker scattering denoted by solid symbols, which means more homogeneous medium, southeast of the aftershock area as shown in Figure 5. This area almost corresponds to the region where the basement rock outcrops (ACTIVE FAULT RESEARCH CENTER, 2004), and this shows a good correlation between the surface geology and the scattering properties estimated in this study.

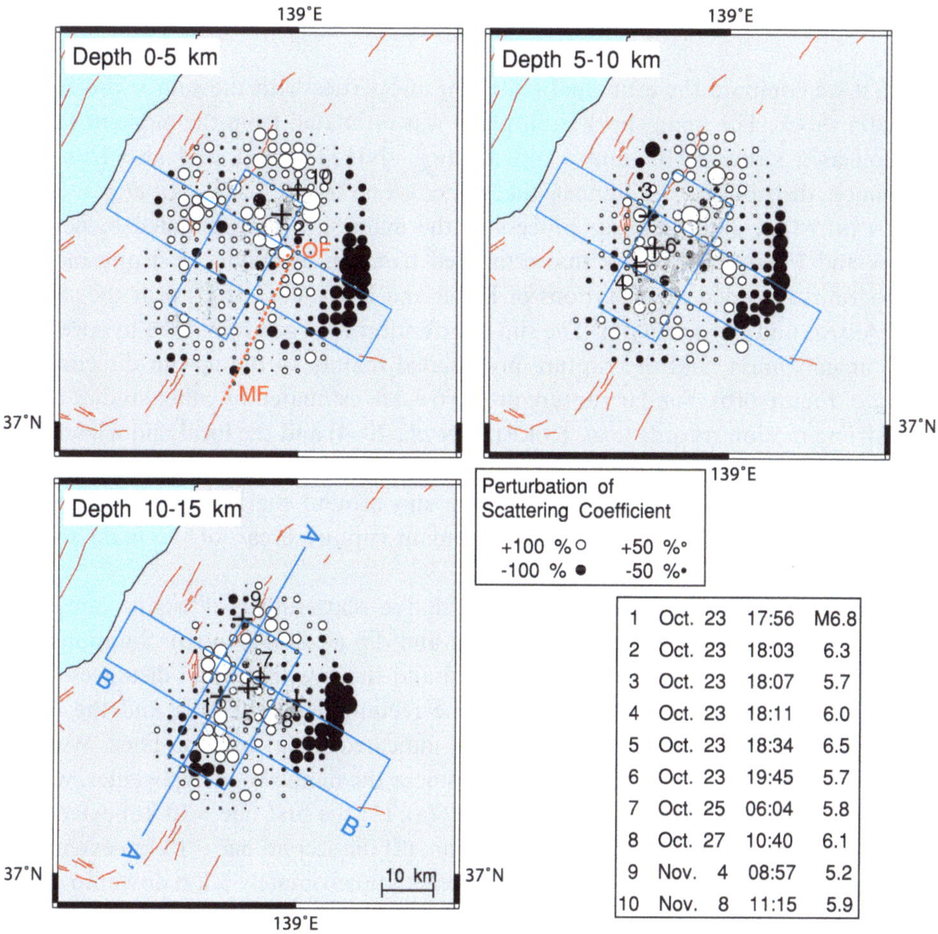

1	Oct.	23	17:56	M6.8
2	Oct.	23	18:03	6.3
3	Oct.	23	18:07	5.7
4	Oct.	23	18:11	6.0
5	Oct.	23	18:34	6.5
6	Oct.	23	19:45	5.7
7	Oct.	25	06:04	5.8
8	Oct.	27	10:40	6.1
9	Nov.	4	08:57	5.2
10	Nov.	8	11:15	5.9

Figure 5

Distribution of relative scattering coefficients at depths from 0–5 km to 10–15 km. Solutions with half-block shifted are superimposed, with the plot of the aftershocks from Oct. 23 to Nov. 8 (light gray dots) in each depth range. The main shock (no. 1) and major large aftershocks (nos. 2–10) are shown by crosses. The line A-A' with a rectangle indicates the approximate projection line of the cross section and the surface projection of the rectangular fault plane, respectively, shown in Figure 6(a). The rectangle with B-B' indicates the area of the cross section shown in Figure 7. Red broken lines represent approximate locations of the Obiro fault (OF) and the Muikamachi fault (MF).

It should be noted that the hypocenters of the main shock (no. 1) and most of the large aftershocks (nos. 2–10), except nos. 8 and 9, seem to be located near the clusters of stronger scatterers stated above (Fig. 5). We consider that these heterogeneous structures in the source region may be related to the earthquake generating process and this will be discussed in the next section.

4. Discussion

First we compare the estimated scattering properties with the source process of the main shock. The main shock fault plane was estimated from the moment tensor solution as a strike ~210° and a dip 50° (e.g., NIED, 2004), and also from the aftershock distribution as almost the same geometry (e.g., SAKAI *et al.*, 2005; SHIBUTANI *et al.*, 2005). Source process of the main shock is estimated by several studies and Figure 6(b) shows that estimated from an inversion of strong motion seismograms recorded at 16 stations of K-net and KiK-net (NIED) near the source area (ASANO and IWATA, 2005). The slip distribution is large around the hypocenter, 3.8 m at maximum, and the rupture propagated mainly to the up-dip direction as shown in Figure 6(b). Similar rupture properties are estimated by other studies using local strong motion records (e.g., KOKETSU *et al.*, 2004) and the local and teleseismic records (YAGI, 2004), and the extent of the large slip area in those studies is almost similar to that within the contour of 0.8 m slip around the hypocenter shown in Figure 6(b), which, therefore, we call the "main rupture area" of the main shock, hereafter.

Figure 6(a) shows the distribution of relative scattering coefficients along the main shock fault plane with the same strike and dip as shown in (b). Solutions are projected along A-A' line shown in Figure 5 and those within 5 km distances from the main shock fault plane are plotted. The rectangular fault area and the main rupture area shown in Figure 6(b) are also indicated in (a) by blue lines. We can recognize three clusters of stronger scatterers near the main shock hypocenter, which is represented by the largest cross in Figure 6(a); (1) the first one ~10 km extending from the hypocenter to the southwest direction, (2) the second one ~10 km extending from the hypocenter to the northeast direction at approximately 3 km downdip-ward from the main shock hypocenter, and (3) the third one ~15 km extending from the second deeper scatterers to the updip direction. Also there exists an area with relatively weaker scattering at ~10 km updip-ward from the main shock hypocenter, extending in an approximately 10 km × 10 km area as indicated by continuous distribution of solutions with solid symbols. Figure 6(c) shows a result of the CB resolution test, in which the result shown in Figure 4(c) is projected onto the main shock fault plane in the same way as in Figure 6(a). The CB pattern is almost well ▶

Figure 6
(a) Distribution of relative scattering coefficients along the fault plane of the main shock, which is shown by the largest cross. The fault plane dips at 50° to the northwest and the projection line A-A' at the depth of the main shock hypocenter is shown in Figure 5. Solutions with resolution greater than 0.2 and within 5 km distances from the fault plane are plotted by open and solid symbols. Aftershocks (Oct. 23-Nov. 8) within 1 km distances from the fault plane are also plotted by red crosses. The heavy solid line (OF) indicates an approximate location of the Obiro fault shown in Figure 5. (b) Slip distribution on the main shock fault plane, with contour interval of 0.8 m (ASANO and IWATA, 2005). The contour line of 0.8 m slip around the hypocenter (a star) and the fault rectangle are also indicated by blue lines in (a). (c) Result of the checkerboard resolution test projected onto the main shock fault plane.

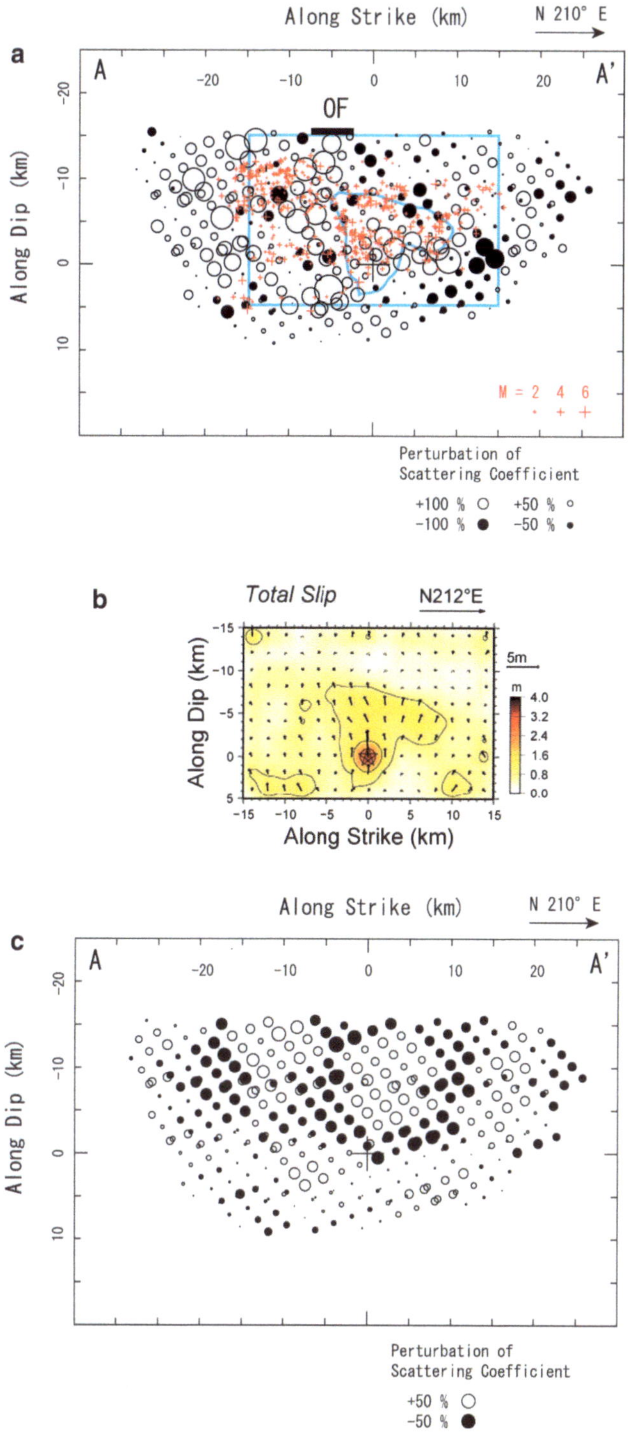
a
Along Strike (km)
N 210° E
A
A'
-20
-10
0
10
20
OF
Along Dip (km)
M = 2 4 6
Perturbation of
Scattering Coefficient
+100 %
+50 %
-100 %
-50 %
b
Total Slip
N212°E
Along Dip (km)
-15
-10
-5
0
5
5m
m
4.0
3.2
2.4
1.6
0.8
0.0
Along Strike (km)
c
Along Strike (km)
N 210° E
A
A'
Along Dip (km)
Perturbation of
Scattering Coefficient
+50 %
-50 %

recovered within the rectangular fault area shown in Figure 6(a), and therefore the scattering properties stated above are considered to be reliable.

The main rupture area indicated in Figure 6(a) seems to be bounded at its northeastern side by the third updipping scatterers stated above. The first scatterers stated above also seem to restrict the lower bound of the main rupture area. The correlation between the main rupture area and the relatively weaker scattering area stated above, however, is not clear. This is partly because of the inaccuracy of the hypocenter of the main shock and also differences in the velocity structure model and the station coverage between the scattering and source process inversions. The blue contour shown in Figure 6(a) is plotted with superposing the location of the main shock hypocenter and therefore, for example, the main rupture area moves to the updip direction if the main shock hypocenter is to be determined shallower. Further analyses will be essential to more thoroughly discuss the relationships between the main shock source process and scattering properties. We can stress, at present, that the large slip area during the main shock rupture seems to be bounded by strong heterogeneous zones with larger scattering coefficients. We assume the possibility that the relatively homogeneous part of the fault-zone acted as an asperity, which is defined as the relatively stronger part on the fault, resulting in a large slip area during

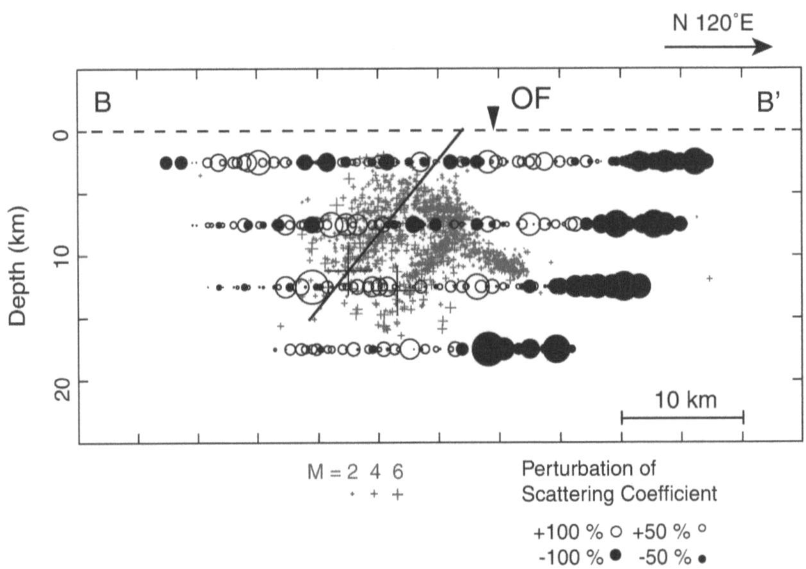

Figure 7
Cross section of relative scattering coefficients, perpendicular to the main shock fault plane, within the rectangle B-B' shown in Figure 5. Aftershocks within the same rectangle are also plotted by small gray crosses. Hypocenters of the main shock and the largest aftershock are represented by two large crosses, for the main shock with its fault plane shown in Figure 6(a). Surface location of the Obiro fault shown in Figure 5 is indicated by a solid triangle (OF).

the event fracturing the entire fault. Similar results were also estimated in the San Andreas fault (NISHIGAMI, 2000) and the 2000 Western Tottori earthquake (ASANO and HASEGAWA, 2004).

In Figure 6(a) we plotted scattering coefficients within 5 km distances from the main shock fault plane, because of the spatial resolution in the present analysis. Figure 7 shows a cross section perpendicular to the main shock fault plane, for solutions inside the rectangle with B-B' indicated in Figure 5. We should be careful that the discussion in this paper has been made with a spatial uncertainty of 5 km. It will be essential to increase the resolution of solutions by adding more data in the inversion in order to make the above discussion more reliable.

We showed that a stronger scattering area exists close to the main shock hypocenter in Figure 6(a). As pointed out in the previous section, hypocenters of large aftershocks also tend to be located near stronger scattering areas (Figure 5). We consider that the rupture initiation is promoted by the existence of localized heterogeneous mediums, such as those with high crack density or high liquid content or low rigidity, in the fault zone or the source area. This kind of heterogeneous bodies may be detected as low velocity zones if they extend in relatively wider areas compared with the seismic wavelength. KATO *et al.* (2005) analyzed the aftershock data observed immediately after the main shock, Oct. 24–27, using a double-difference tomography and estimated a low V_p zone near the main shock hypocenter. This low velocity zone may be correlated with the strong scattering areas around the main shock hypocenter estimated in our present analysis. Existence of low velocity and high Poisson's ratio zone was also estimated below the hypocenter of the M_{JMA} 7.3 Kobe earthquake in 1995 (ZHAO *et al.*, 1996). We analyzed the wave data from earthquakes which occurred on Nov. 1 to. 7, i.e., just before the occurrence of the M 5.9 aftershock on Nov. 8 (no. 10 in Fig. 5). Figure 5 shows that a strong scattering area already existed close to the hypocenter of this event. This result suggests that some of the strong heterogeneities estimated in this study existed before the occurrence of these earthquake sequences and affected their generating process.

The third strong scattering area stated above, existing on the northeastern boundary of the main rupture area, reaches a shallow part of the main shock fault plane as shown in Figure 6(a). It should be noted that the Obiro fault is located around the surface extension of this strong scatterer and also here the Muikamachi fault branches off to the left (Figs. 5, 6(a)). These strong scattterers may have relationships with the deeper extension of the Obiro fault or with the segment boundary between the Obiro and Muikamachi faults. The dip of the main shock fault plane becomes gentle at shallow depth (e.g., KATO *et al.*, 2005), and therefore its surface extension reaches closer to the Obiro and Muikamachi faults than that indicated in Figures 5 and 7. NISHIGAMI and MAMADA (2005) made temporary seismic observations just on the Obiro fault, trying to detect fault-zone trapped waves by the event array analysis. They observed some later phases, not yet identified as fault-zone trapped waves, ~1–2 seconds after direct S arrivals for the wave paths along the third strong scattering area, but not for the paths along the weaker scattering area where the correlation with the main rupture area was discussed. The scattering properties on the main shock fault plane will be examined also from a viewpoint of the low-velocity wave guide.

In the present study we analyzed only the vertical component seismograms. The S coda envelopes observed for local earthquakes usually show similar shapes for the vertical and horizontal components (e.g., SATO and FEHLER, 1998). The S coda envelopes synthesized for a point shear dislocation source, assuming a single scattering model, show that S-to-S wave scattering strongly dominates in the three components for a wide range of lapse times but the partition of coda energy is still controlled by the source radiation pattern, even at long lapse times (SATO, 1984). Also in the present study, we used only one frequency band of 7–15 Hz, i.e., almost dominant frequency range of seismic waves analyzed. Frequency dependency of scattering properties gives information of scale length of heterogeneities as discussed in NISHIGAMI (1997). Further investigation will be necessary for evaluating the coda energy with three component seismograms and also for frequency-dependent scatterer distribution. It will also be essential to increase the number of events and stations used in the analysis in order to increase the resolution of solutions, while decreasing the effect of noises contained in the data as discussed in section 2.

5. *Conclusions*

We estimated a 3-D distribution of relative scattering coefficients in the source region of the 2004 Mid Niigata Prefecture earthquake (M_{JMA} 6.8) by an inversion of coda envelopes from aftershocks. The results of the analyses are summarized as follows:

1. The large slip area during the main shock rupture seems to be bounded by strong heterogeneous zones with larger scattering coefficients.
2. Hypocenters of the main shock and major large aftershocks with M 5–6 classes are located close to stronger scattering areas estimated in the source region.
3. One of the strong scattering areas stated above, close to the hypocenter of the M 5.9 aftershock on Nov. 8, already existed before the occurrence of the event.

We have shown that the heterogeneous structures in the source region seem to have affected the initiation and growth of ruptures of the main shock and major large aftershocks. It will be important to investigate the heterogeneous structure in more detail, not only to understand the generating process of earthquakes in the heterogeneous crust but also to estimate the occurrence and the rupture process of future large events on active faults or plate boundaries with high seismic potential.

Acknowledgements

The author is grateful to the National Research Institute for Earth Science and Disaster Prevention, the Japan Meteorological Agency, and the Kyoto-Kyushu Universities Joint Observation Group for offering the wave data, Dr. T. SHIBUTANI of Kyoto University for offering the JHD hypocenter data, Drs. K. ASANO and T. IWATA of Kyoto University for giving their original figure shown in Figure 6(b), and Dr. Y. Ben-Zion of University of Southern California for inviting me to the special issue publication. Comments by Dr. Z. Peng of University of California, Los Angeles and an anonymous reviewer were very helpful in improving the manuscript. General mapping tool (Wessel and Smith, 1991) was used for drawing part of figures.

REFERENCES

ACTIVE FAULT RESEARCH CENTER, National Institute of Advanced Industrial Science and Technology (2004), *Quick report on the Niigata-ken Chuetsu Earthquake on Oct. 23, 2004*, http://unit.aist.go.jp/actfault/niigata/index.html (in Japanese).

AKI, K., and LEE, W. H. K. (1976), *Determination of three-dimensional velocity anomalies under a seismic array using first P-arrival times from local earthquakes, 1, A homogeneous initial model*, J. Geophys. Res. *81*, 4381–4399.

ASANO, Y., and HASEGAWA, A. (2004), *Imaging the fault zones of the 2000 western Tottori earthquake by a new inversion method to estimate three-dimensional distribution of the scattering coefficient*, J. Geophys. Res. *109*, B06306, doi:10.1029/2003JB002761.

ASANO, K., and IWATA, T. (2005), *Source process of the 2004 Mid Niigata Prefecture earthquake estimated from the waveform inversion of strong motion records*, Abstract of the Annual Meeting of Disas. Prev. Res. Inst., A-01 (in Japanese).

CAPON, J. (1969), *High-resolution frequency-wavenumber spectrum analysis*, Proc. IEEE, *57*, 1408–1418.

GEOGRAPHICAL SURVEY INSTITUTE (2004), *Active fault map in urban areas. Disaster measures for the 2004 Mid Niigata Prefecture earthquake*, http://www.gsi.go.jp/BOUSAI/NIIGATAJISIN/niigatajisin_dansou/niigata_danso_map.html (in Japanese).

KATO, A., KURASHIMO, E., HIRATA, N., SAKAI, S., IWASAKI, T., and KANAZAWA, T. (2005), *Imaging the source region of the 2004 Mid Niigata prefecture earthquake and evolution of a seismogenic thrust-related fault*, Geophys. Res. Lett. doi: 10.1029/ 2005GL022326.

KOKETSU, K., HIKIMA, K., MIYAKE, H., and TANAKA, Y. (2004), *The 2004 Mid Niigata Prefecture earthquake– Strong ground motion and source process–*, http://taro.eri.u-tokyo.ac.jp/saigai/chuetsu/chuetsu.html (in Japanese).

NAGOYA UNIVERSITY, TOYO UNIVERSITY, and AICHI INSTITUTE OF TECHNOLOGY (2005), *Surface earthquake fault with the Mid Niigata prefecture earthquake in 2004*, Report of the Coordinating Committee for Earthquake Prediction, *73*, 387–392 (in Japanese).

NATIONAL RESEARCH INSTITUTE FOR EARTH SCIENCE and DISASTER PREVENTION (2004), *Earthquake Mechanism Information (F-net)*, http://www.fnet.bosai.go.jp/freesia/event/tdmt/20041023085500/update3/index.html.

NEIDELL, N.S., and TANER, M.T. (1971), *Semblance and other coherency measures for multichannel data*, Geophysics *36*, 482–497.

NISHIGAMI, K. (1991), *A new inversion method of coda waveforms to determine spatial distribution of coda scatterers in the crust and uppermost mantle*, Geophys. Res. Lett. *18*, 2225–2228.

NISHIGAMI, K. (1997), *Spatial distribution of coda scatterers in the crust around two active volcanoes and one active fault system in central Japan: Inversion analysis of coda envelope*, Phys. Earth Planet. Inter. *104*, 75–89.

Nishigami, K. (2000), *Deep crustal heterogeneity along and around the San Andreas fault system in central California and its relation to the segmentation*, J. Geophys. Res. *105*, 7983–7998.

Nishigami, K., *Seismological investigation of deep structure of active faults using scattered waves and trapped waves*. In Seismogenic Process Monitoring (eds. Ogasawara, H., Yanagidani, T., and Ando, M.) (Balkema, Rotterdam 2002) pp. 245–256.

Nishigami, K. and Mamada, Y. (2005), *Crustal heterogeneity in the source region of the 2004 Mid Niigata prefecture earthquake–Analyses of scattered waves and fault-zone trapped waves–*, Annuals. Disas. Prev. Res. Inst., Kyoto Univ., No. 48 B, 133–142 (in Japanese with English abstract).

Research Group for Active Faults of Japan, *Active Faults in Japan, Revised edition* (University of Tokyo Press, Tokyo 1991).

Revenaugh, J. (1995), A scattered-wave image of subduction beneath the Transverse Ranges, Science, *268*, 1888–1892.

Sagiya, T., Miyazaki, S., and Tada, T. (2000), *Continuous GPS array and present-day crustal deformation of Japan*, Pure Appl. Geophys. *157*, 2303–2322.

Sakai, S., Hirata, N., Kato, A., Kurashimo, E., Iwasaki, T., and Kanazawa, T. (2005), *Multi-fault system of the 2004 Mid Niigata prefecture earthquake and its aftershocks*, Earth Planets Space *57*, 417–422.

Sato, H. (1977), *Energy propagation including scattering effects: Single isotropic scattering approximation*, J. Phys. Earth *25*, 27–41.

Sato, H. (1984), *Attenuation and envelope formation of three-component seismograms of small local earthquakes in randomly inhomogeneous lithosphere*, J. Geophys. Res. *89*, 1221–1241.

Sato, H. (1988), *Is the single scattering model invalid for the coda excitation at long lapse time?* Pure Appl. Geophys. *128*, 43–47.

Sato, H. (1994), *The relationship between late Cenozoic tectonic events and stress field and basin development in northeast Japan*, J. Geophys. Res. *99*, 22261–22274.

Sato, H., and Fehler, M.C., *Seismic Wave Propagation and Scattering in the Heterogeneous Earth* (Springer-Verlag, New York 1998).

Shibutani, T., Iio, Y., Matsumoto, S., Katao, H., Matsushima, T., Ohmi, S., Takeuchi, F., Uehira, K., Nishigami, K., Enescu, B., Hirose, I., Kano, Y., Kohno, Y., Korenaga, M., Mamada, Y., Miyazawa, M., Tatsumi, K., Ueno, T., Wada, H., and Yhukutake, Y. (2005), *Aftershock distribution of the 2004 Mid Niigata prefecture earthquake*, Earth Planets Space *57*, 545–549.

Tarantola, A., *Inverse Problem Theory: Methods for Data Fitting and Model Parameter Estimation* (Elsevier, New York, 1987), pp. 294–296.

Wessel, P., and Smith, W.H.F. (1991), *Free software helps map and display data*, EOS, trans., Am. Geophys. Union *72*, 441.

Wu, R.S. and Aki, K. (1988), *Introduction: seismic wave scattering in three-dimensionally heterogeneous Earth*, Pure Appl. Geophys. *128*, 1–6.

Yagi, Y. (2004), *Rupture process of the Mid Niigata prefecture earthquake on Oct. 23, 2004 (Tentative version)*, http://iisee.kenken.go.jp/staff/yagi/eq/20041023/Japan20041023-j.html (in Japanese).

Zhao, D., Kanamori, H., Negishi, H., and Wiens, D. (1996), *Tomography of the source area of the 1995 Kobe earthquake: Evidence for fluids at the hypocenter*? Science *274*, 1891–1894.

Zhao, D., Ochi, F., Hasegawa, A., and Yamamoto, A. (2000), *Evidence for the location and cause of large crustal earthquakes in Japan*, J. Geophys. Res. *105*, 13579–13594.

(Received May 4, 2005; accepted September 9, 2005)

Pure appl. geophys. 163 (2006) 617–618
0033–4553/06/030617–3
DOI 10.1007/s00024-005-0028-4

Pure and Applied Geophysics

In Memory of Keiiti Aki, 1930–2005

While editing this volume in honor of Kei Aki, we were greatly saddened to receive the news that he passed away on May 17, 2005, in La Réunion, a French Island in the Indian Ocean. According to Thomas Staudacher, the Director of the Observatoire volcanologique du Piton de la Fournaise, Aki fell in the street coming home from a grocery store on May 13. The next day, he was disoriented and X-rays revealed bleeding in the brain. Aki was sent to the hospital where doctors performed two surgeries to stop the bleeding without success. He fell into a coma, died on May 17, and was buried in the cemetery of the town of Le Tampon on May 18.

Keiiti Aki was born on March 3, 1930, in Yokohama, Japan to a family of engineers with a 100-year tradition of education and openness to the west. Various aspects of his family history, upbringing and scientific influences on his career, described by Kei to one of us, are given in the introduction to the 2003 Pageoph Aki Volume. Here we provide complementary details that shed additional light on his life and career.

When Aki was 19 years old, he was among the 12 students admitted to the Department of Geophysics of the University of Tokyo. It was a very exciting place with world class professors in both Solid and Fluid Earth, including Chuji Tsuboi, Takeshi Nagata, Takeo Matuzawa, and Koji Hidaka. Among Aki's contemporary schoolmates were Syun-iti Akimoto, Syukuro Manabe, and Seiya Uyeda. Kei obtained his B.Sc. and Ph.D. degrees in geophysics in 1952 and 1958, respectively.

Aki's scientific career may be divided chronologically to activities carried in four geographic centers. From 1952 to 1966, Aki was mostly at the University of Tokyo, and visited Caltech as a research fellow. In 1966, Aki moved to the United States to join the faculty of MIT as a Professor of Geophysics, and became a mentor to a large number of students and postdocs. He also spent summers working with colleagues all over the world. In 1984, Aki moved to the University of Southern California as its first W.M. Keck Foundation Professor of Geophysics. He was responsible for the establishment of the Southern California Earthquake Center, where Aki's vision for integrated research on earthquakes has been realized by scientists and engineers in many disciplines working together. In 1995, Aki moved to La Réunion. After making progress in predicting volcanic eruptions, he returned to the problem of earthquake prediction and completed a book manuscript in 2004. During the last five years of his life, Aki annually lectured in Japan as an adviser to the Japanese Association for the Development of Earthquake Prediction. His last project was promoting the "Coda

Club" for collaborative research of using coda waves for earthquake prediction worldwide.

Aki was a pioneer in many fields of seismology, including statistical seismology, earthquake source properties, seismic tomography, coda waves and scattering, fault rupture process, and volcanic tremors. He established scaling laws for a broad range of phenomena in seismology and seismotectonics. He initiated new seismological analyses encompassing the earthquake source, Earth structures and dynamics of geophysical systems. His scientific works were instrumental in paving the way toward achieving a predictive understanding of earthquakes. His leadership in developing new research programs in seismology and his vision for integrated seismological research have been felt around the world. The two-volume treatise, *Quantitative Seismology: Theory and Methods* (co-authored with Paul Richards), has set the standard in seismology teaching and research since its first publication in 1980. A second edition of this classic was published in a single volume in 2002.

In 2004, the 1964 Niigata (Japan) Earthquake Archive at the IRIS Data Center (USA) was established in honor of Kei Aki. He was the first to introduce the concept of seismic moment in 1966, and showed how to measure it using the seismograms of the World-Wide Standardized Seismograph Stations (WWSSN) that recorded the Niigata earthquake. This archive, now under construction, serves as an online website for Aki, so that his many contributions and related materials can be posted for free access by anyone anywhere in the world. The website includes a collection of personal tributes to Kei that were written by former students and colleagues for the 2004 AGU session in his honor and reveal aspects of Kei that are not captured in more formal publications. Interested readers should visit: http://www.iris.edu/seismo/quakes/1964niigata/

Aki published more than 200 papers (see "Publication list of Keiiti Aki" that follows this Memorial), supervised more than 50 Ph.D. dissertations, and received many honors including election to the U.S. National Academy of Sciences (1979), the Medal of the Seismological Society of America (1986), the Bowie Medal of the American Geophysical Union (2004), and the Gutenberg Medal of the European Geosciences Union (2005). Frank Press called Aki the "Dean of world's seismologists – an exemplar of scholar, teacher, and humanitarian of the highest distinction".

We sorely miss Kei Aki not only as a brilliant seismologist and unselfish colleague and teacher, but also as a cheerful and faithful friend.

William H. K. Lee and Yehuda Ben-Zion
October 31, 2005

Pure appl. geophys. 163 (2006) 619–628
0033–4553/06/030619–10
DOI 10.1007/s00024-005-0030-x

Pure and Applied Geophysics

Publication List of Keiiti Aki

Compiled by T. Miyatake[1]

Acknowledgements

I wish to thank Anshu Jin for her "Keiiti Aki's Publication List", which I have used to check against my list. I am grateful to H. Kawase, K. Yomogida, Y. Okada, H. Sato, and Y. Hisada for providing additional information.

References

AKI, K. (1954), *Quantitative prediction of earthquake occurrence as stochastic phenomena*, J. Phys. Earth *2*, 63–69.

AKI, K. (1954), *Thermal processes near the earth's crust*, Zishin II *7*, 65–76, (in Japanese).

AKI, K. (1955), *Relay computer for correlogram analysis*, Zishin II, *8*, 56–58 (in Japanese).

AKI, K. (1955), *Correlogram analysis of seismograms*, Zishin II *8*, 99–107 (in Japanese).

TOMODA, Y. and AKI, K. (1955), *Use of the function sin x/x in gravity problems*, Proc. Japan Acad., *31*, 443–448.

AKI, K. (1956), *Correlogram analyses of seismograms by means of a simple automatic computer*, J. Phys. Earth. *4*, 71–79.

AKI, K. (1956), *Some problems in statistical seismology*, Zishin II, *8*, 205–228 (in Japanese). Note: An English translation by Augustine S. Furumoto, Hawaii Institute of Geophysics, University of Hawaii, March, 1963, has been widely circulated in the United States.

AKI, K. (1956), *Automatic computer of correlation coefficient between signals sent from two seismometers by the use of decatron counting tube*, Zishin II, *9*, 40–46 (in Japanese).

AKI, K. (1957), "Hendogensho" ("Stochastic Phenomena"), Koyama Shoten, Tokyo, (in Japanese).

AKI, K. (1957), *Space and time spectra of stationary stochastic waves, with special reference to microtremors*, Bull. Earthq. Res. Inst. *35*, 415–456.

AKI, K. (1957), *Local concentration of magma as a stationary irreversible process*, Bull. Volcanol. Soc. Japan, Ser. *2*(1), 33–46 (in Japanese).

AKI, K., TSUJIURA, M., HORI, M., and GOTO, K. (1958), *Spectral study of near earthquake waves (1)*, Bull. Earthq. Res. Inst. *36*, 71–98.

[1]Earthquake Research Institute, University of Tokyo, Yayoi, 1-1-1, Bunkyo, Tokyo, 113–0032, Japan
E-mail: miyatake@eri.u-tokyo.ac.jp

AKI, K. and TSUJIURA, M. (1959), *Correlation study of near earthquake waves*, Bull. Earthq. Res. Inst. *37*, 207–232.

AKI, K. (1960), *The use of Love waves for the study of earthquake mechanism*, J. Geophys. Res. *65*, 323–331.

AKI, K. (1960), *Study of earthquake mechanism by a method of phase equalization applied to Rayleigh and Love waves*, J. Geophys. Res. *65*, 729–740.

AKI, K. (1960), *Interpretation of source functions of circum-Pacific earthquakes obtained from long-period Rayleigh waves*, J. Geophys. Res. *65*, 2405–2417.

AKI, K. (1960), *Further study of mechanism of circum-Pacific earthquakes from Rayleigh waves*, J. Geophys. Res. *65*, 4165–4172.

AKI, K. (1961), *Crustal structure in Japan from the phase velocity of Rayleigh waves. Part 1.Use of the network of seismological stations operated by the Japan Meteorological Agency*, Bull. Earthq. Res. Inst. *39*, 255–283.

AKI., K., and PRESS, F. (1961), *Upper mantle structure under oceans and continents from mantle Rayleigh waves*, Geophys. J. Roy. Astron. Soc. *5*, 292–305.

AKI, K. (1962), *Accuracy of the Rayleigh wave method for studying the earthquake mechanism*, Bull. Earthq. Res. Inst., *40*, 91–106.

AKI, K. (1962), *Study of earthquake waves by a seismometer array, Part 1. Aftershocks of the Kitamino earthquake of Aug. 19*, Bull. Earthq. Res. Inst. *40*, 371–389 (in Japanese).

AKI, K. (1962), *A note on crustal structure*, Bull. Earthq. Res. Inst. *40*, 425–430.

KAMINUMA, K. and AKI, K. (1963), *Crustal structure in Japan from the phase velocity of Rayleigh waves (Part 2): Rayleigh waves from the Aleutian shock of March 9, 1957*, Bull. Earthq. Res. Inst. *41*, 217–241.

AKI, K., and KAMINUMA, K. (1963), *Phase velocity of Love waves in Japan, Part 1: Love waves from the Aleutian shock of March 9, 1957*, Bull. Earthq. Res. Inst. *41*, 243–259.

AKI, K., and MATSUMOTO, H. (1963), *Study of earthquake waves by means of a seismometer array, Part 2: A Study of the crustal structure in the Kanto region by the observation of apparent velocities of P waves caused by many small earthquakes*, Bull. Earthq. Res. Inst. *41*, 279–292 (in Japanese).

AKI, K. (1964), *Study of Love and Rayleigh waves from earthquakes with fault plane solutions or with known faulting: Part I*, Bull. Seism. Soc. Am. *54*, 511–527; *Part II*, Bull. Seism. Soc. Am. *54*, 529–558.

MIKUMO, T. and AKI, K. (1964), *Determination of local phase velocity by intercomparison of seismograms from strain and pendulum instruments*, J. Geophys. Res. *69*, 721–731.

AKI, K. (1964), *A note on surface waves from the Hardhat nuclear explosion*, J. Geophys. Res. *69*, 1131–1134.

AKI, K. (1965), *A note on the use of microseisms in determining the shallow structures of the Earth's crust*, Geophysics *30*, 665–667.

AKI, K. (1965), *A computer program for precise determination of focal mechanism of local earthquakes by revising focal depths and crust-mantle structure*, Bull. Earthq. Res. Inst. *43*, 15–22.

AKI, K. (1965), *Accuracy of origin time, epicentre and focal depth of local earthquake determined routinely by the Japan Meteorological Agency*, Bull. Earthq. Res. Inst. *43*, 23–38.

AKI, K. (1965), *Maximum likelihood estimate of b in the formula log* $N=a - bM$ *and its confidence limits*, Bull. Earthq. Res. Inst. *43*, 237–239.

AKI, K., MATUMOTO, H., TSUJIURA, M., and MARUYAMA, T. (1965), *A digital, tele-recorded, long-period seismograph system*, Bull. Earthq. Res. Inst., *43*, 381–397.

AKI, K. (1966), *Generation and propagation of G waves from the Niigata earthquake of June 16, 964, Part 1: A statistical analysis*, Bull. Earthq. Res. Inst. *44*, 23–72.

AKI, K. (1966), *Generation and propagation of G waves from the Niigata earthquake of June 16, 1964, Part 2: Estimation of earthquake moment, released energy, and stress-strain drop from the G wave spectrum*, Bull. Earthq. Res. Inst. *44*, 73–88.

AKI, K. (1966), *Earthquake generating stress in Japan for the years 1961 to 1963 obtained by smoothing the first motion radiation patterns*, Bull. Earthq. Res. Inst. *44*, 447–471.

AKI, K. (1967), *Scaling law of seismic spectrum*, J. Geophys. Res. *72*, 1217–1231.

AKI, K. (1968), *Seismological evidence for soft thin layers in the upper mantle under Japan*, J. Geophys. Res. *73*, 585–594.

AKI, K. (1968), *Seismic displacements near a fault*, J. Geophys. Res. *73*, 5359–5375.

AKI, K. (1968), *Seismicity and seismological method*, Tectonophysics *6*, 41–58.

AKI, K. (1969), *Analysis of the seismic coda of local earthquakes as scattered waves*, J. Geophys. Res. *74*, 6215–6231.

AKI, K., HORI, M., and MATSUMOTO, H. (1969), *Micro-aftershocks observed at a temporary array station on the Kinai peninsula from May 19 to June 7, 1964*, In *The Prince William Sound, Alaska Earthquake of 1964 and Aftershocks*, U.S. Coast and Geodetic Survey *2*, 131–156.

TSAI, Y.B., and AKI, K. (1969), *Simultaneous determination of the seismic moment and attenuation of seismic surface waves*, Bull. Seism. Soc. Am., *59*, 275–287.

AKI, K., REASENBERG, P., DEFAZIO, T., and TSAI, Y.B. (1969), *Near-field and far-field evidence for triggering of an earthquake by the Benham explosion*, Bull. Seism. Soc. Am. *59*, 2197–2207.

WARE, J.A., and AKI, K. (1969), *Continuous and discrete inverse-scattering problems in a stratified elastic medium, 1. Plane waves at normal incidence*, J. Acoust. Soc. Am. *45*, 911–921.

TSAI, Y.B., and AKI, K. (1970), *Source mechanism of the Truckee, California earthquake of September 12, 1966*, Bull. Seism. Soc. Am. *60*, 1199–1208.

AKI, K., DEFAZIO, T., REASENBERG, P., and NUR, A. (1970), *An active experiment with earthquake fault for an estimation of the in-situ stress*, Bull. Seism. Soc. Am. *60*, 1315–1336.

AKI, K. and LARNER, K. (1970), *Surface motion of a layered medium having an irregular interface due to incident plane SH waves*, J. Geophys. Res. *75*, 933–954.

TSAI, Y.B., and AKI, K. (1970), *Precise focal depth determination from amplitude spectra of surface waves*, J. Geophys. Res. *75*, 5729–5743.

AKI, K. (1970), Geophysics, In *Seismic Design for Nuclear Power Plants*, (Hamsen, R.J. Ed). MIT Press, Cambridge, Mass. 91–113.

BOORE, D.M., AKI, K., and TODD, T. (1971), *A two-dimensional moving dislocation model for a strike slip fault*, Bull. Seism. Soc. Am. *61*, 177–194.

BOORE, D.M., AKI, K., and LARNER, K. (1971), *Comparison of two independent methods for the solution of wave scattering problems: response of a sedimentary basin to vertical incident SH waves*, J. Geophys. Res. *76*, 558–569.

TSAI, Y.B., and AKI, K. (1971), *Amplitude spectra of surface waves from small earthquakes and underground nuclear explosions*, J. Geophys. Res. *76*, 3940–3952.

AKI, K. (1972), *Earthquake mechanism*, Tectonophysics *13*, 423–441.

AKI, K. (1972), *Recent results on the mechanism of earthquakes with implications for the prediction and control program*, Tectonophysics *14*, 227–243.

AKI, K. (1972), *Scaling law of earthquake source-time function*, Geophys. J. Roy. Astron. Soc. *31*, 3–25.

AKI, K., and TSAI, Y.B. (1972), *Mechanism of Love wave excitation by explosive source*, J. Geophys. Res. *77*, 1452–1475.

IDA, Y., and AKI, K. (1972), *Seismic source time function of propagating longitudinal shear cracks*, J. Geophys. Res., *77*, 2034–2044.

MADARIAGA, R., and AKI, K. (1972), *Spectral splitting of toroidal free oscillations due to lateral heterogeneity of earth structure*, J. Geophys. Res. *77*, 4421–4431.

DEFAZIO, T., AKI, K., and ALBA, J. (1973), *Solid earth tide and observed change in the in-situ seismic velocity*, J. Geophys. Res. *78*, 1319–1322.

AKI, K. (1973), *Scattering of P waves under the Montana LASA*, J. Geophys. Res. *78*, 1334–1346.

WEIDNER, D., and AKI, K. (1973), *Focal depth and mechanism of mid-ocean ridge earthquakes*, J. Geophys. Res. *78*, 1818–1831.

AKI, K., BOUCHON, M., and REASENBERG, P. (1974), *Seismic source function for an underground nuclear explosion*, Bull. Seism. Soc. Am. *64*, 131–148.

REASENBERG, P., and AKI, K. (1974), *A precise, continuous measurement of seismic velocity for monitoring in situ stress*, J. Geophys. Res. *79*, 399–406.

AKI, K., and CHOUET, B. (1975), *Origin of coda waves: Source, attenuation and scattering effects*, J. Geophys. Res., *80*, 3322–3342.

WARD, R.W., and AKI, K. (1975), *Synthesis of teleseismic P waves from sources near sinking lithospheric slabs*, Bull. Seism. Soc. Am., *65*, 1667–1680.

AKI, K., and LEE, W.H.K. (1976), *Determination of three-dimensional velocity anomalies under a seismic array using first P arrival times from local earthquakes, 1. A homogeneous initial model*, J. Geophys. Res. *81*, 4381–4399.

AKI, K., CHRISTOFFERSSON, A., and HUSEBYE, E.S. (1976), *Three-dimensional seismic structure of lithosphere under Montana LASA*, Bull. Seism. Soc. Am., *66*, 501–524.

HUSEBYE, E.S., CHRISTOFFERSSON, A., AKI, K., and POWELL, C. (1976), *Preliminary results on the three-dimensional seismic structure of the lithosphere under the USGS Central California Seismic Array*, Geophys. J. R. Astron. Soc. *46*, 319–340.

AKI, K., CHRISTOFFERSSON, A., and HUSEBYE, E.S. (1977), *Determination of the three-dimensional seismic structure of the lithosphere*, J. Geophys. Res. *82*, 277–296.

AKI, K. BOUCHON, M., CHOUET, B., and DAS, S. (1977), *Quantitative prediction of strong motion for a potential earthquake fault*, Ann. Geofis. *30*, 341–386.

AKI, K., FEHLER, M., and DAS, S. (1977), *Source mechanism of volcanic tremors: fluid driven crack models and their application to the 1963 Kilauea eruption*, J. Volcanol. Geotherm. Res. *2*, 259–287.

AKI, K. (1977), *Three-dimensional seismic velocity anomalies in the lithosphere*, J. Geophys, *43*, 235–242.

DAS, S., and AKI, K. (1977), *A numerical study of two-dimensional spontaneous rupture propagation*, Geophys. J. Roy. Astron. Soc. *50*, 643–668.

BOUCHON, M., and AKI, K. (1977), *Near-field of a seismic source in a layered medium with irregular interface*, Geophys. J. Roy. Astron. Soc. *50*, 669–684.

DAS, S., and AKI, K. (1977), *Fault plane with barriers: A versatile earthquake model*, J. Geophys. Res., *82*, 565–570.

AKI, K., CHOUET, B., FEHLER, M., ZANDT, G., KOYANAGI, R., COLP, J., and HAY, R.G. (1978), *Seismic properties of a shallow magma reservoir in the Kilauea Iki by active and passive experiments*, J. Geophys. Res. *83*, 2273–2282.

CHOUET, B., AKI, K., and TSUJIURA, M. (1978), *Regional variation of the scaling law of earthquake source spectra*, Bull. Seism. Soc. Am. *68*, 49–79.

FEHLER, M., and AKI, K. (1978), *Numerical study of diffraction of plane elastic waves by a finite crack with application to location of a magma lens*, Bull. Seism. Soc. Am. *68*, 573–598.

MENDIGUREN, J., and AKI, K. (1978), *Source mechanism of the deep Colombian earthquake of 1970 July 31 from the free oscillation data*, Geophys. J. Roy. Astron. Soc. *55*, 539–556.

PATTON, H., and AKI, K. (1978), *Determination of seismic moment tensor using surface waves*, Tectonophysics, *49*, 213–222.

AKI, K. (1978), *Introduction to seismology*, In *Physics of Earthquake*, (Kanamori, H. Ed), Iwanami Shoten (in Japanese).

AKI, K. (1979), *Characterization of barriers on an earthquake fault*, J. Geophys. Res. *84*, 6140–6148.

AKI, K. (1979), *A direction of earthquake prediction study*, Kagaku, *49*(5), 308–313 (in Japanese).

AKI, K. (1979), *Orogeny*, Iwanami Press (in Japanese). Project team for the development of borehole small-size long-period seismometer (S. Ando, Y. Okada, K. AKI, T. Daikubara, Y. Suzuki, and T. Hara), *Development of the Borehole Easy-operation Long-period Seismometer (BELS Type 79) which has a stable period*, Bull. Earthq. Res. Inst. *54*, 135–146, (in Japanese).

AKI, K., and RICHARDS, P.G (1980), *Quantitative Seismology: Theory and Methods*, Freeman, San Francisco, *2*.

AKI, K. (1980), *Attenuation of shear waves in the lithosphere for frequencies from 0.05 to 25 Hz*, Phys. Earth Planet. Inter. *21*, 50–60.

AKI, K. (1980), *Scattering and attenuation of shear waves in the lithosphere*, J. Geophys. Res. *85*, 6496–6504.

BOUCHON, M., and AKI, K. (1980), *Simulation of long-period, near-field motion for the great California earthquake of 1857*, Bull. Seism. Soc. Am. *70*, 1669–1682.

AKI, K. (1980), *SSA presidential address Possibility of seismology in the 1980's*, Bull. Seism. Soc. Am. *70*, 1969–1976.

AKI, K. (1980), *Re-evaluation of stress drop and seismic energy using a new model of earthquake faulting*, In *Source Mechanism and Earthquake Prediction*, (Allegre, C. Ed), 23–50 CNRS.

AKI, K., and KOYANAGI, R. (1981), *Deep volcanic tremors and magma ascent mechanism under Kilauea, Hawaii*, J. Geophys. Res. *86*, 7095–7109.

AKI, K. (1981), *3-D inhomogeneities in the upper mantle*, Tectonophysics *75*, 31–40.

AKI, K. (1981) *A probabilistic synthesis of precursory phenomena*, In *Earthquake Prediction, an International Review*, (Simpson, D.W. and Richards, P.G. Eds). M. Ewing Ser. *4*, 566–574 (American Geophysical Union, Washington, D.C.).

AKI, K. (1981), *Source and scattering effects on the spectra of small local earthquakes*, Bull. Seism. Soc. Am., *71*, 1687–1700.

AKI, K., FEHLER, M., AAMODT, R.L., ALBRIGHT, J.N., POTTER, R.M., PEARSON, C.M., and TESTRE, J.W. (1982), *Interpretation of seismic data from hydraulic fracturing experiments at Fenton Hill, New Mexico, hot dry rock geothermal site*, J. Geophys. Res. *87*, 936–944.

AKI, K. (1982), *Strong motion prediction using mathematical modeling technique*, Bull. Seism. Soc. Am. *72*, 829–841.

AKI, K. (1982), *Three-dimensional seismic inhomogeneities in the lithosphere and asthenosphere: Evidence for de-coupling in the lithosphere and flow in the asthenosphere*, Rev. Geophys. Space Phys. *20*, 161–170.

BOUCHON, M., and AKI, K. (1982), *Strain, tilt, and rotation associated with strong ground motion in the vicinity of earthquake faults*, Bull. Seism. Soc. Am. *72*, 1717–1738.

HORIE, A., and AKI, K. (1982), *The three-dimensional velocity structure beneath the Kanto distinct, Japan*, J. Phys. Earth *30*, 255–281.

PAPAGEORGIOU, A.S., and AKI, K. (1982), *Aspects of the mechanics of earthquake rupture related to the generation of high frequency waves and the prediction of strong ground motion*, Soil Dynamics Earthq. Engineering *1*, 67–74.

AKI, K. (1982), *Scattering and attenuation*, Bull. Seism. Soc. Am. *72*, S319–S330.

AKI, K. (1983), *Strong Motion Seismology*, In *Earthquakes: Observation, Theory and Interpretation*, (Kanamori, H. and Boschi, E. Ed), pp. 223–250, North-Holland Pub. Co., Amsterdam.

CAO, T., and AKI, K. (1983), *Assigning probability gain for precursors of four large Chinese earthquakes*, J. Geophys. Res. *88*, 2185–2190.

PAPAGEORGIOU, A.S., and AKI, K. (1983), *A specific barrier model for the quantitative description of inhomogeneous faulting and prediction of strong ground motion, Parts I and II*, Bull. Seism. Soc. Am. *73*, 693–722, 953–978.

AKI, K. (1984), *Asperities, barriers and characteristics of earthquakes*, J. Geophys. Res. *89*, 5867–5872.

AKI, K. (1984), *Evidence for magma intrusion during the Mammoth Lakes earthquakes of May, 1980 and implications of absence of volcanic (harmonic) tremor*, J. Geophys. Res. *89*, 7689–7696.

AKI, K. (1984), *Prediction of strong motion using physical model of earthquake faulting*, Proc. Eighth World Conf. Earthq. Engineering, Vol. 2, *Ground Motion and Seismicity*, 433–444, Prentice-Hall, Englewood Cliffs, New Jersey.

AKI, K. (1984), *Short period seismology*, J. Computational Physics *54*, 3–17.

BISWAS, N.N., and AKI, K. (1984), *Characteristics of coda waves in central and south-central Alaska*, Bull. Seism. Soc. Am. *74*, 493–509.

CAO, T., and AKI, K. (1984), *Seismicity simulation with a mass-spring model and a displacement hardening-softening friction law*, Pure Appl. Geophys. *122*, 10–24.

NOWACK, R., and AKI, K. (1984), *The two-dimensional Gaussian beam synthetic method: Testing and application*, J. Geophys. Res. *89*, 7797–7819.

AKI, K. (1985), *Theory of earthquake prediction with special reference to monitoring of the quality factor of lithosphere by the coda method*, Earthq. Prediction Res. *3*, 219–230.

PAPAGEORGIOU, A.S., and AKI, K. (1985), *Scaling law of far-field spectra based on observed parameters of specific barrier model*, Pure Appl. Geophys. *123*, 354–374.

WU, R.S., and AKI, K. (1985), *Scattering characteristics of elastic waves by an elastic heterogeneity*, Geophysics *50*, 582–595.

WU, R.S., and AKI, K. (1985), *The fractal nature of the inhomogeneities in the lithosphere evidenced from seismic wave scattering*, Pure Appl. Geophys. *123*, 805–818.

YOMOGIDA, K., and AKI, K. (1985), *Waveform synthesis of surface waves in a laterally heterogeneous Earth by the Gaussian beam method*, J. Geophys. Res. *90*, 7665–7688.

JIN, A., CAO, T., and AKI, K. (1985), *Regional change of coda Q in the oceanic lithosphere*, J. Geophys. Res. *90*, 8651–8659.

WU, R.S., and AKI, K. (1985), *Elastic wave scattering by a random medium and the small-scale inhomogeneities in the lithosphere*, J. Geophys. Res. *90*, 10261–10274.

CAO, T., and AKI, K. (1986), *Seismicity simulation with rate- and state-dependent friction law*, Pure Appl. Geophys. *124*, 447–513.

CAO, T., and AKI, K. (1986), *Effect of slip rate on stress drop*, Pure Appl. Geophys. *124*, 515–529.

JIN, A., and AKI, K. (1986), *Temporal changes in coda Q before the Tangshan earthquake of 1976 and the Haicheng earthquake of 1975*, J. Geophys. Res. *91*, 665–673.

PHILLIPS, W.S., and AKI, K. (1986), *Site amplification of coda waves from local earthquakes in central California*, Bull. Seism. Soc. Am. *76*, 627–648.

LEE, W.H.K., AKI, K., CHOUET, B., JOHNSON, P., MARKS, S., NEWBERRY, J.T., RYALL, A.S., STEWART, S.W., and TOTTINGHAM, D.M. (1986), *A preliminary study of coda Q in California and Nevada*, Bull. Seism. Soc. Am. *76*, 1143–1150.

AKI, K. (1986), *Is Geophysics useful?, Kagaku (Iwanami Shoten)*, *56*(11), 684–689 (in Japanese).

OKUBO, P.G., and AKI, K. (1987), *Fractal geometry in San Andreas fault system*, J. Geophys. Res. *92*, 345–355.

AKI, K. (1987), *Magnitude frequency relation for small earthquakes: A clue for the origin of f_{max} of large earthquakes*, J. Geophys. Res. *92*, 1349–1355.

PENG, J.Y., AKI, K., CHOUET, B., JOHNSON, P., LEE, W.H.K., MARKS, S., NEWBERRY, J.T., RYALL, A.S., STEWART, S.W., and TOTTINGHAM, D.M. (1987), *Temporal changes in coda Q associated with the Round Valley, California, earthquake of November 23, 1984*, J. Geophys. Res. *92*, 3507–3526.

FERRAZZINI, V., and AKI, K. (1987), *Slow waves trapped in a fluid-filled infinite crack: Implication for volcanic tremor*, J. Geophys. Res. *92*, 9215–9223.

KOYANAGI, R.Y., CHOUET, B., and AKI, K. (1987), *Origin of volcanic tremor in Hawaii, Part I data from the Hawaiian Volcano Observatory 1969–1985*, In *Volcanism in Hawaii*, (Decker, R.W., Wright, T.L., and Stauffer, P.H. Eds), U.S. Geol. Surv. Prof. Paper 1350, 1221–1257.

CHOUET, B., KOYANAGI, R.Y., and AKI, K. (1987), *Origin of volcanic tremor in Hawaii, Part II Theory and discussion*, In *Volcanism in Hawaii*, (Decker, R.W., Wright, T.L., and Stauffer, P.H. Eds), U.S. Geol. Surv. Prof. Paper 1350, 1259–1280.

LEARY, P.C., LI, Y.G., and AKI, K. (1987), *Observation and modeling of fault-zone fracture seismic anisotropy*, Geophys. J. Roy. Astron. Soc. *91*, 461–492.

THURBER, C.H., and AKI, K. (1987), *Three-dimensional seismic imaging*, Ann. Rev. Earth Planet Sci. *15*, 115–139.

YOMOGIDA, K., and AKI, K. (1987), *Amplitude and phase data inversion for phase velocity anomalies in the Pacific Ocean basin*, Geophys. J. Roy. Astron. Soc. *88*, 161–204.

AKI, K. (1988), *Separation of source and site effects in acceleration power spectral of major California earthquakes*, Proc. 9th World Conf. Earthq. Engineering *8*, 163–167.

AKI, K. (1988), *Impact of earthquake seismology on the geological community since the Benioff zone*, Geol. Soc. Am. Bull. *100*, 625–629.

AKI, K. (1988), *Local site effects on strong ground motion, Proc. Earthq. Eng. Soil Dynamics II, - Recent advances in ground motion evaluation*, ASCE, Utah, June 27–30, pp. 103–155.

AKI, K. (1988), *Physical theory of earthquakes*, In *Seismic Hazard in Mediterranean Region*, (Bonin, J., Cara, M., Cisternas, M., and Fantechi, R. Ed), pp. 3–34, Kluwer Academic Publ.

JIN, A., and AKI, K. (1988), *Special and temporal correlation between coda Q and seismicity in China*, Bull. Seism. Soc. Am. *78*, 741–769.

WU, R.S., and AKI, K. (1988), *Introduction: seismic scattering in three-dimensionally heterogeneous earth*, Pure Appl. Geophys. *128*, 1–6.

WU, R.S., and AKI, K. (1988), *Multiple scattering and energy transfer of seismic waves -separation of scattering effect from intrinsic attenuation. II. Application of the theory to Hindu-Kush region*, Pure Appl. Geophys. *128*, 49–80.

AKI, K. (1989), *Deterministic and stochastic approaches in seismogram analysis*, In *Observatory Seismology*, (LITEHISER, J.J. Ed), pp. 257–265, University of California Press, Berkeley.

AKI, K. (1989), *Elasticity and wave propagation: Principles*, In *Encyclopedia of Solid Earth Geophysics*, (James, D.E. Ed), pp. 363–367, Van Nostrand Reinhold Co., New York.

BENITES, R., and AKI, K. (1989), *Boundary integral-Gaussian beam method for seismic wave scattering: SH in two-dimensional media*, J. Acoust. Soc. Am. *86*, 375–368.

JIN, A., and AKI, K. (1989), *Spatial and temporal correlation between coda Q^{-1} and seismicity and its physical mechanism*, J. Geophys. Res. *94*, 14041–14059.

KAWASE, H., and AKI, K. (1989), *A study on the response of a soft basin for incident S, P, and Rayleigh waves with special reference to the long duration observed in Mexico City*, Bull. Seism. Soc. Am. *79*, 1361–1382.

CAMPILLO, M., GARIEL, J.C., AKI, K., SANCHEZ-SESMA, F.J. (1989), *Destructive strong ground motion in Mexico city: Source, path and site effects during great 1985 Michoakan earthquake*, Bull. Seism. Soc. Am. *79*, 1718–1735.

AKI, K., and PAPAGEORGIOU, A.S. (1989), *Separation of source and site effects in acceleration power spectra of major California earthquakes*, Proc. 9th World Conf. Earthq. Engineering *8*, 163–167.

AKI, K. (1990), *Haskell's source mechanism papers and their impact on modern seismology*, in *Vincit Veritas: A Portrait of the Life and Work of Norman Abraham Haskell, 1905–1970*, (Ben-Menahem, A. Ed), pp. 42–45, American Geophysical Union, Washington, D.C.

KAWASE, H., and AKI, K. (1990), *Topography effect at the critical SV-wave incidence: Possible explanation of damage pattern by the Whitter Narrows, California earthquake of October 1, 1987*, Bull. Seism. Soc. Am. *80*, 1–22.

BEN-ZION, Y., and AKI, K. (1990), *Seismic radiation from an SH line source in a laterally heterogeneous planar fault zone*, Bull. Seism. Soc. Am. *80*, 971–994.

AKI, K. (1990), Comment on "Microearthquake spectra from the Anza, California, Seismic Network: Site response and source scaling" by A. Frankel, and L. Wennerberg, Bull. Seism. Soc. Am., *80*, 1034–1035.

LI, Y., LEARY, P.C., AKI, K., and MALIN, P.E. (1990) *Seismic trapped modes in the Oroville and San Andreas fault zones*, Science *249*, 763–766.

SU, F., and AKI, K. (1990), *Spatial and temporal variation in coda Q^{-1} associated with the North Palm Springs earthquake of 1986*, Pure Appl. Geophys. *133*, 23–52.

AKI, K. (1991), *Summary of discussion on coda waves at the Istanbul IASPEI Meeting*, Phys. Earth Planet. Inter. *67*, 1–3.

AKI, K., and IRIKURA, K. (1991), *Characterization and mapping of earthquake shaking for seismic zonation*, Proc. Fourth Int. Conf. Seismic Zonation 61–110.

CHEN, X., and AKI, K. (1991), *General coherence functions for amplitude and phase fluctuations in a randomly heterogeneous medium*, Geophys. J. Int. *105*, 155–162.

ZENG, Y., SU, F., and AKI, K. (1991), *Scattering wave energy propagation in a medium with randomly distributed isotopic scatterers, 1. Theory*, J. Geophys. Res. *96*, 607–619.

FERRAZZINI, V., AKI, K., and CHOUET, B. (1991), *Characteristics of seismic waves composing Hawaiian volcanic tremors and gas-piston events observed by a near-source array*, J. Geophys. Res. *96*, 6199–6214.

SU, F. AKI, K., and BISWAS, N.N. (1991), *Discriminating quarry blasts from earthquakes using coda waves*, Bull. Seism. Soc. Am. *81*, 162–178.

CHIN, B.H., and AKI, K. (1991), *Simultaneous determination of source, path and recording site effects on strong ground motion during the Loma Prieta earthquake - a preliminary result on pervasive non-linear site effect*, Bull. Seism. Soc. Am. *81*, 1859–1884.

MAYEDA, K, KOYANAGI, S., and AKI, K. (1991), *Site amplification factor from S wave coda in the Long Valley caldera region, California*, Bull. Seism. Soc. Am. *81*, 2194–2213.

ROBERTS, P.M., AKI, K., and FEHLER, M.C. (1991), *A low-velocity zone in the basement beneath the Valles Caldera, New Mexico*, J. Geophys. Res. *96*(B13), 21583–21596.

ROSA, J.W.C., and AKI, K. (1991), *Global compilation of phase and group velocities of fundamental mode Rayleigh waves in the period range from 20 to 100 sec.*, Revista Brasileira de Geofisica *9*, 223–248.

ROSA, J.W.C., and AKI, K. (1991), *Global compilation of phase and group velocities of fundamental mode Rayleigh waves in the period range from 20 to 100 sec.*, Revista Brasileira de Geofisica *9*, 249–273.

AKI, K. (1992), *Earthquake sources and strong motion prediction*, Proc. IDNDR Int. Symp. Earthq. Disaster Reduction Technol., Tsukuba, Japan, 29–43.

AKI, K. (1992), *Higher order interrelations between seismogenic structures and earthquake processes*, Tectonophysics *211*, 1–12.

SU, F., AKI, K., TENG, T., ZENG, Y., KOYANAGI, S., and MAYEDA, K. (1992), *The relation between site amplification factor and surficial geology in central California*, Bull. Seism. Soc. Am. *82*, 580–602.

AKI, K. (1992), *Scattering conversions P to S versus S to P*, Bull. Seism. Soc. Am. *82*, 1969–1972.

AKI, K. (1992), Seismological and geotechnical studies of local site effect on strong and weak motions, *Proc. of Int. Symposium on the Effects of Surface Geology on Seismic Motion (Balkema)*, *1*, 97–110.

AKI, K. (1992), *State of the art in volcanic seismology*. In *Volcanic Seismology*, (Gasparini, P., Scarpa, R., and Aki, K. Eds), pp. 3–9, Springer-Verlag, Berlin.

BENITES, R., AKI, K., and YOMOGIDA, K. (1992), *Multiple scattering of SH waves in 2-D media with many cavities*, Pure Appl. Geophys. *138*, 353–390.

FERRAZZINI, V., and AKI, K. (1992), *Preliminary results from a field experiment on volcanic events at Kilauea using an array of digital seismographs.* In *Volcanic Seismology*, (Gasparini, P., Scarpa, R., and Aki, K. Eds), pp. 168–189, Springer-Verlag, Berlin.

MAYEDA, K., KOYANAGI, S., HOSHIBA, M., Aki, K., and ZENG, Y. (1992), *A comparative study of scattering, intrinsic and coda Q for Hawaii, Long Valley and central California between 1.5 and 15 Hz*, J. Geophys. Res. *97*, 6643–6659.

AKI, K., and DMOWSKA, R. (1993), *Preface.* In *Relating Geophysical Structures and Processes: The Jeffreys Volume*, (Aki, K. and DMOWSKA, R. Eds), Geophysical Monograph *76*, Am. Geophys. Union.

AKI, K. (1993), *Local site effects on weak and strong ground motion*, Tectonophysics *218*, 93–111.

AKI, K. (1993), *Overview.* In *Seismic Tomography*, (Iyer, H.M. and Hirahara, K. Eds), pp. 1–8, Chapman and Hall, London.

CHEN, X., AL-FARES, M., and AKI, K. (1993), Imaging the shape of an elastic heterogeneity by the T-matrix method, *Proc. Symp. Mathematical methods in Geophysical imaging, San Diego, California*, pp. 109–123, Soc. Photo-Optical Inst. Eng.

JIN, A., and AKI, K. (1993), *Temporal correlation between coda Q^{-1} and seismicity - evidence for a structural unit in the brittle-ductile transition zone*, J. Geodynamics *17*, 95–120.

ZENG, Y., AKI, K., and TENG, T. (1993), *Source inversion of the 1987 Whittier Narrows earthquake, California, using the isochron method*, Bull. Seism. Soc. Am. *83*, 358–377.

KATO, K., AKI, K., and TENG, T. (1993), *3-D simulations of surface wave propagation in the Kanto sedimentary basin, Japan. Part 1: Application of the surface wave Gaussian beam method*, Bull. Seism. Soc. Am. *83*, 1676–1699.

HISADA, Y., AKI, K., and TENG, T. (1993), *3-D simulations of surface wave propagation in the Kanto sedimentary basin, Japan. Part 2: Application of the surface wave BEM*, Bull. Seism. Soc. Am. *83*, 1700–1720.

ZENG, Y., AKI, K., and TENG, T.L. (1993), *Mapping of the high-frequency source radiation for the Loma Prieta earthquake, California*, J. Geophys. Res. *98*(B7), 11981–11994.

AMINZADEH, F., KATZ, S., and AKI, K. (1994), *Adaptive neural nets for generation of artificial earthquake precursors*, IEEE Trans. Geosci. Remote Sensing *32*, 1139–1143.

BENITES, R., and AKI, K. (1994), *Ground motion at mountains and sedimentary basins with vertical seismic velocity gradient*, Geophys. J. Int. *116*, 95–118.

LI, Y.G., VIDALE, J.E., AKI, K., MARONE, C.J., and LEE, W.H.K. (1994), *Fine structure of the Landers fault zone: Segmentation and the rupture process*, Science *265*, 367–370.

LI, Y.G., AKI, K., ADAMS, D., HASEMI, A., and LEE, W.H.K. (1994), *Seismic guided waves trapped in the fault zone of the Landers, California, earthquake of 1992*, J. Geophys. Res. *99*, 11705–11722.

JIN, A., MAYEDA, K., ADAMS, D., and AKI, K. (1994), *Separation of intrinsic and scattering attenuation in southern California using TERRAscope data*, J. Geophys. Res. *99*, 17835–17848.

AKI, K. (1995), (as a member of Working Group on California Earthquake Probabilities), *Seismic hazards in southern California: Probable earthquakes 1994 to 2024*, Bull. Seism. Soc. Am. *85*, 379–439.

AKI, K. (1995), *Earthquake prediction: Societal implications, U. S. National Report to International Union of Geodesy and Geophysics 1991–1994*, Rev. Geophys, Suppl. 243–247.

SU, F., and AKI, K. (1995), *Site amplification factors in central and southern California determined from coda waves*, Bull. Seism. Soc. Am., *85*, 452–466.

KATO, K., AKI, K., and TAKEMURA, M. (1995), *Site amplification from coda waves: Validation and application to S-wave site response*, Bull. Seism. Soc. Am. *85*, 467–477.

KOYANAGI, S., AKI, K., BISWAS, N., and MAYEDA, K. (1995), *Inferred attenuation from site-effect-corrected T phase recorded on the Island of Hawaii*, Pure Appl. Geophys. *144*, 1–16.

ROBERTS, P.M., AKI, K., and FEHLER, M.C. (1995), *A shallow attenuating anomaly inside the ring fracture of Valles caldera, New Mexico*, J. Volcanol. Geotherm. Res. *67*, 79–99.

WANG, R., and AKI, K. (1995), *Introduction*, Pure Appl. Geophys. *145*, 399–401.

AKI, K. (1995), *Interrelation between fault-zone structure and earthquake process*, Pure Appl. Geophys. *145*, 647–676.

AKI, K. (1996), *Scale dependence in earthquake phenomena and its relevance to earthquake prediction*, Proc. Nat. Acad. Sci. USA, *93*, 3740–3747.

CHEN, X., and AKI, K. (1996), *An effective approach to determine the dynamic source parameters*, Pure Appl. Geophys. *146*, 689–696.

GAO, L.S., and AKI, K. (1996), *Effect of finite thickness of scattering layer on coda Q of local earthquakes*, J. Geodynamics *21*, 191–213.

LI, Y.G., ELLSWORTH, W.L., THURBER, C.H., MALIN, P.E., and AKI, K. (1997), *Observations of fault-zone trapped waves excited by explosions at the San Andreas fault, central California*, Bull. Seism. Soc. Am. *87*, 210–221.

LI, Y.G., AKI, K., and VERNON, F.L. (1997), *San Jacinto fault zone guided waves: A discrimination for recently active fault strands near Anza, California*, J. Geophys. Res. *102*(B6), 11689–11701.

YOMOGIDA, K., AKI, K., and BENITES, R. (1997), *Coda Q in two-layer random media*, Geophys, J. Int. *128*, 425–433.

LI, Y.G., VIDALE, J.E., AKI, K., XU, F., and BURDETTE, T. (1998), *Evidence of shallow fault zone strengthening after the 1992 M7.5 Landers, California, earthquake*, Science *279*, 217–219.

LI, Y.G., AKI, K., VIDALE, J.E., and ALVAREZ, M.G. (1998), *A delineation of the Nojima fault ruptured in the M7.2 Kobe, Japan, earthquake of 1995 using fault zone trapped waves*, J. Geophys. Res. *103*(B4), 7247–7263.

AKI, K. (1999), *Scale dependence in earthquake processes and seismogenic structure*. In *1st ACES Workshop Proceedings*, (Mora, P. Ed), pp. 437–440.

LI, Y.G., AKI, K., VIDALE, J.E., and XU, F. (1999), *Shallow structure of the Landers fault zone from explosion-generated trapped waves*, J. Geophys. Res. *104*(B9), 20257–20275.

AKI, K., and JIN, A. (2000), *Seismic monitoring and modeling of earthquake processes for prediction*. In *2nd ACES Workshop Proceedings*, (Matsu'ura, M., Nakajima, L., and Mora, P. Eds), pp. 121–127.

AKI, K. (2000), *Scale-dependence in earthquake processes and seismogenic structures*, Pure Appl. Geophys. *157*, 2249–2258.

LI, Y.G., VIDAL, J.E., AKI, K., and XU, F. (2000), *Depth-dependent structure of Landers fault zone using fault zone trapped waves generated by aftershocks*, J. Geophys. Res. *105*, 6237–6254.

AKI, K., and FERRAZZINI, V. (2000), *Seismic monitoring and modeling of an active volcano for prediction*, J. Geophys. Res. *105*(B7), 16617–16640.

AKI, K., and FERRAZZINI, V. (2001), *Comparison of Mount Etna, Kilauea, and Piton de la Fournaise by a quantitative modeling of their eruption histories*, J. Geophys. Res. *106*(B3), 4091–4102.

AKI, K. (2002), *Synthesis of earthquake science information and its public transfer: A History of the Southern California Earthquake Center*. In *International Handbook of Earthquake and Engineering Seismology*, Lee, W.H.K., Kanamori, H., Jennings, J.C., and Kisslinger, C. Eds), Part A, pp. 39–50, Academic Press, San Diego.

AKI, K. (2002), *Introduction to physical modeling for earthquake prediction*, Zishin Journal *33*, 1–8 (in Japanese).

AKI, K., and RICHARDS, P.G. *Quantitative Seismology*, 2nd ed. (University Science Books, Sausalito, California, 2002).

AKI, K., and LEE, W.H.K. (2003), *Glossary of interest to earthquake and engineering seismologists*. In *International Handbook of Earthquake and Engineering Seismology*, Lee, W.H.K., Kanamori, H., Jennings, J.C., and Kisslinger, C. Eds), Part B, pp.1793–1856, Academic Press, San Diego.

AKI, K. (2003), *Physical model for earthquake prediction*, Chikyu Monthly, *25*, 730–737 (in Japanese).

AKI, K. (2003), *A perspective on the history of strong motion seismology*, Phys. Earth Planet. Inter. *137*, 5–11.

AKI, K. (2003), "Seismology of Earthquake and Volcanic Prediction", Lecture notes for the Seventh International Workshop on Non-linear Dynamics and Earthquake Prediction, Workshop on Volcanic Disaster Mitigation on September 24–26, 2003, Nat. Res. Inst. Earth Sci. Disast. Prev. and Yamanashi Inst. Environ. Sci., 219 pp.

AKI, K. (2004), *A new view of earthquake and volcano precursors, A Special issue of Earth, Planets and Space: the IUGG Hagiwara symposium*, Earth Plan. Space *56*(8), 689–713.

JIN A., AKI, K., and ZHEN, L. (2004), *Seismological evidence for the brittle-ductile interaction hypothesis on earthquake loading*, Earth Plan. Space *56*(8), 823–830.

AKI, K. (2004), *From modern seismology to earthquake prediction engineering, Part 1*, Earthq. Disaster Prevention, *199*, 19–22 (in Japanese).

AKI, K., and RICHARDS P.G., (Japanese translation by K. Uenishi, N. Kame, and H. Aochi), *Quantitative Seismology*, 2nd ed., Kokin Shoin Press, 2004, (in Japanese).

ZALIAPIN, I., JIN, A., LIU, Z., AKI, K., and KEILIS-BOROK, V. (2005), *Temporal (Un)correlations between Coda Q and seismicity: Multiscale trend analysis*, Pure Appl. Geophys. *162*, 827–841.

JIN, A., and AKI, K. (2005), *High-resolution maps of Coda Q in Japan and their interpretation by the brittle-ductile interaction hypothesis*, Earth Plan. Space *57*(5), 403–409.

AKI, K. (2005), *From modern seismology to earthquake prediction engineering, Part 2*, Earthq. Disaster Prevention *202*, 9–16 (in Japanese).

MIX
Papier aus verantwortungsvollen Quellen
Paper from responsible sources
FSC® C105338

If you have any concerns about our products,
you can contact us on
ProductSafety@springernature.com

In case Publisher is established outside the EU,
the EU authorized representative is:
Springer Nature Customer Service Center GmbH
Europaplatz 3, 69115 Heidelberg, Germany

Printed by Libri Plureos GmbH
in Hamburg, Germany